The Facts On File

DICTIONARY of SPACE TECHNOLOGY

Revised Edition

Joseph A. Angelo, Jr.

Facts On File, Inc.

This book is dedicated to my beloved daughter,
Jennifer April Angelo (1975–1993).
Her life, though tragically brief, was filled with such love and compassion
for others that she will always serve as my inspiration. Young people, like
Jennifer, can make this world a better place, and they most certainly
deserve to inherit the stars!

The Facts On File Dictionary of Space Technology
Revised Edition

Copyright © 2004, 1999 by Joseph A. Angelo, Jr.

Facts On File, Inc.
132 West 31st Street
New York NY 10001

Library of Congress Cataloging-in-Publication Data
Angelo, Joseph A.
    The Facts On File dictionary of space technology / Joseph A. Angelo, Jr.—Rev. ed.
        p. cm.
    Rev. ed. of Dictionary of space technology. 1982.
    ISBN 0-8160-5222-0
    1. Astronautics—Dictionaries. I. Angelo, Joseph A. Dictionary of space technology. II.
Facts on File, Inc. III. Title.

TL788.A53 2003
629.4′03—dc21   2003049148

Facts On File books are available at special discounts when purchased in bulk quantities for businesses, associations, institutions, or sales promotions. Please call our Special Sales Department in New York at (212) 967-8800 or (800) 322-8755.

You can find Facts On File on the World Wide Web at http://www.factsonfile.com

Text and cover design by Cathy Rincon
Illustrations by Sholto Ainslie

Printed in the United States of America

MP Hermitage 10 9 8 7 6 5 4 3 2 1

This book is printed on acid-free paper.

CONTENTS

ACKNOWLEDGMENTS

I wish to publicly acknowledge the generous support of the National Aeronautics and Space Administration (NASA), the United States Air Force (USAF), the National Reconnaissance Office (NRO), and the European Space Agency (Washington, D.C. office) during the preparation of this revised edition. Special thanks are also extended to the editorial staff at Facts On File, particularly my editor, Frank K. Darmstadt. The staff at the Evans Library of Florida Tech again provided valuable help. Finally, without the patient support of my wife, Joan, this edition would never have emerged from the chaotic piles of revised manuscript that were quite literally scattered throughout our home.

INTRODUCTION

Tantalizing discoveries made by American robot spacecraft at the end of the 20th century strongly suggest that large quantities of liquid water once existed on the surface of an ancient Mars. If that hypothesis is confirmed, life may have emerged on the Red Planet eons ago. Space exploration in the 21st century can then provide indisputable evidence that the emergence of life is not unique to planet Earth. A space technology–enabled discovery of this magnitude will have profound scientific and philosophical implications. It represents the equivalent of a biological Copernican Revolution—shattering the widely cherished geocentric perception of life as a uniquely terrestrial phenomenon.

Throughout the 21st century, space technology will reveal the secrets of a universe filled with wonder, excitement, and many surprises. Marvelous space-based observatories, like the Hubble Space Telescope and the Chandra X-Ray Observatory, will provide an endless stream of exciting new facts that promote intellectual stimulation for an entire new generation of scientists.

Space technology is also helping human beings rediscover the beauty and fragility of their home planet. The natural circle of life on Earth was recognized and respected by many ancient peoples. Today, powerful new Earth observing spacecraft, like NASA's Aqua and Terra, use remote-sensing technologies to reveal the intricacies of the Earth system—the complex collection of processes within Earth's biosphere that collectively strive to optimize the conditions for life in all its many wonderful forms. Earth-orbiting astronauts and cosmonauts constantly remind us that, when viewed from orbit, our planet no longer displays artificial boundary lines, separating peoples, places, cultures, and religions. What each of these space travelers has personally witnessed is the inspirationally beautiful tapestry of a living planet.

The initial assembly of the *International Space Station* took place in December 1998. The event marked the start of an exciting new era in space travel—an era characterized by increased cooperation among space-faring nations. However, as suggested by the Latin phrase *Ad astra per aspera* (To the stars through hardship and struggle), travel in space is not without risk. On February 1, 2003, the Space Shuttle *Columbia* and its crew of seven perished in a reentry accident. The tragic event reminded us all of the dangers constantly encountered by those brave men and women who dare to boldly venture beyond the boundaries of Earth.

Remaining true to the purpose of the original book, this revised edition presents the basic concepts of space technology, its underlying scientific principles, and the technical characteristics of the engineered systems that have (or will) accomplish some of humankind's most important space missions. Careful attention is also given to the contemporary role of space technology in global security. For example, the swift and successful conclusion of the 2003 Iraqi conflict by the United States and Coalition forces was due in no small measure to the information dominance of the battle space provided by satellite systems.

After working almost four decades in the field of space technology, I still marvel every time I witness a rocket vehicle blasting off from Cape Canaveral and rising into space on a pillar of fire. Each such event reminds me of the profound statement made by the German space visionary Hermann Oberth in response to the question: Why space travel? His succinct and poignant answer: "To make available for life every place where life is possible. To make inhabitable all worlds as yet uninhabitable, and all life purposeful."

Space technology encourages us to view the universe as both a destination and a destiny for all humankind!

—Joseph A. Angelo, Jr.
Cape Canaveral, Florida

A   Symbol for RELATIVE MASS NUMBER.

aberration   1. In optics, a specific deviation from a perfect IMAGE, such as spherical aberration, astigmatism, COMA, curvature of field, and DISTORTION. For example, spherical aberration results in a point of LIGHT (i.e., a point image) appearing as a circular DISK at the FOCAL POINT of a spherical LENS or curved MIRROR. This occurs because the focal points of light rays far from the optical AXIS are different from the focal points of light rays passing through or near the center of the lens or mirror. Light rays passing near the center of a spherical lens or curved mirror have longer focal lengths than do those passing near the edges of the optical system. 2. In ASTRONOMY, the apparent angular displacement of the position of a CELESTIAL BODY in the direction of motion of the observer. This effect is caused by the combination of the velocity of the observer and the VELOCITY OF LIGHT (c). An observer on EARTH would have Earth's ORBITAL VELOCITY around the SUN (V_E), which is approximately 30 km/s. As a result of this effect in the course of a YEAR, the light from a FIXED STAR appears to move in a small ELLIPSE around its mean position on the CELESTIAL SPHERE. The British astronomer James Bradley (1693–1762) discovered this phenomenon in 1728.

abiotic   Not involving living things; not produced by living organisms.

ablation   A form of MASS TRANSFER COOLING that involves the removal of a special surface material (called an ablative material) from a body, such as a REENTRY VEHICLE, a planetary PROBE or a reusable AEROSPACE VEHICLE (such as the SPACE SHUTTLE ORBITER), by MELTING, VAPOR-IZATION, SUBLIMATION, chipping, or other erosive process due to the AERODYNAMIC HEATING effects of moving through a planetary ATMOSPHERE at very high speed. The rate at which ablation occurs is a function of the reentering body's passage through the aerothermal environment, a high-TEMPERATURE environment caused by atmospheric FRICTION. Ablation is also a function of other factors including: (1) the amount of thermal ENERGY (i.e., HEAT) needed to raise the temperature of the ablative material to the ablative temperature (including PHASE CHANGE); (2) the heat blocking action created by BOUNDARY LAYER thickening due to the injection of MASS; and (3) the thermal energy dissipated into the interior of the body by CONDUCTION at the ablative temperature. To promote maximum thermal protection, the ablative material should not easily conduct heat into the reentry body. To minimize mass loss during the ablative cooling process, the ablative material also should have a high value for its effective heat of ablation—a thermophysical property that describes the efficiency with which thermal energy (in JOULES [J]) is removed per unit mass lost or "ablated" (in KILOGRAMS [kg]). Contemporary fiberglass-resin ablative materials can achieve more than 10^7 J/kg thermal energy removal efficiencies through sublimation processes during reentry. Ablative cooling generally is considered to be the least mass-intensive approach to reentry vehicle thermal protection. However, these mass savings are achieved at the expense of heat shield (and possibly reentry vehicle) reusability.

abort   1. To cancel, cut short, or break off an action, operation, or procedure with an aircraft, SPACE VEHICLE, or the like, especially because of equipment failure. For example, the LUNAR landing mission was *aborted* during the *Apollo 13* flight. 2. In defense, failure to accomplish a military mission for any reason other than enemy action. The abort may occur at any point from the initiation of an operation to arrival at the TARGET or destination. 3. An aircraft, space vehicle, or planetary PROBE that aborts. 4. An act or instance of aborting. 5. To cancel or cut short a flight after it has been launched. (See Figure 1.)

abort modes (Space Transportation System) Selection of an ascent ABORT mode may become necessary if there is a failure that affects SPACE SHUTTLE vehicle performance, such as the failure of a main engine or an ORBITAL MANEUVERING SYSTEM (OMS). Other failures requiring early termination of a space shuttle flight, such as a cabin leak, might also require the selection of an abort mode.

There are two basic types of ascent abort modes for space shuttle missions: *intact aborts* and *contingency aborts*. Intact aborts are designed to provide a safe return of the ORBITER vehicle and its crew to a planned landing site. Contingency aborts are designed to permit flight crew survival following more severe failures when an intact abort is not possible. A contingency abort would generally result in a ditch operation.

There are four types of intact aborts: Abort-to-Orbit (ATO), Abort-Once-Around (AOA), Transatlantic-Landing (TAL), and Return-to-Launch-Site (RTLS).

The *Abort-to-Orbit* (ATO) mode is designed to allow the Orbiter vehicle to achieve a temporary ORBIT that is lower than the nominal orbit. This abort mode requires less performance and provides time to evaluate problems and then choose either an early deorbit maneuver or an orbital maneuvering system (OMS) thrusting maneuver to raise the orbit and continue the mission.

The *Abort-Once-Around* (AOA) mode is designed to allow the Orbiter vehicle to fly once around the EARTH and make a normal entry and landing. This abort

Figure 1. A malfunction in the first stage of the Vanguard launch vehicle caused the vehicle to lose thrust after just two seconds, aborting the mission on December 6, 1957. The catastrophic destruction of this rocket vehicle and its small scientific satellite temporarily shattered American hopes of effectively responding to the successful launches of two different Sputnik satellites by the former Soviet Union at the start of the Space Age in late 1957. *(Courtesy of U.S. Navy)*

mode generally involves two OMS thrusting sequences, with the second sequence being a deorbit maneuver. The atmospheric entry sequence would be similar to a normal mission entry.

The *Transatlantic-Landing* (TAL) mode is designed to permit an intact landing on the other side of the Atlantic Ocean at emergency landing sites in either Morocco, the Gambia, or Spain. This abort mode results in a BALLISTIC TRAJECTORY that does not require an OMS burn. (See Figure 2.)

Finally, the *Return-to-Launch-Site* (RTLS) mode involves flying downrange to dissipate PROPELLANT and then turning around under power to return the Orbiter vehicle, its crew, and its PAYLOAD directly to a landing at or near the KENNEDY SPACE CENTER.

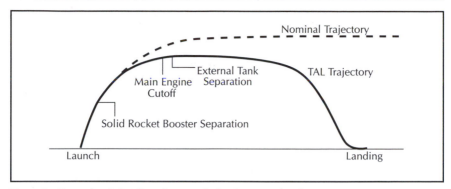

Figure 2. Transatlantic landing abort mode for the space shuttle.

The type of failure (e.g., loss of one main engine) and the time of the failure determine which type of abort mode should be selected. For example, if the problem is a SPACE SHUTTLE MAIN ENGINE (SSME) failure, the flight crew and MISSION CONTROL CENTER (MCC) (at the NASA JOHNSON SPACE CENTER in Houston, Texas) would select the best option available at the time when the main engine fails. *See also* SPACE TRANSPORTATION SYSTEM.

abrasion The loss of surface material due to frictional FORCES.

absentee ratio With respect to a hypothetical CONSTELLATION of orbiting space weapon platforms, the ratio of the number of platforms not in position to participate in a battle to the number that are. In contemporary SPACE DEFENSE studies, typical absentee ratios for postulated KINETIC ENERGY WEAPON systems were approximately 10 to 30, depending on the ORBITS of the SPACE PLATFORMS and the hypothesized space battle scenarios.

absolute   magnitude (symbol: M) The measure of the brightness (or APPARENT MAGNITUDE) that a STAR would have if it were hypothetically located at a reference distance of 10 PARSECS (10 pc), about 32.6 LIGHT-YEARS, from the SUN.

absolute pressure In engineering, the total PRESSURE exerted on the boundary

(i.e., walls) of a system. This term is used to indicate the pressure (normal component of FORCE per unit area exerted by a FLUID on a system boundary) above the hypothetical ABSOLUTE zero value of pressure that is theoretically obtainable in empty space. The ATMOSPHERIC PRESSURE is defined as the pressure exerted on a system boundary by a planetary ATMOSPHERE (location dependent). The term *(positive) gage pressure* is used to identify the difference for a particular system between the absolute pressure and the atmospheric pressure, which is generally used as the reference. However, in VACUUM system technology work on EARTH, the atmospheric pressure exceeds the absolute pressure of the system, so the terms *vacuum* and *negative gage pressure* (NGP) are often used to describe the amount by which the atmospheric pressure exceeds the absolute pressure of the vacuum system.

In   traditional   engineering   units, absolute pressure is usually expressed in pounds-force per square inch-absolute (psia), while gage pressure is designated in pounds-force per square inch-gage (psig). The PASCAL (Pa) is the unit of pressure in SI UNITS.

absolute   temperature TEMPERATURE value relative to ABSOLUTE ZERO, which corresponds to 0 K, or −273.15°C (after the Swedish astronomer Anders Celsius [1701–1744]). In SI UNITS, the absolute temperature   values   are   expressed   in

KELVINS (K), a unit named in honor of the Scottish physicist Baron William Thomson Kelvin (1827–1907). In the traditional engineering unit system, absolute temperature values are expressed in DEGREES Rankine (R), named after the Scottish engineer William Rankine (1820–1872). See also ABSOLUTE TEMPERATURE.

absolute zero The TEMPERATURE at which molecular motion vanishes and an object has no thermal ENERGY (or HEAT). From THERMODYNAMICS, absolute zero is the lowest possible temperature, namely zero degrees KELVIN (0 K).

absorbed dose (symbol: D) When ionizing radiation passes through matter, some of its ENERGY is imparted to the matter. The amount of energy absorbed per unit MASS of irradiated material is called the absorbed dose. The traditional unit of absorbed dose is the RAD (an acronym for *radiation absorbed dose*), while the SI UNIT for absorbed dose is called the GRAY (symbol: Gy). One gray is defined as one JOULE (J) of energy deposited per KILOGRAM (kg) of irradiated matter, while one rad is defined as 100 ergs of energy deposited per gram of irradiated matter.

The traditional and SI units of absorbed dose are related as follows:
$$100 \text{ rad} = 1 \text{ Gy}$$
The absorbed dose is often loosely referred to in radiation protection activities as "dose" (although this use is neither precise nor correct) and is also frequently confused with the term DOSE EQUIVALENT (symbol: H).

absorptance (absorptivity) In HEAT TRANSFER by THERMAL RADIATION, the absorptance (commonly used symbol: α) of a body is defined as the ratio of the incident radiant ENERGY absorbed by the body to the total radiant energy falling upon the body. For the special case of an ideal BLACKBODY, all radiant energy incident upon this blackbody is absorbed, regardless of the WAVELENGTH or direction. Therefore, the absorptance for a blackbody has a value of unity, that is, $\alpha_{blackbody} = 1$. All other real-world SOLID objects have an absorptance of less than one. *Compare with* REFLECTANCE and TRANSMITTANCE.

absorption 1. In general, the taking up or assimilation of one substance by another; for example, a process in which one or more soluble components of a GAS mixture are dissolved in a LIQUID. 2. In HEAT TRANSFER, the process by which incident radiant ENERGY is absorbed and converted into other forms of energy. 3. In VACUUM technology, the process of a gas entering the interior of a SOLID. 4. In modern physics, a nuclear interaction in which an incident NEUTRON disappears as a free particle upon interacting with an atomic NUCLEUS.

absorption line A gap, dip, or dark line occurring in a SPECTRUM caused by the absorption of ELECTROMAGNETIC RADIATION (radiant ENERGY) at a specific WAVELENGTH by an absorbing substance, such as a planetary ATMOSPHERE or a monatomic GAS. *See also* ABSORPTION SPECTRUM.

absorption spectrum The array of ABSORPTION LINES and bands that results from the passage of ELECTROMAGNETIC RADIATION (i.e., radiant ENERGY) from a continuously emitting high-TEMPERATURE source through a selectively absorbing medium that is cooler than the source. The absorption spectrum is characteristic of the absorbing medium, just as an EMISSION SPECTRUM is characteristic of the radiating source.

An absorption spectrum formed by a monatomic GAS (e.g., HELIUM) exhibits discrete dark lines, while one formed by a polyatomic gas (e.g., CARBON DIOXIDE, CO_2) exhibits ordered arrays (bands) of dark lines that appear regularly spaced and very close together. This type of ABSORPTION is often referred to as line absorption. Line spectra occur because the atoms of the absorbing gas are making transitions between specific energy levels. In contrast, the absorption spectrum formed by a selectively absorbing LIQUID or SOLID is generally continuous in nature, that is, there is a continuous WAVELENGTH region over which RADIATION is absorbed. *See also* ELECTROMAGNETIC SPECTRUM.

abundance of elements (in the universe)
STELLAR SPECTRA provide an estimate of the *cosmic abundance* of ELEMENTS as a percentage of the total MASS of the UNIVERSE. The 10 most common elements are HYDROGEN (H) at 73.5%, of the total mass, HELIUM (He) at 24.9%, oxygen (O) at 0.7%, carbon (C) at 0.3%, iron at 0.15%, neon (Ne) at 0.12%, nitrogen (N) at 0.10%, silicon (Si) at 0.07%, magnesium (Mg) at 0.05%, and sulfur (S) at 0.04%.

accelerated life test(s)
The series of test procedures for a SPACECRAFT or AEROSPACE system that approximate in a relatively short period of time the deteriorating effects and possible failures that might be encountered under normal, long-term space mission conditions. Accelerated life tests help aerospace engineers detect critical design flaws and material incompatibilities (for example, excessive wear or FRICTION) that eventually might affect the performance of a spacecraft component or subsystem over its anticipated operational lifetime. *See also* LIFE CYCLE.

acceleration (usual symbol: a)
The rate at which the VELOCITY of an object changes with time. Acceleration is a VECTOR quantity and has the physical dimensions of length per unit time to the second power (for example, METERS per SECOND per second, or m/s^2).

acceleration of gravity
The local ACCELERATION due to GRAVITY on or near the surface of a PLANET. On EARTH, the acceleration due to gravity (symbol: g) of a free-falling object has the standard value of $9.80665 \ m/s^2$ by international agreement. According to legend, the famous Italian scientist GALILEO GALILEI simultaneously dropped a large and small cannonball from the top of the Tower of Pisa to investigate the acceleration of gravity. As he anticipated, each object fell to the ground in exactly the same amount of time (neglecting air resistance)—despite the difference in their MASSes. During the APOLLO PROJECT, ASTRONAUTS repeated a similar experiment on the surface of the Moon, dramatically demonstrating the universality of physical laws and honoring Galileo's scientific genius. It was Galileo's pioneering investigation of the physics of FREE FALL that helped SIR ISAAC NEWTON unlock the secrets of motion of the mechanical UNIVERSE.

accelerator
A device used in nuclear physics for increasing the VELOCITY and ENERGY of charged elementary PARTICLES, such as ELECTRONS or PROTONS, through the application of ELECTROMAGNETIC FORCES.

accelerometer
An INSTRUMENT that measures ACCELERATION or gravitational FORCES capable of imparting acceleration. Frequently used on SPACE VEHICLES to assist in guidance and navigation and on PLANETARY PROBES to support scientific data collection.

acceptance test(s)
In the AEROSPACE industry, the required formal tests conducted to demonstrate the acceptability of a unit, component, or system for delivery. These tests demonstrate performance to purchase specification requirements and serve as quality-control screens to detect deficiencies of workmanship and materials.

accretion
The gradual accumulation of small PARTICLES of GAS and dust into larger material bodies, mostly due to the influence of GRAVITY. For example, in the early stages of stellar formation, matter begins to collect or accrete into a NEBULA (a giant INTERSTELLAR cloud of gas and dust). Eventually STARS are born in this nebula. When a particular star forms, small quantities of residual matter may collect into one or more PLANETS that ORBIT the new star. *See also* ACCRETION DISK.

accretion disk
The whirling DISK of inflowing (or infalling) material from a

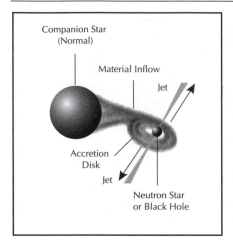

Figure 3. The inflow of material from a companion star, creating an accretion disk around a neutron star or black hole.

normal stellar companion that develops around a massive COMPACT BODY, such as a NEUTRON STAR or a BLACK HOLE. The conservation of ANGULAR MOMENTUM shapes this disk, which is often accompanied by a pair of very high-speed material jets that depart in opposite directions perpendicular to the plane of the disk. (See Figure 3.) *See also* ACCRETION.

accumulator In general, a device or mechanism that stores up or accumulates something. For example, in hydraulics, an accumulator stores FLUID under PRESSURE.

acoustic Arising from or associated with SOUND. For example, an *acoustic wave.*

acoustic absorber An array of ACOUSTIC resonators distributed along the wall of a COMBUSTION CHAMBER, designed to prevent OSCILLATORY COMBUSTION by increasing DAMPING in the ROCKET ENGINE system.

acoustics The scientific study of SOUND, including its production, propagation, and interaction with materials.

acoustic velocity The speed at which SOUND WAVES travel in a particular medium.

Also called the *speed of sound.* For example, sound travels at approximately 331 METERS per SECOND through AIR that has a TEMPERATURE of 0° Celsius and a PRESSURE of 1 ATMOSPHERE.

acquisition 1. The process of locating the ORBIT of a SATELLITE or the TRAJECTORY of a SPACE PROBE, so that mission control personnel can track the object and collect its TELEMETRY DATA. 2. The process of pointing an ANTENNA or TELESCOPE so that it is properly oriented to gather TRACKING or telemetry data from a satellite or space probe. 3. The process of searching for and detecting a potentially threatening object in space. An *acquisition sensor* is designed to search a large area of space and to distinguish potential targets from other objects against the background of space.

acquisition and tracking radar A RADAR system that locks onto a strong SIGNAL and tracks the object reflecting the signal.

acronym A word formed from the first letters of a name, such as HST, which means the HUBBLE SPACE TELESCOPE. It is also a word formed by combining the initial parts of a series of words, such as lidar, which means LIGHT detection and ranging. Acronyms are frequently used in the AEROSPACE industry, space technology, and ASTRONOMY.

activation analysis A method for identifying and measuring chemical ELEMENTS in a sample of material. The sample is first made radioactive by bombarding (i.e., irradiating) it with NEUTRONS, PROTONS, or other nuclear PARTICLES. The newly formed radioactive atoms in the sample then experience radioactive DECAY, giving off characteristic NUCLEAR RADIATIONS, such as GAMMA RAYS of specific ENERGY levels, that reveal what kinds of ATOMS are present and possibly how many.

active 1. In general, transmitting a SIGNAL—as, for example, an *active* SATELLITE. 2. In SURVEILLANCE and REMOTE SENSING, an adjective applied to actions or equipment that emit signals or ENERGY capable of being detected. 3. In physics and nuclear technology, radioactive—as, for example, an *active* ISOTOPE of the ELEMENT cesium.

active control The automatic activation of various control functions and equipment onboard an AEROSPACE VEHICLE or SATELLITE. For example, to achieve active attitude control of a satellite, the satellite's current ATTITUDE is measured automatically and compared with a reference or desired value. Any significant difference between the satellite's current attitude and the reference or desired attitude produces an ERROR SIGNAL, which is then used to initiate appropriate corrective maneuvers by onboard ACTUATORS. Since both the measurements and the automatically imposed corrective maneuvers will not be perfect, the active control CYCLE usually continues through a number of iterations until the difference between the satellite's actual and desired attitude is within preselected, tolerable limits.

active defense The employment of limited offensive action and counterattacks to deny a contested area or position to the enemy. *Compare with* PASSIVE DEFENSE.

active discrimination In BALLISTIC MISSILE DEFENSE (BMD), the illumination of a potential TARGET with ELECTROMAGNETIC RADIATION in order to determine from the characteristics of the reflected radiation whether this object is a genuine threat object (e.g., an enemy REENTRY VEHICLE or POST-BOOST VEHICLE) or just a DECOY. RADAR and LASER RADAR systems are examples of active discrimination devices.

active galactic nucleus (AGN) The central region of a distant (active) GALAXY that appears to be a point-like source of intense X-RAY or GAMMA RAY emissions. Astrophysicists speculate that the AGN is caused by the presence of a centrally located, super-heavy BLACK HOLE accreting nearby matter. *See also* ACCRETION.

active galaxies Collectively, those unusual celestial objects, including QUASARS, BL LAC OBJECTS, and SEYFERT GALAXIES, that have extremely energetic central regions, called active galactic nuclei (AGN). These emit enormous amounts of ELECTROMAGNETIC RADIATION, from RADIO WAVES to X-RAYS and GAMMA RAYS.

active homing guidance A system of HOMING GUIDANCE wherein a MISSILE carries within itself both the source for illuminating the TARGET and the receiver for detecting the SIGNAL reflected by the target. Active homing guidance systems also can be used to assist SPACE SYSTEMS in RENDEZVOUS and DOCKING operations.

active microwave instrument A MICROWAVE instrument, such as a RADAR ALTIMETER, that provides its own source of illumination. For example, by measuring the RADAR returns from the ocean or sea, a radar altimeter on a SPACECRAFT can be used to deduce WAVE height, which is an indirect measure of surface wind speed. *See also* REMOTE SENSING.

active remote sensing A REMOTE SENSING technique in which the SENSOR supplies its own source of ELECTROMAGNETIC RADIATION to illuminate a TARGET. A SYNTHETIC APERTURE RADAR (SAR) system is an example.

active satellite A SATELLITE that transmits a SIGNAL, in contrast to a passive (dormant) satellite.

active sensor A SENSOR that illuminates a TARGET, producing return secondary RADIATION that is then detected in order to track and possibly identify the target. A LIDAR is an example of an active sensor.

active Sun The name scientists give to the collection of dynamic SOLAR phenomena, including SUNSPOTS, SOLAR FLARES, and PROMINENCES, associated with intense variations in the SUN's magnetic activity. *Compare with* QUIET SUN.

active tracking system A system that requires the addition of a TRANSPONDER or

TRANSMITTER on board a SPACE VEHICLE or MISSILE to repeat, transmit, or retransmit information to the TRACKING equipment.

activity The rate of disintegration or DECAY of a RADIOACTIVE ISOTOPE. *See also* RADIOACTIVITY.

actuator A servomechanism that supplies and transmits ENERGY for the operation of other mechanisms, systems, or process control equipment.

acute radiation syndrome (ARS) The acute organic disorder that follows exposure to relatively severe doses of IONIZING RADIATION. A person will initially experience nausea, diarrhea, or blood cell changes. In the later stages loss of hair, hemorrhaging, and possibly death can take place. *Radiation dose equivalent* values of about 4.5 to 5 SIEVERT (450 to 500 REM) will prove fatal to 50% of the exposed individuals in a large general population. Also called *radiation sickness.*

adapter Any device used or designed primarily to fit or adjust one component to another; for example, a fitting to join two pipes that have different threads or different diameters.

adapter skirt A flange or extension on a LAUNCH VEHICLE stage or SPACECRAFT section that provides a means of fitting on another stage or section.

adaptive control system A CONTROL SYSTEM that continuously monitors the dynamic response of the system being controlled and then automatically adjusts critical system parameters to satisfy preassigned response criteria, thereby producing the same response over a wide range of environmental conditions.

adaptive optics Optical systems that can be modified (e.g., by controlling the shape of a MIRROR) to compensate for distortions. An example is the use of information from a BEAM of LIGHT passing through the ATMOSPHERE to compensate for the distortion experienced by another

beam of light on its passage through the atmosphere. Adaptive optics systems are used in observational ASTRONOMY to eliminate the "twinkling" of STARS and in BALLISTIC MISSILE DEFENSE to reduce the dispersive effect of the atmosphere on LASER beam weapons. At visible and near-infrared WAVELENGTHS, the ANGULAR RESOLUTION of EARTH-based TELESCOPES with APERTURES greater than 10 to 20 cm is limited by turbulence in Earth's atmosphere rather than by the inherent, DIFFRACTION-limited IMAGE size of the system. Large telescopes can be equipped with adaptive optics systems to compensate for atmospheric turbulence effects and can achieve imaging on scales that approach the diffraction limit. These adaptive optics systems continuously measure the WAVE front errors resulting from atmospheric turbulence. Then, using a pointlike reference source situated above the distorting layers of Earth's atmosphere, compensation is achieved by rapidly adjusting a deformable optical element located in or near a pupil plane of the optical system.

additive A material or substance added to something else for a specific purpose. For example, a substance can be added to a ROCKET PROPELLANT to achieve a more even rate of combustion.

adhesion Molecular attraction that holds the surfaces of two substances together.

adiabatic A process or phenomenon that takes place without gain or loss of thermal ENERGY (HEAT); in THERMODYNAMICS, a process in which heat is neither added to nor removed from the system involved.

adiabatic wall In a thermodynamic system, a boundary that is a perfect HEAT insulator (i.e., neither transports nor absorbs thermal ENERGY); in BOUNDARY LAYER theory, the wall condition in which the TEMPERATURE gradient at the wall is zero.

adsorbent A material that takes up GASES or LIQUIDS by ADSORPTION. For example, activated carbon is a highly adsorbent form of carbon used to remove odors and toxic substances from liquid or gaseous streams.

adsorption The adhesion of a thin film of LIQUID or GAS to the surface of a SOLID substance. The solid does not combine chemically with the adsorbed substance.

Advanced X-Ray Astrophysics Facility-Imaging (AXAF-I) *See* CHANDRA X-RAY OBSERVATORY.

Aegis A surface ship-launched MISSILE used by the U.S. Navy as part of a totally integrated shipboard weapon system that combines computers, RADARS, and missiles to provide a defense umbrella for surface shipping. The system is capable of automatically detecting, TRACKING, and destroying a hostile air-launched, sea-launched, or land-launched weapon. In the late 1960s the U.S. Navy developed the *Advanced Surface Missile System* (ASMS) and then renamed this system Aegis after the shield of the god Zeus in Greek mythology. The Aegis system provides effective defense against antiship CRUISE MISSILES and human-crewed enemy aircraft in all environmental conditions.

aeolian Pertaining to, carried by, or caused by the wind; for example, aeolian sand dunes on the surface of MARS. Also called eolian.

aerial 1. (adjective) of or pertaining to the AIR, ATMOSPHERE, or aviation. 2. (noun) an ANTENNA.

aero- A prefix that means of or pertaining to the AIR, the ATMOSPHERE, aircraft, or flight through the atmosphere of a PLANET.

aeroassist The use of the thin, upper regions of a PLANET's ATMOSPHERE to provide the LIFT or DRAG needed to maneuver a SPACECRAFT. Near a planet with a SENSIBLE ATMOSPHERE, aeroassist allows a spacecraft to change direction or to slow down without expending PROPELLANT from the CONTROL ROCKET system. (See Figure 4.)

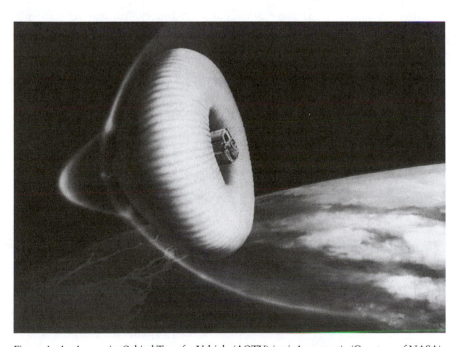

Figure 4. An Aeroassist Orbital Transfer Vehicle (AOTV) (artist's concept). *(Courtesy of NASA)*

aerobraking The use of a specially designed SPACECRAFT structure to deflect rarefied (very low DENSITY) AIRFLOW around a spacecraft, there by supporting AEROASSIST maneuvers in the vicinity of a PLANET. Such maneuvers reduce the spacecraft's need to perform large propulsive burns when making orbital changes near a planet. In 1993 NASA's MAGELLAN MISSION became the first planetary exploration spacecraft to use aerobraking as a means of changing its ORBIT around the target planet (VENUS).

aerodynamic force The LIFT (L) and/or DRAG (D) forces exerted by a moving GAS upon a body completely immersed in it. Lift acts in a direction normal to the flight path, while drag acts in a direction parallel and opposite to the flight path. (See Figure 5.) The aerodynamic forces acting upon a body flying through the ATMOSPHERE of a PLANET are dependent on the VELOCITY of the object, its geometric characteristics (i.e., size and shape), and the thermophysical properties of the atmosphere (e.g., TEMPERATURE and DENSITY) at flight ALTITUDE. For low flight speeds, lift and drag are determined primarily by the ANGLE OF ATTACK (α). At high (i.e., SUPERSONIC) speeds, these forces become a function of both the angle of attack and the MACH NUMBER (M). For a typical BALLISTIC MISSILE, the angle of attack is usually very low (i.e., $\alpha < 1°$). *See also* AIRFOIL; REENTRY VEHICLE.

aerodynamic heating Frictional surface heating experienced by an AEROSPACE VEHICLE, SPACE SYSTEM, or REENTRY VEHICLE as it enter the upper regions of a planetary ATMOSPHERE at very high VELOCITY. Peak aerodynamic heating generally occurs in STAGNATION POINT regions on the object, such as on the leading edge of a WING or on the blunt surfaces of a NOSE CONE. Special thermal protection is needed to prevent structural damage or destruction. NASA's SPACE SHUTTLE ORBITER vehicle, for example, uses thermal protection tiles to survive the intense aerodynamic heating environment that occurs during reentry and landing.

aerodynamic missile A MISSILE that uses AERODYNAMIC FORCES to maintain its flight path, generally employing propulsion guidance.

aerodynamics The branch of science that deals with the motion of GASES (especially AIR) and with the FORCES acting on SOLID bodies when they move through gases or when gases move against or around solid bodies.

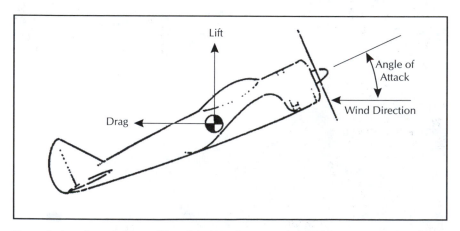

Figure 5. Aerodynamic forces (lift and drag) acting on an aircraft flying at a certain angle of attack.

aerodynamic skip An atmospheric entry ABORT caused by entering a PLANET'S ATMOSPHERE at too shallow an ANGLE. Much like a stone skipping across the surface of a pond, this condition results in a TRAJECTORY back out into SPACE rather than downward toward the planet's surface.

aerodynamic throat area The effective flow area of a NOZZLE'S THROAT; generally, the effective flow area is less than is the geometric flow area because the flow is not uniform.

aerodynamic vehicle A craft that has lifting and control surfaces to provide stability, control, and maneuverability while flying through a PLANET'S ATMOSPHERE. For AERODYNAMIC MISSILES, the flight profile both during and after THRUST depends primarily upon AERODYNAMIC FORCES. A GLIDER or an AIRPLANE is capable of flight only within a SENSIBLE ATMOSPHERE, and such vehicles rely on aerodynamic forces to keep them aloft.

aeronautics The science of flight within the ATMOSPHERE of a PLANET; the engineering principles associated with the design and construction of craft for flight within the atmosphere; the theory and practice of operating craft within the atmosphere.

aeropause A region of indeterminate limits in a PLANET'S upper ATMOSPHERE considered a boundary between the denser (sensible) portion of the atmosphere and OUTER SPACE.

aerosol A very small dust PARTICLE or droplet of LIQUID (other than water or ice) in a PLANET'S ATMOSPHERE, ranging in size from about 0.001 MICROMETER (μm) to larger than 100 micrometers (μm) in radius. Terrestrial aerosols include smoke, dust, haze, and fumes. They are important in EARTH'S atmosphere as nucleation sites for the condensation of water droplets and ice crystals, as participants in various chemical cycles, and as absorbers and scatterers of SOLAR RADIATION. Aerosols

influence the Earth's radiation budget (i.e., the overall balance of incoming versus outgoing radiant ENERGY), which in turn influences the climate on the surface of the planet. *See also* GLOBAL CHANGE.

aerospace A term derived from *aeronautics* and *space* meaning of or pertaining to EARTH'S atmospheric envelope and OUTER SPACE beyond it. These two separate physical entities are taken as a single realm for activities involving LAUNCH, guidance, control, and recovery of vehicles and systems that can travel through and function in both physical regions. For example, NASA's SPACE SHUTTLE ORBITER is called an AEROSPACE VEHICLE because it operates both in the ATMOSPHERE and in outer space. *See also* AERONAUTICS; ASTRONAUTICS.

aerospace control operations The employment of AEROSPACE forces, supported by ground and naval forces, as appropriate, to achieve military objectives in vital aerospace areas. Such operations include destruction of enemy aerospace and surface-to-air forces, interdiction of enemy aerospace operations, protection of vital air lines of communication, and establishment of local military superiority in areas of AIR and SPACE operations.

aerospace defense 1. An inclusive term encompassing AIR DEFENSE and SPACE DEFENSE. 2. All defensive measures designed to destroy attacking enemy aircraft, MISSILES, and/or SPACE VEHICLES after they leave EARTH'S surface, or to nullify or reduce the effectiveness of attacks by such systems.

aerospace ground equipment (AGE) All the support and test equipment needed on EARTH'S surface to make an AEROSPACE system or SPACECRAFT function properly during in its intended SPACE MISSION.

aerospace medicine The branch of medical science that deals with the effects of flight upon the human body. The treatment of SPACE SICKNESS (space adaptation syndrome) falls within this field.

aerospace technology transfer The generation of innovation in virtually all branches of science and technology as a result of the pursuit of AEROSPACE technology goals and applications. Specifically, the secondary ("spin-off") applications of aerospace technologies in such diverse fields as health and longevity, sports, wildlife research, pollution control, search and rescue operations, archaeology, and many other areas that help improve the quality of life on EARTH.

aerospace vehicle A vehicle capable of operating both within EARTH's SENSIBLE (measurable) ATMOSPHERE and in OUTER SPACE. The SPACE SHUTTLE ORBITER vehicle is an example.

aerospike nozzle A ROCKET NOZZLE design that allows combustion to occur around the periphery of a spike (or center plug). The THRUST-producing, hot-exhaust flow is then shaped and adjusted by the AMBIENT (atmospheric) PRESSURE. Also called a *plug nozzle* or a *spike nozzle*.

aerozine A LIQUID FUEL for ROCKETS consisting of a mixture of HYDRAZINE (N_2H_4) and unsymmetrical dimethylhydrazine (UDMH), which has the chemical formula: $(CH_3)_2NNH_2$.

afterbody Any companion body (usually JETTISONed, expended HARDWARE) that trials a SPACECRAFT following LAUNCH and contributes to the SPACE DEBRIS problem. Any expended portion of a LAUNCH VEHICLE or ROCKET that enters EARTH's ATMOSPHERE unprotected behind a returning NOSE CONE or SPACE CAPSULE that is protected against the AERODYNAMIC HEATING. Finally, it is any unprotected, discarded portion of a SPACE PROBE or SPACECRAFT that trials behind the protected probe or LANDER SPACECRAFT as either enters a PLANET's atmosphere to accomplish a MISSION.

afterburner A device used for increasing the THRUST of a JET engine by burning additional FUEL in the uncombined oxygen present in the TURBINE exhaust GASES.

afterburning 1. In ROCKETRY, the characteristic of some ROCKET motors to BURN irregularly for some time after the main burning and THRUST generation have ceased. 2. In AERONAUTICS, the process of fuel injection and COMBUSTION in the exhaust GASES of a turbojet engine (aft, or to the rear, of the TURBINE). The function of an AFTERBURNER, a device for augmenting the thrust of a JET engine by burning additional FUEL in the uncombined oxygen in the exhaust gases from the turbine.

aft flight deck That part of the SPACE SHUTTLE ORBITER cabin on the upper deck where PAYLOAD controls can be located. *See also* SPACE TRANSPORTATION SYSTEM.

aft skirt The component of the SPACE SHUTTLE SOLID ROCKET BOOSTER (SRB) used for mounting the THRUST VECTOR CONTROL system and for supporting the aft cluster of the four BOOSTER SEPARATION MOTORS. *See also* SPACE TRANSPORTATION SYSTEM.

Agena A versatile, UPPER STAGE ROCKET that supported numerous American military and civilian SPACE MISSIONs in the 1960s and 1970s. One special feature of this LIQUID PROPELLANT system was its in-space engine restart capability. The Agena was originally developed by the UNITED STATES AIR FORCE for use in combination with THOR or ATLAS first STAGEs. Agena A, the first version of this upper stage, was followed by Agena B, which had a larger FUEL capacity and engines that could restart in space. The later Agena D was standardized to provide a LAUNCH VEHICLE for a variety of military and NASA PAYLOADs. For example, NASA used the Atlas-Agena vehicles to launch large EARTH-orbiting SATELLITEs as well as LUNAR and INTERPLANETARY space PROBEs; Thor-Agena vehicles launched scientific satellites, such as the *ORBITING GEOPHYSICAL OBSERVATORY* (OGO), and applications satellites, such as NIMBUS METEOROLOGICAL SATELLITES. In the GEMINI PROJECT the Agena D vehicle, modified to suit specialized requirements of space RENDEZVOUS and DOCKING maneuvers, became the Gemini Agena Target Vehicle (GATV). (See Figure 6.)

age of the Moon The elapsed time, usually expressed in days, since the last new MOON. *See also* PHASES OF THE MOON.

agglutinate(s) Common type of PARTICLE found on the MOON consisting of small rock, mineral, and glass fragments impact-bonded together with glass.

air The overall mixture of GASes that make up EARTH's ATMOSPHERE, primarily nitrogen (N_2) at 78% (by volume), oxygen (O_2) at 21%, argon (Ar) at 0.9%, and carbon dioxide (CO_2) at 0.03%. Sometimes AEROSPACE engineers use this word for the breathable gaseous mixture found inside the crew compartment of a SPACE VEHICLE or in the PRESSURIZED HABITABLE ENVIRONMENT of a SPACE STATION.

airborne laser (ABL) A high-priority program within the MISSILE DEFENSE AGENCY (MDA) to place a weapons-class chemical oxygen–iodine LASER (COIL) aboard a modified Boeing 747-400 series freighter aircraft, called the YAL-1A Attack Laser. The powerful infrared WAVELENGTH (1.315 MICROMETERS) laser would destroy a SCUD-like enemy BALLISTIC MISSILE shortly after it was launched, causing debris from the missile's chemical, biological, or nuclear WARHEAD to fall short of the intended TARGET. Depending on battle zone deployment scenarios, if an ABL engaged the enemy missile early enough during BOOST PHASE, debris from that missile's warhead might even fall on enemy territory.

airborne optical adjunct In BALLISTIC MISSILE DEFENSE (BMD), a set of SENSORS designed to detect, track, and discriminate an incoming WARHEAD. These sensors, typically optical or infrared devices, would be flown onboard a long-endurance AERIAL platform (either human-crewed or UNMANNED) stationed high above the clouds. *See also* AIRBORNE LASER; UNMANNED AERIAL VEHICLE.

airborne particulates Total suspended matter found in EARTH's ATMOSPHERE as

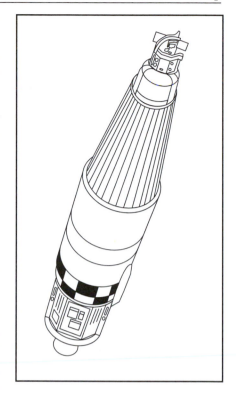

Figure 6. Agena upper stage vehicle configured for the SNAPSHOT mission flown by the U.S. government in 1965. During this mission, the Agena successfully placed its payload, an experimental SNAP-10A space nuclear reactor (the small, wastebasket-sized structure shown in the upper portion of the drawing), into a near-circular, 1,300-kilometer altitude orbit around Earth. Once at this orbit, the reactor was commanded to start up and operated at approximately 500 watts (electric) for 45 days.

SOLID pieces or LIQUID droplets. Airborne particulates include windblown dust, emissions from industrial processes, smoke from the COMBUSTION of wood and coal, and the exhaust of motor vehicles. *See also* AEROSOL.

air breakup The breakup of a REENTRY BODY or ROCKET (e.g., scientific SOUNDING ROCKET) after REENTRY into the ATMOSPHERE. Air breakup sometimes is accomplished by detonating a small explosive charge inside the reentry body or rocket. The air breakup process can retard the fall of certain pieces, thereby increasing the chances of recovery of selected components (such as a scientific INSTRUMENT CAPSULE) by reducing the severity of ground IMPACT.

air-breather An AERODYNAMIC VEHICLE propelled by the COMBUSTION of FUEL that is oxidized by AIR taken in from the ATMOS-PHERE; an *air-breathing vehicle*. CRUISE MIS-SILES and JET aircraft are examples of air-breathers, while ROCKETs carry their own supply of OXIDIZER and operate indepen-dently of the surrounding environment.

airburst An explosion of a bomb or PRO-JECTILE above the surface, as distinguished from an explosion on contact with the sur-face or after surface penetration.

aircraft A vehicle designed to be sup-ported by the AIR, either by the dynamic action of the air upon the surfaces of the vehicle or else by its own buoyancy. Air-craft include fixed-wing airplanes, GLID-ERS, helicopters, free and captive balloons, and airships.

air defense All defensive measures designed to destroy enemy aircraft or MIS-SILES in EARTH's ATMOSPHERE or to nullify or reduce the effectiveness of such attacks.

airflow A flow or stream of AIR. Airflow can occur in a WIND TUNNEL and in the intake portion of a JET engine; relative air-flow occurs past the WING or other com-ponent of a moving aircraft. Airflow is usually measured in terms of MASS or VOL-UME per unit time.

airfoil A WING, FIN, or CANARD designed to provide an AERODYNAMIC FORCE (e.g., LIFT) when it moves through the AIR (on EARTH) or through the SENSI-BLE ATMOSPHERE of a PLANET (such as MARS or VENUS) or of TITAN, the largest MOON of SATURN. (See Figure 7.)

Air Force Space Command *See* UNITED STATES AIR FORCE; UNITED STATES STRATEGIC COMMAND.

airframe The assembled structural and aerodynamic components of an aircraft or ROCKET vehicle that support the different systems and subsystems integral to the vehicle. The term *airframe,* a carryover from aviation technology, is still appropri-ate for rocket vehicles, since a major func-tion of the airframe is performed during the rocket's flight within the ATMOSPHERE.

airglow A faint general luminosity in the EARTH's sky, most apparent at night. As a relatively steady (visible) radiant emis-sion from the upper ATMOSPHERE, it is dis-tinguished from the sporadic emission of

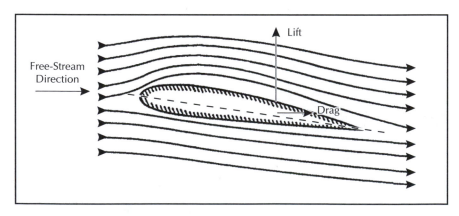

Figure 7. Basic forces generated by an airfoil in an airstream.

Figure 8. A U.S. Air Force B-52 bomber launches a supersonic Hound Dog air-to-surface missile during a demonstration test flight in 1959. *(Courtesy of U.S. Air Force)*

AURORA. Airglow is a CHEMILUMINESCENCE that occurs during recombination of ionized atmosphere ATOMS and MOLECULES that have collided with high-energy PARTICLES and ELECTROMAGNETIC RADIATION, primarily from the SUN. Airglow includes emissions from the oxygen molecule (O_2), the nitrogen molecule (N_2), the hydroxyl radical (OH), atomic oxygen (O), and sodium (Na).

air launch The process of launching a GUIDED MISSILE or ROCKET from an aircraft while it is in flight. For example, during AERIAL combat a pilot may choose to launch an AIR-TO-AIR MISSILE at an enemy aircraft. (See Figure 8.)

air launched cruise missile (ALCM) The UNITED STATES AIR FORCE developed the AGM-86B/C as a guided, air-launched, surface attack CRUISE MISSILE. This small, SUBSONIC winged MISSILE is powered by a turbofan JET engine. Launched from aircraft, such as the B-52 bomber, it can fly complicated routes to a TARGET through the use of modern information technology, such interaction between its inertial navigation computer system and the GLOBAL POSITIONING SYSTEM (GPS).

airlock A small chamber with "airtight" doors that is capable of being pressurized and depressurized. The airlock serves as a passageway for crew members and equipment between places at different PRESSURE levels—for example, between a SPACECRAFT's pressurized crew cabin and OUTER SPACE.

airplane A heavier-than-air, fixed-wing vehicle capable of flight through the ATMOSPHERE. AERODYNAMIC FORCES are used in conjunction with a powerplant (e.g., JET engine or propeller) to achieve motion through the AIR.

airscoop A hood or open end of an air DUCT that projects into the airstream around an aircraft or MISSILE in such a way as to use the motion of the vehicle to capture AIR for use in an engine, a ventilator, and so on.

air sounding The act of measuring atmospheric phenomena or determining atmospheric conditions at ALTITUDE, especially by means of SENSORS carried by balloons or SOUNDING ROCKETS.

air-to-air missile (AAM) A MISSILE launched from an airborne vehicle at a

TARGET above the surface of EARTH (i.e., at a target in the AIR). One example is the Advance Medium Range, Air-to-Air Missile (AMRAAM), also called the AIM-120, developed by the UNITED STATES AIR FORCE and Navy in conjunction with several NATO allies. AMRAAM serves as a new generation follow-on to the SPARROW (AIM-7) air-to-air missile. The AIM-120 missile is faster, smaller, and lighter and has improved capabilities against low-altitude TARGETS.

air-to-ground missile (AGM) *See* AIR-TO-SURFACE MISSILE.

air-to-surface missile (ASM) A MISSILE launched from an airborne platform at a surface TARGET. Sometimes referred to as an *AGM*, or *air-to-ground missile*. One example is the AGM-130A missile, equipped with a television and imaging infrared seeker that allows an airborne weapon system officer to observe and/or steer the missile as it attacks a ground target. Developed by the UNITED STATES AIR FORCE, the AGM-130A supports high- and low-ALTITUDE precision strikes at relatively safe standoff ranges. *See also* AIR LAUNCHED CRUISE MISSILE; HELLFIRE MISSILE.

air-to-underwater missile (AUM) A MISSILE launched from an airborne vehicle toward an underwater TARGET, such as a submarine.

albedo The ratio of the ELECTROMAGNETIC RADIATION (such as visible LIGHT) reflected by the surface of a nonluminous object to the total amount of electromagnetic radiation incident upon that surface. The albedo is usually expressed as a percentage. For example, the planetary albedo of EARTH is about 30%. This means that approximately 30% of the total SOLAR RADIATION falling upon Earth is reflected back to OUTER SPACE. With respect to the MOON, a high albedo indicates a light-colored region, while a low albedo indicates a dark region. LUNAR MARIA ("seas") have a low albedo; LUNAR HIGHLANDS have a high albedo.

algorithm A special mathematical procedure or rule for solving a particular type of problem.

alien In the context of advanced space technology and science fiction, an EXTRATERRESTRIAL; an inhabitant (presumably intelligent) of another world.

alien life-form (ALF) A general, thought at present hypothetical, expression for EXTRATERRESTRIAL LIFE, especially life that exhibits some degree of intelligence. *See also* EXOBIOLOGY; EXTREMOPHILE.

all-inertial guidance The guidance of a ROCKET vehicle or MISSILE entirely by use of inertial devices; the equipment used for this. *See also* INERTIAL GUIDANCE SYSTEM.

alloy A substance that has metallic properties and is composed of two or more chemical ELEMENTS, at least one of which is an elemental metal. For example, brass is an alloy of zinc (Zn) and copper (Cu), while steel is an alloy of iron (Fe), carbon (C), and often small amounts of other elements.

Almagest The Arabic name (meaning "the greatest") for the collection of ancient Greek astronomical and mathematical knowledge written by PTOLEMY in about 150 C.E. and translated by Arab astronomers about 820 C.E. This compendium included the 48 ancient Greek CONSTELLATIONS upon which today's astronomers base the modern systems of constellations.

Alpha Centauri The closest STAR system, about 4.3 LIGHT-YEARS away. It is actually a triple-star system, with two stars orbiting around each other and a third star, called Proxima Centauri, revolving around the pair at some distance. In various CELESTIAL configurations, Proxima Centauri becomes the closest known star to our SOLAR SYSTEM—approximately 4.2 light-years away.

alphanumeric (*alphabet* plus *numeric*) Including letters and numerical digits, for example, the term, JEN75WX11.

alpha particle (symbol: α) A positively-charged atomic PARTICLE emitted by certain RADIONUCLIDES. It consists of two NEUTRONS and two PROTONS bound together and is identical to the NUCLEUS of a helium-4 ($^4_2$ He) ATOM. Alpha particles are the least penetrating of the three common types of IONIZING RADIATION (alpha particle, BETA PARTICLE, and GAMMA RAY). They are stopped easily by materials such as a thin sheet of paper or even a few centimeters of AIR.

altazimuth mounting A TELESCOPE mounting that has one AXIS pointing to the ZENITH.

altimeter An INSTRUMENT for measuring the height (ALTITUDE) above a PLANET's surface; generally reported relative to a common planetary reference point, such as sea level on EARTH. A barometric altimeter is an instrument that displays the height of an aircraft or AEROSPACE VEHICLE above a specified PRESSURE datum; a radio/radar altimeter is an instrument that displays the distance between an aircraft or SPACECRAFT DATUM and the surface vertically below as determined by a reflected radio/RADAR transmission.

altitude 1. (ASTRONOMY) The ANGLE between an observer's horizon and a TARGET CELESTIAL BODY. The altitude is 0° if the object is on the horizon and 90° if the object is at ZENITH (directly overhead). 2. (spacecraft) In SPACE VEHICLE navigation, the height above the mean surface of the reference celestial body. Note that the *distance* of a space vehicle or SPACECRAFT from the reference celestial body is taken as the distance from the center of the object.

altitude chamber A chamber within which the AIR PRESSURE and TEMPERATURE can be adjusted to simulate different conditions that occur at different ALTITUDES; used for crew training, experimentation, and AEROSPACE system testing. Also called *hypobaric chamber.*

altitude engine ROCKET ENGINE that is designed to operate at high-ALTITUDE conditions.

ambient The surroundings, especially of or pertaining to the environment around an aircraft, AEROSPACE VEHICLE, or other aerospace object (e.g., NOSE CONE or PLANETARY PROBE). The natural conditions found in a given environment, such as the ambient PRESSURE or ambient TEMPERATURE on the surface of a PLANET. For example, a planetary probe on the surface of VENUS must function in an infernolike environment where the ambient temperature is about 480° C (753 K).

Amor group A collection of NEAR-EARTH ASTEROIDS that cross the ORBIT of MARS but do not cross the orbit of EARTH. This ASTEROID group acquired its name from the 1-kilometer diameter Amor asteroid, discovered in 1932 by the Belgian astronomer Eugène Delporte (1882–1955).

amorphotoi Term used by the early Greek astronomers to describe the spaces in the night sky populated by dim STARS between the prominent groups of stars making up the ancient CONSTELLATIONS. It is Greek for *unformed.*

amorphous In planetary geology, a rock that lacks crystal structure; without form.

ampere (symbol: A) The SI UNIT of electric CURRENT, defined as the constant current that, if maintained in two straight parallel conductors of infinite length, of negligible circular cross sections, and placed 1 meter apart in a VACUUM, would produce a FORCE between these conductors equal to 2×10^{-7} NEWTONS per METER of length. The unit is named after the French physicist André M. Ampère (1775–1836).

amplifier A device capable of reproducing an input electrical or ELECTROMAGNETIC RADIATION SIGNAL with increased intensity, or GAIN. The ENERGY required to increase the intensity of the input signal is drawn from an external source. If the output signal is a linear function of the input signal, the device is called a *linear amplifier;* otherwise, it is called a *nonlinear amplifier.*

amplitude Generally, the maximum value of the displacement of a WAVE or other periodic phenomenon from a reference (average) position. Specifically, in ASTRONOMY and ASTROPHYSICS, the overall range of BRIGHTNESS (from maximum MAGNITUDE to minimum magnitude) of a VARIABLE STAR.

amplitude modulation (AM) In TELEMETRY and communications, a form of modulation in which the AMPLITUDE of the CARRIER WAVE is varied, or "modulated," about its unmodulated value. The amount of modulation is proportional to the amplitude of the SIGNAL wave. The FREQUENCY of the carrier wave is kept constant. *Compare with* FREQUENCY MODULATION.

analog (device) A device that depicts values by a continuously variable physical property, such as voltage, PRESSURE, or position. An analog representation of a SIGNAL can vary continuously over a range, while a digital representation of the same signal is restricted to a discrete set of numbers. A mercury-filled THERMOMETER is an example of a simple analog device. As the TEMPERATURE varies, the mercury moves within the device continuously to indicate all new temperature values (within the range of temperatures for the particular thermometer).

analog computer A computing device that processes continuously variable physical (analog) data, such as voltage or PRESSURE variations. Specialized analog computers are used in selected scientific and industrial applications. However, the vast majority of modern computers are digital devices that process discrete data, such as binary numbers. A slide rule is an example of a simple analog computer. *See also* DIGITAL COMPUTER.

analog-to-digital converter (ADC) A device that transforms continuously variable analog data or SIGNALs into discrete, digitized signals. In the ADC device, the incoming analog signal is sampled, digitized, and encoded.

ancient astronaut theory The (unproven) hypothesis that EARTH was visited in the past by a race of intelligent EXTRATERRESTRIAL beings who were exploring this portion of the MILKY WAY GALAXY.

ancient constellations The collection of approximately 50 CONSTELLATIONS drawn up by ancient astronomers and recorded by PTOLEMY, including such familiar constellations as the signs of the ZODIAC, Ursa Major (the Great Bear), Boötes (the Herdsman), and Orion (the Hunter).

androgynous interface A nonpolar interface; one that physically can join with another of the same design; literally, having both male and female characteristics.

android A term from science fiction describing a robot with near-human form or features.

Andromeda galaxy The Great Spiral Galaxy (or M31) in the CONSTELLATION of Andromeda, about 2.2 million LIGHT-YEARS (670 KILOPARSECS) away. It is the most distant object visible to the NAKED EYE and is the closest SPIRAL GALAXY to the MILKY WAY GALAXY.

anechoic chamber A test enclosure especially designed for experiments in acoustics. The interior walls of the chamber are covered with special materials (typically SOUND-absorbing, pyramid-shaped surfaces) that absorb sufficiently well the sound incident upon the walls, thereby creating an essentially "sound-free" condition in the FREQUENCY range(s) of interest.

angle The inclination of two intersecting lines to each other, measured by the ARC of a circle intercepted between the two lines forming the angle. There are many types of angles. An *acute angle* is less than 90°; a *right angle* is precisely 90°; an *obtuse angle* is greater than 90° but less that 180°; and a *straight angle* is 180°.

angle of attack The angle (commonly used symbol: α) between a reference line fixed with respect to an AIRFRAME and a line in the direction of movement of the body. *See also* AERODYNAMIC FORCE.

angle of incidence The ANGLE at which a ray of LIGHT (or other type of ELECTROMAGNETIC RADIATION) impinges on a surface. This angle is usually measured between the direction of propagation and a perpendicular to the surface at the point of incidence and, therefore, has a value between 0° and 90°.

angle of reflection The ANGLE at which a reflected ray of LIGHT (or other type of ELECTROMAGNETIC RADIATION) leaves a reflecting surface. This angle is usually measured between the direction of the outgoing ray and a perpendicular to the surface at the point of reflection. For a plane mirror the angle of reflection is equal to the ANGLE OF INCIDENCE.

angstrom (symbol: Å) A unit of length used to indicate the WAVELENGTH of ELECTROMAGNETIC RADIATION in the visible, near-infrared, and near-ultraviolet portions of the ELECTROMAGNETIC SPECTRUM. Named after the Swedish physicist Anders Jonas Ångstrom (1814–1874), who quantitatively described the SUN's SPECTRUM in 1868. One angstrom equals 0.1 NANOMETER (10^{-10}m).

angular acceleration (symbol: α) The time rate of change of ANGULAR VELOCITY (ω).

angular frequency (symbol: ω) The FREQUENCY of a periodic quantity expressed as ANGULAR VELOCITY in RADIANS per second. It is equal to the frequency (in HERTZ or CYCLES per second) times 2π radians per cycle.

angular measure Units of ANGLE generally expressed in terms of DEGREES (°), ARC MINUTES ('), and ARC SECONDS ("), where 1° of angle equals 60', and 1' equals 60". A full circle contains 360°, or 2π RADIANS. In the case of the KEPLERIAN ELEMENT called the *right ascension of the ascending node* (Ω), angular measure is

expressed in terms of hours, minutes, and seconds. This particular ORBITAL ELEMENT describes the rotation of the ORBITAL PLANE from the reference AXIS and helps uniquely specify the position and path of a SATELLITE (natural or human-made) in its ORBIT as a function of time.

angular momentum (symbol: L) A measure of an object's tendency to continue rotating at a particular rate around a certain AXIS. It is defined as the product of the ANGULAR VELOCITY (ω) of the object and its moment of INERTIA (I) about the axis of rotation; that is, L = I ω.

angular resolution The angle between two points that can just be distinguished by a DETECTOR and a collimator. The normal human eye has an angular resolution of about 1 ARC MINUTE (1'). A collimator is a device (often cylindrical in form) that produces a certain size BEAM of PARTICLES or LIGHT.

angular velocity (symbol: ω) The change of ANGLE per unit time; usually expressed in RADIANS per second.

anisotropic Exhibiting different properties along axes in different directions; an anisotropic RADIATOR would, for example, emit different amounts of RADIATION in different directions as compared to an ISOTROPIC radiator, which would emit radiation uniformly in all directions.

annihilation radiation Upon collision, the conversion of an atomic PARTICLE and its corresponding ANTIPARTICLE into pure ELECTROMAGNETIC ENERGY (called *annihilation radiation*). For example, when an ELECTRON (e^-) and POSITRON (e^+) collide, the minimum annihilation radiation released consists of a pair of GAMMA RAYS, each of approximately 0.511 million ELECTRON VOLTS (MeV) energy.

annual parallax (symbol: π) The PARALLAX of a STAR that results from the change in the position of a reference observing point during EARTH's annual

REVOLUTION around the SUN. It is the maximum angular displacement of the star that occurs when the star-Sun-Earth ANGLE is 90° (as illustrated in Figure 9). Also called the *heliocentric parallax*.

annular nozzle   A NOZZLE with a ring-shaped (annular) THROAT formed by an outer wall and a center body wall.

anode   The positive ELECTRODE in a system (e.g., an electrolytic cell or a discharge tube) whereby ELECTRONs leave the system.

anomalistic period   The time interval between two successive PERIGEE passages of a SATELLITE in ORBIT about its PRIMARY BODY. For example, the term *anomalistic month* defines the mean time interval between successive passages of the MOON through its closest point to EARTH (perigee), about 27.555 days.

anomaly   1. (AEROSPACE operations) A deviation from the normal or anticipated result. 2. (ASTRONOMY) The ANGLE used to define the position (at a particular time) of a celestial object, such as a PLANET or ARTIFICIAL SATELLITE in an ELLIPTICAL ORBIT about its PRIMARY BODY. The *true anomaly* of a planet is the angle (in the direction of the planet's motion) between the point of closest approach (i.e., the PERIHELION), the focus (the SUN), and the planet's current orbital position.

antenna   A device used to detect, collect, or transmit RADIO WAVES. A RADIO TELESCOPE is a large receiving antenna, while many SPACECRAFT have both a DIRECTIONAL ANTENNA and an OMNIDIRECTIONAL ANTENNA to transmit (DOWNLINK) TELEMETRY and to receive (UPLINK) instructions.

antenna array   A group of ANTENNAS coupled together into a system to obtain directional effects or to increase sensitivity. *See also* VERY LARGE ARRAY.

anthropic principle   The controversial hypothesis in modern COSMOLOGY that suggests that the UNIVERSE evolved in just the right way after the Big Bang event to allow for the emergence of human life.

antiballistic missile (ABM) system   A MISSILE system designed to intercept and destroy a strategic offensive BALLISTIC MISSILE or its deployed REENTRY VEHICLES. *See also* BALLISTIC MISSILE DEFENSE.

antiextrusion ring   Ring installed on the low-PRESSURE side of a seal or packing to prevent extrusion of the sealing material; sometimes called a *backup ring*.

antimatter   Matter in which the ordinary nuclear PARTICLES (such as ELECTRONS, PROTONS, and NEUTRONS) are replaced by their corresponding antiparticles—that is, POSITRONS, antiprotons, antineutrons, and so on. Sometimes called *mirror matter*. Normal matter and antimatter mutually annihilate each other upon contact and are converted into pure energy, called ANNIHILATION RADIATION.

antimissile missile (AMM)   A MISSILE launched against a hostile missile in flight.

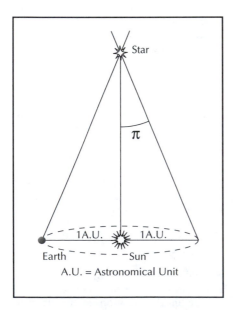

Figure 9. Annual parallax.

For example, a PATRIOT missile being fired against an incoming SCUD missile during Desert Storm operations in the Persian Gulf War of 1991. *See also* BALLISTIC MISSILE DEFENSE.

antiparticle Elementary atomic PARTICLES have corresponding real (or hypothetical) antiparticles. The antiparticle has equal MASS but opposite electric charge (or other property, as in the case of the NEUTRON and antineutron) of its normal matter twin. The antiparticle of the ELECTRON is the POSITRON; of the PROTON, the antiproton; and so on. However, the PHOTON is its own antiparticle. When a particle and its corresponding antiparticle collide, they are converted into ENERGY in a process called *annihilation*. *See also* ANTIMATTER; ANNIHILATION RADIATION.

antipodal Located at the opposite point on a CELESTIAL BODY. For example, EARTH's North Pole is antipodal to its South Pole.

antiradiation missile (ARM) A MISSILE that homes passively on a RADIATION source, such as an enemy RADAR SIGNAL.

antirotation device Mechanical device (e.g., a key) used in rotating machinery to prevent rotation of one component relative to an adjacent component.

antisatellite (ASAT) operations Defensive and offensive operations designed to neutralize, disrupt, or threaten hostile SATELLITES or important satellite control elements, such as a GROUND RECEIVING STATION or a relay satellite system.

antisatellite (ASAT) spacecraft A SPACECRAFT designed to destroy other SATELLITES in space. An ASAT spacecraft could be deployed in space disguised as a peaceful satellite that quietly lurks as a secret hunter/killer satellite, awaiting instructions to track and attack its prey.

antisatellite (ASAT) weapon A weapon system designed to destroy SATELLITES in space. The ASAT weapon may be launched from the ground or an aircraft or may be based in space. The TARGET satellite may be destroyed by nuclear or conventional explosion, collision at high velocity (i.e., a KINETIC ENERGY WEAPON), or DIRECTED ENERGY WEAPON (DEW).

antislosh baffle A device provided in a PROPELLANT tank of a LIQUID-FUEL ROCKET to dampen unwanted liquid motion, or sloshing, during flight. This device can take many forms, including flat ring, truncated CONE, and vane.

antisubmarine rocket (ASROC) A surface-ship-launched, ROCKET-propelled, nuclear depth charge or homing torpedo.

antivortex baffles Assemblies installed in the PROPELLANT tanks of LIQUID-FUEL ROCKETS to prevent GAS from entering the ROCKET ENGINE(s). The baffles minimize the rotating action as the propellants flow out the bottom of the tanks. Without these baffles, the rotating propellants would create a vortex similar to a whirlpool in a bathtub drain.

apareon That point on a MARS-centered ORBIT at which a SATELLITE or SPACECRAFT is at its greatest distance from the PLANET. The term is analogous to the term APOGEE.

apastron 1. The point in a body's ORBIT around a STAR at which it is at a maximum distance from the star. 2. That point in the orbit of one member of a BINARY STAR SYSTEM at which the stars are farthest apart. *Compare with* PERIASTRON.

aperture 1. The opening in front of a TELESCOPE, camera, or other optical instrument through which LIGHT passes. 2. The diameter of the OBJECTIVE of a telescope or other optical instrument. 3. Concerning a unidirectional ANTENNA, that portion of a plane surface near the antenna, perpendicular to the direction of maximum radiation, through which the major portion of the ELECTROMAGNETIC RADIATION passes.

aperture synthesis A resolution-improving technique in RADIO ASTRONOMY that uses a variable-APERTURE radio INTERFEROMETER to mimic the full-DISH size of a huge RADIO TELESCOPE.

apex The direction in the sky toward which the SUN and its system of PLANETS appear to be moving relative to the local STARS. Also called the *solar apex*, it is located in the CONSTELLATION of Hercules.

aphelion The point in an object's ORBIT around the SUN that is most distant from the Sun. *Compare with* PERIHELION.

Aphrodite Terra A large, fractured highland region near the EQUATOR of VENUS.

apoapsis That point in an ORBIT farthest from the orbited central body or center of attraction. *See also* ORBITS OF OBJECTS IN SPACE.

apocynthion That point in the LUNAR ORBIT of a SPACECRAFT or SATELLITE launched from EARTH that is farthest from the MOON. *Compare with* APOLUNE.

apogee 1. In general, the point at which a MISSILE TRAJECTORY or a SATELLITE ORBIT is farthest from the center of the gravitational field of the controlling body or bodies. 2. Specifically, the point that is farthest from EARTH in a GEOCENTRIC orbit. The term is applied to both the orbit of the MOON around Earth as well as to the orbits of ARTIFICIAL SATELLITES around Earth. At apogee, the VELOCITY of the orbiting object is at a minimum. To enlarge or "circularize" the orbit, a spacecraft's thruster (i.e., the APOGEE MOTOR) is fired at apogee, giving the SPACE VEHICLE and its PAYLOAD the necessary increase in velocity. Opposite of PERIGEE. 3. The highest ALTITUDE reached by a SOUNDING ROCKET launched from Earth's surface. *See also* ORBITS OF OBJECTS IN SPACE.

apogee motor A SOLID-PROPELLANT ROCKET motor that is attached to a SPACECRAFT and fired when the deployed space-craft is at the APOGEE of an initial (relatively low-ALTITUDE) PARKING ORBIT around EARTH. This firing establishes a new orbit farther from Earth or permits the spacecraft to achieve ESCAPE VELOCITY. Also called *apogee kick motor* or *apogee rocket*.

apojove The farthest point in an ORBIT around the PLANET JUPITER.

Apollo group A collection of NEAR-EARTH ASTEROIDS that have PERIHELION distances of 1.017 ASTRONOMICAL UNITS (AU) or less, taking them across the ORBIT of EARTH around the SUN. This group acquired its name from the first ASTEROID to be discovered, Apollo, in 1932 by the German astronomer Karl Reinmuth (1892–1979).

Apollo Lunar Surface Experiments Package (ALSEP) Scientific devices and equipment placed on the MOON by the APOLLO PROJECT ASTRONAUTs and left there to transmit data back to EARTH. Experiments included the study of METEORITE IMPACTS, LUNAR surface characteristics, seismic activity on the Moon, SOLAR WIND interaction, and analysis of the very tenuous lunar ATMOSPHERE.

Apollo Project The American effort in the 1960s and early 1970s to place ASTRONAUTS successfully on the surface of the MOON and return them safely to EARTH. The project was launched in May 1961 by President John F. Kennedy (1917–1963) in response to a growing space technology challenge from the former Soviet Union. Managed by NASA, the *Apollo 8* mission sent the first three humans to the vicinity of the Moon in December 1968. The *Apollo 11* mission involved the first human landing on another world (July 20, 1969). *Apollo 17,* the last LUNAR landing mission under this project, took place in December 1972. (See table.) The project is often considered one of the greatest technical accomplishments in all human history. (See Figure 10.) *See also* GEMINI PROJECT; LUNAR ORBITER; MERCURY PROJECT; MOON; RANGER PROJECT; SURVEYOR PROJECT.

Apollo Project Summary

Spacecraft Name	Crew	Date	Flight time (hours, minutes, seconds)	Revolutions	Remarks
Apollo 7	Walter H. Schirra Donn Eisele Walter Cunningham	10/11–22/68	260:8:45	163	First crewed Apollo flight demonstrated the spacecraft, crew, and support elements. All performed as required.
Apollo 8	Frank Borman James A. Lovell, Jr. William Anders	12/21–27/68	147:00:41	10 rev. of Moon	History's first crewed flight to the vicinity of another celestial body.
Apollo 9	James A. McDivitt David R. Scott Russell L. Schweikart	3/3–13/69	241:00:53	151	First all-up crewed Apollo flight (with Saturn V and command, service, and lunar modules). First Apollo extravehicular activity. First docking of command service module with lunar module (LM).
Apollo 10	Thomas P. Stafford John W. Young Eugene A. Cernan	5/18–26/69	192:03:23	31 rev. of Moon	Apollo LM descended to within 14.5 km of Moon and later rejoined command service module. First rehearsal in lunar environment.
Apollo 11	Neil A. Armstrong Michael Collins Edwin E. Aldrin, Jr.	7/16–24/69	195:18:35	30 rev. of Moon	First landing of a person on the Moon. Total stay time: 21 hr., 36 min.
Apollo 12	Charles Conrad, Jr. Richard F. Gordon, Jr Alan L. Bean	11/14–24/69	244:36:25	45 rev. of Moon	Second crewed exploration of the Moon. Total stay time: 31 hr., 31 min.
Apollo 13	James A. Lovell, Jr. John L. Swigert, Jr. Fred W. Haise, Jr.	4/11–17/70	142:54:41	—	Mission aborted because of service module oxygen tank failure.
Apollo 14	Alan B. Shepard, Jr. Stuart A. Roosa Edgar D. Mitchell	1/31–2/9/71	216:01:59	34 rev. of Moon	First crewed landing in and exploration of lunar highlands. Total stay time: 33 hr., 31 min.
Apollo 15	David R. Scott Alfred M. Worden James B. Irwin	7/26–8/7/71	295:11:53	74 rev. of Moon	First use of lunar roving vehicle. Total stay time: 66 hr., 55 min.
Apollo 16	John W. Young Thomas K. Mattingly II Charles M. Duke, Jr.	4/16–27/72	265:51:05	64 rev. of Moon	First use of remote-controlled television camera to record liftoff of the lunar module ascent stage from the lunar surface. Total stay time: 71 hr., 2 min.
Apollo 17	Eugene A. Cernan Ronald E. Evans Harrison H. Schmitt	12/7–19/72	301:51:59	75 rev. of Moon	Last crewed lunar landing and exploration of the Moon in the Apollo Program returned 110 kg of lunar samples to Earth. Total stay time: 75 hr.

Source: NASA.

Figure 10. *Apollo 12* astronauts Charles Conrad, Jr., and Alan L. Bean on the lunar surface, November 1969. *(Courtesy of NASA)*

Apollo-Soyuz Test Project (ASTP)
Joint United States–former Soviet Union space mission (July 1975) that centered on the RENDEZVOUS and DOCKING of the *Apollo 18* SPACECRAFT (three ASTRONAUT crew: Thomas Stafford, Vance Brand, and Donald [Deke] Slayton) and the *Soyuz 19* spacecraft (two COSMONAUT crew: Aleskey Leonov and Valeriy Kubasov). *See also* APOLLO PROJECT; SOYUZ SPACECRAFT.

apolune That point in an ORBIT around the MOON of a SPACECRAFT launched from the lunar surface that is farthest from the Moon. *Compare with* PERILUNE.

aposelene The farthest point in an ORBIT around the MOON. *See also* APO-CYNTHION; APOLUNE.

apparent In ASTRONOMY, observed. True values are reduced from apparent (observed) values by eliminating those factors, such as REFRACTION and flight time, that can affect the observation.

apparent diameter The observed diameter (but not necessarily the actual diameter) of a CELESTIAL BODY. Usually expressed in degrees, minutes, and seconds of arc. Also called ANGULAR DIAMETER.

apparent magnitude (symbol: *m*) The brightness of a STAR (or other CELESTIAL BODY) as measured by an observer on EARTH. Its value depends on the star's intrinsic BRIGHTNESS (LUMINOSITY), how far away it is, and how much of its LIGHT has been absorbed by the intervening INTER-

STELLAR MEDIUM. *See also* ABSOLUTE MAGNITUDE; MAGNITUDE.

apparent motion The observed motion of a heavenly body across the CELESTIAL SPHERE, assuming the EARTH is at the center of the celestial sphere and is standing still (stationary).

approach The maneuvers of a SPACECRAFT or AEROSPACE VEHICLE from its normal orbital position (STATION-KEEPING position) toward another orbiting spacecraft for the purpose of conducting RENDEZVOUS and DOCKING operations.

appulse 1. The APPARENT near APPROACH of one CELESTIAL BODY to another. For example, the apparent close approach (on the CELESTIAL SPHERE) of a PLANET or ASTEROID to a STAR, without the occurrence of an occultation. 2. A penumbral ECLIPSE of the MOON.

apsis (plural: apsides) In CELESTIAL MECHANICS, either of the two orbital points nearest (*periapsis*) or farthest (*apoapsis*) from the center of gravitational attraction. The apsides are called the PERIGEE and APOGEE for an ORBIT around EARTH and the PERIHELION and APHELION for an orbit around the SUN. The straight line connecting these two points is called the *line of apsides* and represents the major AXIS of an ELLIPTICAL ORBIT. *See also* ORBITS OF OBJECTS IN SPACE.

***Aqua* spacecraft** An advanced EARTH OBSERVING SATELLITE placed into POLAR ORBIT by NASA on May 4, 2002. The primary role of *Aqua,* as its name implies (Latin for "water"), is to gather information about changes in ocean circulation and how clouds and surface water processes affect Earth's climate. Equipped with six STATE OF THE ART instruments, *Aqua* will collect data on global precipitation, evaporation, and the cycling of water on a planetary basis. *See also* GLOBAL CHANGE; *TERRA* SPACECRAFT.

arc 1. In mathematics, a part of a curved line, such as a portion of a circle.

2. In physics, a luminous glow that appears when an electric current passes through ionized AIR or GAS. 3. In planetary science, an auroral arc.

arc discharge A luminous, gaseous electrical discharge in which charge transfer occurs continuously along a narrow channel of high ION DENSITY. An ARC discharge requires a continuous source of electric potential difference across the terminals of the arc. The intense IONIZATION needed to maintain the high CURRENT is provided by the evaporation of the ELECTRODES.

archaeological astronomy Scientific investigation concerning the astronomical significance of ancient structures and sites, such as Stonehenge in the United Kingdom.

arc-jet engine An electric ROCKET engine in which the PROPELLANT GAS is heated by passing through an electric ARC. In general, the CATHODE is located axially in the arc-jet engine, and the ANODE is formed into a combined plenum chamber/constrictor/NOZZLE configuration. The main problems associated with arc-jet engines are electrode erosion and low overall efficiency. Electrode erosion occurs as a result of the intense heating experienced by the electrodes and can seriously limit the thruster lifetime. The electrode erosion problem can be reduced by careful design of the electrodes. Overall arc-jet engine efficiency is dominated by "frozen-flow" losses, which are the result of dissociation and IONIZATION of the propellant. These "frozen-flow" losses are much more difficult to reduce since they are dependent on the THERMODYNAMICS of the flow and heating processes. A high-power arc-jet engine with EXHAUST VELOCITY values between 8×10^3 and 2×10^4 METERS per second is an attractive option for propelling an ORBITAL TRANSFER VEHICLE. *See also* ELECTRIC PROPULSION.

arc minute One-sixtieth (1/60th) of a DEGREE of ANGLE. This unit of angle is associated with precise measurements of motions and positions of CELESTIAL objects as occurs in the science of ASTROMETRY.

$$1° = 60 \text{ arc min} = 60'$$

See also ARC SECOND.

arc second One-three thousand six hundredth (1/3, 600th) of a DEGREE of ANGLE. This unit of angle is associated with very precise measurements of stellar motions and positions in the science of ASTROMETRY.

$$1' \text{ (arc min)} = 60 \text{ arc sec} = 60''$$

See also ARC MINUTE.

area ratio Ratio of the geometric flow area of the NOZZLE exit to the geometric flow area of the nozzle THROAT; also called *expansion area ratio* or simply *expansion ratio*. *See also* NOZZLE.

Arecibo Observatory The world's largest radio/radar TELESCOPE, with a 305-m (1,000-ft) diameter DISH. It is located in a large, bowl-shaped natural depression in the tropical jungle of Puerto Rico. When it operates as a RADIO WAVE receiver, the giant RADIO TELESCOPE can listen for SIGNALS from celestial objects at the farthest reaches of the UNIVERSE. As a RADAR transmitter/receiver, it assists astronomers and planetary scientists by bouncing signals off the MOON, off nearby PLANETS and their satellites, and even off layers of EARTH's IONOSPHERE. *See also* RADAR ASTRONOMY; RADIO ASTRONOMY.

areocentric With MARS as the center.

areodesy The branch of science that determines by careful observation and precise measurement the size and shape of the PLANET MARS and the exact position of points and features on the surface of the planet. *Compare with* GEODESY.

Ares The planet MARS. This name for the fourth PLANET in our SOLAR SYSTEM is derived from the ancient Greek word Ares (Αρεσ), their name for the mythological god of war (who was called Mars by the Romans). Today the term *Ares* is seldom used for Mars, except in combined forms, such as AREOCENTRIC and AREODESY.

argument of periapsis The argument (angular distance) of PERIAPSIS from the ASCENDING NODE. *See also* ORBITS OF OBJECTS IN SPACE.

Ariane The Ariane family of LAUNCH VEHICLES evolved from a European desire, first expressed in the early 1960s, to achieve and maintain an independent access to space. Early manifestation of the efforts that ultimately resulted in the creation of Arianespace (the international company that now markets the Ariane launch vehicles) included France's Diamant launch vehicle program (with operations at Hammaguir, Algeria, in the Sahara Desert) and the Europa launch vehicle program, which operated in the Woomera range in Australia before moving to Kourou, French Guiana, in 1970. These early efforts eventually yielded the first Ariane flight on December 24, 1979. That mission, called L01, was followed by 10 more Ariane-1 flights over the next six years. The initial Ariane vehicle family (Ariane-1 through Ariane-4) centered on a three-STAGE launch vehicle configuration with evolving capabilities. The Ariane-1 vehicle gave way in 1984 to the more powerful Ariane-2 and Ariane-3 vehicle configurations. These configurations, in turn, were replaced in June 1988 with the successful launch of the Ariane-4 vehicle—a launcher that has been called Europe's "space workhorse." Ariane-4 vehicles are designed to orbit SATELLITES with a total mass value of up to 4,700 kg (10,350 lbm). The various launch vehicle versions differ according to the number and type of strap-on boosters and the size of the fairings.

On October 30, 1997, a new, more powerful launcher, called Ariane-5, joined the Ariane family, when its second qualification flight (called V502 or Ariane 502) successfully took place at the GUIANA SPACE CENTER in Kourou. Twenty-seven minutes into this flight, the Maqsat-H and Maqsat-B platforms, carrying INSTRUMENTS to analyze the new launch vehicle's performance, and the TeamSat technology satellite were injected into orbit. The Ariane-5 launch vehicle is an advanced "two-stage" system, consisting of a powerful

LIQUID HYDROGEN/LIQUID OXYGEN-fueled main engine (called the Vulcain) and two strap-on SOLID PROPELLANT ROCKETs. It is capable of placing a satellite payload of 5,900 kg (13,000 lbm) (dual satellite launch) or 6,800 kg (14,950 lbm) (single satellite launch) into geostationary transfer orbit (GTO), and approximately 20,000 kg (44,000 lbm) into low EARTH orbit (LEO). (See Figure 11.)

However, the development of the Ariane-5 has not been without some technical challenge. On June 4, 1996, the maiden flight of the Ariane-5 launch vehicle ended in a spectacular failure. Approximately 40 seconds after initiation of the flight sequence, at an altitude of about 3,700 m (12,140 ft), a guidance system error caused the vehicle to suddenly veer off its flight path, break up, and explode—as the errant vehicle properly self-destructed. Then, on December 12, 2002, approximately three minutes after lift-off, the more powerful (10-tonne) Ariane-5 vehicle involved in Flight 157 experienced another catastrophic failure. A leak in the Vulcain 2 NOZZLE's cooling circuit during ascent appears to have produced a nozzle failure that then led to a loss of control over the launch vehicle's TRAJECTORY.

The last flight of the Ariane-4 system is scheduled for 2003. Beyond that point, the continuously improving Ariane-5 PROPULSION SYSTEM will serve as the main launch vehicle for the European Community, placing a wide variety of payloads into low Earth orbit, into GEOSTATIONARY EARTH ORBIT (GEO), and on INTERPLANE-TARY trajectories.

Ariane vehicles are launched from Kourou, French Guiana, in South America. Kourou was chosen in part because it is close to Earth's EQUATOR, which makes the site ideal for launching most satellites. Following the first firing of a Diamant rocket by CNES (the French Space Agency) in 1970, Europe decided to use Kourou for its Europa launch vehicles. In 1975 the European Space Agency (ESA) took over the existing European facilities at the Guiana Space Center to build the ELA-1 complex (the Ensemble de Lancement Ariane, or Ariane Launch Complex)

for its Ariane-1, 2, and 3 launch vehicles. The ESA next created the ELA-2 complex for the Ariane-4 vehicle. The Ariane-5 vehicle is launched from ELA-3, a modern launch complex whose facilities spread over 21 km^2 (8.1 mi^2). In all, the "European spaceport" at Kourou covers 960 km^2 (370.6 mi^2) and has a workforce of more than 1,000 persons. *See also* CENTRE NATIONAL D'ETUDES SPATIALES (CNES); EUROPEAN SPACE AGENCY.

Armstrong, Neil A. (b. 1930) American ASTRONAUT who served as the commander for NASA's *Apollo 11* LUNAR-landing mission in July 1969. As he became the first human being to set foot on the MOON (July 20, 1969), he uttered these historic words: "That's one small step for a man, one giant leap for mankind." *See also* APOLLO PROJECT.

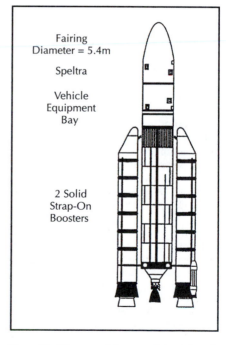

Figure 11. The general features of the Ariane-5 "two-stage" launch vehicle, consisting of a powerful liquid hydrogen/liquid oxygen–fueled Vulcain main engine and two strap-on solid propellant rockets.

articulated Segmented or jointed and thereby able to accommodate motion.

artificial gravity Simulated GRAVITY conditions established within a SPACECRAFT, SPACE STATION, or SPACE SETTLEMENT. Rotating the space system about an AXIS creates this condition, since the CENTRIFUGAL FORCE generated by the rotation produces effects similar to the FORCE of gravity within the vehicle. This technique was first suggested by the Russian ASTRONAUTICS pioneer KONSTANTIN TSIOLKOVSKY at the start of the 20th century.

artificial intelligence (AI) Information-processing functions (including thinking and perceiving) performed by machines that imitate (to some extent) the mental activities performed by the human brain. Advances in AI will allow "very smart" ROBOT SPACECRAFT to explore distant alien worlds with minimal human supervision. *See also* ROBOTICS; TELEOPERATION.

artificial satellite A human-made object, such as a SPACECRAFT, placed in ORBIT around EARTH or other CELESTIAL BODY. SPUTNIK 1 was the first artificial satellite to be placed in orbit around Earth.

ascending node That point in the ORBIT of a CELESTIAL BODY when it travels from south to north across a reference plane, such as the equatorial plane of the CELESTIAL SPHERE or the plane of the ECLIPTIC. Also called the *northbound node*. *Compare with* DESCENDING NODE.

asteroid A small, solid, rocky object that orbits the SUN but is independent of any major PLANET. Most asteroids (or minor planets) are found in the main ASTEROID BELT. The largest asteroid is Ceres, about 1,000 km in diameter and discovered in 1801 by the Italian astronomer Giuseppe Piazzi (1746–1826). EARTH-CROSSING ASTEROIDS (ECAs) or NEAR-EARTH ASTEROIDS (NEAs), have ORBITS that take them near or across EARTH's orbit around the SUN and are divided into the ATEN, APOLLO, and AMOR GROUPS. Some scientists have postulated that the impact of an Earth-crossing asteroid may have been responsible for the disappearance of the dinosaurs and many other animal species on Earth some 65 million years ago. Asteroids have been classified on the basis of their spectral characteristics—that is, by how much LIGHT of different colors they reflect. The three major classifications are: C-type (carbonaceous), S-type (silicaceous), and M-type (metallic). *See also* EXTRATERRESTRIAL CATASTROPHE THEORY; NEAR-EARTH ASTEROID RENDEZVOUS (NEAR) MISSION; TROJAN GROUP.

asteroid belt The region of OUTER SPACE between the ORBITS of MARS and JUPITER that contains the great majority of the ASTEROIDS. These minor planets or planetoids have ORBITAL PERIODS of between three and six years and travel around the SUN at distances of between 2.2 to 3.3 ASTRONOMICAL UNITS (AUs).

asteroid detection system (ADS) A proposed planetary defense observation network consisting of multispectral ground-based and space-based SENSORS for the SURVEILLANCE, detection, TRACKING, and characterization of (natural) space objects (i.e., ASTEROIDS and COMETS) that might pose a major threat if they were to collide with EARTH. This advanced system also would include a central facility to collect data from all the sensors in the network, maintain a current DATABASE of all known space objects, and disseminate collected information to appropriate authorities.

asteroid mining The ASTEROIDS, especially EARTH-CROSSING ASTEROIDS, might serve as "space warehouses" from which future space workers could extract certain critical materials needed in space-based industrial activities and in the construction and operation of large space habitats. On some asteroids scientists expect to find water (trapped), organic compounds, and metals. In time, it may be far more efficient to mine an Earth-crossing asteroid than to lift the very same raw materials from EARTH's surface. Therefore, some space visionaries now suggest that asteroid mining could play a major role in the evolution of a permanent human civilization in space.

asteroid negation system (ANS) A proposed planetary defense system capable of intercepting any (natural) space object (i.e., an ASTEROID or COMET) that was determined to be a threat to EARTH. The system would accomplish this negation task in sufficient time to deflect the space object from its collision course with Earth or else fragment it into smaller pieces that would not pose a major planetary threat. Deflection and/or fragmentation could be accomplished with a variety of means, ranging from the detonation of nuclear explosives to the use of high SPECIFIC IMPULSE thrusters, KINETIC ENERGY PROJECTILES, or DIRECTED ENERGY devices.

astro- A prefix that means "STAR" or (by extension) OUTER SPACE or CELESTIAL; for example, ASTRONAUT, ASTRONAUTICS, and ASTROPHYSICS.

astrobiology The search for and study of living organisms found on CELESTIAL bodies beyond EARTH. *See also* EXOBIOLOGY.

astrobleme A geological structure (often eroded) produced by the HYPERVELOCITY IMPACT of a METEOROID, COMET, or ASTEROID.

astrochimp(s) The nickname given to the primates used during the early U.S. space program. In the MERCURY PROJECT (the first U.S. crewed space program), these astrochimps were used to test SPACE CAPSULE and LAUNCH VEHICLE system hardware prior to its commitment to human flight. For example, on January 31, 1961, a 17 kg (37 lbm) chimpanzee named Ham was launched from CAPE CANAVERAL AIR FORCE STATION by a REDSTONE ROCKET on a suborbital flight test of the Mercury Project SPACECRAFT. During this mission, the PROPULSION SYSTEM developed more THRUST than planned, and the simian space traveler experienced overacceleration. After recovery Ham appeared to be in good physical condition. (See Figure 12.)

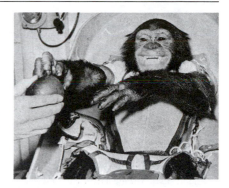

Figure 12. Astrochimp Ham reaches out for an apple after his successful suborbital rocket flight on January 31, 1961. *(Courtesy of NASA)*

However, when shown another space capsule, his reactions made it quite clear to his handlers that this particular astrochimp wanted no further role in the space program.

On November 29, 1961, another astrochimp, a 17-kg (37 lbm) chimpanzee named Enos was launched successfully from Cape Canaveral by an ATLAS rocket on the final orbital qualification flight test of the Mercury Project spacecraft. During the flight, Enos performed psychomotor duties and upon recovery was found to be in excellent physical condition.

astrodynamics The application of CELESTIAL MECHANICS, PROPULSION SYSTEM theory, and related fields of science and engineering to the problem of carefully planning and directing the TRAJECTORY of a SPACE VEHICLE.

astrolabe Instrument used by ancient astronomers to measure the ALTITUDE of a STAR.

astrology Attempt by many early astronomers to forecast future events on EARTH by observing and interpreting the relative positions of the FIXED STARS, the SUN, the PLANETS, and the MOON. Such mystical stargazing was a common activity in most ancient societies, was enthusiastically practiced in western Europe up through the 17th century, and lingers today as daily horoscopes. At the dawn of the scientific revolution, GALILEO GALILEI taught a required university course on medical astrology, and JOHANNES KEPLER earned a living as a court astrologer. The popular "science" of astrology is based on the unscientific hypothesis that the motion of CELESTIAL bodies controls and influences human lives and TERRESTRIAL events. *See also* ZODIAC.

astrometric binary A BINARY STAR SYSTEM in which irregularities in the PROPER MOTION (wobbling) of a visible STAR imply the presence of an undetected companion.

astrometry Branch of ASTRONOMY that involves the very precise measurement of the motion and position of CELESTIAL bodies.

astronaut Within the American space program, a person who travels in OUTER SPACE; a person who flies in an AEROSPACE VEHICLE to an ALTITUDE of more than 80 km (50 mi). The word comes from a combination of two ancient Greek words that literally mean "STAR" (*astro*) "sailor or traveler" *(naut). Compare with* COSMONAUT.

astronautics The branch of engineering science dealing with spaceflight and the design and operation of SPACE VEHICLES.

astronomical Of or pertaining to ASTRONOMY or to observations of CELESTIAL objects and phenomena.

Astronomer Royal The honorary title created in 1675 by King Charles II and given to a prominent English astronomer. Up until 1971, the Astronomer Royal also served as the director of the ROYAL GREENWICH OBSERVATORY. John Flamsteed (1646–1719) was the first English astronomer to hold this position. He served from 1675 until his death in 1719.

astronomical unit (AU) A convenient unit of distance defined as the semimajor axis of EARTH'S ORBIT around the SUN. One AU, the average distance between Earth and the Sun, is equal to approxi-

mately 149.6×10^6 km (approximately 92.9×10^6 mi), or 499.01 light-seconds.

astronomy Branch of science that deals with CELESTIAL bodies and studies their size, composition, position, origin, and dynamic behavior. *See also* ASTROPHYSICS; COSMOLOGY.

astrophotography The use of photographic techniques to create IMAGES of CELESTIAL bodies. Astronomers are now replacing LIGHT-sensitive photographic emulsions with CHARGED-COUPLED DEVICES (CCDs) to create digital images in the visible, infrared, and ultraviolet portions of the ELECTROMAGNETIC SPECTRUM.

astrophysics The branch of physics that investigates the nature of STARS and star systems. It provides the theoretical principles enabling scientists to understand ASTRONOMICAL observations. ASTRONOMY addresses fundamental questions that have puzzled human beings since our primitive beginnings. What is the nature of the UNIVERSE? How did it begin, how is it evolving, and what will be its eventual fate? As important as these questions are, there is another motive for astronomical studies. Since the 17th century, when SIR ISAAC NEWTON's studies of CELESTIAL MECHANICS helped him formulate the three basic laws of motion and the universal law of GRAVITATION, the sciences of astronomy and physics have become intertwined. Astrophysics provides the theoretical framework for understanding astronomical observations. Astrophysicists also predict new phenomena such as BLACK HOLES—often before such unusual objects are actually discovered by astronomers. The vast laboratory of OUTER SPACE makes it possible to investigate large-scale physical processes that cannot be duplicated in a TERRESTRIAL laboratory.

Modern astrophysics has within its reach the ability to bring about one of the greatest scientific achievements ever— a unified understanding of the total evolutionary scheme of the universe. This remarkable revolution in astrophysics is happening thanks to the confluence of two streams of technical development: REMOTE SENSING and space technology. With remote sensing, scientists have developed sensitive INSTRUMENTS capable of detecting and analyzing RADIATION across the entire range of the ELECTRO-MAGNETIC (EM) SPECTRUM. Space technology enables astrophysicists to place these sensitive instruments above EARTH's ATMOSPHERE, where they can sense electromagnetic SIGNALS over nearly the entire spectrum (i.e., from the primordial COSMIC MICROWAVE BACKGROUND of the big bang explosion to the energetic GAMMA RAY emissions associated with compact bodies, such as NEUTRON STARS and black holes). Due to the absorbing nature of Earth's atmosphere, ground-based observatories are restricted to a rather narrow range of observable WAVELENGTHS. *See also* BIG BANG THEORY; *CHANDRA X-RAY OBSERVATORY*; *COMPTON GAMMA RAY OBSERVATORY*; COSMIC BACKGROUND EXPLORER; COSMIC RAYS; COSMOLOGY; GAMMA RAY ASTRONOMY; *HUBBLE SPACE TELESCOPE*; INFRARED ASTRONOMY; ULTRAVIOLET ASTRONOMY; X-RAY ASTRONOMY.

asymmetric Lacking a mirror-image construction on both sides of a dividing line.

Aten group A collection of NEAR-EARTH ASTEROIDS that cross the ORBIT of EARTH but whose average distances from the SUN lie inside Earth's orbit. This ASTEROID group acquired its name from the 0.9-km (0.6-mi) diameter asteroid Aten, discovered in 1976 by the American astronomer Eleanor Kay Helin (née Francis).

Atlas A family of versatile LIQUID-FUEL ROCKETs originally developed by General Bernard Schriever (b. 1913) of the UNITED STATES AIR FORCE in the late 1950s as the first operational American INTERCONTINENTAL BALLISTIC MISSILE (ICBM). Evolved and improved Atlas LAUNCH VEHICLEs now serve many government and commercial space transportation needs. The modern Atlas ROCKET vehicle fleet, as developed and marketed by the Lockheed Martin Company, included three basic families: the Atlas II (IIA and IIAS), the

Atlas III (IIIA and IIIB), and the Atlas V (400 and 500 series). The Atlas II family was capable of lifting PAYLOADS ranging in MASS from approximately 2,800 kg (6,200 lbm) to 3,700 kg (8,100 lbm) to a geostationary TRANSFER ORBIT (GTO). The Atlas III family of launch vehicles can lift payloads of up to 4,500 kg (9,900 lbm) to a GTO, while the more powerful Atlas V family can lift payloads of up to 8,650 kg (19,100 lbm) to a GTO.

The family of Atlas II vehicles evolved from the Atlas I family military MISSILES and COLD WAR era SPACE LAUNCH VEHICLEs. The transformation was brought about by the introduction of higher THRUST engines and longer PROPELLANT tanks for both STAGES. The Atlas III family incorporated the pressure-stabilized design of the Atlas II vehicle but used a new single-stage Atlas engine built by NPO Energomash of Russia. The RD-180 is a throttleable ROCKET ENGINE that uses LIQUID OXYGEN and KEROSENE propellants and provides approximately 3,800 kilonewtons (860,000 pounds-force [lbf]) of thrust at sea level. The Atlas V uses the RD-180 main engine in a Common Core Booster first-stage configuration with up to five strap on SOLID PROPELLANT ROCKETS and a CENTAUR UPPER STAGE. The first Atlas V vehicle successfully lifted off from CAPE CANAVERAL AIR FORCE STATION on August 21, 2002. Later that year an important chapter in PROPULSION SYSTEM history closed when the last Atlas II vehicle was successfully launched from Cape Canaveral Air Force Station on December 4, 2002.

atmosphere 1. Gravitationally bound gaseous envelope that forms an outer region around a PLANET or other CELESTIAL BODY. 2. (cabin) Breathable environment inside a SPACE CAPSULE, AEROSPACE VEHICLE, SPACECRAFT, or SPACE STATION. 3. (EARTH's) Life-sustaining gaseous envelope surrounding Earth. Near sea level it contains the following composition of GASES (by volume): nitrogen 78%, oxygen 21%, argon 0.9%, and carbon dioxide 0.03%. There are also lesser amounts of many other gases, including water vapor and human-generated chemical pollutants.

Earth's electrically neutral atmosphere is composed of four primary layers: troposphere, stratosphere, mesosphere, and thermosphere. Life occurs in the troposphere, the lowest region, which extends up to about 16 km (9.94 mi) ALTITUDE. It is also the place within which most of Earth's weather occurs. Extending from about 16 km altitude to 48 km (29.8 mi) is the stratosphere. The vast majority (some 99%) of the AIR in Earth's atmosphere is located in these two regions. Above the stratosphere, from approximately 48 km (29.8 mi) to 85 km (52.8 mi) is the mesosphere. In this region, TEMPERATURE decreases with altitude and reaches a minimum of –90°C at about 85 km (the mesopause). Finally, the uppermost layer of the atmosphere is called the thermosphere and extends from the mesopause out to approximately 1,000 km (621 mi). In this region, temperature rises with altitude as oxygen and nitrogen MOLECULES absorb ULTRAVIOLET RADIATION from the SUN. This heating process creates ionized ATOMS and molecules and forms the layers of the IONOSPHERE, which actually starts at about 60 km (37.3 mi). Earth's upper atmosphere is characterized by the presence of electrically charged gases, or PLASMA. The boundaries of and structure within the ionosphere vary considerably according to SOLAR ACTIVITY. Finally, overlapping the ionosphere is the MAGNETOSPHERE, which extends from approximately 80 km (49.7 mi) to 65,000 km (40,400 mi) on the side of Earth toward the Sun and trails out more than 300,000 km (186,540 mi) on the side of Earth away from the Sun. The geomagnetic field of Earth plays a dominant role in the behavior of charged PARTICLES in the magnetosphere.

atmospheric braking The action of slowing down an object entering the ATMOSPHERE of EARTH or another PLANET from space by using the aerodynamic DRAG exerted by the atmosphere. See also AEROBRAKE.

atmospheric compensation The physical distortion or modification of the components of an optical system to compensate for the distortion of LIGHT WAVES as they pass through the ATMOSPHERE and the optical system.

atmospheric drag With respect to EARTH-orbiting SPACECRAFT, the retarding FORCE produced on a SATELLITE, AEROSPACE VEHICLE, or spacecraft by its passage through upper regions of Earth's ATMOSPHERE. This retarding force drops off exponentially with ALTITUDE and has only a small effect on spacecraft whose PERIGEE is higher than a few hundred KILOMETERS. For spacecraft with lower perigee values, the cumulative effect of atmospheric drag eventually will cause them to reenter the denser regions of Earth's atmosphere (and be destroyed) unless they are provided with an onboard PROPULSION SYSTEM that can accomplish periodic reboost.

atmospheric entry The penetration of any planetary ATMOSPHERE by an object from OUTER SPACE; specifically, the penetration of EARTH's atmosphere by a crewed or uncrewed SPACE CAPSULE, AEROSPACE VEHICLE, or SPACECRAFT. Also called *entry* and sometimes *reentry* (although typically the object is making its initial return or entry after a flight in space).

atmospheric pressure The PRESSURE (FORCE per unit area) at any point in a PLANET's ATMOSPHERE due solely to the WEIGHT of the atmospheric GASes above that point.

atmospheric probe The special collection of scientific INSTRUMENTS (usually released by a MOTHER SPACECRAFT) for determining the PRESSURE, composition, and TEMPERATURE of a PLANET's ATMOSPHERE at different ALTITUDES. An example is the probe released by NASA's GALILEO PROJECT SPACECRAFT in December 1995. As it plunged into the JOVIAN atmosphere, the probe successfully transmitted its scientific data to the Galileo spacecraft (the mother spacecraft) for about 58 minutes.

atmospheric refraction REFRACTION resulting when a ray of radiant ENERGY passes obliquely through an ATMOSPHERE.

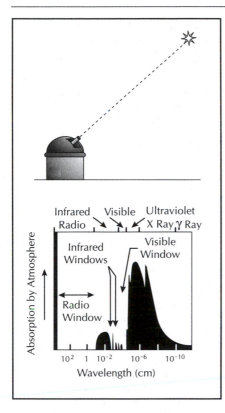

Figure 13. Earth's atmosphere absorbs much of the electromagnetic radiation arriving from sources in outer space. The atmosphere contains only a few wavelength bands (called atmospheric windows) that transmit most of the incident radiation within that band to a detector on the planet's surface.

atmospheric window A WAVELENGTH interval within which a PLANET'S ATMOSPHERE is transparent (that is, easily transmits) ELECTROMAGNETIC RADIATION. (See Figure 13.)

atom A tiny PARTICLE of matter (the smallest part of an ELEMENT) indivisible by chemical means. It is the fundamental building block of the chemical elements. The elements, such as HYDROGEN (H), HELIUM (He), carbon (C), iron (Fe), lead (Pb), and uranium (U), differ from one another because they consist of different types of atoms. According to (much sim-

plified) modern atomic theory, an atom consists of a dense inner core (the NUCLEUS) that contains PROTONS and NEUTRONS and a cloud of orbiting ELECTRONS. Atoms are electrically neutral, with the number of (positively charged) protons being equal to the number of (negatively charged) electrons.

atomic clock A precise device for measuring or standardizing time that is based on periodic vibrations of certain ATOMS (cesium) or MOLECULES (ammonia). Widely used in military and civilian SPACECRAFT, as, for example, the GLOBAL POSITIONING SYSTEM (GPS).

atomic mass The MASS of a neutral ATOM of a particular NUCLIDE usually expressed in ATOMIC MASS UNITS (amu). *See also* MASS NUMBER.

atomic mass unit (amu) One-twelfth (1/12) the MASS of a neutral ATOM of the most abundant ISOTOPE of carbon, carbon-12.

atomic number (symbol: Z) The number of PROTONS in the NUCLEUS of an ATOM and also its positive charge. Each chemical ELEMENT has its characteristic atomic number. For example, the atomic number for carbon is 6, while the atomic number for uranium is 92.

atomic weapon An explosive weapon in which the ENERGY is produced by nuclear FISSION or FUSION. *See also* NUCLEAR WEAPON.

atomic weight The MASS of an ATOM relative to other atoms. At present, the most abundant ISOTOPE of the ELEMENT carbon, namely carbon-12, is assigned an atomic weight of exactly 12. As a result, 1/12 the mass of a carbon-12 atom is called one ATOMIC MASS UNIT, which is approximately the mass of one PROTON or one NEUTRON. Also called *relative atomic mass*.

atomize To divide a LIQUID into extremely minute PARTICLES, either by impact with a JET of steam or compressed AIR or by passage through some mechanical device, such as an INJECTOR.

attached payload 1. With respect to the SPACE SHUTTLE, any PAYLOAD that is launched in the ORBITER, remains structurally and functionally attached throughout the flight, and is landed in the Orbiter. 2. With respect to a SPACE STATION, a payload located on the base truss outside the pressurized MODULES.

attenuation The decrease in intensity (strength) of an electromagnetic wave as it passes through a transmitting medium. This loss is due to absorption of the incident ELECTROMAGNETIC RADIATION (EMR) by the transmitting medium or to scattering of the EMR out of the path of the detector. Attenuation does not include the reduction in EMR wave strength due to geometric spreading as a consequence of the INVERSE SQUARE LAW. The attenuation process can be described by the equation:

$$I = I_0 e^{-\alpha x}$$

where I is the FLUX (PHOTONS OR PARTICLES per unit area) at a specified distance (x) into the attenuating medium, I_0 is the initial flux, α is the attenuation coefficient (with units of inverse length), and e is a special mathematical function, the irrational number 2.71828. . . . The expression e^x is called the exponential function.

attitude The position of an object as defined by the inclination of its axes with respect to a frame of reference. The orientation of a SPACE VEHICLE (e.g., a SPACE-CRAFT or AEROSPACE VEHICLE) that is either in motion or at rest, as established by the relationship between the vehicle's axes and a reference line or plane. Attitude is often expressed in terms of PITCH, ROLL, and YAW.

attitude control system The onboard system of computers, low-THRUST ROCK-ETS *(thrusters),* and mechanical devices (such as a MOMENTUM wheel) used to keep a SPACECRAFT stabilized during flight and to precisely point its instruments in some desired direction. Stabilization is achieved by spinning the spacecraft or by using a three-AXIS active approach that maintains the spacecraft in a fixed, reference ATTITUDE by firing a selected combination of thrusters when necessary.

Stabilization can be achieved by spinning the spacecraft, as was done on the *PIONEER 10* and *11* missions to the outer SOLAR SYSTEM. In this approach, the gyroscopic action of the rotating spacecraft MASS is the stabilizing mechanism. PROPULSION SYSTEM thrusters are fired to make any desired changes in the spacecraft's spin-stabilized attitude.

Spacecraft also can be designed for active three-axis stabilization, as was done on the *VOYAGER 1* and *2* spacecraft, which explored the outer solar system and beyond. In this method of stabilization, small propulsion system thrusters gently nudge the spacecraft back and forth within a DEADBAND of allowed attitude error. Another method of achieving active three-axis stabilization is to use electrically powered REACTION wheels, which are also called momentum wheels. These massive wheels are mounted in three orthogonal axes onboard the spacecraft. To rotate the spacecraft in one direction, the proper wheel in the opposite direction is spun. To rotate the vehicle back, the wheel is slowed down. Excessive momentum, which builds up in the system due to internal FRICTION and external FORCES, occasionally must be removed from the system; this usually is accomplished with propulsive maneuvers.

Either general approach to spacecraft stabilization has basic advantages and disadvantages. Spin stabilized vehicles provide a continuous "sweeping motion" that is generally desirable for fields and PARTI-CLE INSTRUMENTS. However, such spacecraft may then require complicated systems to despin ANTENNAS or optical instruments that must be pointed at TAR-GETS in space. Three-axis controlled spacecraft can point antennas and optical instruments precisely (without the necessity for despinning), but these craft then may have to perform rotation maneuvers to use their fields and particle science instruments properly.

attitude jet   A JETSTREAM used to correct or change the ATTITUDE of an aircraft, SPACECRAFT, or AEROSPACE VEHICLE; the NOZZLE that directs this jetstream.

audible sound   SOUND containing FREQUENCY components lying between 15 HERTZ (a low rumble) and 20,000 hertz (a shrill, high-pitched whistle)—the hearing limits of the human ear.

aurora   Visible glow in a PLANET's upper ATMOSPHERE (IONOSPHERE) caused by interaction of the planet's MAGNETOSPHERE and PARTICLES from the SUN (SOLAR WIND). On EARTH, the aurora borealis (or Northern lights) and the aurora australis (or southern lights) are visible manifestations of the magnetosphere's dynamic behavior. At high LATITUDES, disturbances in Earth's geomagnetic field accelerate trapped particles into the upper atmosphere, where they excite nitrogen MOLECULES (red emissions) and oxygen ATOMS (red and green emissions). Auroras also occur on JUPITER, SATURN, URANUS, and NEPTUNE.

autoignition   Self-ignition or spontaneous combustion of PROPELLANT.

automated payload   PAYLOAD that is supported by an uncrewed SPACECRAFT capable of operating independently of the SPACE SHUTTLE ORBITER vehicle. Automated payloads are detached from the orbiter during the operational phase of their flights.

automatic pilot   Equipment that automatically stabilizes the ATTITUDE of a vehicle about its PITCH, ROLL, and YAW axes. Also called *autopilot;* sometimes referred as "George"—as in "let George (the autopilot) fly the vehicle."

auxiliary power unit (APU)   A power unit carried on a SPACECRAFT or AEROSPACE VEHICLE that supplements the main source of electric power on the craft.

auxiliary stage   1. A small propulsion unit that may be used with a PAYLOAD. One or more of these units may be used to provide the additional VELOCITY required to place a payload in the desired ORBIT or TRAJECTORY. 2. A PROPULSION SYSTEM used to provide midcourse trajectory corrections, braking maneuvers, and/or orbital adjustments.

avionics   The contraction of *avi*ation and electr*onics*. The term refers to the application of electronics to systems and equipment used in AERONAUTICS and ASTRONAUTICS.

Avogadro's hypothesis   The volume occupied by a MOLE of GAS at a given PRESSURE and TEMPERATURE is the same for all gases. This hypothesis also can be expressed as: Equal volumes of all gases measured at the same temperature and pressure contain the same number of MOLECULES. The Italian physicist Count Amedeo Avogadro (1776–1856) first proposed this hypothesis in 1811. In a strict sense, it is valid only for IDEAL GASes.

Avogadro's number (symbol: NA)   The number of MOLECULES per MOLE of any substance.

$$N_A = 6.022 \times 10^{23} \text{ molecules/mole}$$

Also called the *Avogadro constant.*

axis (plural: axes)   Straight line about which a body rotates (axis of rotation) or along which its CENTER OF GRAVITY moves (axis of translation). Also, one of a set of reference lines for a coordinate system, such as the x-axis, y-axis, and z-axis in CARTESIAN COORDINATES.

azimuth   The horizontal direction or bearing to a CELESTIAL BODY measured in DEGREES clockwise from north around a TERRESTRIAL observer's horizon. On EARTH, azimuth is 0° for an object that is due north, 90° for an object due east, 180° for an object due south, and 270° for an object due west. *See also* ALTITUDE (ASTRONOMY).

B

background radiation The NUCLEAR RADIATION that occurs in the natural environment, including COSMIC RAYS and radiation from naturally radioactive ELEMENTS found in the air, the soil, and the human body. Sometimes referred to as *natural radiation background*. The term also may mean IONIZING RADIATION (natural and human-made) that is unrelated to a particular experiment or series of measurements.

backlobe A RADIATION (RADIO FREQUENCY) LOBE whose AXIS makes an ANGLE of approximately 180 DEGREES with respect to the axis of the major lobe of the ANTENNA.

backout The process of undoing tasks that have already been completed during the COUNTDOWN of a LAUNCH VEHICLE, usually in reverse order.

backscattering In general, the SCATTERING of RADIATION in a direction generally opposite to the direction of the incident radiation. In RADAR SURVEILLANCE, that portion of the radar BEAM scattered directly back toward the source by the terrain surface.

backup 1. A unit or item kept available to replace one that fails to perform satisfactorily. 2. An item or system under development (e.g., a PLANETARY PROBE) intended to perform the same general functions as another item or system under development.

backup crew A crew of ASTRONAUTS or COSMONAUTS trained to replace the prime crew, if necessary, on a particular space MISSION.

baffle 1. In the COMBUSTION CHAMBER of a LIQUID ROCKET ENGINE, an obstruc-

tion used to prevent COMBUSTION INSTABILITY by maintaining uniform PROPELLANT mixtures and equalizing PRESSURES. 2. In the fuel tank of a liquid rocket system, an obstruction used to prevent SLOSHING of the propellant. 3. In an optical or MULTISPECTRAL SENSING system, a barrier or obstruction used to prevent stray (unwanted) RADIATION from reaching a sensitive DETECTOR element.

Baikonur Cosmodrome A major LAUNCH SITE for the space program of the former Soviet Union and later the Russian Federation. The complex is located just east of the Aral Sea in Kazakhstan (now an independent republic). Also known as the Tyuratam launch site during the COLD WAR, the Soviets launched *SPUTNIK 1* (1957), the first ARTIFICIAL SATELLITE, and COSMONAUT YURI GAGARIN, the first human to travel into OUTER SPACE (1961), from this location.

ballistic flyby Unpowered flight similar to a bullet's TRAJECTORY, governed by GRAVITY and by the body's previously acquired VELOCITY.

ballistic missile A MISSILE that is propelled by ROCKET engines and guided only during the initial (THRUST-producing) phase of its flight. In the nonpowered and nonguided phase of its flight, it assumes a BALLISTIC TRAJECTORY similar to that of an artillery shell. After thrust termination, REENTRY VEHICLES (RVs) can be released, and these RVs also follow free-falling (ballistic) trajectories toward their targets. *Compare with* CRUISE MISSILE; GUIDED MISSILE.

ballistic missile defense (BMD) A defense system designed to protect a territory from incoming BALLISTIC MISSILES in both strategic and theater tactical roles. A BMD system usually is conceived of as having several independent layers designed to destroy attacking ballistic missiles or their WARHEADS at any or all points in their trajectories, from LAUNCH until just before IMPACT at the TARGETS.

The phases of a typical ballistic missile TRAJECTORY are shown in Figure 14. During the BOOST PHASE, the ROCKET engines accelerate the missile and its warhead PAYLOAD through and out of the ATMOSPHERE and provide intense, highly specific observables. A POST-BOOST PHASE, or BUS deployment phase, occurs next, during which multiple warheads (or perhaps a single warhead) and PENETRATION AIDS are released from a post-boost vehicle. In the midcourse phase, the warheads and penetration aids travel on trajectories above the atmosphere; they reenter the atmosphere in the TERMINAL PHASE, where ATMOSPHERIC DRAG affects them.

An effective ballistic missile defense system capable of engaging the missile attack all along its flight path must perform certain key functions, including: (1) promptly and reliably warning of an attack and initiating the defense; (2) continuously TRACKING all threatening objects from the beginning to the end of their trajectories; (3) efficiently intercepting and destroying the BOOSTER or POST-BOOST VEHICLE (PBV); (4) efficiently discriminating between enemy warheads and DECOYS through filtering of lightweight penetration aids; (5) efficiently and economically intercepting and destroying enemy warheads (REENTRY VEHICLES) in the midcourse phase of their flight; (6) intercepting and destroying the enemy warheads (reentry vehicles) at the outer reaches of the atmosphere during the terminal phase of their flight; and (7) effectively coordinating all of the defensive system's components through BATTLE MANAGEMENT, communications, and DATA PROCESSING.

Figure 15 presents a concept for ballistic missile defense during the boost phase of the enemy's ballistic missile flight. An essential requirement is a global, full-time SURVEILLANCE capability to detect a ballistic missile attack, to define its destination and intensity, to determine targeted areas, and to provide data to guide boost-phase interceptors and post-boost vehicle tracking systems. Attacks may range from a single or perhaps a few missiles to a massive simultaneous launch. For every enemy booster destroyed during this phase, the number of potentially hostile objects to be identified and sorted out by the remaining elements of a multilayered (or multitiered) ballistic missile defense system will be significantly reduced. An early defensive response also will minimize the numbers

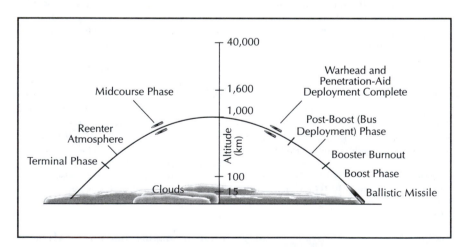

Figure 14. Major phases of a typical ballistic missile flight trajectory.

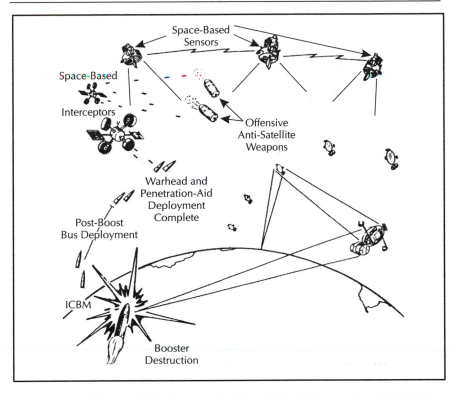

Figure 15. General scenario for ballistic missile defense during the boost phase of a ballistic missile's flight.

of deployed penetration aids. Space-based SENSORS would be used to detect and define the ballistic missile attack. SPACE-BASED INTERCEPTORS would protect these sensor systems from enemy ANTISATELLITE (ASAT) WEAPONS and, as a secondary mission, also would be used to attack the enemy ballistic missiles. For the boost phase defense scenario depicted in Figure 15, nonnuclear, direct-impact projectiles (i.e., KINETIC ENERGY WEAPONS) are being used against the enemy's offensive weapons (i.e., ASAT weapons and warheads) while DIRECTED ENERGY WEAPONS (i.e., HIGH-ENERGY LASERS) are being used to destroy the enemy's boosters early in the flight trajectory.

Similarly, Figure 16 illustrates a generalized ballistic missile defense scenario that might occur during the midcourse phase. Intercept outside the atmosphere during

the midcourse phase of a missile attack would require the defense to cope with decoys designed to attract interceptors and to exhaust the defensive forces. However, continuing discrimination of nonthreatening objects and continuing attrition of reentry vehicles would reduce the pressure on the terminal phase missile defense system. Engagement times are longer during the midcourse phase than during other phases of a ballistic missile trajectory. Figure 16 depicts space-based sensors that can discriminate among warheads, decoys, debris, and the interceptors that the defense has committed to the battle. The nonnuclear, direct-impact (hit to kill) projectiles race toward warheads that the space-based sensors have identified as credible targets.

Finally, Figure 17 illustrates a ballistic missile defense scenario during the terminal

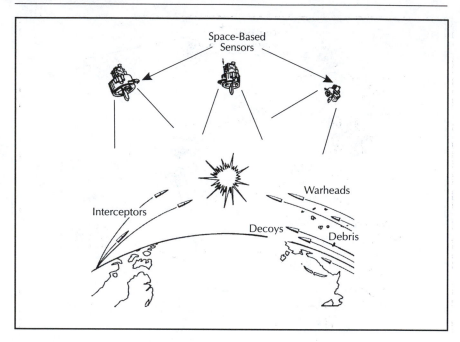

Figure 16. General scenario for ballistic missile defense during the midcourse phase of a ballistic missile's flight.

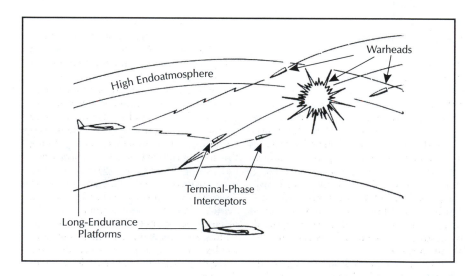

Figure 17. General scenario for ballistic missile defense during the terminal phase of a ballistic missile's flight.

phase of the attack. This phase is essentially the final line of defense. Threatening objects include warheads shot at but not destroyed, objects not previously detected, and decoys that have neither been discriminated against nor destroyed. The interceptors of the terminal phase defensive system must now deal with all of these threatening objects. An AIRBORNE OPTICAL ADJUNCT (i.e., airborne sensor platform) is depicted here. Surviving enemy reentry vehicles are detected in the late EXOATMOSPHERIC portion of their flight with sensors onboard the long-endurance (crewed or robotic) aerial platforms. Then terminal defense interceptors, illustrated here as nonnuclear, direct-impact projectiles, are guided to these remaining warheads and destroy them as they travel through the upper portions of the atmosphere.

Since the end of the COLD WAR (1989), the increased proliferation of ballistic missile systems and weapons of mass destruction throughout the world has forced the United States to consider the timely development of a ballistic missile defense system that can engage all classes of ballistic missile threats, including and especially the growing threat from rogue nations working in concert with terrorist groups. Consequently, in January 2003, President George W. Bush directed the U.S. Department of Defense to begin fielding initial missile defense capabilities in the 2004–2005 period to meet the near-term ballistic missile threat to the American homeland, deployed military forces, and territories of friendly or allied nations. As planned by the MISSILE DEFENSE AGENCY, up to 20 ground-based interceptors capable of destroying attacking intercontinental ballistic missiles during the midcourse phase of their trajectories will be located at Fort Greely, Alaska (16 interceptors), and VANDENBERG AIR FORCE BASE, California (4 interceptors). In addition, up to 20 sea-based interceptors will be deployed on existing AEGIS ships. Their mission is to intercept short- and medium-range ballistic missiles in the midcourse phase of flight. These systems are being complemented by the deployment of air-transportable PATRIOT Advanced Capability-

3 (PAC-3) systems to intercept short- and medium-range ballistic missiles.

The evolutionary American approach to missile defense includes development of the THEATER HIGH ALTITUDE AIR DEFENSE (THAAD) system to intercept short- and medium-range missiles at high ALTITUDE and an AIRBORNE LASER (ABL) aircraft to destroy a ballistic missile in the boost phase. Finally, the program includes the development and testing of space-based defenses, particularly space-based kinetic energy (hit to kill) interceptors and advanced target tracking SATELLITES. However, BMD is a technically challenging undertaking and can be likened to stopping an incoming high-VELOCITY rifle bullet with another rifle bullet.

ballistic missile submarine (SSBN) A nuclear-powered submarine armed with long-range strategic or INTERCONTINENTAL BALLISTIC MISSILES. The U.S. Navy currently operates 18 Ohio-class/Trident ballistic missile submarines. Each nuclear-powered submarine carries 18 nuclear-armed TRIDENT MISSILES.

ballistic trajectory The path an object (that does not have lifting surfaces) follows while being acted upon by only the FORCE of GRAVITY and any resistive AERODYNAMIC FORCES of the medium through which it passes. A stone tossed into the AIR follows a ballistic trajectory. Similarly, after its propulsive unit stops operating, a ROCKET vehicle describes a ballistic trajectory.

balloon-type rocket A LIQUID FUEL ROCKET, such as the early versions of the ATLAS, that requires the PRESSURE of its PROPELLANTS (or other GASes) within to give it structural integrity.

band A range of (RADIO WAVE) frequencies. Alternatively, a closely spaced set of SPECTRAL LINES that is associated with the ELECTROMAGNETIC RADIATION (EMR) characteristic of some particular atomic or molecular ENERGY levels.

band-pass filter A WAVE FILTER that has a single transmission BAND extending from a lower cutoff FREQUENCY greater than zero to a finite upper cutoff frequency.

bandwidth   The number of HERTZ (cycles per second) between the upper and lower limits of a FREQUENCY band.

bar   A unit of PRESSURE in the centimeter-gram-second (cgs) system.

1 bar = 10^6 dynes/cm^2 = 10^5 newtons/m^2 (PASCALS) = 750 mm Hg

The term *millibar* (symbol: mb) is encountered often in METEOROLOGY.

1 millibar (mb) = 10^{-3} bar = 100 pascals

barbecue mode   The slow roll of an orbiting AEROSPACE VEHICLE or SPACECRAFT to help equalize its external TEMPERATURE and to promote a more favorable HEAT (thermal ENERGY) balance. This maneuver is performed during certain missions. In OUTER SPACE, SOLAR RADIATION is intense on one side of a SPACE VEHICLE, while the side opposite the SUN can become extremely cold.

barn (symbol: b)   A unit of area used in expressing nuclear CROSS SECTIONS.

1 barn (b) = 10^{-24} cm^2 = 10^{-28} m^2

Barnard's star   A RED DWARF STAR approximately six LIGHT-YEARS from the SUN, making it the fourth-nearest star to the SOLAR SYSTEM. Discovered in 1916 by the American astronomer Edward E. Barnard (1857–1923), it has the largest PROPER MOTION (some 10.3 seconds of ARC per YEAR) of any known star.

barred spiral galaxy   A type of SPIRAL GALAXY that has a bright bar of STARS across the central regions of the GALACTIC NUCLEUS.

barrier cooling   In ROCKETRY, the use of a controlled mixture ratio near the wall of a COMBUSTION CHAMBER to provide a film of low-TEMPERATURE GASes to reduce the severity of gas-side heating of the chamber.

barycenter   The CENTER OF MASS of a system of masses at which point the total MASS of the system is assumed to be con-

centrated. In a system of two PARTICLES or two CELESTIAL bodies (that is, a binary system), the barycenter is located somewhere on a straight line connecting the geometric center of each object, but closer to the more massive object. For example, the barycenter for the EARTH-MOON system is located about 4,700 km from the center of Earth—a point actually inside Earth, which has a radius of about 6,400 km.

base   1. In mathematics, the number of different symbols in a number system; for example, 2 is the base in the binary system, while 10 is the base in the decimal system. Also, the number that, when raised to a specific POWER, has a LOGARITHM equal to that power. For example, if the number 10 is raised to the power 4, it is equal to 10,000; the number 4 is then called the common logarithm of 10,000 to the base 10. 2. In chemistry, any of a large class of compounds that react with acids to form water and a salt. 3. In AEROSPACE, a major facility or central location for EXTRATERRESTRIAL OPERATIONS, such as a LUNAR BASE, a MARS BASE, or a SPACE BASE.

base heating   In AEROSPACE, the heating of the aft (rear) portion of an AEROSPACE VEHICLE or SPACECRAFT. For example, as the SPACE SHUTTLE ascends through the ATMOSPHERE, THERMAL RADIATION from the expanding main engine and SOLID ROCKET BOOSTER exhaust PLUMES can preferentially HEAT the aft portion of the vehicle.

baseline   In general, any line that serves as the basis for measurement of other lines. Often in AEROSPACE usage, a reference case, such as a *baseline design,* a *baseline schedule,* or a *baseline budget.*

basin (impact)   A large, shallow, lowland area in the CRUST of a TERRESTRIAL PLANET formed by the IMPACT of an ASTEROID or COMET.

battery   An electrochemical ENERGY storage device that serves as a source of direct

CURRENT or voltage, usually consisting of two or more electrolytic cells that are joined together and function as a single unit. The earliest Russian and American EARTH-orbiting SPACECRAFT (e.g., *SPUTNIK 1* and *EXPLORER 1* depended on batteries for all their electric POWER. In general, however, batteries are acceptable as the sole source of electric power only on spacecraft or probes with missions of very short duration—hours, perhaps days, or at the most weeks in length. SOLAR PHOTOVOLTAIC CONVERSION is used in combination with rechargeable batteries on the vast majority of today's spacecraft. The rechargeable batteries provide electrical power during "dark times" when the solar arrays cannot view the SUN. For example, the NICKEL-CADMIUM BATTERY has been the common energy storage companion for SOLAR CELL power supply systems on many spacecraft. Specific energy densities (i.e., energy per unit mass) of about 10 WATT-hours per KILOGRAM are common at the 10% to 20% depths of discharge used to provide CYCLE LIFE. The energy storage

subsystem is usually the largest and heaviest part of a solar cell–rechargeable battery space power system.

Storable batteries can be used as the sole power supply for expendable PLANETARY PROBES required to operate in harsh planetary environments. For example, Figure 18 shows a typical high-TEMPERATURE battery with a service life of about five minutes that can operate at a temperature of 430° Celsius. Activated when needed by a PYROTECHNIC device, this type of high-temperature battery could power a probe that briefly functioned in the high-temperature surface environment of MERCURY or VENUS. Unless extraordinary design features were built into such a probe, its INSTRUMENTS and supporting subsystems most likely would succumb to the elevated temperature environment within a short time.

battle management In general, the analysis of data on the state of a battle and decisions regarding weapons aiming and allocation. Subtasks include command and

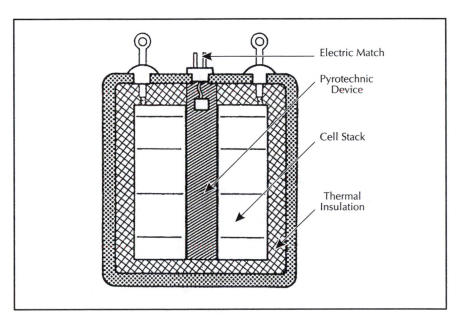

Figure 18. Diagram of a typical high-temperature battery with a service life of about five minutes and an operating temperature of 430° Celsius.

communication, kill assessment, maintaining knowledge of the state and positions of all elements of the defense system, and calculation of target track files. With respect to BALLISTIC MISSILE DEFENSE (BMD), the set of instructions and rules and the corresponding HARDWARE controlling the operation of a BMD system. SENSORS and interceptors are allocated by the BMD system, and updated battle results are presented to the individuals in command for analysis and (as necessary) human intervention.

baud (rate) A unit of signaling speed. The baud rate is the number of electronic SIGNAL changes or data symbols that can be transmitted by a communications channel per second. Named after J. M. Baudot (1845–1903), a French telegraph engineer.

beacon A light, group of lights, electronic apparatus, or other type of signaling device that guides, orients, or warns aircraft, SPACECRAFT, or AEROSPACE VEHICLES in flight. For example, a crash locator beacon emits a distinctive, characteristic SIGNAL that allows search-and-rescue (SAR) teams to locate a downed aircraft, aerospace vehicle, or aborted or reentered spacecraft or PROBE.

beam A narrow, well-collimated stream of PARTICLES (such as ELECTRONS or PROTONS) or ELECTROMAGNETIC RADIATION (such as, GAMMA RAYS or PHOTONS) that are traveling in a single direction.

beamed energy propulsion An advanced PROPULSION SYSTEM concept in which a remote POWER source (e.g., the SUN or a high-powered LASER) supplies thermal ENERGY to an ORBITAL TRANSFER VEHICLE (OTV). Two beamed energy concepts that have been suggested: solar thermal propulsion and laser thermal propulsion. *Solar thermal propulsion* makes use of the Sun and does not require an onboard energy source. The solar thermal ROCKET would harvest raw sunlight with concentrators or MIRRORS. Then, the collected SOLAR RADIATION would be focused to HEAT a PROPELLANT, such as HYDROGEN. Once at

high TEMPERATURE, the gaseous propellant would flow through a conventional CONVERGING-DIVERGING NOZZLE to produce THRUST. Similarly, *laser thermal propulsion* would use a remotely located high-power laser to beam energy to an orbital transfer vehicle to heat a propellant, such as hydrogen. The resultant high temperature gas is then discharged through a thrust-producing NOZZLE. Such advanced propulsion systems might eventually support the transport of cargo between various destinations in CISLUNAR space and possibly even from the surface of the MOON to low LUNAR ORBIT.

beam rider A MISSILE guided to its TARGET by a BEAM of ELECTROMAGNETIC RADIATION, such as a RADAR beam or a LASER beam.

becquerel (symbol: Bq) The SI UNIT of RADIOACTIVITY. One becquerel corresponds to one nuclear transformation (disintegration) per second. Named after the French physicist A. Henri Becquerel (1852–1908), who discovered radioactivity in 1896. This unit is related to the CURIE (the traditional unit for radioactivity) as follows: 1 curie (Ci) = 3.7×10^{10} becquerels (Bq).

bel (symbol: B) A logarithmic unit (n) used to express the ratio of two power levels, P_1 and P_2. Therefore, n (bels) = $\log_{10}$ (P_2/P_1). The decibel (symbol: dB) is encountered more frequently in physics, ACOUSTICS, TELECOMMUNICATIONS, and ELECTRONICS, where 10 decibels = 1 bel. This unit honors the American inventor Alexander Graham Bell (1847–1922). *Compare with* NEPER.

bell nozzle A NOZZLE with a circular opening for a THROAT and an axisymmetric contoured wall downstream of the throat that gives this type of nozzle a characteristic bell shape.

bellows A thin-walled, circumferentially corrugated cylinder that can be elongated or compressed longitudinally and, when integrated into a DUCT assembly, can accommodate duct movements by deflection of the corrugations (convolutions); also used in tanks to provide positive expulsion of FLUID and in fluid flow systems to isolate a REGULATOR, VALVE, or similar component.

Belt of Orion The line of three bright STARS (Alnilam, Alnitak, and Mintaka) that form the belt of Orion, a very conspicuous CONSTELLATION on the EQUATOR of the CELESTIAL SPHERE. It honors the great hunter in Greek mythology.

bent-pipe communications An AEROSPACE industry expression (jargon) for the use of relay stations to achieve non–LINE OF SIGHT (LOS) transmission links.

Bernal sphere A large, spherical space habitat first proposed by the British physicist and writer John Desmond Bernal (1910–1971) in his 1929 work *The World, the Flesh, and the Devil,* in which he boldly speculated about the colonization of OUTER

SPACE. Bernal's concept of spherical space habitats influenced both early SPACE STATION designs and more recent SPACE SETTLEMENT concepts. (See Figure 19.)

berthing The joining of two orbiting SPACECRAFT using a MANIPULATOR or other mechanical device to move one into contact (or very close proximity) with the other at a selected interface. For example, NASA ASTRONAUTS have used the SPACE SHUTTLE'S REMOTE MANIPULATOR SYSTEM to carefully berth a large FREE-FLYING SPACECRAFT (like the *HUBBLE SPACE TELESCOPE*) onto a special support fixture located in the ORBITER'S PAYLOAD BAY during an on-orbit servicing and repair mission. *See also* DOCKING; RENDEZVOUS.

beta decay RADIOACTIVITY in which an atomic NUCLEUS spontaneously decays and emits two subatomic PARTICLES: a BETA PARTICLE (β) and a NEUTRINO (v). In beta-minus

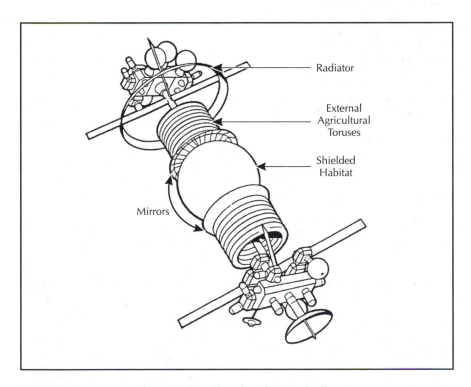

Figure 19. Large space settlement concept based on the Bernal sphere.

(β⁻) decay, a NEUTRON in the transforming (parent) nucleus becomes a PROTON, and a negative beta particle and an antineutrino are emitted. The resultant (daughter) nucleus has its ATOMIC NUMBER (Z) increased by one (thereby changing its chemical properties), while the total ATOMIC MASS (A) remains the same as that of the parent nucleus. In beta-plus (β⁺) decay, a proton is converted into a neutron, and a positive beta particle (POSITRON) is emitted along with a neutrino. Here, the atomic number of the resultant (daughter) nucleus is decreased by one, a process that also changes its chemical properties.

Decimal Number	Binary Notation
0	0
1	1
2	10
3	11
4	100
5	101
6	110
7	111
8	1000
9	1001
10	1010

beta particle (β) The negatively-charged subatomic PARTICLE emitted from the atomic NUCLEUS during the process of BETA DECAY. It is identical to the ELECTRON. *See also* POSITRON.

big bang (theory) Contemporary theory in COSMOLOGY concerning the origin of the UNIVERSE. It suggests that a very large, ancient explosion, called the initial singularity, started space and time of the present universe, which has been expanding ever since. The big bang event is now thought to have occurred about 15 to 20 billion years ago. ASTROPHYSICAL observations, especially discovery of the COSMIC (MICROWAVE) BACKGROUND RADIATION in 1964, tend to support this theory.

big crunch (hypothesis) Within the CLOSED UNIVERSE model of COSMOLOGY, the postulated end state that occurs after the present UNIVERSE expands to its maximum physical dimensions and then collapses in on itself under the influence of GRAVITATION, eventually reaching an infinitely dense end point, or SINGULARITY.

binary digit (bit) Only two possible values (or digits) in the binary number system, namely 0 or 1. Binary notation is a common TELEMETRY (information) encoding scheme that uses binary digits to represent numbers and symbols. For example, DIGITAL COMPUTERS use a sequence of bits, such as an eight-bit-long BYTE (*binary digit eight*), to create a more complex unit of information.

binary notation A numeric system that uses only two different characters, usually 0 and 1. (See table.) Because the numbers 0 and 1 can be represented easily by the "Off" and "On" conditions of an electric circuit, binary notation is widely used in DIGITAL COMPUTERS.

binary (double) star system (s) A pair of STARS that orbit around a common CENTER OF MASS and are bound together by their mutual GRAVITATION. By convention, the star that is nearest the center of mass is called the primary, while the other (smaller) star of the system is called the companion. Binary star systems can be classified further as visual binaries, eclipsing binaries, spectroscopic binaries, and astrometric binaries. A more recent class of binary star system, called X-RAY binary, was discovered during the 1970s through space-based X-ray observations. *See also* X-RAY ASTRONOMY.

binding energy The minimum ENERGY required to dissociate a NUCLEUS (or system of PARTICLES) into its component NUCLEONS (i.e., its NEUTRONS and PROTONS). Also called *total binding energy*. For a particle in a system (e.g., a neutron or proton in a nucleus), the minimum net energy required to remove that particle from its system is sometimes called the *separation energy*. The *electron binding energy* is the energy required to remove an ELECTRON from an ATOM or MOLECULE. Nuclear binding energies have values on the order of millions of ELECTRON VOLTS (MeVs), while chemical binding energies (that is, electron binding energies) have values on the order of tens of electron volts (eVs).

bioclean room Any enclosed area where there is control over viable and nonviable particulates in the AIR with TEMPERATURE, HUMIDITY, and PRESSURE control as required to maintain specified standards. A viable PARTICLE is a particle that will reproduce to form observable colonies when placed on a specified culture medium and incubated according to optimum environmental conditions for growth after collection. A nonviable particle will not reproduce to form observable colonies when placed on a specified culture medium and incubated according to optimum environmental conditions for growth after collection. A bioclean room is operated with emphasis on minimizing airborne viable and nonviable particle generation or concentrations to specified levels. Particle size is expressed as the apparent maximum linear dimension or diameter of the particle, usually in MICROMETERS (μm) or MICRONS.

There are three clean room classes based on total (viable and nonviable) particle count, with the maximum number of particles per unit volume or horizontal surface area being specified for each class. The "cleanest," or most stringent, airborne particulate environment is called a *Class 100 Bioclean Room* (called a *Class 3.5 Room* in the SI UNIT SYSTEM). In this type of bioclean room, the particle count may not exceed a total of 100 particles per cubic foot (3.5 particles per LITER) of a size 0.5 micrometer and larger; the viable particle count may not exceed 0.1 per cubic foot (0.0035 per liter), with an average value not to exceed 1,200 per square foot (12,900 per square METER) per week on horizontal surfaces. The *Class 10,000 Bioclean Room* (called a *Class 350 Room* in SI unit system) is the next cleanest environment, followed by a *Class 100,000 Bioclean Room* (called a *Class 3,500* Room in the SI unit system). The table summarizes the air cleanliness conditions for each type of clean room.

In AEROSPACE applications, bioclean rooms, or "clean rooms," as they are frequently called, are used to manufacture, assemble, test, disassemble, and repair delicate SPACECRAFT SENSOR systems, electronic components, and certain mechanical subsystems. Special protective clothing, including gloves, smocks (often called "bunny suits"), and head and foot coverings are worn by aerospace workers to reduce the level of dust and contamination in clean rooms. A Class 10,000 Clean Room facility is typically the cleanliness level encountered in the assembly and testing of large spacecraft. If a PLANETARY PROBE is being prepared for a trip to a PLANET such as MARS, care is also taken in the clean room to ensure that "hitchhiking TERRESTRIAL microorganisms" are

Bioclean Room Air Cleanliness Levels			
Class English System (metric system)	Maximum Number of Particles per Cu Ft 0.5 Micron and Larger (per liter)	Maximum Number of Viable Particles per Cu Ft (per liter)	Average Number of Viable Particles per Sq Ft per Week (per m² per week)
100 (3.5)	100 (3.5)	0.1 (0.0035)	1,200 (12,900)
10,000 (350)	10,000 (350)	0.5 (0.0176)	6,000 (64,600)
100,000 (3,500)	100,000 (3,500)	2.5 (0.0884)	30,000 (323,000)

Source: NASA.

brought to the minimum population levels consistent with planetary quarantine protocols, thereby avoiding the potential of forward contamination of the TARGET alien world. *See also* EXTRATERRESTRIAL CONTAMINATION.

biogenic elements Those ELEMENTS generally considered by scientists (astrobiologists) as essential for all living systems, including HYDROGEN (H), carbon (C), nitrogen (N), oxygen (O), sulfur (S), and phosphorous (P). The availability of the chemical compound water (H_2O) is also considered necessary for life both here on EARTH and possibly elsewhere in the UNIVERSE. *See also* ASTROBIOLOGY.

biopak A container for housing a living organism (e.g., an insect, fish, or small mammal) in a habitable environment and for recording biological activities during flight in OUTER SPACE.

bioregenerative life support system *See* LIFE SUPPORT SYSTEM.

biosensor A SENSOR used to provide information about a life process. Developments in miniaturization and information technologies are making possible biosensors that are only a few thousandths of a centimeter wide. Smaller biosensors have two distinct advantages: (1) They can be placed in areas that were previously inaccessible, and (2) they are fast since life process measurements can be made in situ (e.g., in the human being or in the field) instead of at a central laboratory. New generations of biosensors will take advantage of their common heritage with information technology to integrate logic functions into the same tiny package with the biosensors. Integrating logic with sensors results in more useful devices. For example, a glucose sensor with integrated logic functions might be designed to respond differently to measured blood sugar levels at different times of the day.

biosphere The life zone of a planetary body; for example, that part of EARTH inhabited by living organisms. On this PLANET, the biosphere includes portions of the ATMOSPHERE, the HYDROSPHERE, the CRYOSPHERE, and surface regions of the solid Earth. *See also* ECOSPHERE; GLOBAL CHANGE.

biota The fauna and flora (animal and plant life) of a given area.

biotelemetry The remote measurement of life functions. data from biosensors attached to an ASTRONAUT or COSMONAUT are sent back to EARTH (as TELEMETRY) for the purposes of space crew health monitoring and evaluation by medical experts and mission managers. During a strenuous EXTRAVEHICULAR ACTIVITY (EVA), for example, medical specialists at mission control would use biotelemetry to monitor an astronaut's heartbeat and respiration rate.

bipropellant rocket A ROCKET that uses two unmixed (uncombined) liquid chemicals as its FUEL and OXIDIZER. The two chemical PROPELLANTs flow separately into the rocket's COMBUSTION CHAMBER, where they are combined and combusted to produce high-TEMPERATURE, THRUST-generating GASes. The combustion gases then exit the rocket system through a suitably designed NOZZLE.

bipropellant valve A VALVE incorporating both FUEL and OXIDIZER valving (flow control) units driven by a common ACTUATOR.

bird A popular AEROSPACE industry expression (jargon) for a ROCKET, MISSILE, SATELLITE, or SPACECRAFT.

birth-to-death tracking In BALLISTIC MISSILE DEFENSE, the ability to track a BALLISTIC MISSILE and its PAYLOAD (e.g., SATELLITES, REENTRY VEHICLES, and/or DECOYs that simulate these objects) from LAUNCH until each object is intercepted and destroyed or completes its mission (i.e., impacting on a TARGET or deploying a POST-BOOST VEHICLE).

bistatic radar A RADAR system that has transmitters and receivers at two separate locations; this is a special case of multistatic radar. The reasons for this separation may include enhanced SURVIVABILITY and the ability to overcome the COUNTERMEASURE of RETROREFLECTORS.

bit   A BINARY DIGIT, the basic unit of information (either 0 or 1) in BINARY NOTATION.

blackbody   A perfect emitter and perfect absorber of ELECTROMAGNETIC RADIATION. According to PLANCK'S RADIATION LAW, the radiant ENERGY emitted by a blackbody is a function only of the ABSOLUTE TEMPERATURE of the emitting object.

black box   A unit or subsystem (often involving an electronic device) of a SPACECRAFT or AEROSPACE VEHICLE that is considered only with respect to its input and output characteristics, without any specification of its internal elements.

black dwarf   The cold remains of a WHITE DWARF STAR that no longer emits visible RADIATION or a nonradiating ball of INTERSTELLAR GAS that has contracted under GRAVITATION but contains too little MASS to initiate nuclear FUSION.

black hole   An incredibly compact, gravitationally collapsed MASS from which nothing (light, matter, or any other kind of information) can escape. Astrophysicists believe that a black hole is the natural end product when a massive STAR dies and collapses beyond a certain critical dimension, called the SCHWARZSCHILD RADIUS. Once the massive star shrinks to this critical radius, its gravitational ESCAPE VELOCITY is equal to the SPEED OF LIGHT, and nothing can escape from it. Inside this radius, called the *event horizon,* lies an extremely dense point mass (SINGULARITY).

In late November 2002, scientists using the *CHANDRA X-RAY OBSERVATORY* collected data suggesting the presence of two supermassive black holes orbiting each other in the same GALAXY (NCG 6240). The observatory was able to "see" them because the black holes are surrounded by hot swirling vortices of matter called ACCRETION DISKS. Such disks are strong sources of X RAYS. Several hundred million years from now these black holes will merge. This merger will be a spectacular COSMIC event that unleashes intense RADIATION and GRAVITATION WAVES and leaves behind an even larger black hole.

blackout   1. In communications, a fadeout, due to environmental factors, of radio and TELEMETRY transmission between ground stations and AEROSPACE VEHICLES traveling at high speeds in the ATMOSPHERE. It is caused by ATTENUATION of the transmission SIGNAL as it passes through the ionized boundary layer (PLASMA sheath) and SHOCK WAVE regions generated by the high VELOCITY vehicle. 2. In defense, the disabling of RADAR by means of a NUCLEAR EXPLOSION. The intense ELECTROMAGNETIC ENERGY released generates a large background that obscures signals and renders many types of radar useless for minutes or longer. 3. In AEROSPACE MEDICINE, a physical condition in which vision is temporarily obscured by a blackness accompanied by a dullness of certain other senses. This is brought on by decreased blood pressure in the head and a consequent lack of oxygen. The condition may occur when pulling out of a high-speed dive in a plane.

blastoff   The moment a ROCKET or AEROSPACE VEHICLE rises from its LAUNCH PAD under full THRUST. *See also* LIFTOFF.

blazar   A variable EXTRAGALACTIC object (possibly a high-speed JET from an ACTIVE GALACTIC NUCLEUS) that exhibits very dynamic, sometimes violent behavior. *See also* BL LAC (BL LACERTAE) OBJECT.

bl lac (bl lacertae) object   A class of EXTRAGALACTIC objects thought to be the active centers of faint elliptical galaxies that vary considerably in brightness over very short periods of time (typically hours, days, or weeks). Scientists further speculate that a very high-speed (RELATIVISTIC) JET is emerging from such an object straight at an observer on EARTH.

blockhouse (block house) A reinforced-concrete structure, often built partially underground, that provides protection against blast, HEAT, and possibly an ABORT explosion during ROCKET launchings.

blowdown system A closed PROPELLANT/pressurant system that decays in ULLAGE PRESSURE level as the propellant is consumed and ullage volume is increased.

blue giant A massive, very-high-LUMINOSITY STAR with a surface TEMPERATURE of about 30,000 K that has exhausted all its HYDROGEN THERMONUCLEAR fuel and left the MAIN SEQUENCE.

blueshift When a CELESTIAL object (such as a distant GALAXY) approaches an observer at high VELOCITY, the ELECTROMAGNETIC RADIATION it emits in the visible portion of the SPECTRUM appears shifted toward the blue (higher FREQUENCY, shorter WAVELENGTH) region. *Compare with* REDSHIFT. *See also* DOPPLER SHIFT.

boattail The aft (rear) end of a ROCKET containing the PROPULSION SYSTEM and its INTERFACE with vehicle TANKAGE.

boiloff The loss of a CRYOGENIC PROPELLANT, such as LIQUID OXYGEN or LIQUID HYDROGEN, due to vaporization. This happens when the TEMPERATURE of the cryogenic propellant rises slightly in the PROPELLANT tank of a ROCKET being prepared for LAUNCH. The longer a fully fueled rocket vehicle sits on its LAUNCH PAD, the more significant the problem of boiloff becomes.

bolide A brilliant METEOR, especially one that explodes into fragments near the end of its TRAJECTORY in EARTH'S ATMOSPHERE.

bolometer A sensitive INSTRUMENT that measures the intensity of incident radiant (thermal) ENERGY, usually by means of a thermally sensitive electrical resistor. The change in electrical resistance is an indication of the amount of radiant energy reaching the exposed material.

Boltzmann constant (symbol: k) The physical constant that describes the relationship between ABSOLUTE TEMPERATURE and the KINETIC ENERGY of the ATOMS or MOLECULES in a PERFECT GAS. It equals 1.380658×10^{-23} JOULES per KELVIN (J/K) and is named after the Austrian physicist Ludwig Boltzmann (1844–1906).

Bond albedo The fraction of the total amount of ELECTROMAGNETIC RADIATION (such as the total amount of LIGHT) falling upon a nonluminous spherical body that is reflected in all directions by that body. The Bond albedo is measured or calculated over all WAVELENGTHs and is named after the American astronomer George Phillips Bond (1825–1865).

boost 1. To LAUNCH or to push along during a portion of flight, as to boost a RAMJET to flight speed by means of a ROCKET, or a SPACECRAFT or rocket vehicle boosted to ALTITUDE with a (BOOSTER) rocket. 2. The first portion of the TRAJECTORY of a BALLISTIC MISSILE. 3. Additional POWER, PRESSURE, or FORCE supplied by a BOOSTER ENGINE; as, for example, a HYDRAULIC boost, or the extra propulsion given a flying vehicle during LIFTOFF, climb, or another phase of its flight, as with a booster engine. 4. To supercharge.

booster 1. The ROCKET that places a BALLISTIC MISSILE in its TRAJECTORY toward a TARGET or that launches a SATELLITE or SPACECRAFT into ORBIT. 2. An auxiliary or initial PROPULSION SYSTEM that travels with a MISSILE or aircraft and that may or may not separate from the parent craft when its IMPULSE has been delivered. A booster system may contain, or consist of, one or more units. 3. A HIGH EXPLOSIVE (HE) element sufficiently sensitive as to be actuated by small explosive elements in a FUSE or primer and powerful enough to cause DETONATION of the main explosive charge.

booster clustering Locating BOOSTERS relatively near one another (typically within hundreds of kilometers) in order to force a space-based BALLISTIC MISSILE DEFENSE

(BMD) system to a higher ABSENTEE RATIO and, therefore, to increase the required number of SPACE-BASED INTERCEPTORS.

booster decoy A ROCKET that would imitate the early phase of BOOSTER PLUME and TRAJECTORY in order to draw fire from an opponent's BALLISTIC MISSILE DEFENSE (BMD) system but that would actually not be an armed (i.e., WARHEAD-equipped) INTERCONTINENTAL BALLISTIC MISSILE (ICBM). A booster decoy would cost substantially less than a fully equipped ICBM.

booster engine An engine, especially a BOOSTER ROCKET, that adds its THRUST to the thrust of a SUSTAINER ENGINE.

booster pump A PUMP in a FUEL system, HYDRAULIC system, or the like, used to provide additional or auxiliary PRESSURE when needed or to provide an initial pressure differential before entering a main pump, as in pumping HYDROGEN near the boiling point.

booster rocket A ROCKET motor, either SOLID or LIQUID PROPELLANT, that assists the main propulsive system (called the SUSTAINER ENGINE) of a LAUNCH VEHICLE during some part of its flight.

booster separation motor (BSM) The small, SOLID-PROPELLANT ROCKETS that serve as the separation motors that translate, or move, the SPACE SHUTTLE's large SOLID ROCKET BOOSTERS away from the ORBITER vehicle's still-thrusting main engines and the EXTERNAL TANK.

boostglide vehicle An AEROSPACE VEHICLE (part aircraft, part SPACECRAFT) designed to fly to the limits of the SENSIBLE ATMOSPHERE, then to be boosted by ROCKETs into OUTER SPACE, returning to EARTH by gliding under aerodynamic control.

boost phase In general, that portion of the flight of a BALLISTIC MISSILE or LAUNCH VEHICLE during which the BOOSTER and SUSTAINER ENGINES operate. In BALLISTIC MISSILE DEFENSE (BMD), the first portion of a ballistic missile's TRAJECTORY, during which it is being powered by its ROCKET engines. During this period, which typically lasts from three to five minutes for an INTERCONTINENTAL BALLISTIC MISSILE (ICBM), the MISSILE reaches an ALTITUDE of about 200 km (124 mi), whereupon powered flight ends and the ICBM begins to dispense its REENTRY VEHICLES. The other portions of the missile flight, including the midcourse phase and the REENTRY PHASE, take up the remainder of an ICBM's flight total time of about 25 to 30 minutes.

bootstrap Refers to a self-generating or self-sustaining process. An AEROSPACE example is the operation of a LIQUID-PROPELLANT ROCKET ENGINE in which (during MAIN STAGE operation) the GAS GENERATOR is fed by PROPELLANTs from the TURBOPUMP SYSTEM, and the turbopump (in turn) is driven by hot gases from the generator system. Of course, the operation of such a system must be started by outside power or propellant sources. However, when its operation is no longer dependent on these external sources, the rocket engine is said to be in "bootstrap" operation. Similarly, in computer science, bootstrap refers to a sequence of commands or instructions the execution of which causes additional commands or instructions to be loaded and executed until the desired computer program is completely loaded into storage.

boundary layer The layer of FLUID in the immediate vicinity of a bounding surface; in FLUID MECHANICS, the layer of fluid affected primarily by the VISCOSITY of the fluid. This general term refers ambiguously to the LAMINAR BOUNDARY LAYER, the turbulent boundary layer, the planetary boundary layer, or the surface boundary layer. In AERODYMANICS, the boundary layer thickness often is measured from the surface to a point at which the AIRFLOW has 99% of the FREE STREAM VELOCITY.

bow shock 1. A SHOCK WAVE in front of a body, such as an AEROSPACE VEHICLE or an AIRFOIL. 2. The INTERFACE formed where the electrically charged SOLAR WIND encounters an obstacle in OUTER SPACE, such as a PLANET's MAGNETIC FIELD or ATMOSPHERE. *See also* MAGNETOSPHERE.

braking ellipses A series of ELLIPSES, decreasing in size due to aerodynamic DRAG, followed by an orbiting SPACECRAFT as it encounters and enters a planetary ATMOSPHERE. *See also* AEROBRAKE.

Brayton cycle A thermodynamic CYCLE that represents the idealized behavior of a gaseous WORKING FLUID in a TURBINE type of HEAT ENGINE. The idealized Brayton cycle is described in Figure 20. In this description, each stage of the thermodynamic process is assumed to have been completed before the next stage is initiated. However, in an actual Brayton cycle engine, there is a gradual rather than a sharp transition from one stage to the next; consequently, the sharp points in this idealized figure actually would be rounded off. This cycle, developed by the American engineer George Brayton (1830–1892), can be either a closed cycle (in which the

working gas is constantly recycled) or an open cycle (in which the working gas, like AIR, passes through the system just once).

breadboard An assembly of preliminary circuits or parts used to prove the feasibility of a device, circuit, system, or principle without regard to the final configuration or packaging of the parts.

breakaway disconnect Separable connector that is disengaged by the separation FORCE as the LAUNCH VEHICLE rises from the LAUNCH PAD or as a STAGE separates from a lower stage. Also called a *rise-off* or *staging disconnect*.

breakoff phenomenon The feeling of isolation that sometimes occurs during high-ALTITUDE flight or space flight. The traveler can feel totally separated and detached from EARTH and human society. Also called the *breakaway phenomenon* and the *Shimanagashi syndrome*.

bremsstrahlung (German for breaking radiation) ELECTROMAGNETIC RADIATION

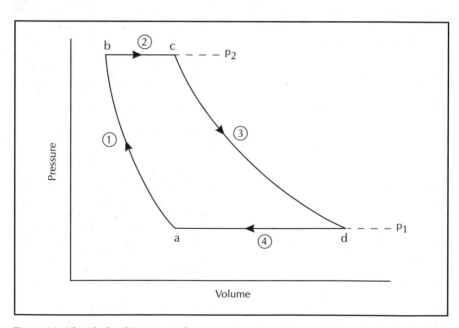

Figure 20. The idealized Brayton cycle.

emitted as energetic PHOTONS (that is, as X RAYS) when a fast-moving charged PARTICLE (usually an ELECTRON) loses ENERGY upon being accelerated and deflected by the electric field surrounding a positively charged atomic NUCLEUS. X rays produced in ordinary X-ray machines are an example of bremsstrahlung.

brennschluss (German for combustion termination) The cessation of burning in a ROCKET, resulting from consumption of the PROPELLANTS, from deliberate shutoff, or from other causes; the time at which this cessation occurs.

bridgewire Resistance wire, attached to the leads of an ELECTROEXPLOSIVE DEVICE (EED), whose function is to convert the electrical firing SIGNAL into thermal ENERGY adequate to ignite the prime charge of an IGNITER.

brightness In BALLISTIC MISSILE DEFENSE (BMD), the amount of POWER that can be delivered on a TARGET per unit SOLID ANGLE by a DIRECTED ENERGY WEAPON.

British thermal unit (BTU) A unit quantity of ENERGY, especially thermal energy (HEAT). Defined as the amount of thermal energy (heat) needed to increase the TEMPERATURE of 1 pound of water 1 degree Fahrenheit (F) at normal ATMOSPHERIC PRESSURE. In general, the initial temperature of the water is specified as 39° F and the final temperature as 40° F—a physical condition in which water has its maximum DENSITY.

1 BTU = 1055 joules

brown dwarf A very low LUMINOSITY, substellar (almost a STAR) CELESTIAL OBJECT that contains starlike material (that is, HYDROGEN and HELIUM) but has too low a MASS (typically 1% to 10% of a SOLAR MASS) to allow its core to initiate thermonuclear FUSION (hydrogen burning).

buffeting The beating of an aerodynamic structure or surface by unsteady flow, gusts, and the like; the irregular shaking or OSCILLATION of a vehicle component owing to turbulent AIR or separated flow.

bulge of the Earth The extra extension of EARTH's EQUATOR, caused by the CENTRIFUGAL FORCE of Earth's ROTATION, which slightly flattens the spherical shape of Earth. This bulge causes the planes of SATELLITE ORBITS inclined to the equator (but not POLAR ORBITS) to slowly rotate around Earth's AXIS.

bulkhead A dividing wall in a ROCKET, SPACECRAFT, or aircraft FUSELAGE, at right ANGLES to the LONGITUDINAL AXIS of the structure, that serves to strengthen, divide, or help shape the structure. A bulkhead sometimes is designed to withstand PRESSURE differences.

burble A separation or breakdown of the LAMINAR FLOW past a body; the eddying or TURBULENT FLOW resulting from this. Burble occurs over an AIRFOIL operating at an ANGLE OF ATTACK greater than the angle of maximum LIFT, resulting in a loss of lift and an increase of DRAG.

burn In AEROSPACE operations, the firing of a ROCKET ENGINE. For example, the "third burn" of the SPACE SHUTTLE's ORBITAL MANEUVERING SYSTEM (OMS) engines would mean the third time during a space shuttle flight that the OMS engines had been fired. The *burn time* is simply the length of a thrusting period.

burning surface The surface of a SOLID PROPELLANT GRAIN that is not restricted from burning at any given time during propellant COMBUSTION.

burnout The moment in time or the point in a ROCKET's TRAJECTORY when COMBUSTION of FUELS in the engine is terminated. This usually occurs when all the PROPELLANTS are consumed.

burnout angle The ANGLE between the local vertical and the VELOCITY VECTOR at termination of THRUST (i.e., at BURNOUT).

burnout velocity The VELOCITY attained by a MISSILE, ROCKET, or AEROSPACE VEHICLE at the termination of THRUST (i.e., at BURNOUT).

burn rate Literally, the rate at which a SOLID PROPELLANT burns—that is, the rate of recession of a burning propellant surface, perpendicular to that surface, at a specified PRESSURE and GRAIN TEMPERATURE; with respect to solid propellant grain design, the rate at which the WEB decreases in thickness during rocket motor operation. The web is the minimum thickness of a solid-propellant rocket grain from the initial IGNITION surface to the insulated case wall or to the intersection of another burning surface at the time when the burning surface experiences a major change. Also called *burning rate*. See also ROCKET.

burn time In general, the length of a THRUST-producing period for a ROCKET ENGINE. For a SOLID-PROPELLANT ROCKET, the interval from attainment of a specified initial fraction of maximum thrust or PRESSURE level to WEB BURNOUT. The web is the minimum thickness of a solid propellant rocket grain from the initial surface to the insulated case wall or to the intersection of another burning surface at the time when the burning surface undergoes a major change. See also ROCKET.

burnup 1. (AEROSPACE technology) The vaporization and disintegration of a SATELLITE, SPACECRAFT, or ROCKET or any of their components by AERODYNAMIC HEATING upon entry into a planetary ATMOSPHERE. 2. (nuclear energy technol-

ogy) A measure of NUCLEAR REACTOR fuel consumption. It can be expressed as (a) the percentage of nuclear fuel ATOMS that have undergone FISSION, or (b) the amount of ENERGY produced per MASS (or WEIGHT) unit of fuel in the reactor (e.g., megawatt days per metric TON [MWd/t]).

burst disk Passive physical barrier in a FLUID system that blocks the flow of fluid until ruptured by (excessive) fluid PRESSURE.

bus The ROCKET-propelled final stage of an INTERCONTINENTAL BALLISTIC MISSILE (ICBM) that, after BOOSTER BURNOUT, places WARHEADS and (possibly) DECOYS on ballistic trajectories toward their TARGETs. This is also called the POST-BOOST VEHICLE (PBV).

bus deployment phase In BALLISTIC MISSILE DEFENSE (BMD), the portion of a MISSILE flight during which multiple WARHEADS are deployed on different paths to different TARGETS. The warheads on a single INTERCONTINENTAL BALLISTIC MISSILE (ICBM) are carried on a platform, called the BUS (or POST-BOOST VEHICLE), which has small ROCKET motors to move the bus slightly from its original flight path. This phase is also called the POST-BOOST PHASE.

byte (*binary digit eight*) A basic unit of information or data consisting of eight BINARY DIGITS (BITS). The information storage capacity of a computer system often is defined in terms of kilobytes (kb), megabytes (Mb), and even gigabytes (Gb). One kilobyte corresponds to 2^{10}, or 1,024 bytes, while 1 megabyte corresponds to 2^{20}, or 1,048,576 bytes.

caldera A large volcanic depression, more or less circular in form and much larger than the included volcanic vents. A caldera may be formed by three basic geologic processes: explosion, collapse, or erosion. *See also* OLYMPUS MONS.

calendar year The time interval that is used as the basis for an annual calendar. For example, the *Gregorian calendar year* contains an average of 365.2425 days, so that each calendar year contains 365 days, with an extra day being added every fourth calendar year to make a *leap year* (which contains 366 days).

calibration The process of translating the SIGNALS collected by a measuring INSTRUMENT into something that is scientifically useful. The calibration procedure generally involves comparing measurements made by an instrument with a standard or set of reference data. This process usually removes most of the errors caused by instabilities in the instrument or in the environment through which the signal has traveled.

Callisto The second-largest MOON of JUPITER and the outermost of the four GALILEAN SATELLITES.

calorie (symbol: cal) A unit of thermal ENERGY (HEAT) originally defined as the amount of energy required to raise 1 g of water through 1°C. This energy unit (often called a *small calorie*) is related to the JOULE as follows: 1 cal = 4.1868 J. Scientists use the term *kilocalorie* (1,000 small calories) for one *big calorie* when describing the energy content of food.

calorimeter A device in or with which precise measurements of thermal energy (heat) flow can be made.

Caloris basin A very large, ringed impact BASIN about 1,300 km across on MERCURY.

camber Curvature of the mean line on an airfoil; the distance from the point of greatest curvature to the chord.

canali The Italian word for *channels* used in 1877 by the Italian astronomer Giovanni Schiaparelli (1835–1910) to describe natural surface features he observed on MARS. Subsequent pre–space age investigators, including the American astronomer Percival Lowell (1855–1916), took the Italian word quite literally as meaning *canals* and sought additional evidence of an intelligent civilization on Mars. Since the 1960s, many SPACECRAFT have visited Mars, dispelling such popular speculations and revealing no evidence of any Martian canals constructed by intelligent beings.

canard 1. Pertaining to an aerodynamic vehicle in which the horizontal surfaces used for trim and control are placed forward of the main lifting surface. 2. The horizontal trim and control surfaces in such an arrangement.

candela (symbol: Cd) The SI UNIT of LUMINOUS INTENSITY. It is defined as the luminous intensity in a given direction of a source that emits monochromatic light at a FREQUENCY of 540×10^{12} HERTZ and whose radiant intensity in that direction is 1/683 WATT per STERADIAN. The physical basis for this unit is the luminous intensity of a BLACKBODY RADIATOR at a TEMPERATURE of 2046 K (the temperature at which molten platinum solidifies). *See also* LUMEN.

cannibalize The process of taking functioning parts from a nonoperating

SPACECRAFT or LAUNCH VEHICLE and installing these salvaged parts in another spacecraft or launch vehicle in order to make the latter operational.

Cape Canaveral Air Force Station (CCAFS) The region on Florida's east central coast from which the United States Air Force and NASA have launched more than 3,000 ROCKETs since 1950. Cape Canaveral Air Force Station is the major East Coast LAUNCH SITE for the Department of Defense, while the adjacent NASA KENNEDY SPACE CENTER is the SPACEPORT for the fleet of SPACE SHUTTLE vehicles. The station covers a 65-square-km (about 25-square-mi) area. Much of the land is now inhabited by large populations of animals, including deer, alligators, and wild boar. The nerve center for CCAFS and the entire Eastern Range is the range control center, from which all launches, as well as the status of range resources, are monitored. The range safety function also is performed in the range control center. *See also* VANDENBERG AIR FORCE BASE.

capillary A tube or pipe of very narrow diameter. A FLUID, such as water, will rise or move through a capillary as a result of the fluid's SURFACE TENSION. This process is called *capillarity*. On EARTH, for example, capillarity is a very important process in living plants, since it is responsible for the transport of water throughout the plant against the forces of terrestrial GRAVITY. *See also* HEAT PIPE.

capital satellite A highly valued or costly satellite, as distinct from an expensive decoy satellite.

capsule 1. In general, a boxlike container, component, or unit that often is sealed. 2. A small, sealed, pressurized cabin that contains an acceptable environment for human crew, animals, or sensitive equipment during extremely high-ALTITUDE flight, space flight, or emergency escape. As these pressurized cabins became larger, such as the APOLLO PROJECT capsule, the term SPACECRAFT replaced the term *capsule*

in AEROSPACE use. 3. An enclosed container carried on a ROCKET or spacecraft that supports and safely transports a PAYLOAD, experiment, or INSTRUMENT intended for recovery after flight.

captive firing The firing of a ROCKET PROPULSION system at full or partial THRUST while the rocket is restrained in a test stand facility. Usually engineers instrument the propulsion system to obtain test data that verify rocket design and demonstrate performance levels. Sometimes called a *holddown test*.

capture 1. In CELESTIAL MECHANICS, the process by which the CENTRAL FORCE field of a PLANET or STAR overcomes by gravitational attraction the VELOCITY of a passing celestial object (such as a COMET or tiny PLANETOID) or SPACECRAFT and brings that body under the control of the central force field. 2. In AEROSPACE operations, that moment during the RENDEZVOUS and DOCKING of two spacecraft when the capture mechanism engages and locks the two orbiting vehicles together. 3. In SPACE SHUTTLE operations, the event of the REMOTE MANIPULATOR SYSTEM (RMS) and effector making contact with and firmly attaching to a PAYLOAD grappling fixture. A payload is considered captured any time it is firmly attached to the remote manipulator system. 4. In nuclear physics, a process in which an atomic or a nuclear system acquires an additional PARTICLE, such as the capture of ELECTRONs by positive IONs or the capture of NEUTRONs by nuclei.

carbon dioxide (symbol: CO_2) A colorless, odorless, noncombustible GAS present in EARTH'S ATMOSPHERE. Carbon dioxide is formed by the combustion of carbon and carbon compounds (e.g., fossil fuels, wood, and alcohol), by respiration, and by the gradual oxidation of organic matter in the soil. Carbon dioxide is removed from the atmosphere by green plants (during photosynthesis) and by absorption in the oceans. Since it is a *greenhouse gas,* the increasing presence of CO_2 in Earth's atmosphere could have significant environmental consequences, such as global warming. The pre–industrial revolution atmospheric concentration of CO_2 is estimated to have been about 275 parts per million by volume (ppmv). Environmental scientists now predict that if current trends in the rise in atmospheric CO_2 concentration continue (currently about 340 to 350 ppmv), then sometime after the middle of the 21st century the atmospheric concentration of CO_2 will reach *twice* its pre–industrial revolution value. This condition might severely alter Earth's environment through global warming, polar ice cap melting, and increased desertification. *See also* GLOBAL CHANGE; MISSION TO PLANET EARTH.

cargo The total complement of payloads (one or more) on any one space shuttle flight. Cargo includes everything contained in the *orbiter cargo bay* plus other equipment, hardware, and consumables located elsewhere in the orbiter vehicle that are unique to the user and are not carried as part of the basic orbiter payload support. *See also* CARGO BAY.

cargo bay The unpressurized midpart of the space shuttle orbiter vehicle fuselage behind the cabin aft (rear) bulkhead where most payloads are carried. Its maximum usable payload envelope is 18.3 meters (m) (60 feet [ft]) long and 4.6 m (15 ft) in diameter. Hinged doors extend the full length of the cargo bay. *See also* SPACE TRANSPORTATION SYSTEM.

Carnot cycle An idealized reversible thermodynamic CYCLE for a theoretical HEAT ENGINE, first postulated in 1824 by the French engineer Nicolas Sadi Carnot (1746–1832). As shown in Figure 21, the Carnot cycle consists of two ADIABATIC (no ENERGY transfer as HEAT) stages, alternating with two isothermal (constant TEMPERATURE) stages. In stage 1, an ideal WORKING FLUID experiences adiabatic compression along path *ab*. Since no thermal energy (heat) is allowed to enter or leave the system, the mechanical work of compression causes the temperature of the working fluid to increase from T_1

(the sink temperature) to T_2 (the source temperature). During stage 2 of the Carnot cycle, the working fluid experiences isothermal expansion. While the working fluid expands in traveling along path *bc,* heat is transferred to it from the source at temperature T_2. Then in traveling along path *cd* during stage 3, the working fluid undergoes adiabatic expansion. Travel along this path essentially completes the "power stroke" of the cycle, and the work of expansion causes the temperature of the working fluid to fall from T_2 to T_1. Finally, in stage 4, the working fluid undergoes isothermal compression while traveling along path *da*. Mechanical work is performed on the working fluid, and an equivalent amount of heat is rejected to the sink (environment), which is at the lower temperature, T_1. The working fluid is now back to its initial state and is ready to repeat the cycle again.

The thermal efficiency (η_{th}) of the Carnot cycle is given by the expression

$$\eta_{th} = 1 - (T_1/T_2)$$

where T_1 is the sink (lower) temperature and T_2 is the source (higher) temperature expressed on an absolute scale (i.e., temperature in kelvins or degrees Rankine). The thermal efficiency of the Carnot cycle is the best possible efficiency of any heat engine operating between temperatures T_1 and T_2. This fact is sometimes referred to as the *Carnot principle*. Therefore, for maximum efficiency in a reversible heat engine, the source temperature (T_2) should be as high as possible (within the temperature limits of the materials involved), while the sink or environmental reservoir temperature (T_1) should be as low as possible. Heat engines operating on EARTH must reject thermal energy to the surrounding terrestrial environment. Therefore, T_1 typically is limited to about 300 K (depending on the local environmental conditions). Heat engines operating in space have the opportunity (depending on RADIATOR design) to reject thermal energy to the coldest regions of OUTER SPACE (about 3 K). Although the Carnot cycle is a postulated, idealized cycle, it is very useful

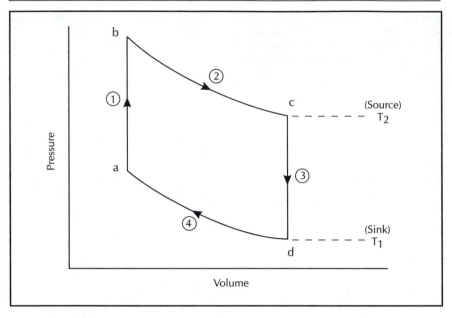

Figure 21. Pressure-volume relationships for an ideal working fluid undergoing the four stages of the Carnot cycle—an ideal, reversible heat engine cycle.

in understanding the performance limits of real heat engines that operate using somewhat less than idealized BRAYTON CYCLE or RANKINE CYCLE.

carrier wave (CW) In TELECOMMUNI-CATIONS, an ELECTROMAGNETIC WAVE intended for MODULATION. This wave is transmitted at a specified FREQUENCY and AMPLITUDE. Information then is superimposed on this carrier wave by making small changes in (i.e., modulating) either its frequency or its amplitude.

Cartesian coordinates A coordinate system, developed by the French mathematician René Descartes (1596–1650), in which locations of points in space are expressed by reference to three mutually perpendicular planes, called *coordinate planes*. The three planes intersect in straight lines called the *coordinate axes*. The distances and the axes are usually marked (x, y, z), and the origin is the (zero) point at which the three axes intersect.

case 1. In general, the outer portion or a metallic material. 2. In rocketry, the structural envelope for the solid propellant in a solid-propellant rocket motor.

case bonding Cementing of the solid propellant to the motor case through the insulation by use of a thin layer of adhesive (the liner). *See also* ROCKET; SOLID-PROPELLANT ROCKET.

Cassegrain telescope A compound REFLECTING TELESCOPE in which a small CONVEX (secondary) MIRROR reflects the convergent BEAM from the PARABOLIC primary mirror through a hole in the primary mirror to an EYEPIECE in the back of the primary mirror. Designed in 1672 by the somewhat mysterious Frenchman Guillaume Cassegrain (c. 1629–1693), it has become the most widely used reflecting telescope in ASTRONOMY.

***Cassini* mission** The *Cassini* mission was launched successfully by a Titan IV–Centaur vehicle on October 15, 1997, on a joint NASA and European Space Agency (ESA) project to Saturn, its major moon Titan, and its complex system of other moons. The *Cassini* orbiter spacecraft's mission consists of delivering a probe, called Huygens (provided by ESA), to Titan and then remaining in orbit around Saturn for detailed studies of the planet and its rings and satellites. (See Figure 22) The principal objectives of this mission are to: (1) determine the three-dimensional structure and dynamical behavior of the rings; (2) determine the composition of the satellite surfaces and the geological history of each object; (3) determine the nature and origin of the dark material on Iapetus' leading hemisphere; (4) measure the three-dimensional structure and dynamical behavior of the Saturnian magnetosphere; (5) study the dynamical behavior of Saturn's atmosphere at cloud level; (6) study the time variability of Titan's clouds and hazes; and (7) characterize Titan's surface on a regional scale. The *Cassini* spacecraft's instrumentation consists of: a radar mapper, a charge-coupled-device (CCD) imaging system, a visual/infrared mapping spectrometer, a composite infrared spectrograph, a cosmic dust analyzer, radio and plasma wave experiments, an ultraviolet imaging spectrograph, a magnetospheric imaging instrument, a magnetometer, and an ion/neutral mass spectrometer. Telemetry from the communications antenna will be used to make observations of the atmospheres of Titan and Saturn and to measure the gravity fields of the planet and its satellites. The spacecraft is named in honor of the Italian-born French astronomer Giovanni Domenico Cassini (1625–1712), who was the first director of the Royal Observatory in Paris.

The *Cassini* mission is taking a similar tour of the solar system as did the Galileo mission—namely the Venus-Venus-Earth-Jupiter Gravity Assist (VVE-JGA) trajectory. As currently planned, the Cassini spacecraft will arrive in the

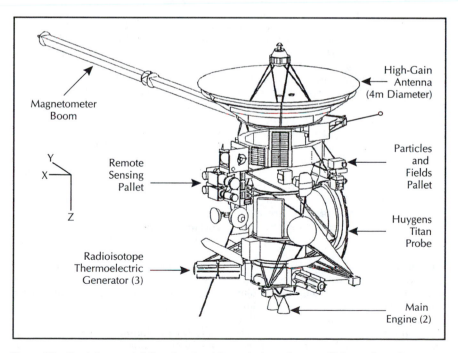

Figure 22. *Cassini* spacecraft line drawing (shown in interplanetary flight configuration).

Saturnian system in June 2004. Shortly after entering orbit around Saturn, the Huygens probe will separate from the Cassini orbiter spacecraft and begin its entry into the atmosphere of Titan. This atmospheric probe carries six science instruments. Should it survive impact with the surface of Titan, it may even be able to transmit some science data from the surface. The Huygens probe is sponsored by the European Space Agency and is named after the Dutch physicist and astronomer Christiaan Huygens (1629–1695), who first described the nature of Saturn's rings and discovered its major moon, Titan. The Cassini spacecraft, which receives its electric power (over 625 watts projected at end of mission) from three radioisotope thermoelectric generators (RTGs), is then expected to make at least 30 loose elliptical orbits of Saturn, with each orbit being optimized for a different set of Saturnian system observations. *See also* GALILEO PROJECT; SATURN.

catalyst A substance that changes (increases) the rate of a chemical reaction without being consumed or changed by the reaction. A *negative catalyst* is a substance that retards a chemical reaction.

catapult A power-activated device that provides an auxiliary source of thrust to a missile or aircraft; it combines the functions of directing and accelerating the missile during its travel on the catapult and serves the same functions for a missile as does a gun tube for a shell.

cathode The negative electrode in an electron (discharge) tube or electrolytic cell through which a primary stream of electrons enters a system. *Compare with* ANODE.

cavitation The formation of bubbles (or cavities) in a fluid (liquid) that occurs whenever the static pressure at any point in the fluid becomes less than the fluid vapor pressure. In the flow of a liquid, the lowest pressure reached is the vapor pressure. If the fluid velocity is further increased, the flow condition changes and

cavities are formed. The formation of these cavities (or vapor regions) alters the fluid flow path and therefore the performance of hydraulic machinery and devices, such as pumps. The collapse of these bubbles in downstream regions of high pressure creates local pressure forces that may result in pitting or deformation of any solid surface near the cavity at the time of collapse. Cavitation effects are most noticeable with high-speed hydraulic machinery, such as a liquid propellant rocket engine's turbopumps.

celestial Of or pertaining to the heavens.

celestial body Any aggregation of matter in space constituting a unit for astronomical study, such as the Sun, Moon, a planet, comet, asteroid, star, nebula, and so on. Also called a *heavenly body*.

celestial guidance The guidance of a missile or other aerospace vehicle by reference to celestial bodies. *See also* GUIDANCE SYSTEM.

celestial-inertial guidance The process of directing the movements of an aerospace vehicle or spacecraft, especially in the selection of a flight path, by an inertial guidance system that also receives inputs from observations of celestial bodies. *See also* GUIDANCE SYSTEM.

celestial latitude (symbol: β) With respect to the CELESTIAL SPHERE, the angular distance of a CELESTIAL BODY from 0° to 90° north (considered positive) or south (considered negative) of the ECLIPTIC.

celestial longitude (symbol: λ) With respect to the CELESTIAL SPHERE, the angular distance of a CELESTIAL BODY from 0° to 360° measured eastwards along the ECLIPTIC to the intersection of the body's circle of celestial longitude. The VERNAL EQUINOX is taken as 0°.

celestial mechanics The branch of science that investigates the dynamic relationships among bodies of the solar system and deals with the relative motions of celestial bodies under the influence of gravitational fields. *See also* ORBITS OF OBJECTS IN SPACE.

celestial navigation The process of directing an aircraft, AEROSPACE vehicle, or SPACECRAFT from one point to another by reference to celestial bodies of known coordinates. Celestial navigation usually refers to the process as accomplished by a human operator. The same process accomplished automatically by a machine usually is called *celestial guidance* or sometimes *automatic celestial navigation*. *See also* GUIDANCE SYSTEM.

celestial sphere To create a consistent coordinate system for the heavens, early astronomers developed the concept of a celestial sphere. It is an imaginary sphere of very large radius with EARTH as its center and on which all observable CELESTIAL bodies are assumed projected. The rotational AXIS of Earth intersects the north and south poles of the celestial sphere. An extension of Earth's equatorial plane cuts the celestial sphere and forms a great circle, called the *celestial equator*. The direction to any STAR or other celestial body can then be plotted in two dimensions on the inside of this imaginary sphere using CELESTIAL LATITUDE and the CELESTIAL LONGITUDE.

Celsius temperature scale The widely used relative TEMPERATURE scale, originally developed by the Swedish scientist Anders Celsius (1701–1744), in which the range between two reference points (ice at 0° and boiling water at 100°) is divided into 100 equal units, or degrees.

Centaur (rocket) A powerful and versatile UPPER STAGE ROCKET originally developed by the United States in the 1950s for use with the ATLAS LAUNCH VEHICLE. Engineered by the German-American rocket engineer Krafft Ehricke (1917–1984), it was the first American rocket to use LIQUID HYDROGEN as its PROPELLANT. Centaur has supported many important military and scientific missions, including the CASSINI MISSION (launched on October 15, 1997) to SATURN.

Centaurs A group of unusual CELESTIAL objects residing in the outer SOLAR SYSTEM, such as CHIRON, that exhibit a dual ASTEROID-COMET nature. Named after the centaurs in Greek mythology, who were half-human and half-horse beings.

centi- (symbol: c) The prefix in the SI unit system that denotes one hundredth (1/100).

centimeter (cm) One hundredth of a meter (i.e., 0.01 meter).
2.54 cm = 1 inch (US) (exactly)

centimeter-gram-second system (cgs-system) A system of units based on the centimeter (cm) as the unit of length, the gram (g) as the unit of mass, and the second (s) as the unit of time. In the great majority of contemporary technical literature, the cgs-system has been superseded by SI units. *See also* SI UNITS.

center-of-gravity For a rigid body, that point about which the resultant of the gravitational torques of all the particles of the body is zero. If the mass of a rigid body can be considered as concentrated at a central point, called the CENTER-OF-MASS, then the "weight" of this body (i.e., its response to a gravitational field) also may be considered concentrated at this point, such that the center-of-mass is also the center-of-gravity.

center-of-mass That point in a given body, or in a system of two or more bodies acting together in respect to another body (e.g., the Earth-Moon system orbiting around the Sun), which represents the mean position of matter in the bodies or system of bodies. Also called *barycenter*.

central force A force that for the purposes of computation can be considered to be concentrated at one central point with its intensity at any other point being a function of the distance from the central point. For example, gravitation is considered as a central force in CELESTIAL MECHANICS.

central processing unit (CPU) The computational and control unit of a computer—the device that functions as the "brain" of a computer system. The central processing unit (CPU) interprets and executes instructions and transfers information within the computer. Microprocessors, which have made possible the personal computer "revolution," contain single-chip CPUs, while the CPUs in large mainframe computers and many early minicomputers contain numerous circuit boards (each packed full of integrated circuits).

Centre National d'Etudes Spatiales (CNES) The public body responsible for all aspects of French space activity including LAUNCH VEHICLES and SPACECRAFT. CNES has four main centers: Headquarters (Paris), the Launch Division at Evry, the Toulouse Space Center, and the Guiana Space Center (LAUNCH SITE) in Kourou, French Guiana (South America). *See also* ARIANE; EUROPEAN SPACE AGENCY.

centrifugal force A reaction FORCE that is directed opposite to a CENTRIPETAL FORCE, such that it points out along the radius of curvature away from the center of curvature.

centripetal force The central (inward-acting) FORCE on a body that causes it to move in a curved (circular) path. Consider a person carefully whirling a stone secured by a strong (but lightweight) string in a circular path at a constant speed. The string exerts a radial "tug" on the stone, which is called the centripetal force. Now as the stone keeps moving in a circle at constant speed, the stone also exerts a reaction force on the string, which is called the CENTRIFUGAL FORCE. It is equal in magnitude but opposite in direction to the centripetal force exerted by the string on the stone.

Cepheid variable A type of very bright SUPERGIANT STAR that exhibits a regular pattern of changing its brightness as a function of time. The period of this pulsation pattern is directly related to the star's intrinsic brightness, so modern astronomers can use Cepheid variables to determine astronomical distances.

Ceres The first and largest (940-km diameter) ASTEROID to be found. It was discovered on January 1, 1801, by the Italian astronomer and monk Giuseppe Piazzi (1746–1826).

CETI An acronym that stands for "communication with extraterrestrial intelligence." *Compare with SETI. See also* SEARCH FOR EXTRATERRESTRIAL INTELLIGENCE.

chaff Radar confusion reflectors, which consist of thin, narrow metallic strips of various lengths and frequency responses; usually dropped from aircraft or expelled from shells, rockets, or spacecraft as a radar countermeasure.

chain reaction A reaction that stimulates its own repetition. In nuclear physics, a *fission chain reaction* starts when a fissile nucleus (such as uranium-235 or plutonium-239) absorbs a neutron and splits, or "fissions," releasing additional neutrons. A fission chain reaction is self-sustaining when the number of neutrons released in a given time equals or exceeds the number of neutrons lost by escape from the system or by nonfission (parasitic) capture. *See also* FISSION (NUCLEAR).

***Challenger* accident** NASA's SPACE SHUTTLE *Challenger* was launched from Complex 39-B at the KENNEDY SPACE CENTER on January 28, 1986, as part of the STS 51-L mission. At approximately 74 seconds into the flight, an explosion occurred that caused the loss of the AEROSPACE VEHICLE and its entire crew, including ASTRONAUTS Francis R. Scobee, Michael J. Smith, Ellison S. Onizuka, Judith A. Resnik, Ronald E. McNair, S. Christa Corrigan McAuliffe, and Gregory B. Jarvis.

chamber filling interval Time period from complete ignition of the solid-propellant grain to achievement of equilibrium burning pressure. *See also* SOLID-PROPELLANT ROCKET.

chamber pressure (symbol: P_c) The pressure of gases within the combustion chamber of a rocket engine. *See also* ROCKET.

chamber volume (symbol: V_c) The volume of a rocket's combustion chamber, including the convergent portion of the nozzle up to the throat. *See also* ROCKET.

Chandrasekhar, Subrahmanyan (aka: Chandra) (1910–1995) Indian-American astrophysicist who made important contributions to the theory of stellar evolution, especially the role of the WHITE DWARF STAR as the last stage of evolution of many stars that are about the MASS of the SUN. He shared the 1983 Nobel Prize in physics for his theoretical studies of the physical processes important to the structure and evolution of stars. In July 1999, NASA successfully launched an Advanced X-Ray Astrophysics Facility that was renamed the *CHANDRA X-RAY OBSERVATORY* (CXO) in his honor.

Chandrasekhar limit In the 1920s CHANDRASEKHAR SUBRAHMANYAN used RELATIVITY theory and QUANTUM MECHANICS to show that if the MASS of a DEGENERATE STAR is more than about 1.4 SOLAR MASSES (a maximum mass called the *Chan-*

drasekhar limit), it will not evolve into a WHITE DWARF STAR, but rather it will continue to collapse under the influence of GRAVITY and become either a NEUTRON STAR, a BLACK HOLE, or else blow itself apart in a SUPERNOVA explosion.

Chandra X-ray Observatory (CXO) One of NASA's major orbiting astronomical observatories, launched in July 1999 and named after SUBRAHMANYAN CHANDRASEKAR. It was previously called the Advanced X-Ray Astrophysics Facility (AXAF). This EARTH-orbiting facility studies some of the most interesting and puzzling X-RAY sources in the UNIVERSE, including emissions from active galactic nuclei, exploding STARS, NEUTRON STARS, and matter falling into BLACK HOLES. (See Figure 23.)

change detection The practice in remote sensing applications of comparing two digitized images of the same scene that have been acquired at different times. By comparing the intensity differences (either graytone or natural color differences) between corresponding pixels in the time-separated images of the scene, interesting information concerning change and activity (i.e., vegetation growth, urban sprawl, migrating environmental stress in crops, etc.) can be detected quickly in a

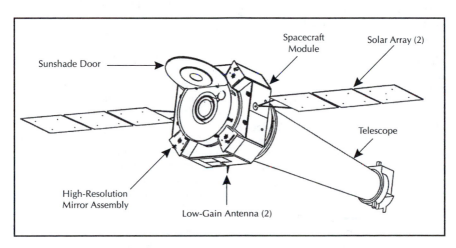

Figure 23. *Chandra X-ray Observatory.*

semiautomated, quasi-empirical fashion. *See also* REMOTE SENSING.

channel construction Use of machined grooves in the wall of the nozzle or combustion chamber of a rocket to form coolant passages. *See also* ROCKET.

chaos The branch of physics (mechanics) that studies unstable systems.

chaotic orbit The ORBIT of a CELESTIAL BODY that changes in a highly unpredictable manner, usually when a small object, such as an ASTEROID or COMET, passes close to a massive PLANET (such as SATURN or JUPITER) or the SUN. For example, the chaotic orbit of CHIRON is influenced by both URANUS and Saturn.

chaotic terrain A planetary surface feature (first observed on MARS in 1969) that is characterized by a jumbled, irregular assembly of fractures and blocks of rock.

charge coupled device (CCD) An electronic (solid state) device containing a regular array of SENSOR elements that are sensitive to various types of ELECTROMAGNETIC RADIATION (e.g., LIGHT) and emit ELECTRONS when exposed to such radiation. The emitted electrons are collected and the resulting charge analyzed. CCDs are used as the light-detecting component in modern television cameras and TELESCOPES.

charged particle An ION; an elementary PARTICLE that carries a positive or negative electric charge, such as an ELECTRON, a PROTON, or an ALPHA PARTICLE.

charged particle detector A device that counts and/or measures the energy of charged particles; frequently found as part of the scientific instrument payload on planetary flyby and deep-space exploration missions.

Charon The large (about 1,200 km in diameter) MOON of PLUTO discovered in 1978 by the American astronomer James Walter Christy. It orbits at a mean distance of 20,000 km and keeps one face permanently towards the PLANET.

charring ablator An ablation material characterized by the formation of a carbonaceous layer at the heated surface that impedes thermal energy (heat) flow into the material by virtue of its insulating and radiation heat transfer characteristics. *See also* ABLATION.

chaser spacecraft The SPACECRAFT or AEROSPACE VEHICLE that actively performs the key maneuvers during orbital RENDEZVOUS and DOCKING/BERTHING operations. The other SPACE VEHICLE serves as the TARGET and remains essentially PASSIVE during the encounter.

chasm A canyon or deep linear feature on a PLANET's surface.

checkout The sequence of actions (such as functional, operational, and calibration tests) performed to determine the readiness of a SPACECRAFT or LAUNCH VEHICLE to perform its intended mission.

chemical energy Energy liberated or absorbed in the process of a chemical reaction. In such a reaction, energy losses or gains usually involve only the outermost electrons of the atoms or ions of the system undergoing change; here a chemical bond of some type is established or broken without disrupting the original atomic or ionic identities of the constituents. Chemical changes, according to the nature of the substance entering into the change, may be induced by thermal (thermochemical), light (photochemical), or electric (electrochemical) energies.

chemical fuel A fuel that depends on an oxidizer for combustion or the development of thrust, such as liquid or solid rocket fuel or internal-combustion-engine fuel; as distinguished from nuclear fuel (i.e., uranium-235).

chemical laser A laser that uses chemical reactions to pump energy into a lasting medium, thereby creating the inverted state population needed for lasing. An example of a high-power chemical laser is the hydrogen fluoride (HF) laser—a laser in which a hydrogen-fluorine reaction produces lasing in a hydrogen-fluoride medium.

chemical pressurization The pressurization of propellant tanks in a rocket by means of high-pressure gases developed by the combustion of a fuel and oxidizer or by the decomposition of a substance.

chemical rocket A ROCKET that uses the combustion of a chemical FUEL in either solid or liquid form to generate THRUST. The chemical fuel requires an OXIDIZER to support combustion.

Cheyenne Mountain Air Force Station (CMAFS) During the COLD WAR, a hollowed-out mountain in the Colorado Rockies functioned as the U.S. nerve center that would sound the first alarm of a Soviet attack against North America. Since the end of the cold war, the activities at Cheyenne Mountain have not diminished. Rather, they have evolved in response to the realities and needs of a quickly changing world. The original NORTH AMERICAN AIR DEFENSE COMMAND (NORAD) Combat Operations Center (COC) has transformed into the Cheyenne Mountain Operations Center (CMOC).

Today, Cheyenne Mountain is known as Cheyenne Mountain Air Force Station (CMAFS) and hosts four commands: the North American Aerospace Defense Command (NORAD), the United States Northern Command (USNORTHCOM), the UNITED STATES STRATEGIC COMMAND (USSTRATCOM), and AIR FORCE SPACE COMMAND (AFSPC). Cheyenne Mountain serves as the command center for both NORAD and USNORTHCOM. It is the central collection and coordination center for a worldwide system of SATELLITES, RADARS, and SENSORS that provide early warning of any MISSILE, air, or space threat to North America.

Supporting the NORAD mission, CMOC provides warning of BALLISTIC MISSILE or air attacks against North America, assists the air sovereignty mission of the United States and Canada, and, if necessary, serves as the focal point for AIR DEFENSE operations to counter enemy bombers and CRUISE MISSILES. In addition, CMOC provides theater ballistic missile warning for U.S. and allied forces. In support of USSTRATCOM, Cheyenne Mountain provides a day-to-day picture of precisely what is in space and where it is located. SPACE CONTROL OPERATIONS include protection, prevention, and negation functions supported by the surveillance of space.

Cheyenne Mountain is one of the most unique installations in the world. Apart from the fact that it is housed about 600 meters (2,000 ft) underground, CMOC is also different from most military units because it is a joint and binational military organization comprised of more than 200 men and women from the U.S. Army, Navy, Marine Corps, and Air Force, as well as the Canadian Forces. Operations are conducted in seven centers crewed 24 hours a day, 365 days a year. These centers are the Air Warning Center, the Missile Warning Center, the Space Control Center, the Operational Intelligence Watch, the Systems Center, the Weather Center, and the Command Center.

chilldown Cooling all or part of a cryogenic (very cold) ROCKET engine system from AMBIENT (room) TEMPERATURE down to cryogenic temperature by circulating CRYOGENIC PROPELLANT (FLUID) through the system prior to engine start.

Chiron An unusual CELESTIAL BODY in the outer SOLAR SYSTEM with a CHAOTIC ORBIT that lies almost entirely between the ORBITS of SATURN and URANUS. This massive ASTEROID-sized object has a diameter of about 200 kilometers and is the first object placed in the CENTAURS group because it also has a detectable COMA, a feature characteristic of COMETS.

choked flow A flow condition in a DUCT or pipe such that the flow upstream of a certain critical section (like a NOZZLE or VALVE) cannot be increased by further reducing downstream PRESSURE.

chord line A straight line through the centers of curvature of the leading and trailing edges of an airfoil section. This line generally is used as a reference line from which the ordinates and angles of an airfoil are measured. *See also* AIRFOIL; WING.

chromatic aberration A phenomenon that occurs in a refracting optical system because LIGHT of different WAVELENGTHS (colors) is refracted (bent) by a different amount. As a result, a simple LENS will give red light a longer FOCAL LENGTH than blue light.

chromosphere The reddish layer in the SUN's ATMOSPHERE located between the PHOTOSPHERE (the apparent SOLAR surface) and the base of the CORONA. It is the source of solar PROMINENCES.

chronic radiation dose A dose of ionizing radiation received either continuously or intermittently over a prolonged period of time.

Chryse Planitia A large plain on MARS characterized by many ancient channels that could have once contained flowing surface water. Landing site for NASA's *VIKING 1* LANDER SPACECRAFT in July 1976.

chuffing The characteristic of some rockets to burn intermittently and with an irregular noise.

chugging A form of combustion instability that occurs in a LIQUID-PROPELLANT ROCKET ENGINE. It is characterized by a pulsing operation at a fairly low FREQUENCY.

circadian rhythms An organism's day/night cycle of living; a regular change in physiological function occurring in approximately 24-hour cycles.

circularize To change an elliptical orbit into a circular one, usually by firing a rocket motor when the space vehicle is at apogee. *See also* APOGEE; APOGEE MOTOR.

circular velocity At any specific distance from the primary (i.e., central body), the orbital velocity required to maintain a constant-radius orbit.

circumferential seal Seal composed of a continuous ring or of one or more segmented rings whose sealing surface is parallel to the centerline of the flow passage. Also called *radial seal.*

circumlunar Around the Moon; a term generally applied to trajectories.

circumsolar space Around the Sun; heliocentric (Sun-centered) space.

circumstellar Around a star—as opposed to *interstellar,* between the stars.

cislunar Of or pertaining to phenomena, projects, or activities happening in the region of OUTER SPACE between EARTH and the MOON. It comes from the Latin word *cis,* meaning "on this side," and *lunar,* which means "of or pertaining to the Moon." Therefore, it means "on this side of the Moon."

Clarke orbit A GEOSTATIONARY ORBIT; named after the British writer and engineer Sir Arthur C. Clarke (b. 1917), who first proposed in 1945 the use of this special ORBIT around EARTH for COMMUNICATIONS SATELLITES.

clean room A controlled work environment for SPACECRAFT and AEROSPACE systems in which dust, TEMPERATURE, and HUMIDITY are carefully controlled during the fabrication, assembly, and/or testing of critical components. *See also* BIOCLEAN ROOM.

Clementine A joint project between the Department of Defense and NASA. The objective of the mission, also called the Deep Space Program Science Experiment, was to test sensors and spacecraft components under extended exposure to the space environment and to make scientific observations of the Moon and the near-Earth asteroid 1620 Geographos. The observations included imaging at various wave-lengths, including ultraviolet (UV) and infrared (IR), laser altimetry, and charged particle measurements. The *Clementine* craft was launched on January 25, 1994, from Vandenburg Air Force Base in California aboard a Titan II-G rocket. After two Earth flybys, lunar insertion was achieved on February 21, 1994. Lunar mapping took place over approximately two months, in two parts. The first part consisted of a five-hour elliptical polar orbit with a perilune of about 400 kilometers (249 miles) at 28 degrees south latitude. After one month of mapping, the orbit was rotated to a perilune of 29 degrees north latitude, where it remained for one more month. After leaving lunar orbit, a malfunction in one of the onboard computers occurred on May 7, 1994. This malfunction caused a thruster to fire until it had used up all its propellant, leaving the spacecraft spinning at about 80 revolutions per minute (RPM) with no spin control. This made the planned continuation of the mission, a flyby of the asteroid 1620 Geographos, impossible. The spacecraft remained in geocentric orbit and continued testing the spacecraft components until the end of mission.

clevis A fitting with a U-shaped end for attachment to the end of a pipe or rod.

climate change The long-term fluctuations in temperature, precipitation, wind, and all other aspects of Earth's climate. External processes, such as solar-irradiance variations, variations of Earth's orbital parameters (e.g., eccentricity, precession, and inclination), lithosphere motions, and volcanic activity, are factors in climatic variation. Internal variations of the climate system also produce fluctuations of sufficient magnitude and variability to explain observed climate change through the feedback processes that interrelate the components of the climate system. *See also* GLOBAL CHANGE.

clock An electronic circuit, often an integrated circuit, that produces high-frequency timing signals. A common application is synchronization of the operations performed by a computer or microprocessor-based system. Typical clock rates in microprocessor circuits are in the megahertz range, with 1 megahertz (1 MHz) corresponding to 1 million cycles per second (10^6 cps). A *spacecraft clock* is usually part of the command and data handling subsystem. It meters the passing time during the life of the spacecraft and regulates nearly all activity within the spacecraft. It may be very simple (e.g., incrementing every second and bumping its value up by one), or it may be much more complex (with several main and subordinate fields of increasing resolution). In aerospace operations, many types of commands that are uplinked to the spacecraft are set to begin execution at specific spacecraft clock counts. In downlinked telemetry, spacecraft clock counts (which indicate the time a telemetry frame was created) are included with engineering and science data to facilitate processing, distribution, and analysis.

closed ecological life support system (CELSS) A system that provides for the maintenance of life in an isolated living chamber or facility through complete reuse of the materials available within the chamber or facility. This is accomplished, in particular, by means of a cycle in which exhaled carbon dioxide, urine, and other waste matter are converted chemically or by photosynthesis into oxygen, water, and food. On a grand (macroscopic) scale, the planet Earth itself is a closed ecological system; on a more modest scale, a "self-sufficient" space base or space settlement also would represent another example of a closed ecological system. In this case, however, the degree of "closure" for the space base or space settlement would be determined by the amount of makeup materials that had to be supplied from Earth.

Material recycling in a life support system can be based on physical and chemical processes, can be biological in nature, or can be a combination of both.

Chemical and physical systems are designed more easily than biological systems, but provide little flexibility or adaptability to changing needs. A life support system based solely on physical and chemical methods also would be limited because it would still require resupply of food and some means of waste disposal. A *bioregenerative* life support system incorporates biological components in the creation, purification, and renewal of life support elements. Plants and algae are used in food production, water purification, and oxygen release. While the interactions of the biomass with the environment are very complex and dynamic, creating a fully closed ecological system—one that needs no resupply of materials (although energy can cross its boundaries, as does sunlight into Earth's biosphere)—appears possible and even essential for future, permanently inhabited human bases and settlements within the solar system. Life scientists call a human-made closed ecological system that involves a combination of chemical, physical, and biological processes a *closed ecological life support system* (CELSS). This type of closed ecological system is sometimes called a *controlled ecological life support system.*

closed loop Term applied to an electrical or mechanical system in which the output is compared with the input (command) signal, and any discrepancy between the two results in corrective action by the system elements.

closed-loop RF check In aerospace launch operations, the radio frequency (RF) check of the command destruct system without open-air RF transmission. This is generally accomplished by hardline from the transmitter to the receiver. *See also* COMMAND DESTRUCT.

closed system 1. In thermodynamics, a system in which no transfer of mass takes place across its boundaries. Compare with OPEN SYSTEM. 2. A closed ecological system. *See also* CLOSED ECOLOGICAL LIFE SUPPORT SYSTEM. 3. In mathematics, a sys-

tem of differential equations and supplementary conditions such that the values of all the unknowns (dependent variables) of the system are mathematically determined for all values of the independent variables (usually space and time) to which the system applies. 4. A system that constitutes a feedback loop so that the inputs and controls depend on the resulting output.

closed universe The model in COSMOLOGY that assumes the total MASS of the UNIVERSE is sufficiently large that one day the galaxies will slow down and stop expanding because of their mutual gravitational attraction. At that time, the universe will have reached its maximum size, and then GRAVITATION will make it slowly contract, ultimately collapsing to a single point of infinite density (sometimes called the BIG CRUNCH). Also called *bounded universe model. Compare with* OPEN UNIVERSE.

close encounter (CE) An interaction with an unidentified flying object (UFO). *See also* UNIDENTIFIED FLYING OBJECT.

closest approach 1. The point in time and space when two planets or other celestial bodies are nearest to each other as they orbit about the Sun or other primary. 2. The point in time and space when a scientific spacecraft on a flyby encounter trajectory is nearest the target celestial object. 3. The place or time of such an event.

closing rate The speed at which two bodies approach each other—as, for example, two spacecraft "closing" for an orbital rendezvous and/or docking operation.

cluster Two or more rocket motors bound together so as to function as one propulsion unit. *See also* ROCKET.

cluster of galaxies An accumulation of galaxies that lie within a few million LIGHT-YEARS of one another and are bound by GRAVITATION. Galactic clusters can occur with just a few member galaxies (say 10 to 100), such as the LOCAL GROUP, or they can occur in great groupings involving thousands of galaxies.

coasting flight   In rocketry, the flight of a rocket between burnout or thrust cutoff of one stage and ignition of another, or between burnout and summit altitude or maximum horizontal range.

coated optics   Optical elements (lenses, prisms, etc.) that have their surfaces covered with a thin transparent film to minimize light loss due to reflection.

coaxial injector   Type of liquid rocket engine injector in which one propellant surrounds the other at each injection point.

cold-flow test   The thorough testing of a LIQUID-PROPELLANT ROCKET ENGINE without actually firing (igniting) it. This type of test helps AEROSPACE engineers verify the performance and efficiency of a PROPULSION SYSTEM, since all aspects of PROPELLANT flow and conditioning, except combustion, are examined. Tank pressurization, propellant loading, and propellant flow into the COMBUSTION CHAMBER (without ignition) are usually included in a cold flow test. *Compare with* HOT FIRE TEST.

coldsoak   The exposure of a system or equipment to low temperature for a long period of time to ensure that its temperature is lowered to that of the surrounding atmosphere or operational environment.

cold war   The ideological conflict between the United States and the former Soviet Union from approximately 1946 to 1989, involving rivalry, mistrust, and hostility just short of overt military action. The tearing down of the Berlin Wall in November 1989 generally is considered as the (symbolic) end of the cold war period.

collimator   A device for focusing or confining a BEAM of PARTICLES or ELECTROMAGNETIC RADIATION, such as X-RAY PHOTONS.

color   A quality of LIGHT that depends on its WAVELENGTH. The *spectral color* of

emitted light corresponds to its place in the SPECTRUM of the rainbow. *Visual light,* or *perceived color,* is the quality of light emission as recognized by the human eye. Simply stated, the human eye contains three basic types of light-sensitive cells that respond in various combinations to incoming spectral colors. For example, the color brown occurs when the eye responds to a particular combination of blue, yellow, and red light. Violet light has the shortest wavelength, while red light has the longest wavelength. All the other colors have wavelengths that lie in between.

Columbia **accident**   While gliding back to EARTH on February 1, 2003, after a very successful 16-day scientific research mission in LOW EARTH ORBIT, NASA's SPACE SHUTTLE *Columbia* experienced a catastrophic reentry accident and broke apart at an ALTITUDE of about 63 kilometers over Texas. Traveling through the upper ATMOSPHERE at approximately 18 times the SPEED OF SOUND, the ORBITER vehicle disintegrated, taking the lives of its seven crewmembers: six American ASTRONAUTS (Rick D. Husband, William C. McCool, Michael P. Anderson, Kalpana Chawla, Laurel Blair Salton Clark, and David M. Brown) and the first Israeli astronaut (Ilan Ramon). Disaster struck the STS 107 mission when *Columbia* was just 15 minutes from its landing site at the KENNEDY SPACE CENTER in Florida. Preliminary investigations indicate that a severe heating problem occurred in *Columbia*'s left wing, but a month after the catastrophe experts remained baffled by the root cause of the fatal accident.

coma   1. The gaseous envelope that surrounds the nucleus of a comet. *See also* COMET. 2. In an optical system, a result of spherical aberration in which a point source of light (not on axis) has a blurred, comet-shaped image. *See also* ABERRATION.

combustion   Burning, or rapid oxidation, accompanied by the release of energy in the form of heat and light. The combustion of hydrogen, however, emits radiation outside of the visible spectrum—making a "hydrogen flame" essentially invisible to the naked eye.

combustion chamber The part of a ROCKET engine in which the combustion of chemical PROPELLANTS takes place at high PRESSURE. The combustion chamber and the diverging section of the NOZZLE make up a rocket's THRUST chamber. Sometimes called the *firing chamber* or simply the *chamber*.

combustion efficiency The efficiency with which rocket fuel is burned, often expressed as the ratio of the actual energy released by the combustion of a unit mass of fuel to the potential chemical energy stored in a unit mass of fuel. *See also* ROCKET.

combustion instability Unsteadiness or abnormality in the combustion of fuel, as may occur in a rocket engine. *See also* ROCKET.

combustion product Gaseous or solid materials produced during the burning (oxidation) of a fuel or combustible substance.

comet A dirty ice "rock" consisting of dust, frozen water, and gases that orbits the SUN. As a comet approaches the inner SOLAR SYSTEM from deep space, solar radiation causes its frozen materials to vaporize (sublime), creating a COMA and a long TAIL of dust and IONS. Scientists think these icy PLANETESIMALS are the remainders of the primordial material from which the OUTER PLANETS were formed billions of years ago. *See also* KUIPER BELT and OORT CLOUD.

Comet Halley (1P/Halley) The most famous PERIODIC COMET. Named after the British mathematician and astronomer Edmond Halley (1656–1742), who successfully predicted its 1758 return. Reported since 240 B.C.E., this COMET reaches PERIHELION approximately every 76 years. During its most recent inner SOLAR SYSTEM appearance, an international fleet of five different SPACECRAFT, including the *GIOTTO* SPACECRAFT, performed scientific investigations that sup-

ported the dirty ice rock model of a comet's NUCLEUS.

command A signal that initiates or triggers an action in the device that receives the signal. In computer operations, also called an *instruction*.

command and control system The facilities, equipment, communications, procedures, and personnel essential to a commander for planning, directing, and controlling operations of assigned forces pursuant to the missions assigned.

command destruct An intentional action leading to the destruction of a ROCKET or MISSILE in flight. Whenever a malfunctioning vehicle's performance creates a safety hazard on or off the rocket test range, the range safety officer sends the command destruct signal to destroy it.

commander In AEROSPACE operations involving human spaceflight, the crew person who has ultimate responsibility for the safety of the personnel onboard the SPACECRAFT or aerospace vehicle and has authority throughout the flight to deviate from the flight/mission plan, procedures, and personnel assignments as necessary to preserve crew safety or vehicle integrity.

command guidance A guidance system wherein information (including the latest data/intelligence about a target) is transmitted to the in-flight missile from an outside source, causing the missile to traverse a directed flight path. *See also* GUIDANCE SYSTEM.

command module (CM) The part of the Apollo spacecraft in which the astronauts lived and worked; it remained attached to the service module (SM) until reentry into EARTH's atmosphere, when the SM was jettisoned and the command module (CM) brought the lunar astronauts back safely to the surface of Earth. *See also* APOLLO PROJECT.

communications satellite A SATELLITE that relays or reflects ELECTROMAGNETIC

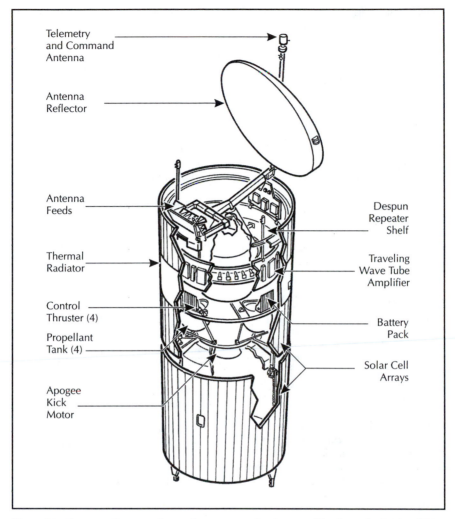

Figure 24. Cutaway diagram of a modern communications satellite, here the Boeing 376 spacecraft configuration. On orbit, this 2.13-meter (7-ft) diameter satellite extends to a full vertical configuration (as shown) of about 6.7 meters (22 ft). Numerous versions of this commercial communications satellite have been placed into service in geosynchronous orbit.

SIGNALS between two (or more) communications stations. An *active communications satellite* receives, regulates, and retransmits electromagnetic signals between stations, while a *passive communications satellite* simply reflects (or bounces) signals between stations. In 1945 the British writer and engineer Sir Arthur C. Clarke (b. 1917) proposed placing communications satellites in GEOSTATIONARY ORBIT around EARTH. Numerous active communications satellites now maintain a global telecommunications infrastructure. (See Figure 24.)

compact body A small, very dense CELESTIAL BODY that represents the end product of stellar evolution: a WHITE DWARF, a NEUTRON STAR, or a BLACK HOLE.

companion body A NOSE CONE, protective SHROUD, last-STAGE ROCKET, or PAYLOAD separation hardware that ORBITS EARTH along with an operational SATELLITE or SPACECRAFT. Companion bodies contribute significantly to a growing SPACE DEBRIS population in LOW EARTH ORBIT.

complex Aerospace jargon for *launch complex;* for example, Complex 39A at the NASA Kennedy Space Center.

composite materials Structural materials of metals, ceramics, or plastics with built-in strengthening agents that may be in the form of filaments, foils, powders, or flakes of a different compatible material.

composite propellant A solid-rocket propellant consisting of a fuel (typically organic resins or plastics) and an oxidizer (i.e., nitrates or perchlorates) intimately mixed in a continuous solid phase. Neither the fuel nor the oxidizer would burn without the presence of the other. *See also* ROCKET.

compressed liquid In THERMODYNAMICS, a liquid that exists at a pressure greater than the saturation pressure for its temperature or at a temperature lower than the saturation temperature corresponding to its pressure. Also called a *subcooled liquid.*

compressible flow In fluid mechanics, flow in which density changes in the fluid cannot be neglected.

compressor A device for compressing air or other fluid. Compressors can be distinguished by the manner in which they handle the fluid being compressed, such as an *axial-flow compressor, centrifugal compressor, single-entry compressor,* and *supersonic compressor.* Compressors also can be identified by the number of stages they possess, such as *single-stage compressor* or *multistage compressor.*

Compton, Arthur Holly (1892–1962) American physicist who shared the 1927 Nobel Prize in physics for his pioneering work on the scattering of high-energy PHOTONS by ELECTRONS. This phenomenon, called the COMPTON EFFECT, is the foundation of many GAMMA RAY detection techniques used in high-energy ASTROPHYSICS. NASA named the large gamma ray observatory launched in April 1991 the COMPTON GAMMA RAY OBSERVATORY (CGRO) to honor his scientific accomplishments.

Compton effect Scattering of high-energy photons (either X rays or gamma rays) by free electrons. In this process the electron gains energy and recoils, and the scattered photon loses some of its energy (which has been transferred to the scattered electron) and increases in wavelength. First observed in 1923 by the American physicist A. H. Compton (1892–1962). *See also* GAMMA RAYS; X RAY.

Compton Gamma Ray Observatory (CGRO) A major NASA orbiting astrophysical observatory dedicated to GAMMA RAY ASTRONOMY. The CGRO was placed in ORBIT around EARTH in April 1991. At the end of its useful scientific mission, flight controllers intentionally commanded the massive (16,300 kg) SPACECRAFT to perform a DE-ORBIT BURN. This caused it to reenter and safely crash in June 2000 in a remote region of the Pacific Ocean. The spacecraft was named in honor of the American physicist ARTHUR HOLLY COMPTON.

concave lens (or mirror) A lens or mirror with an inward curvature.

concentration For a solution, the quantity of dissolved substance per unit quantity of solvent. Concentration can be expressed in several ways. The *mass concentration,* for example, is defined as the mass of solute per unit volume of solvent and has as typical units (i.e., kilograms per cubic meter [kg/m^3] or grams per cubic centimeter [g/cm^3]). The *molal concentration,* on the other hand, is defined as the amount of substance dissolved per unit mass of substance and has the units mol/kg (i.e., moles per kilogram).

condensation The phase change process by which a vapor becomes a liquid. The opposite of *evaporation*.

condensation trail A visible cloud streak, usually brilliantly white in color, that trails behind a missile or other vehicle in flight under certain conditions. There are three general types of condensation trails: the *aerodynamic* type, caused by reduced pressure of the air in certain areas as it flows past the missile or aircraft; the *convection* type, caused by the rising of air warmed by an engine; and the *engine-exhaust* (or exhaust-moisture) type, formed by the ejection of water vapor into a cold atmosphere. Also called a *contrail* or *vapor trail.*

conductance (symbol: G) The reciprocal of electrical resistance (for a direct current circuit); generally measured in siemens (S), the SI UNIT of electrical conductance; previously measured in *mho* ("ohm" spelled backward) or reciprocal ohm.

conduction (thermal) The transport of HEAT (thermal ENERGY) through an object by means of a TEMPERATURE difference from a region of higher temperature to a region of lower temperature. For SOLIDS and LIQUID metals, thermal conduction is accomplished by the migration of fast-moving ELECTRONS, while atomic and molecular collisions support thermal conduction in GASES and other liquids. *Compare with* CONVECTION.

conductivity 1. In general, the ability to transmit, as electricity, thermal energy, sound and so on. 2. *Thermal conductivity* is a measure of a substance's ability to transport (via conduction) thermal energy (heat). Typical SI UNITS are joules per second per degree Kelvin per meter ($J \ s^{-1} \ m^{-1}$). Metals, such as copper and silver, have high values of thermal conductivity. 3. *Electrical conductivity* is a measure of a substance's ability to transport electric charge. It is the reciprocal of the material's resistivity. Electrical conductivity is generally expressed by the SI units siemens per meter ($S \ m^{-1}$). Metals, such as copper and silver, also have high values of electrical conductivity.

cone A geometric configuration having a circular bottom and sides tapering off to an apex (as in *nose cone*).

conic section A curve formed by the intersection of a plane and a right circular CONE. Also called *conic*. The conic sections are the ELLIPSE, the *parabola,* and the hyperbola—all curves that describe the paths of bodies moving in space. The *circle* is simply an ellipse with an ECCENTRICITY of zero. (See Figure 25.)

conjunction The alignment of two bodies in the SOLAR SYSTEM so that they have the same CELESTIAL LONGITUDE as seen from EARTH (that is, when they appear closest together in the sky). For example, a SUPERIOR PLANET forms a *superior conjunction* when the SUN lies between it and Earth. An INFERIOR PLANET (either VENUS or MERCURY) forms an *inferior conjunction* when it lies directly between the Sun and Earth and a *superior conjunction* when it lies directly behind the Sun.

conservation of angular momentum The principle of physics that states that absolute angular momentum is a property which cannot be created or destroyed but can only be transferred from one physical system to another through the action of a net torque on the system. As a consequence, the absolute angular momentum of an isolated physical system remains constant.

conservation of charge A principle of physics that states that for an isolated system, the total net charge is a constant; that is, charge can neither be created nor destroyed.

conservation of energy The principle of physics that states that the total energy of an isolated system remains constant if no interconversion of mass and energy takes place within the system. Also called the *First Law of Thermodynamics.*

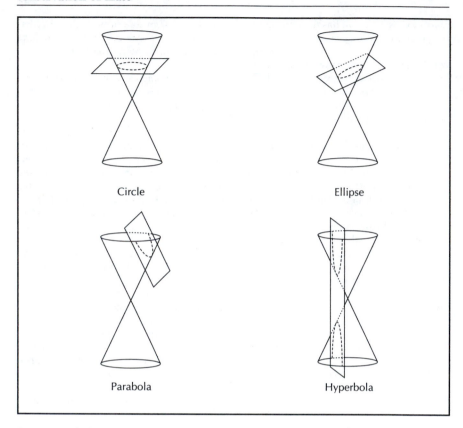

Figure 25. The basic conic sections.

conservation of mass The principle of classical (Newtonian) physics that states that mass can neither be created nor destroyed but only transferred from one volume to another. *See also* CONSERVATION OF MASS AND ENERGY.

conservation of mass and energy From special relativity and Einstein's famous mass-energy equivalence formula ($E = \Delta m\ c^2$), this conservation principle states that for an isolated system, the sum of the mass and energy remains constant, although interconversion of mass and energy can occur within the system.

conservation of momentum The principle of physics that states that in the absence of external forces, absolute momentum is a property that cannot be created or destroyed. *See also* NEWTON'S LAWS OF MOTION.

conservation principle A principle or law of physics that states that the total quantity or magnitude of a certain physical property of a system (e.g., mass, charge, or momentum) remains unchanged even though there may be exchanges of that property between components within the system. Scientists consider such conservation principles to be universally true; so, for example, the conservation of momentum principle applies both on EARTH as well as on MARS.

console A desk-like array of controls, indicators, and video display devices for monitoring and controlling AEROSPACE operations, such as the CHECKOUT,

COUNTDOWN, and LAUNCH of a ROCKET. During the critical phases of a space MISSION, the console becomes the central place from which to issue commands to or at which to display information concerning an AEROSPACE VEHICLE, a deployed PAYLOAD, an EARTH-orbiting SPACECRAFT, or a PLANETARY PROBE. The MISSION CONTROL CENTER generally contains clusters of consoles (each assigned to specific monitoring and control tasks). Depending on the nature and duration of a particular space mission, operators will remain at their consoles continuously or only work there intermittently.

constellation 1. (AEROSPACE) A term used to collectively describe the number and orbital disposition of a set of SATELLITES, such as the constellation of GLOBAL POSITIONING SYSTEM (GPS) satellites. 2. (ASTRONOMY) An easily identifiable configuration of the brightest STARS in a moderately small region of the night sky. Originally, constellations, such as Orion the Hunter and Ursa Major (the Great Bear), were named by early astronomers after heroes and creatures from various ancient cultures and mythologies. 3. In BALLISTIC MISSILE DEFENSE, the number and orbital disposition of a set of space-based weapons forming part of a defensive system.

consumption rate The average quantity of an item consumed or expended during a given time interval, expressed in quantities by the most appropriate unit of measurement per applicable stated basis. For example, on a 300-day-duration human mission to the planet Mars, each crew member might consume an average of 2 kilograms of food each day (24-hour period). The food consumption rate for this mission could then be expressed as: 2 kg/(day-crew member).

continuously crewed spacecraft A SPACECRAFT that has accommodations for continuous habitation (human occupancy) during its MISSION. The *INTERNATIONAL SPACE STATION* (*ISS*) is an example. Sometimes (though not preferred) called a *continuously manned spacecraft.*

continuously habitable zone (CHZ) The region around a STAR in which one or several PLANETs can maintain conditions appropriate for the emergence and sustained existence of life. One important characteristic of a planet in the CHZ is that its environmental conditions support the retention of significant amounts of LIQUID water on the planetary surface.

continuous-wave laser (CW) A laser system in which the coherent light is generated continuously rather than at fixed time intervals. Also called a *CW laser.*

continuous wave radar (CW) A family of radar systems that transmits either modulated or unmodulated continuous waves. The simplest form of this type of radar system transmits a single frequency and detects moving targets by means of the DOPPLER EFFECT.

contraction ratio For a rocket engine, the ratio of the area of the combustion chamber at its maximum diameter to the area of the nozzle throat. *See also* ROCKET.

contrail A long, narrow cloud (usually bright white) produced by high-flying aircraft and missiles in the upper atmosphere. *See also* CONDENSATION TRAIL.

contravane A vane that reverses or neutralizes rotation of a flow. Also called a *countervane.*

controller In general, a device that converts an input signal from the controlled variable (temperature, pressure, fluid level, or fluid flow rate) to a valve actuator input (pneumatic, hydraulic, electrical, or mechanical) to vary the valve position to provide the required correction of the controlled variable.

control mass In THERMODYNAMICS the mass inside a closed system; for example, the working fluid inside a sealed piston-cylinder system.

control rocket A low-THRUST ROCKET, such as a RETROROCKET or a VERNIER

ENGINE, used to guide, to change the ATTI-TUDE of, or to make small corrections in the VELOCITY of an AEROSPACE VEHICLE, SPACE-CRAFT, or EXPENDABLE LAUNCH VEHICLE.

control system (missile) A system that serves to maintain attitude stability and to correct deflections. *See also* GUID-ANCE SYSTEM.

control vane A movable vane used for control; especially a movable air vane or exhaust jet vane on a rocket, used to control flight attitude.

control volume In thermodynamics, a defined region in space across the boundaries of which matter and energy can flow.

convection 1. In heat transfer, mass motions within a fluid resulting in the transport and mixing of the properties of that fluid. The up-and-down drafts in a fluid heated from below in a gravitational environment. Because the density of the heated fluid is lowered, the warmer fluid rises (natural convection); after cooling, the density of the fluid increases, and it tends to sink. Compare to CONDUCTION and RADIATION. 2. In planetary science, atmospheric or oceanic motions that are predominately vertical and that result in the vertical transport and mixing of atmospheric or oceanic properties. Because the most striking meteorological features result if atmospheric convective motion occurs in conjunction with the rising current of air (i.e., updrafts), convection sometimes is used to imply only upward vertical motion.

converging-diverging (CD) nozzle A THRUST-producing flow device for expanding and accelerating hot exhaust GASES from a ROCKET engine. A properly designed NOZZLE efficiently converts the THERMAL ENERGY of combustion into KINETIC ENERGY of the combustion product gases. In a SUPERSONIC converging-diverging nozzle, the hot gas upstream of the nozzle THROAT is at SUBSONIC VELOCITY (that is, the MACH NUMBER (M) < 1), reaches sonic velocity (the SPEED OF SOUND, for which

$M = 1$) at the throat of the nozzle, and then expands to supersonic velocity ($M > 1$) downstream of the nozzle throat region while flowing through the diverging section of the nozzle. *See also* DE LAVAL NOZZLE.

converging lens (or mirror) A LENS (or MIRROR) that refracts (or reflects) a parallel BEAM of LIGHT, making it converge at a point (called the principal focus). A converging mirror is a CONCAVE MIRROR, while a converging lens is generally a CONVEX LENS that is thicker in the middle than at its edges. *Compare with* DIVERGING LENS.

convex lens (or mirror) A lens or mirror with an outward curvature. *Compare* CONCAVE LENS.

cook-off The deflagration or detonation of ordnance by the absorption of thermal energy (heat) from its environment.

coolant 1. In general, a gas or liquid used to remove thermal energy (heat) from some device or system, as, for example, a rocket's combustion chamber or nozzle throat wall. 2. Specifically, a substance (liquid or gas) circulated through the core of a nuclear reactor to remove or transfer thermal energy (heat). Common coolants are water, air, carbon dioxide (CO_2), liquid sodium (Na), and sodium-potassium alloy (NaK).

coolant tube In astronautical engineering, relatively small-diameter thin-walled conduit attached to or forming the wall of a regeneratively cooled combustion chamber or nozzle and carrying propellant to cool the wall.

cooperative target A three-AXIS stabilized orbiting object that has signaling devices to support RENDEZVOUS and DOCKING/capture operations by a CHASER SPACECRAFT.

co-orbital Sharing the same or very similar ORBIT. For example, during a REN-DEZVOUS operation the CHASER SPACE-CRAFT and its COOPERATIVE TARGET are said to be co-orbital.

coordinated universal time *See* UNI-VERSAL TIME (UT).

coordinates Linear or angular quantities that designate the position that a point occupies in a given reference frame or system. Also used as a general term to designate the particular kind of reference frame or system, such as *plane rectangular coordinates* or *spherical coordinates*.

Copernican system The theory of planetary motions, proposed by the Polish astronomer NICHOLAS COPERNICUS, in which all PLANETs (including EARTH) move in *circular* ORBITs around the SUN, with the planets closer to the Sun moving faster. In this system, the hypothesis of which helped trigger the scientific revolution of the 16th and 17th centuries, Earth was viewed not as an immovable object at the center of the UNIVERSE (as in the GEOCENTRIC PTOLE-MAIC SYSTEM) but rather as a planet orbiting the Sun between VENUS and MARS. Early in the 17th century, JOHANNES KEPLER showed that while Copernicus's HELIOCENTRIC hypothesis was correct, the planets actually moved in (slightly) ELLIPTI-CAL ORBITs around the Sun.

Copernicus, Nicholas (Nicolaus) (1473–1543) Polish astronomer and church official who triggered the scientific revolution of the 17th century with his book *On the Revolution of Celestial Spheres*. When published in 1543 while he lay on his deathbed, this book overthrew the PTOLE-MAIC SYSTEM by boldly suggesting a HELIO-CENTRIC model for the SOLAR SYSTEM in which EARTH and all the other PLANETs moved around the SUN. His heliocentric model—possibly derived from the long-forgotten ideas of the ancient Greek mathematician and astronomer Aristarchus of Samos (c. 320–c. 250 B.C.E.)—caused much technical, political, and social upheaval before finally displacing two millennia of Greek GEOCENTRIC COSMOLOGY.

Copernicus (spacecraft) NASA's *Copernicus* spacecraft was launched on August 21, 1972. This mission was the third in the Orbiting Astronomical Obser-vatory (OAO) program and the second successful spacecraft to observe the celestial sphere from above Earth's atmosphere. An ultraviolet (UV) telescope with a spectrometer measured high-resolution spectra of stars, galaxies, and planets with the main emphasis being placed on the determination of interstellar absorption lines. Three X-ray telescopes and a collimated proportional counter provided measurements of celestial X-ray sources and interstellar absorption between 1 and 100 angstroms (Å) wavelength. Also called the Orbiting Astronomical Observatory-3 (OAO-3), its observational mission life extended from August 1972 through February 1981—some nine and one-half years. The spacecraft was named in honor of the famous Polish astronomer NICHOLAS COPERNICUS (1473–1543).

core 1. (planetary) The high-density, central region of a PLANET. 2. (stellar) The very-high-TEMPERATURE, central region of a STAR. For MAIN SEQUENCE stars, FUSION processes within the core burn HYDROGEN. For stars that have left the main sequence, nuclear fusion processes in the core involve HELIUM and oxygen.

coriolis effect(s) 1. The physiological effects (e.g., nausea, vertigo, dizziness, etc.) felt by a person moving radially in a rotating system, such as a rotating space station. 2. The tendency for an object moving above Earth (e.g., a missile in flight) to turn to the right in the Northern Hemisphere and to the left in the Southern Hemisphere relative to Earth's surface. This effect arises because Earth rotates and is not, therefore, an inertial reference frame.

corona The outermost region of a STAR. The SUN's corona consists of low-DENSITY clouds of very hot GASes (> 1 million K) and ionized materials.

coronal mass ejection (CME) A high-speed (10 to 1,000 km/s) ejection of matter from the SUN's CORONA. A CME travels through space disturbing the SOLAR WIND and giving rise to GEOMAGNETIC STORMS when the disturbance reaches EARTH.

Corporal An early medium-range (about 120-kilometer [75-mile]) liquid-propellant, surface-to-surface guided missile developed and deployed by the U.S. Army.

cosmic Of or pertaining to the UNIVERSE, especially that part outside EARTH's ATMOSPHERE. This term frequently appears in the Russian (former Soviet Union) space program as the equivalent to *space* or ASTRO-, such as *cosmic station* (versus SPACE STATION) or COSMONAUT (versus ASTRONAUT).

Cosmic Background Explorer (COBE) A NASA SPACECRAFT placed in ORBIT around EARTH in November 1989. It successfully measured the SPECTRUM and intensity distribution of the COSMIC MICROWAVE BACKGROUND (CMB).

cosmic dust Fine microscopic particles drifting in outer space. *See also* INTERPLANETARY DUST.

cosmic microwave background (CMB) The background of MICROWAVE RADIATION that permeates the UNIVERSE and has a BLACKBODY TEMPERATURE of about 2.7 K. Sometimes called the *primal glow,* scientists believe it represents the remains of the ancient fireball in which the universe was created. *See also* BIG BANG.

cosmic rays Extremely energetic PARTICLES (usually bare atomic nuclei) that move through OUTER SPACE at speeds just below the SPEED OF LIGHT and bombard EARTH from all directions. Their existence was discovered in 1912 by the Austrian-American physicist Victor Hess (1883–1964). HYDROGEN nuclei (PROTONS) make up the highest percentage of the cosmic ray population (approximately 85%), but these particles range over the entire periodic table of ELEMENTS. *Galactic cosmic rays* are samples of material from outside the SOLAR SYSTEM and provide direct evidence of phenomena that occur as a result of explosive processes in STARS throughout the MILKY WAY GALAXY. *Solar cosmic rays* (mostly protons and ALPHA PARTICLES) are ejected from the SUN during SOLAR FLARE events.

cosmic ray astronomy The branch of high-energy ASTROPHYSICS that uses COSMIC RAYS to provide information on the origin of the chemical ELEMENTS through NUCLEOSYNTHESIS during stellar explosions.

cosmodrome In the Soviet and later Russian space program, a place that launches space vehicles and rockets. The three main cosmodromes in the Russian space program are: the Plesetsk Cosmodrome, which is situated south of Archangel in the northwest corner of Russia; the Kapustin Yar Cosmodrome, on the banks of the Volga River southeast of Moscow; and the first and main cosmodrome (the site which launched *Sputnik 1*), the Baikonur Cosmodrome, located just east of the Aral Sea in Kazakhstan (now an independent country). The Baikonur Cosmodrome has also been referred to as *Tyuratam*.

cosmological principle The hypothesis that the expanding UNIVERSE is ISOTROPIC and homogeneous. In other words, there is no special location for observing the universe, and all observers anywhere in the universe would see the same recession of distant galaxies.

cosmology The study of the origin, evolution, and structure of the UNIVERSE. Contemporary cosmology centers on the BIG BANG hypothesis. This theory states that about 15 to 20 billion (10^9) years ago, the universe began in a great explosion and has been expanding ever since. In the OPEN UNIVERSE (or STEADY-STATE UNIVERSE) model, scientists postulate that the universe is infinite and will continue to expand forever. In the more widely accepted CLOSED UNIVERSE model, the total MASS of the universe is assumed sufficiently large to stop its expansion eventually and then start contracting by GRAVITATION, leading ultimately to a BIG CRUNCH. In the *flat universe model,* the expansion gradually comes to a halt. However, instead of collapsing, the universe achieves an equilibrium condition, with expansion FORCES precisely balancing the forces of gravitational contraction. Contemporary observations indicate that

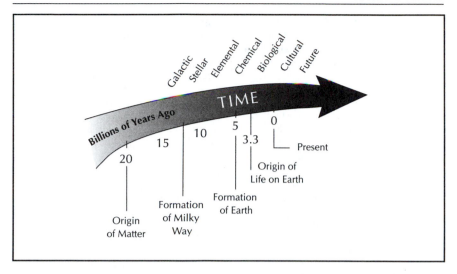

Figure 26. The scenario of cosmic evolution provides a grand synthesis for the long series of alterations of matter that have given rise to our galaxy, our Sun, our planet, and ourselves.

the universe is not only expanding but that the rate at which it expands might also be increasing. (See Figure 26.)

cosmonaut The name used by Russia (formerly the Soviet Union) for its space travelers or "astronauts."

Cosmos The general name applied to a large variety of Soviet and later Russian spacecraft, ranging from military satellites to scientific spacecraft investigating the upper atmosphere of Earth. *Cosmos 1* was launched in March 1962; since then well over 2,000 Cosmos satellites have been sent into space. Also called *Kosmos.*

coulomb (symbol: C) The SI UNIT of electric charge. The quantity of electric charge transported in one second by a current of one ampere. Named after the French physicist Charles de Coulomb (1736–1806).

count 1. In aerospace operations, to proceed from one point in a countdown to another, normally by calling a number to signify the point reached; to proceed in a countdown, as in *T minus 20 seconds and counting.* 2. With respect to the detection of ionizing radiation, one pulse of current or voltage from a detector, indicating the passage of a photon or particle through the detector.

countdown The step-by-step process that leads to the LAUNCH of a ROCKET or AEROSPACE VEHICLE. A countdown takes place in accordance with a specific schedule, with zero being the go, or activate, time.

counterattack Attack by part or all of a defending force against an attacking enemy force, for such specific purposes as regaining lost ground or cutting off or destroying enemy advance units, and with the general objective of denying to the enemy the attainment of his/her purpose in attacking.

counter-countermeasures Measures taken by the defense to defeat offensive countermeasures.

countermeasures A tactic used by the offense to oppose defensive measures or by the defense to oppose offensive measures. Nondestructive countermeasures are those intended to nullify the capability of the opposing system by means other than direct attack. Defense suppression countermeasures include means of attacking defensive system elements.

counterpermeation A pressurant (i.e., an inert gas, such as nitrogen) often is used in a rocket's propellant tank to provide the pressure necessary to displace the liquid propellant from the tank and drive it (as a liquid) into the rocket's combustion chamber. Permeation occurs when the inert gas (pressurant) flows from its storage compartment through a porous (i.e., permeable) membrane and interacts with the liquid propellant in the tank. *Counterpermeation* involves the reverse process—that is, the simultaneous (and undesirable) migration of vaporized propellant and inert pressurant back across the permeable membrane into the pressurant's storage compartment. This can occur when the liquid propellant in a rocket's fuel or oxidizer tank becomes heated (possibly due to a loss of the tank's insulation and/or the excessive aerodynamic heating of the tank's surface during launch ascent). Some of the heated liquid propellant will then vaporize. Depending on the amount of heating, the type of propellant, and the overall design of the propellant tank, the propellant's vapor pressure eventually could exceed the pressure of the pressurant and counterpermeation will occur. *See also* PROPELLANT; ROCKET.

course 1. The line of flight taken by a rocket, aircraft, or aerospace vehicle. 2. A predetermined or intended route; a direction to be followed that is measured with respect to a geographic reference direction.

Crab nebula The SUPERNOVA remnant of an exploding STAR observed in 1054 by Chinese astronomers. It is about 6,500 LIGHT-YEARS away in the CONSTELLATION Taurus and contains a PULSAR that flashes optically.

craft Any vehicle or device designed to fly through air or space; an aircraft or spacecraft.

crater 1. A bowl-shaped topographic depression with steep slopes; a volcanic orifice. Craters are formed naturally by two general processes: *impact* (as from an asteroid or meteoroid) and *eruptive* (as from a volcanic eruption). 2. The pit, depression, or cavity formed in the surface of Earth by an explosion (chemical explosives or nuclear). It may range from saucer-shaped to conical, depending largely on the depth of burst. 3. The depression resulting from high-speed solid-particle impacts on a rigid material as, for example, a meteoroid or space debris particle impact on the skin of a spacecraft.

creep The slow (but continuous) permanent deformation of material caused by a constant tensile or compressive load that is less than the load necessary for the material to give way under pressure (i.e., to yield); some time is required to induce creep, and the process is accelerated at elevated temperatures.

crew activity planning The analysis and development of activities to be performed in space flight by the human crew, resulting in a timeline of these activities together with the necessary procedures and crew reference data to accomplish flight objectives.

crew module A three-section pressurized working, living, and stowage compartment in the forward portion of the space shuttle orbiter vehicle. It consists of the flight deck, the middeck/equipment bay, and an airlock. *See also* SPACE TRANSPORTATION SYSTEM.

crew-tended spacecraft A spacecraft that has accommodations for only temporary habitation during its mission. Sometimes called a *man-tended spacecraft.* *Compare with* CONTINUOUSLY CREWED SPACECRAFT.

critical diameter A diameter of a solid rocket motor such that it can sustain a detonation.

critical hold/scrub point In aerospace operations, that time in the countdown when an unplanned hold normally would be expected to cause a scrubbed launch attempt.

critical ignition pressure Pressure below which propellants cannot be ignited.

criticality The state of a nuclear reactor when it is sustaining a fission chain reaction. See also NUCLEAR REACTOR.

critical mass The minimum amount of fissile material capable of supporting a fission chain reaction under precisely specified conditions.

critical mass density (symbol: Ω) The mean DENSITY of matter within the UNIVERSE (considered as a whole) that cosmologists consider necessary if GRAVITATION is to eventually halt its expansion. If the universe does not contain sufficient MASS (that is, if Ω <1), then the universe will continue to expand forever. If Ω > 1, then the universe has enough mass to stop its expansion eventually and to start an inward collapse under the influence of gravitation. If the critical mass density is just right (that is, Ω = 1), then the universe is considered flat and a state of equilibrium will exist in which the outward FORCE of expansion is precisely balanced by the inward force of gravitation. See also COSMOLOGY.

critical point A fluid's critical point occurs at the highest temperature where liquid and vapor phases can coexist. Above this temperature the fluid, no matter what its density, can have only one phase. In thermodynamics, the critical point is represented as the termination point of the liquid-vapor coexistence line in the pressure-temperature diagram for a particular substance.

critical pressure 1. In thermodynamics, the pressure of a fluid at the critical point—that is, the highest pressure under which the liquid and gaseous phases of a substance can coexist. 2. In rocketry, the pressure in the nozzle throat for which the isentropic (constant entropy) mass-flow rate is a maximum.

critical speed (symbol: V^*) The flow velocity (V^*) at which the Mach number (M) is unity. It is used as a convenient reference velocity in the basic equations of high-speed, compressible fluid flow. See also MACH NUMBER.

critical temperature The temperature above which a substance cannot exist in a liquid state, regardless of the pressure. See also CRITICAL POINT.

critical velocity In rocketry, the speed of sound at conditions prevailing at the nozzle throat. Also called the *throat velocity* and *critical throat velocity*. See also ROCKET.

cross section (common symbol: σ [sigma]) 1. In general, the area of a plane surface that is formed when a solid is cut; often this cut is made at right angles to the longest axis of the solid object. 2. In aerospace engineering, the physical (geometric) area presented by an aircraft, aerospace vehicle, or missile system; the frontal (or "head-on") cross section is the surface area that would be observed if the vehicle was flying directly toward the observer. 3. In aerospace operations, the apparent surface area of an aircraft, aerospace vehicle, rocket, or spacecraft that can be detected by a remote sensing system; for example, an aircraft's *radar cross section*. Stealth (low-observables) technology now is being used to reduce a military aircraft's radar cross section. 4. In physics, a measure of the probability that a nuclear reaction will occur. Usually measured in *barns,* it is the apparent (or effective) area presented by a target nucleus (or particle) to an oncoming particle or other nuclear radiation, such as a gamma ray photon. See also BARN.

cruise missile A guided missile traveling within the atmosphere at aircraft speeds and, usually, low altitude whose trajectory is preprogrammed. It is capable of achieving high accuracy in striking a distant target. It is maneuverable during flight, is constantly propelled, and, therefore, does not follow a ballistic trajectory. Cruise missiles may be armed with nuclear weapons or with conventional warheads (i.e., high explosives). *See also* TOMAHAWK.

cruise phase   In space flight operations involving a scientific spacecraft, the *cruise phase* is bounded at the beginning by the launch phase and at the end by the encounter phase. After launch the scientific spacecraft is commanded to configure itself for cruise. Appendages that might have been stowed for launch (e.g., an antenna system) can now be deployed either fully or to some intermediate cruise position. Telemetry is analyzed to determine the state of health of the spacecraft and to determine how well it survived the launch. The trajectory is fine-tuned to prepare the spacecraft for its scientific encounter mission. During the cruise phase, ground system upgrades can be made and appropriate tests conducted, spacecraft flight software modifications can be implemented and also tested by the space flight operations facility. Finally, as the spacecraft nears its celestial target, scientific instruments are powered on (if necessary) and calibrated.

crust   Outermost, and least dense, solid layer of a planet.

cryogenic   Fluids or conditions at low temperatures, usually below-150° Celsius (123 Kelvin).

cryogenic liquid   Liquified gas at very low temperature, such as liquid oxygen (LO_2 or LOX), liquid hydrogen (LH_2), liquid nitrogen, or liquid argon.

cryogenics   The branch of science dealing with very low temperatures, applications of these low-temperature environments, and methods of producing them. This science may be concerned with practical engineering problems, such as producing, transporting, and storing the metric ton quantities of liquid oxygen (LO_2) or liquid hydrogen (LH_2) needed as chemical rocket propellants; or it may help a physicist investigate some basic properties of matter at extremely low temperatures. Cryogenic researchers work with temperature environments down to one-millionth of a degree

Kelvin (10^{-6} K)—essentially at the physical threshold of absolute zero.

cryogenic materials   Metals and alloys that are usable in systems or structures operating at very low temperatures. These materials usually possess improved physical properties at such temperatures.

cryogenic propellant   A rocket fuel, oxidizer, or propulsion fluid that is liquid only at very low temperatures. Liquid hydrogen (LH_2) and liquid oxygen (LO_2) are examples of cryogenic propellants.

cryogenic pump   A pump that uses CRYOPUMPING to create a vacuum. *See also* CRYOPUMP.

cryogenic seal   A seal that must function effectively at temperatures below about –150° Celsius. (123 Kelvin).

cryogenic system   A system in which the temperature falls below –150° Celsius (123 Kelvin).

cryogenic temperature   Generally, temperatures below –150° Celsius (C) (123 Kelvin [K]); sometimes the boiling point of liquid nitrogen (LN_2), namely –195° (77.4 K) is used as the "cryogenic temperature" threshold; in other instances, the temperature associated with the boiling point of liquid helium (LHe), namely –269° C (4.2 K), is considered as the threshold of the cryogenic temperature regime.

cryopump   An exposed surface refrigerated to cryogenic temperature for the purpose of pumping gases in a vacuum chamber by condensing the residual chamber gases and maintaining the condensate at a temperature such that the equilibrium vapor pressure is equal to or less than the desired ultimate pressure in the chamber. *See also* CRYOPUMPING.

cryopumping   The process of removing gases from an enclosure by condensing the gases on surfaces at cryogenic temperatures.

cryosphere   The portion of Earth's climate system consisting of the world's ice masses and snow deposits. These include the continental ice sheets, mountain glaciers, sea ice, surface snow cover, and lake and river ice. Changes is snow cover on the land surfaces generally are seasonal and closely tied to the mechanics of atmospheric circulation. The glaciers and ice sheets are closely related to the global hydrologic cycle and to variations of sea level. Glaciers and ice sheets typically change volume and extent over periods ranging from hundreds to millions of years. *See also* GLOBAL CHANGE.

curie (symbol: C or Ci) The traditional unit used to describe the intensity of radioactivity of a sample of material. One curie is equal to 3.7×10^{10} disintegrations per second. (This is approximately the rate of decay of one gram of radium). Named after the Curies, Pierre and Marie, who discovered radium in 1898. *See also* BECQUEREL.

Curie point The temperature above which a ferromagnetic material (e.g., iron or nickel) becomes substantially nonmagnetic.

current (symbol: I) The flow of electric charge through a conductor. The ampere (symbol: A) is the SI UNIT of electric current.

cutoff (or cut-off) 1. An act or instance of shutting something off; specifically, in rocketry, an act or instance of shutting off the propellant flow in a rocket, or of stopping the combustion of the propellant. Compare with BURNOUT. 2. Something that shuts off or is used to shut off (e.g., a fuel cutoff valve). 3. Limiting or bounding, as, for example, in *cutoff frequency.*

cutoff velocity The velocity attained by a missile or rocket at the point of cutoff.

cycle Any repetitive series of operations or events; the complete sequence of values of a periodic quantity that occur during a period—for example, one complete wave. The *period* is the duration of one cycle, while the *frequency* is the rate of repetition of a cycle. The *hertz (Hz)* is the SI UNIT of frequency; and 1 hertz equals 1 cycle per second.

cycle life The number of times a unit may be operated (e.g., opened and closed) and still perform within acceptable limits.

Cygnus X-1 The strong X-RAY source in the CONSTELLATION Cygnus that scientists believe comes from a BINARY STAR SYSTEM, consisting of a SUPERGIANT star orbiting a BLACK HOLE companion. Gas drawn off the supergiant star emits X rays as it is intensely heated while falling into the black hole.

cylindrical coordinates A system of curvilinear coordinates in which the position of a point in space is determined by (a) its perpendicular distance from a given line; (b) its distance from a selected reference plane perpendicular to this line; and (c) its angular distance from a selected reference line when projected onto this plane. The coordinates thus form the elements of a cylinder and by convention are written r, θ, and z, where r is the radial distance from the cylinder's axis z and θ is the angular position from a reference line in a cylindrical cross section normal to z. Also called *polar coordinates.*

The relationships between the cylindrical coordinates and the rectangular Cartesian coordinates (x, y, z) are:
$$x = r \cos q; \; y = r \sin q; \; z = z$$

cylindrical grain A solid-propellant grain in which the internal cross section is a circle. *See also* ROCKET.

D

damping The suppression of oscillations usually because energy is being expended by the oscillating system to overcome friction or other types of resistive forces.

dark matter Matter in the UNIVERSE that cannot be observed directly because it emits very little or no ELECTROMAGNETIC RADIATION. Scientists infer its existence through secondary phenomena such as gravitational effects and suggest that it may make up about 90% of the total MASS of the universe. Also called *missing mass*. *See also* COSMOLOGY.

dark nebula A cloud of INTERSTELLAR dust and GAS sufficiently dense and thick that the light from more distant STARs and celestial bodies (behind it) is obscured. The HORSEHEAD NEBULA in the CONSTELLATION Orion is an example of a dark nebula.

dart configuration A configuration of an aerodynamic vehicle in which the control surfaces are at the tail of the vehicle. *Compare with* CANARD.

database A collection of interrelated or independent data items stored together to serve one or more applications.

data fusion The technique in which multivariate data from multiple sources are retrieved and processed as a single, unified entity. A significant set of a priori databases is crucial to the effective functioning of the data fusion process. For example, this technique might be used in supporting the development of an evolving "world model" for a robotic lunar rover, which is teleoperated from a control station on Earth. Inputs from the robot vehicle's sensors would be blended, or "fused," with existing lunar environment databases to support a more intelligent exploration strategy and more efficient field operations.

data handling subsystem The onboard computer responsible for the overall management of a spacecraft's activity is usually the same computer that maintains timing; interprets commands from Earth; collects, processes, and formats the telemetry data that is to be returned to Earth; and manages high-level fault protection and saving routines. This spacecraft computer often is referred to as the command and data handling subsystem.

data link 1. The means of connecting one location to another for the purpose of transmitting and receiving data. 2. A communications link suitable for the transmission of data. 3. Any communications channel used to transmit data from a sensor to a computer, a readout device, or a storage device.

data processing (DP) A general term describing the systematic application of procedures—electrical, mechanical, optical, computational, and so on—whereby data are changed from one form to another. Sometimes used to describe the overall work performed by computers. Also called *automatic data processing (ADP)* and *electronic data processing (EDP)*.

data reduction The transformation of raw or observed data into more compact, ordered, or useful information.

data smoothing The mathematical process of fitting a smooth curve to a dispersed set of data points.

datum 1. A single unit of information; the singular of *data*. 2. Any numerical or geometrical quantity or set of such quantities that may serve as reference or base for other quantities. Where the concept is geometric, the preferred plural form is *datums*—as in *two geodetic datums have been employed.*

dazzling The temporary blinding of a sensor by overloading it with an intense signal of electromagnetic radiation, as, for example, from a laser.

deadband In general, an intentional feature in a guidance and control system that prevents a flight- path error from being corrected until that error exceeds a specified magnitude. With respect to the space shuttle orbiter vehicle, that attitude and rate control region in which no orbiter reaction control subsystem (RCS) or vernier correction forces are being generated.

dead spot In a control system, a region about the neutral control position where small movements of the actuator do not produce any response in the system.

deboost A RETROGRADE (opposite-direction) burn of one or more low-THRUST ROCKETS or an AEROBRAKE maneuver that lowers the ALTITUDE of an orbiting SPACECRAFT.

debris JETTISONed human-made materials, discarded LAUNCH VEHICLE components, and derelict or nonfunctioning SPACECRAFT in ORBIT around EARTH. *See also* SPACE DEBRIS.

debug To isolate and remove malfunctions from a hardware system, subsystem, or component; to correct mistakes in computer software.

decade 1. A group or series of ten; for example, a period of 10 years. 2. The interval between any two quantities having the ratio 10:1.

decay (orbital) The gradual lessening of both the APOGEE and PERIGEE of an orbiting object from its PRIMARY BODY. For example, the orbital decay process for ARTIFICIAL SATELLITES and SPACE DEBRIS often results in their ultimate fiery plunge back into the denser regions of the EARTH's ATMOSPHERE.

decay (radioactive) The spontaneous transformation of one RADIONUCLIDE into a different NUCLIDE or into a different ENERGY state of the same nuclide. This natural process results in a decrease (with time) of the number of original radioactive ATOMS in a sample and involves the emission from the nucleus of ALPHA PARTICLES, BETA PARTICLES, or GAMMA RAYS. *See also* RADIOACTIVITY; RADIOISOTOPE.

decay heat The thermal energy (heat) released by the decay of radionuclides.

decay time 1. In nuclear physics, the time required for a radioactive substance to decay to a stable state. 2. In aerospace operations, the approximate lifetime of an orbiting object in a nonstable orbit. The decay time reflects the tendency of a satellite (or piece of space debris) to lose orbital velocity due to the influences of atmospheric drag and gravitational forces. A decaying object eventually impacts the surface of Earth (or other planet) or else burns up in the planet's atmosphere. The decay time often is measured in terms of revolutions per day per day (rev/day/day).

deceleration The act or process of moving, or causing to move, with decreasing speed; negative acceleration.

deceleration parachute A parachute attached to a craft and deployed to slow the craft, especially during landing. Also called a *drogue parachute*. *See also* DROGUE PARACHUTE.

deci- (symbol: d) A prefix meaning multiplied by one-tenth (10^{-1}); for example, *decibel*.

decibel (symbol: dB) 1. One-tenth of a bel. 2. A dimensionless measure of the ratio of two powers, sound intensities, or voltages. It is equal to ten times the common logarithm (i.e., the logarithm to the base 10) of the power ratio, P_2/P_1. Namely,

n (decibels) = $10 \log_{10} (P_2/P_1)$

Since the bel (B) is an exceptionally large unit, the decibel is encountered frequently in physics and engineering.

declination (symbol: δ) For a CELESTIAL BODY viewed on the CELESTIAL SPHERE, the angular distance north (0° to 90° positive) or south (0° to 90° negative) of the celestial EQUATOR.

decoy 1. In general, an object or phenomenon that is intended to deceive enemy surveillance devices or mislead enemy evaluation of the threat. 2. In ballistic missile defense (BMD), a device that is constructed to look and behave like a nuclear-weapon-carrying warhead, but that is far less costly, much less massive, and can be deployed in large numbers to complicate defenses. *See also* BALLISTIC MISSILE DEFENSE.

decrement A decrease in the value of a variable. *Compare with* INCREMENT.

dedicated Serving a single function; for example, a *dedicated* battery is a power source serving a single load, such as a special ordnance circuit.

deep space The region of outer space at altitudes greater than 5,600 kilometers (approximately 3,500 miles) above Earth's surface.

Deep Space Communications Complex (DSCC) One of three NASA Deep Space Network (DSN) tracking sites at Goldstone, California; Madrid, Spain; and Canberra, Australia. These facilities are approximately equally spaced around Earth for continuous tracking of deep space vehicles. *See also* DEEP SPACE NETWORK.

Deep Space Network (DSN) NASA's global network of ANTENNAS that serve as the RADIO WAVE communications link to distant INTERPLANETARY SPACECRAFT and PROBES, transmitting instructions to them and receiving data from them. Large radio

Figure 27. Front view of the 70-meter diameter antenna at Goldstone, California. The Goldstone Deep Space Communications Complex, located in the Mojave Desert in California, is one of three complexes that make up NASA's Deep Space Network (DSN). *(Digital image courtesy of NASA/JPL)*

antennas of the DSN's three DEEP SPACE COMMUNICATIONS COMPLEXes are located in Goldstone, California; near Madrid, Spain; and near Canberra, Australia—providing almost continuous contact with a SPACECRAFT in DEEP SPACE as EARTH rotates on its AXIS. (See Figure 27.)

deep space probe A SPACECRAFT designed for exploring DEEP SPACE, especially to the vicinity of the MOON and beyond. This includes LUNAR PROBES MARS probes, OUTER PLANET probes, SOLAR probes, and so on.

Defense Meteorological Satellite Program (DMSP) A highly successful family of WEATHER SATELLITES operating in POLAR ORBIT around EARTH that have provided important environmental data to serve American defense and civilian needs for more than two decades. Two operational DMSP SPACECRAFT orbit Earth at an ALTITUDE of about 832 kilometers (516 mi) and scan an area 2,900 kilometers (1,800 mi) wide. Each SATELLITE scans the globe in approximately 12 hours. Using their primary SENSOR, called the Operational Linescan System (OLS), DMSP satellites take visual and infrared imagery of cloud cover.

Military weather forecasters use these imagery data to detect developing weather patterns anywhere in the world. Such data are especially helpful in identifying, locating, and determining the severity of thunderstorms, hurricanes, and typhoons.

Besides the Operational Linescan System, DMSP satellites carry sensors that measure atmospheric moisture and TEMPERATURE levels, X-RAYS, and ELECTRONS that cause AURORAS. These satellites also can locate and determine the intensity of auroras, which are ELECTROMAGNETIC phenomena that can interfere with radar system operations and long-range ELECTROMAGNETIC COMMUNICATIONS. Starting in December 1982, the Block 5D-2 series spacecraft have been launched from VANDENBERG AIR FORCE BASE, California. Later generations of the DMSP spacecraft contain many improvements over earlier models, including new sensors with increased capabilities and a longer life span.

Each DMSP spacecraft is placed in a SUN-SYNCHRONOUS (polar) ORBIT at a nominal altitude of 832 kilometers (516 mi). A sun-synchronous orbit is a special polar orbit that allows a satellite's sensors to maintain a fixed relation to the SUN—a feature that is especially useful for meteorological satellites. Each day, a satellite in sun-synchronous orbit passes over a certain area on Earth at the same local time.

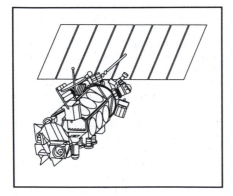

Figure 28. The polar-orbiting Defense Meteorological Satellite Program (DMSP) spacecraft.

One way to characterize sun-synchronous orbits is by the time the spacecraft cross the EQUATOR. Equator crossings (called NODES) occur at the same local time each day, with the descending crossings occurring 12 hours (local time) from the ascending crossings. The terms "AM" and "PM" polar orbiters denote satellites with morning and afternoon equator crossings, respectively. The Block 5D-2 DMSP spacecraft measure 3.64 meters (11.9 ft) in length and are 1.21 meters (3.97 ft) in diameter. Each spacecraft has an on-orbit MASS of 830 kilograms (1,830 lbm) and a design life of about four years. (See Figure 28.)

Starting in 2001, upgraded Block 5D-3 versions of spacecraft with improved sensor technology became available for launch. The UNITED STATES AIR FORCE will maintain its support of the DMSP program until about 2010, corresponding to the projected end of life of the final DMSP satellite. Thereafter, military weather requirements will be filled by a triagency (Department of Defense [DOD], NASA, and Department of Commerce [DOC]) National Polar Orbiting Environmental Satellite System (NPOESS). In May 1994, the president directed that the Departments of Defense and Commerce converge their separate polar orbiting weather satellite programs. Consequently, the command, control, and communications for existing DOD satellites was combined with the control for the National Oceanic and Atmospheric Administration (NOAA) weather satellites. In June 1998, the DOC assumed primary responsibility for flying both DMSP and NOAA's DMSP-cloned civilian polar orbiting weather satellites. The DOC will continue to manage both weather satellite systems, providing essential environmental sensing data to American military forces until the new converged national polar-orbiting environmental satellite system (NPOESS) becomes operational in about 2010.

Defense Satellite Communications System (DSCS)

An evolving family of MILITARY SATELLITES that provide worldwide, responsive wideband and antijam communications in support of U.S. strategic and tactical information transfer needs. DSCS satellites orbit EARTH at an ALTITUDE of approximately 35,900 kilometers (22,300 mi) above the EQUATOR. The satellites are super-high-frequency (SHF) systems capable of providing worldwide secure voice and data transmission. DSCS supports high-priority communications, such as the exchange of wartime information between defense officials and battlefield commanders. The COMMUNICATIONS SATELLITE system can also transmit space operations and early warning data to various systems and military users.

The first of the operational Phase II DSCS systems was launched in 1971. Their characteristic two-DISH ANTENNAS concentrated electronic BEAMS on small areas of Earth's surface but had limited adaptability in comparison to the newer Phase III satellites. The continuously evolving Phase III DSCS SPACECRAFT has a MASS of approximately 1,170 kilograms (2,580 lbm) and a design life of about 10 years. Its rectangular body is 1.83 meters (6 ft) by 1.83 meters (6 ft) by 2.13 meters (7 ft) in a stowed configuration and reaches and 11.6 meters (38 ft) span with its SOLAR PANELS deployed on ORBIT.

The UNITED STATES AIR FORCE began launching the more advanced DSCS Phase III satellites in 1982 and currently operates a CONSTELLATION of 10 satellites in GEOSYNCHRONOUS ORBIT. DSCS users include the National Command Authority, the White House, the Defense Information System Network (DISN), the Air Force Satellite Control Network, and the Diplomatic Telecommunications Service.

Defense Support Program (DSP) The family of MISSILE surveillance SATELLITES operated by the UNITED STATES AIR FORCE since the early 1970s. Placed in GEOSYN-CHRONOUS ORBIT around EARTH, these military SURVEILLANCE SATELLITES can detect missile launches, space launches, and NUCLEAR DETONATIONS occurring around the world. Air force units at Peterson Air Force Base in Colorado operate DSP satellites and report warning information via communications links to the NORTH

AMERICAN AEROSPACE DEFENSE COMMAND (NORAD) and the UNITED STATES STRATEGIC COMMAND early-warning centers within CHEYENNE MOUNTAIN AIR FORCE STATION.

The Defense Support Program is a space-based INFRARED satellite system that provides global coverage and warning of BALLISTIC MISSILE launches, nuclear detonations, and other events. The primary (infrared) SENSOR of each DSP satellite supports near–real time detection and reporting of missile launches against the United States and/or allied forces, interests, and assets worldwide. Other sensors on each satellite support the near–real time detection and reporting of ENDOAT-MOSPHERIC (0–50 km), EXOATMOSPHERIC (50–300 km), and DEEP SPACE (> 300 km) nuclear detonations worldwide.

DSP satellites use an infrared sensor to detect HEAT from missile and BOOSTER plumes against Earth's background SIGNAL. The program started with the first launch of a DSP satellite in the early 1970s. Since that time, DSP satellites have provided an uninterrupted early-warning capability. Developed as a strategic warning system, the DSP's effectiveness in providing tactical warning was demonstrated during the Persian Gulf conflict (from August 1990 through February 1991). During Operation Desert Storm, DSP detected the launch of Iraqi SCUD missiles and provided timely warning to civilian populations and United Nations coalition forces in Israel and Saudi Arabia.

Developments in the DSP have enabled it to provide accurate, reliable data in the face of tougher mission requirements, such as greater numbers of TARGETS, smaller signature targets, and advanced COUNTERMEASURES. Through several upgrade programs, DSP satellites have exceeded their design lives by some 30%. As the satellites' capabilities have grown, so has their MASS and POWER levels. In the early years, the DSP satellite had a mass of about 900 kilograms (2,000 lbm), and its SOLAR paddles generated 400 WATTS of electric ENERGY. New-generation DSP satellites have a mass of approximately 2,270 kilograms (5,000 lbm) and improved Solar Panels that generate more than 1,400 watts of electric energy. The newer DSP satellites are about 6.7 meters (22 ft) in diameter and 10 meters (32.8 ft) high with the solar paddles deployed.

Other space technology developments in the program have led to techniques that have benefited other Department of Defense (DOD) MILITARY SATELLITE systems. For example, the addition of a reaction wheel removed unwanted orbital MOMENTUM from DSP vehicles. The spinning motion of this reaction wheel serves as a countering FORCE on the satellite's movements. This *zero-sum momentum* approach permits precise ORBIT control with a minimum expenditure of attitude control fuel. As a result of DSP experience, reaction wheels have been added to other DOD space systems, including the DEFENSE METEOROLOGICAL SATELLITE PROGRAM (DMSP), the GLOBAL POSITIONING SYSTEM (GPS), and DEFENSE SATELLITE COMMUNICATION SYSTEM (DSCS) satellites.

defensive satellite (DSAT) weapon A proposed space-based weapon system that is intended to defend other orbiting satellites by destroying attacking antisatellite (ASAT) weapons.

deflagration A very rapid combustion process in which large quantities of gas and (thermal) energy are released. It is sometimes accompanied by flame, sparks, or spattering of burning particles. Although a deflagration can be classed as an explosion, the term generally implies the burning process of a substance with self-contained oxygen in which the reaction zone advances into the unreacted material at less than sonic velocity.

deflector A plate or baffle that diverts something in its movement or flow, as: (a) a plate that projects into the airstream on the underside of an airfoil to divert the airflow into a slot; (b) a conelike device placed or fastened on the launch site beneath a vertically launched rocket to deflect the exhaust gases to the sides; (c) any of several different devices used on jet engines to reverse or divert the exhaust

gases; (d) a baffle to deflect and mingle fluids prior to combustion.

degassing The deliberate removal of gas from a material, usually by application of thermal energy (heat) under high-vacuum conditions.

degenerate star A STAR that has collapsed to a high-DENSITY condition, such as a WHITE DWARF or a NEUTRON STAR.

degree (usual symbol: °) A term that has commonly been used to express units of certain physical quantities, such as angles and temperatures. The ancient Babylonians are believed to be the first people to have subdivided the circle into 360 parts, or "degrees," thereby establishing the use of the degree in mathematics as a until of angular measurement.

degrees of freedom (DOF) A mode of motion, either angular or linear, with respect to a coordinate system, independent of any other mode. A body in motion has six possible degrees of freedom, three linear (sometimes called x-, y-, and z-motion with reference to linear [axial] movements in CARTESIAN COORDINATES) and three angular (sometimes called PITCH, YAW, and ROLL with reference to angular movements).

Deimos The tiny, irregularly shaped (about 12 km average diameter) outer MOON of MARS, discovered in 1877 by the American astronomer Asaph Hall (1829–1907).

De Laval nozzle A flow device that efficiently converts the ENERGY content of a hot, high-PRESSURE GAS into KINETIC ENERGY. Originally developed by the Swedish engineer Carl Gustaf de Laval (1845–1913) for use in certain steam TURBINES, this versatile CONVERGING-DIVERGING NOZZLE is now used in practically all modern ROCKETs. The device constricts the outflow of the high-pressure (combustion) gas until it reaches the VELOCITY of sound (at the NOZZLE's THROAT) and then expands the exiting gas to very high velocities.

delayer   In rocketry, a substance mixed in with solid rocket propellants to decrease the rate of combustion.

Delta (launch vehicle)   A versatile family of American two- and three-STAGE LIQUID-PROPELLANT, EXPENDABLE LAUNCH VEHICLES (ELVs) that uses multiple strap-on BOOSTER ROCKETS in several configurations. The family of Delta ROCKETS has successfully launched more than 225 U.S. and foreign SATELLITES, earning it the nickname *space workhorse vehicle.*

The evolving family of Delta rockets began in 1959, when NASA awarded a contract to Douglas Aircraft Company (now part of the Boeing Company) to produce and integrate 12 SPACE LAUNCH VEHICLES. The first Delta vehicle used components from the UNITED STATES AIR FORCE'S THOR MISSILE for its first stage and from the United States Navy's VANGUARD PROJECT rocket for its second stage. On May 13, 1960, the inaugural flight of the Delta I rocket from CAPE CANAVERAL AIR FORCE STATION successfully placed the ECHO 1 COMMUNICATIONS (relay) SATELLITE (actually, a large, inflatable sphere) into ORBIT. The original Delta booster was 17 meters (56 ft) long and 2.4 meters (8 ft) in diameter. It used one Rocketdyne MB-3 Block II ROCKET ENGINE (also called the LR-79-NA11 engine) that burned LIQUID OXYGEN (LOX) and KEROSENE to produce 667,000 NEWTONS (N) (150,000 pounds-FORCE [lbf]) of THRUST at LIFTOFF. Two small VERNIER ENGINES created 4,400 newtons (1,000 lbf) of thrust each, using the same PROPELLANTS as the main engine. Continued improvements resulted in an evolutionary series of vehicles with increasing PAYLOAD capabilities.

The Delta II became a medium capacity vehicle used by the Air Force to launch satellites of the GLOBAL POSITIONING SYSTEM. The first stage of the Delta II rocket was 3.65 meters (12 ft) longer than previous vehicles in the family. Nine more powerful SOLID-PROPELLANT ROCKET MOTORS also encircled the first stage.

In the late 1990s, the Boeing Company introduced the Delta III family of rockets to serve space transportation needs within the expanding commercial COMMUNICATIONS SATELLITE market. With a payload delivery capacity of 3,800 kilograms (8,400 lbm) delivered to a geosynchronous TRANSFER ORBIT (GTO), the Delta III essentially doubled the performance of the Delta II rocket. The first stage of the Delta III is powered by a Boeing RS-27A rocket engine assisted by two vernier rocket engines that help control ROLL during main engine BURN. The vehicle's second stage uses a Pratt and Whitney RL 10B-2 engine that burns CRYOGENIC PROPELLANTS.

On November 20, 2002, a Delta IV rocket lifted off from Cape Canaveral Air Force Station, successfully delivering the W5 Eutelsat commercial communications satellite to a geosynchronous transfer orbit. This event was the inaugural launch of the latest family of Delta launch vehicles. Delta IV rockets combine new and mature launch vehicle technologies and can lift medium to heavy payloads into space. The Delta IV vehicle also represents part of the United States Air Force's *evolved expendable launch vehicle program.* The new rocket uses the new Boeing Rocketdyne-built RS-68 liquid hydrogen and liquid oxygen main engine, an engine capable of generating 2,891,000 newtons (650,000 lbf). Assembled in five vehicle configurations, the new Delta IV rocket family is capable of delivering payloads that range between 5,845 kilograms (12,890 lbm) and 13,130 kilogram (28,950 lbm) to a geosynchronous transfer orbit.

delta-V (symbol: ΔV)   VELOCITY change; a useful numerical index of the maneuverability of a SPACECRAFT or ROCKET. This term often represents the maximum change in velocity that a space vehicle's PROPULSION SYSTEM can provide. It is typically described in terms of kilometers per second (km/s) or METERS per second (m/s).

delta wing   A triangularly shaped wing of an aircraft.

deluge collection pond   A facility at a launch site into which water used to cool the flame deflector is flushed as the rocket begins its ascent.

demodulation The process of recovering the modulating wave from a modulated carrier.

density (usual symbol: ρ) The MASS of a substance per unit volume at a specified TEMPERATURE.

de-orbit burn A retrograde (opposite-direction) ROCKET engine firing by which a SPACE VEHICLE's VELOCITY is reduced to less than that required to remain in ORBIT around a CELESTIAL BODY.

deploy 1. Of a parachute, to release so as to let it fill out or to unfold and fill out. 2. Of a payload, to remove the payload from a stowed or berthed position in the space shuttle orbiter's cargo bay and to release the payload to a position that is free of the orbiter.

descending node That point in the ORBIT of a CELESTIAL BODY when it travels from north to south across a reference plane, such as the equatorial plane of the CELESTIAL SPHERE or the plane of the ECLIPTIC. Also called the southbound node. *Compare with* ASCENDING NODE.

destruct (missile) The deliberate action of destroying a MISSILE or ROCKET vehicle after it has been launched but before it has completed its course. The range safety officer executes the destruct command when a missile or rocket veers off its intended (plotted) course or functions in a way so as to become a hazard. *See also* COMMAND DESTRUCT.

destruct line On a rocket test range, a boundary line on each side of the downrange course beyond which a rocket or missile cannot fly without being destroyed under command destruct procedures, or a line beyond which the impact point cannot pass.

detachment A particular state of isolation in which a person is separated or detached from his/her accustomed behavioral environment by extraordinary physical and/or psychological distances (e.g., a "foreign" astronaut visiting another nation's space station for a long period of time). This condition may compromise the person's performance.

detector A device that can sense and report on radiation (electromagnetic or nuclear) originating from a remote object.

detent A releasable element used to restrain a part before or after its motion. Detents are commonly found in arming mechanisms; for example, the safing pins in safe and arm (S&A) devices often use a spring-loaded detent to secure the pin in the S&A device.

deterrence The prevention of a war or other undesirable acts by military posture threatening unacceptable consequences to an aggressor.

detonating cord A flexible fabric tube containing a filler of high-explosive (HE) material that is to be initiated by an electroexplosive device; often used in missile and rocket (abort) destruct or stage separation functions.

detonation An exothermic (energy-liberating) chemical reaction that propagates with such rapidity that the rate of advance of the reaction zone into the unreacted explosive material exceeds sonic velocity (i.e., the velocity of sound in the material). The rate of advance of the reaction zone is called the *detonation velocity*. When this rate of advance attains such a value that it will continue without diminution through the unreacted material, it is called a *stable detonation velocity*. When the detonation velocity is equal to or greater than the stable detonation velocity of the explosive material, the reaction is called a *high-order detonation*. When it is lower, the reaction is called a *low-order detonation*.

detonator An explosive device (usually an electroexplosive device) that is the first device in an explosive train. It is designed to transform an input signal (generally electrical) into an explosive reaction.

deuterium (usual symbol: D or $^2_1$H) A nonradioactive ISOTOPE of HYDROGEN whose NUCLEUS contains one NEUTRON and one PROTON. It is sometimes called *heavy hydrogen* because the deuterium nucleus is twice as heavy as that of ordinary hydrogen. *See also* TRITIUM.

device (nuclear) A *nuclear explosive* used for tests or experiments. The term is used to distinguish these explosives from nuclear weapons, which are packaged units ready for transportation or use by military forces.

Dewar flask A double-walled container with the interspace evacuated of GAS (AIR) to prevent the contents from gaining or losing thermal ENERGY (HEAT). Named for the Scottish physical chemist Sir James Dewar (1842–1923), large modern versions of this device are used to store CRYO-GENIC PROPELLANTS.

dewetting Phenomenon in solid propellants in which the binder (fuel) breaks free from the embedded oxidizer and metal particles.

dew point The temperature to which air must be cooled for saturation to occur, exclusive of air pressure or moisture content change. At that temperature dew begins to form, and water vapor condenses into liquid.

dexterity Of a human, skill in using the hands or body; of a robot, flexibility in the use of a manipulator arm or its end effector.

diamonds The patterns of SHOCK WAVES (PRESSURE discontinuities) often visible in a ROCKET engine's exhaust. These patterns resemble a series of diamond shapes placed end to end.

diaphragm 1. A thin membrane that can be used as a seal to prevent fluid leakage or as an actuator to transform an applied pressure into a linear force. 2. A positive expulsion device used to expel propellant from a tank in "zero-gravity" (zero-g) conditions. 3. In optics, the physical device that controls the amount of light entering an optical system.

diffraction The spreading out of ELECTROMAGNETIC RADIATION (such as LIGHT WAVES) as they pass by the edge of a body or through closely spaced parallel scratches in the surface of a diffraction grating. For example, when a RAY of white light passes over a sharp opaque edge (like that of a razor blade), it is broken up into its rainbow SPECTRUM of colors.

diffuse background The cosmic microwave radiation background in the 3×10^8 to 3×10^{11} hertz region of the electromagnetic spectrum. It pervades all of space and is believed to be a remnant of the initial Big Bang that marked the birth of the universe. This microwave radiation was first discovered in 1965 and has since been measured carefully by scientific spacecraft, including the NASA *Cosmic Background Explorer* (COBE) satellite. Also called *microwave background radiation* or *cosmic background radiation. See also* BIG BANG THEORY; COSMIC BACKGROUND EXPLORER.

diffuser A specially designed duct, sometimes equipped with stationary guide vanes, that decreases the velocity of a working fluid (i.e., air) and also increases the fluid's pressure. Used in jet engines and wind tunnels to convert low-pressure, high-velocity air flow into low-velocity air flow at a higher pressure. *See also* WIND TUNNEL.

diffuse radiation Electromagnetic radiation propagating in many different directions through a given small volume of space; the opposite of *specular radiation.*

diffuse reflector Any surface that reflects incident electromagnetic radiation (e.g., incoming rays of light) in a multiplicity of directions. This occurs because of the relative "roughness" of the surface (due to surface irregularities and material inhomogeneities) with respect to the wavelength of the incoming radiation. A diffuse reflector is the opposite of a specular reflector, such as a mirror.

digit A single character or symbol in a number system. For example, the binary system has two digits, 0 and 1; while the decimal system has ten digits, 0 through 9.

digital computer The most common type of computer in use today; one that processes data that has been converted into binary notation. *Compare with* ANALOG COMPUTER.

digital image processing Computer processing of the digital number (DN) values assigned to each pixel in an image. For example, all pixels in a particular image with a digital number value within a certain range might be assigned a special "color" or might be changed in value some arbitrary amount to ease the process of image interpretation by a human analyst. Furthermore, two images of the same scene taken at different times or at different wavelengths might have the digital number values of corresponding pixels computer manipulated (e.g., subtracted) to bring out some special features. This is a digital image processing technique called differencing or *change detection. See also* REMOTE SENSING.

digital transmission A technique in telecommunications that sends the signal in the form of one of a discrete number of codes (e.g., in binary code, either 0 or 1). The information content of the signal is concerned with discrete states of this signal, such as the presence or absence of voltage or a contact in a closed or open position.

digitize To express an analog measurement in discrete units, such as a series of binary digits. For example, an image or photograph can be scanned and *digitized* by converting lines and shading (or color) into combinations of appropriate digital values for each pixel in the image or photograph.

dipole A system composed of two equal but opposite electric charges that are separated but still relatively close together. The *dipole moment* (μ) is defined as the product of one of the charges and the distance between them. Some molecules, called *dipole molecules,* have effective centers of positive and negative charge that are permanently separated.

dipole antenna A half-wave (dipole) antenna typically consists of two straight, conducting metal rods each one-quarter of a wavelength long that are connected to an alternating voltage source. The electric field lines associated with this antenna configuration resemble those of an electric dipole. (An electric dipole consists of a pair of opposite electric charges [+q and −q] separated by a distance [d].) The dipole antenna is commonly used to transmit (or receive) radio-frequency signals below 30 megahertz (MHz).

direct ascent nonnuclear antisatellite (DANNASAT) weapon A ground-based rocket with a homing, nonnuclear warhead designed to destroy enemy satellites.

direct ascent nuclear antisatellite (DANASAT) weapon A ground-based rocket with a nuclear warhead designed to destroy enemy satellites.

direct broadcast satellite (DBS) A class of COMMUNICATIONS SATELLITE, usually placed in GEOSTATIONARY ORBIT, that receives broadcast SIGNALs (such as television programs) from points of origin on EARTH and then amplifies, encodes, and retransmits these signals to individual end users scattered throughout some wide area or specific region. Many American households now receive hundreds of television channels directly from space by means of small (less than 0.5 m in diameter), roof-top SATELLITE DISHes that are equipped to decode DBS transmissions. *See also* COMMUNICATIONS SATELLITE.

direct conversion The conversion of thermal ENERGY (HEAT) or other forms of energy (such as sunlight) directly into electrical energy without intermediate conversion into mechanical WORK—that is, without the use of the moving components as found in a conventional electric generator system. The main approaches for converting heat directly into electricity include THERMOELECTRIC CONVERSION, THERMIONIC CONVERSION, and MAGNETOHYDRODYNAMIC CONVERSION. SOLAR ENERGY is directly converted into electrical energy by means of SOLAR CELLS (PHOTOVOLTAIC CONVERSION.) Batteries and FUEL CELLS directly convert chemical energy into electrical energy. *See also* RADIOISOTOPE THERMOELECTRIC GENERATOR.

directed energy   Energy in the form of elementary atomic particles, tiny pellets (i.e., kinetic energy weapon devices), or focused beams of electromagnetic radiation (e.g., laser or microwave beams) that can be sent long distances at, or nearly at, the speed of light.

directed energy weapon (DEW)   A device that uses a tightly focused beam of very intense energy, either in the form of electromagnetic radiation (e.g., light from a laser) or in the form of elementary atomic particles, to kill its target. The DEW device delivers this lethal amount of energy at or near the speed of light. Also called a *speed-of-light weapon.*

directional antenna   An ANTENNA that radiates or receives RADIO-FREQUENCY (RF) signals more efficiently in some directions than in others. A collection of antennas arranged and selectively pointed for this purpose is called a *directional antenna array.*

direct readout   The information technology capability that allows ground stations on EARTH to collect and interpret the data messages (TELEMETRY) being transmitted from SATELLITES.

disarming   In aerospace operations, the normal method of returning ordnance systems or components to a safe status.

disconnect   Short for quick-disconnect—a separable connector characterized by two

separable halves, an interface seal, and, usually, a latch-release locking mechanism; it can be separated without the use of tools in a very short time.

discrete   Composed of distinct elements.

dish   Aerospace jargon used to describe a parabolic radio or radar antenna, whose shape is roughly that of a large soup bowl.

disk   1. (ASTRONOMY) The visible surface of the SUN (or any other CELESTIAL BODY) seen in the sky or through a TELESCOPE. 2. (of a GALAXY) The flattened, wheel-shaped region of STARS, GAS, and dust that lies outside the central region (NUCLEUS) of a galaxy.

distortion   1. In general, the failure of a system (typically optical or electronic) to transmit or reproduce the characteristics of an input signal with exactness in the output signal. 2. An undesired changed in the dimensions or shape of a structure; for example, the *distortion* of a cryogenic propellant tank because of extreme temperature gradients.

diurnal   Having a period of, occurring in, or related to a day; daily.

diverging lens (or mirror)   A LENS (or MIRROR) that refracts (or reflects) a parallel BEAM of LIGHT into a diverging beam. A diverging lens is generally a CONCAVE LENS, while a diverging mirror is a CONVEX MIRROR. *Compare with* CONVERGING LENS.

docking   The act of physically joining two orbiting SPACECRAFT. This is usually accomplished by independently maneuvering one spacecraft (the CHASER SPACECRAFT) into contact with the other (the TARGET SPACECRAFT) at a chosen physical interface. For spacecraft with human crews, a DOCKING MODULE assists in the process and often serves as a special passageway (AIRLOCK) that permits HATCHes to be opened and crew members to move from one spacecraft to the other without the use of a SPACESUIT and without losing cabin PRESSURE.

docking interface The area of contact between two docking mechanisms.

docking mechanism A mechanism that performs functions to connect one spacecraft to another in a docking operation.

docking module A structural element that provides a support and attachment interface between a docking mechanism and a spacecraft. For example, the special component added to the U.S. Apollo spacecraft so that it could be joined with the Russian Soyuz spacecraft in the Apollo-Soyuz Test Project; or the component carried in the cargo bay of the U.S. space shuttle so that it could be joined with the Russian Mir space station. *See also* APOLLO-SOYUZ TEST PROJECT; DOCKING.

doffing The act of removing wearing apparel or other apparatus, such as a SPACESUIT.

dogleg A directional turn made in a LAUNCH VEHICLE's ascent TRAJECTORY to produce a more favorable ORBIT INCLINATION or to avoid passing over a populated (no-fly) region.

donning The act of putting on wearing apparel or other apparatus, such as a SPACESUIT.

Doppler effect The phenomenon evidenced by the change in the observed frequency of a sound or radio wave caused by a time rate of change in the effective length of the path of travel between the source and the point of observation. *See also* DOPPLER SHIFT.

Doppler imaging The use of RADAR or LASER radar to produce reflected Doppler-shifted electromagnetic signals from different parts of an object. This technique can provide an image of the object if it is spinning or tumbling. Since the DOPPLER SHIFT depends on the VELOCITY of the object with respect to the observer, reflections from those parts of the object receding from the observer will have different shifts from those moving toward the observer.

Doppler radar Any form of radar that detects motion relative to a reflecting surface by measuring the frequency shift of reflected electromagnetic energy due to the motion of the observer or of the reflecting surface. a Doppler radar system can differentiate between fixed and moving targets by detecting the apparent change in frequency of the reflected electromagnetic wave due to motion of the target.

Doppler shift The apparent change in the observed FREQUENCY and WAVELENGTH of a source due to the relative motion of the source and an observer. If the source is approaching the observer, the observed frequency is higher and the observed wavelength is shorter. This change to shorter wavelengths is often called the BLUESHIFT. If the source is moving away from the observer, the observed frequency will be lower and the wavelength will be longer. This change to longer wavelengths is called the REDSHIFT. Named after the Austrian physicist Christian Johann Doppler (1803–1853), who discovered this phenomenon in 1842 by observing sounds. (See Figure 29.)

dose In radiation protection, a general term describing the amount of energy

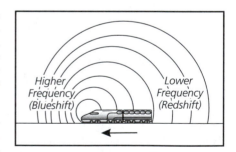

Figure 29. Sound from a train approaching an observer will appear to increase in frequency. Conversely, sound will appear to decrease in frequency as the train goes away (recedes) from an observer. This apparent shift in frequency due to the relative motion of an object and an observer is called the Doppler shift.

delivered to a given volume of matter, a particular body organ, or a person (i.e., a *wholebody dose*) by ionizing radiation. *See also* ABSORBED DOSE.

dose equivalent (symbol: H)   In radiation protection, the product of absorbed dose (D), quality factor (QF), and any other modifying factors used to characterize and evaluate the biological effects of ionizing radiation doses received by human beings (or other living creatures). The traditional unit of dose equivalent is the *rem* (an acronym for *roentgen equivalent man*). The *sievert (Sv)* is the SI special unit for dose equivalent. These units are related as follows:

100 rem = 1 sievert (Sv)

See also SI UNITS.

double-base propellant   A solid ROCKET PROPELLANT that uses two unstable compounds, such as nitrocellulose and nitroglycerin. These unstable compounds contain enough chemically bonded OXIDIZER to sustain combustion.

downlink   The TELEMETRY signal received at a GROUND RECEIVING STATION from a SPACECRAFT or SPACE PROBE.

downrange   A location away from the LAUNCH SITE but along the intended flight path (TRAJECTORY) of a MISSILE or ROCKET flown from a rocket range. For example, the rocket vehicle tracking station on Ascension Island in the south Atlantic Ocean is far downrange from the launch sites at CAPE CANAVERAL AIR FORCE STATION (CCAFS) in Florida.

downtime   A period during which equipment is not operating; this outage can be due to planned maintenance or unplanned component failure.

downweight   Landing weight; specifically space shuttle (STS—Space Transmittal System) payloads and all items required by specific payloads.

drag (symbol: D)   1. A retarding force acting upon a body in motion through a fluid, parallel to the direction of motion of the body. It is a component of the total forces acting on the body. 2. Atmospheric resistance to the orbital motion of a spacecraft. The effect of atmospheric drag is to cause the satellite's orbit to "decay," or spiral downward. A satellite of very high mass, very low cross-sectional area, and in very high orbit will not be affected greatly by drag. However, a high cross-sectional area satellite of low mass in a low-altitude orbit will be affected very strongly by drag. For Earth orbits above about 200 kilometers (124 miles), the satellite's a altitude decreases very slowly due to drag; but below 150 kilometers (93.2 miles), the orbit decays very rapidly. In fact, at lower orbital altitudes, drag is the major space environment condition affecting satellite lifetime. For example, the abandoned U.S. Skylab space station came crashing back to Earth in July 1979 due to relentless action of atmospheric drag.

drag parachute   1. A drogue parachute. 2. Any of several types of parachutes attached to high-performance aircraft or aerospace vehicles that can be deployed, usually during landings, to decrease speed and also, under certain flight conditions, to control and stabilize the vehicle.

drift   1. In ballistics, a shift in projectile direction due to gyroscopic action that results from gravitational and atmospherically induced torques on the spinning projectile. 2. The lateral divergence from the prescribed flight path of an aircraft, rocket, or aerospace vehicle, due primarily to the effect of crosswind. 3. The angular deviation of the spin axis of a gyro from a fixed reference in space. 4. A slow change in frequency of a radio transmitter. 5. A gradual movement in one direction of an instrument's pointer. 6. With respect to semiconductor materials, the movement of carriers (i.e., holes and electrons) in an electric field. Also called *drift mobility* or *carrier mobility*.

drogue   A spacecraft docking receptacle that captures the probe of another spacecraft. The device usually is shaped like a funnel or cone and serves as the "female"-type connector into which the probe ("male"-type connector) fits.

drogue parachute A small parachute specifically used to pull a larger parachute out of stowage; a small parachute used to slow down a descending SPACE CAPSULE, AEROSPACE VEHICLE, or high-performance airplane.

drogue recovery A type of recovery system used for space capsules after initial atmospheric entry, involving the deployment of one or more small parachutes to diminish speed, to reduce aerodynamic heating, and to stabilize the vehicle so that larger recovery parachutes can be deployed safely at lower altitudes without too great an opening shock.

drone 1. In general, a mechanism or vehicle that can operate without direct human guidance. 2. In aerospace, an uncrewed air vehicle that is controlled remotely or automatically. Drones often are used as test targets for missile and fighter aircraft weapon systems.

drop tower A tall tower (about 90 meters [294 feet] high or more) in which experimental packages are carefully dropped under "free-fall" conditions. A well-designed drop tower can provide low-gravity conditions for about 3 to 5 seconds—enough time to perform useful preliminary experiments simulating (briefly) the microgravity or weightlessness conditions encountered in an orbiting spacecraft. A catch tube system at the bottom of the drop tower decelerates and recovers the experimental package. Also called a *low-gravity drop tower* and a *drop tube*. *See also* MICROGRAVITY.

dry cycle Operation of a valve or similar component without propellant or test fluid in the flow passages.

dry emplacement A LAUNCH SITE that has no provision for water cooling of the pad during the LAUNCH of a ROCKET. *Compare with* WET EMPLACEMENT.

dry-film lubricant Material that reduces rolling or sliding friction between mating surfaces by coating one or both of the surfaces with a slippery film, often permanently bonded to the surface; also called *solid lubricant*.

dry run In aerospace operations, a practice exercise or rehearsal for a specific mission task or launch operation; a practice launch without propellant loaded in the rocket vehicle.

dry weight The weight of a missile or rocket vehicle without its fuel. This term, especially appropriate for liquid-propellant rockets, sometimes is considered to include the payload.

dual propellant loading An aerospace operation involving the transfer of both fuel and oxidizer.

duct A tube or passage that confines and conducts a fluid; for example, a passage for the flow of air to the compressor of a gas turbine.

ducted fan A fan enclosed in a duct.

duct propulsion A means of propelling a vehicle by ducting a surrounding fluid through an engine, adding momentum to it by mechanical or thermal means, and ejecting the fluid to obtain a reactive force. *Compare* ROCKET PROPULSION.

dud An explosive munition that has not been armed as intended or that has failed to explode after being armed.

dune In planetary geology, a low hill or bank of drifted sand.

duplexer A device that permits a single antenna system to be used for both transmitting and receiving. (Should not be confused with *diplexer*—a device that permits an antenna to be used simultaneously or separately by two transmitters.)

dust detector A direct-sensing science instrument that measures the velocity, mass (typical range 10^{-16} g to 10^{-6} g), flight direction, charge (if any), and number of dust particles striking the instrument, carried by some spacecraft.

duty cycle In general, the percentage of time a system is functioning; for example, in the case of a pulsed radar system, the pulse repetition rate times the pulse width—or, simply, the percentage of time the radar is transmitting.

dwarf galaxy A small, often elliptical galaxy containing a million (10^6) to perhaps a billion (10^9) stars. The Magellanic Clouds, our nearest galactic neighbors, are examples of dwarf galaxies. *See also* GALAXY; MAGELLANIC CLOUDS.

dwarf star Any STAR that is a MAIN-SEQUENCE star, according to the HERTZSPRUNG-RUSSELL DIAGRAM. Most stars found in the GALAXY, including the SUN, are of this type and are from 0.1 to about 100 SOLAR MASSes in size. However, when scientists use the term *dwarf star,* they are NOT referring to WHITE DWARFS, BROWN DWARFS, or BLACK DWARFS, which are celestial bodies that are not in the collection of main-sequence stars.

Dynamic Isotope Power System (DIPS) A proposed device for producing electric power for spacecraft and planetary rover systems that uses the thermal energy (heat) generated by a quantity of radioactive material (i.e., the radioisotope plutonium-238) as a power source and then converts this thermal energy to electric energy by means of a dynamic heat engine (e.g., a closed Brayton cycle). *See also* SPACE NUCLEAR POWER.

dynamic pressure (common symbol: Q or q) 1. The pressure exerted by a fluid, such as air, by virtue of its motion; it is equal to $\frac{1}{2} \rho V^2$, where ρ is the density and V is the velocity of the fluid. 2. The pressure exerted on a body by virtue of its motion through a fluid, as, for example, the pressure exerted on a rocket vehicle as it flies through the atmosphere. The condition of maximum dynamic pressure experienced by an ascending rocket vehicle is often called *max-Q.*

dynamics The branch of science that studies the motion of bodies under the influence of forces, especially forces that originate outside the body under consideration.

dynamic seal A mechanical device used to minimize leakage of fluid from the flow-stream region of a fluid-system component when there is relative motion of the sealing interfaces.

dynamic stability The characteristics of a body, such as an aircraft, spacecraft, or aerospace vehicle, that causes it, when disturbed from an original state of steady flight or motion, to damp the oscillations induced by the disturbance and gradually return to its original state.

Dyna-Soar (Dynamic Soaring) An early U.S. Air Force space project from 1958 to 1963 that involved a crewed boost–glide orbital vehicle that was to be sent into ORBIT by an EXPENDABLE LAUNCH VEHICLE, perform its military mission, and return to EARTH using wings to glide through the ATMOSPHERE during REENTRY (in a manner similar to NASA's SPACE SHUTTLE). The project was canceled in favor of the civilian (NASA) human spaceflight program involving the MERCURY PROJECT, GEMINI PROJECT, and APOLLO PROJECT. Also called the *X-20 Project.*

dyne (symbol: d) A unit of force in the centimeter-gram-second (c.g.s.) system equal to the force required to accelerate a one-gram mass one centimeter per second per second; that is, 1 dyne = 1 gm-cm/sec^2. *Compare* NEWTON.

dysbarism A general aerospace medicine term describing a variety of symptoms within the human body caused by the existence of a pressure differential between the total ambient pressure and the total pressure of dissolved and free gases within the body tissues, fluids, and cavities. For example, increased ambient pressure, as accompanies a descent from higher altitudes, might cause painful distention of the eardrums.

early warning satellite A military space-craft that has the primary mission of detect-ing and reporting the launch or approach of unknown weapons or weapon systems. For example, a hostile missile launch might be detected and reported through the use of special infrared (IR) sensors onboard this type of surveillance satellite. *See also* DEFENSE SUPPORT PROGRAM.

Earth Earth is the third planet from the SUN and the fifth largest in the SOLAR SYS-TEM. Our PLANET circles its parent STAR at an average distance of 149.6 million kilome-ters (93 million miles). Earth is the only planetary body in the solar system currently known to support life. The table presents some of the physical and dynamic proper-ties of Earth as a planet in the solar system.

From space, our planet is characterized by its blue waters and white clouds, which cover a major portion of it. Earth is sur-rounded by an ocean of air, consisting of 78% nitrogen and 21% oxygen; the remainder is argon, neon, and other gases. The standard atmospheric pressure at sea level is 101,325 newtons per square meter (14.7 pounds per square inch). Sur-face temperatures range from a maximum of about 60° Celsius (C) (140° Fahrenheit [F]) in desert regions along the equator to a minimum of minus 90°C (minus 130°F) in the frigid polar regions. In between, however, surface temperatures are gener-ally much more benign.

Earth's rapid spin and molten nickel-iron core give rise to an extensive magnetic field. This magnetic field, together with the atmosphere, shields us from nearly all of the harmful charged-particle and ultravio-let radiation coming from the Sun and cos-mic sources. Furthermore, most meteors burn up in Earth's protective atmosphere

before they can strike the surface. Earth's nearest celestial neighbor, the MOON, is its only natural satellite. *See also* EARTH'S TRAPPED RADIATION BELTS; GLOBAL CHANGE; MISSION TO PLANET EARTH.

Earth-based telescope A TELESCOPE located on the surface of EARTH. *Compare with* the HUBBLE SPACE TELESCOPE.

Earth-crossing asteroid (ECA) An ASTEROID whose orbit now crosses EARTH's ORBIT or will at some time in the future cross Earth's orbit, as its orbital path evolves under the influence of perturba-tions from JUPITER and the other PLANETS.

Earthlike planet An EXTRASOLAR PLANET that is located in an ECOSPHERE and has planetary environmental condi-tions that resemble the terrestrial BIOS-PHERE—especially a suitable ATMOSPHERE a temperature range that permits the reten-tion of large quantities of liquid water on the PLANET'S surface, and a sufficient quantity of energy striking the planet's sur-face from the parent STAR. These suitable environmental conditions could permit the chemical evolution and the development of carbon-based life as we know it on EARTH. The planet also should have a mass some-what greater than 0.4 Earth masses (to per-mit the production and retention of a breathable atmosphere) but less than about 2.4 Earth masses (to avoid excessive surface gravity conditions).

Earth-observing spacecraft A SATEL-LITE in ORBIT around EARTH that has a spe-cialized collection of SENSORS capable of monitoring important environmental vari-ables. This is also called an environmental satellite or a green satellite. Data from

Dynamic and Physical Properties of the Planet Earth	
Radius	
Equatorial	6,378 km
Polar	6,357 km
Mass	5.98×10^{24} kg
Density (average)	5.52 g/cm^3
Surface area	5.1×10^{14} m^2
Volume	1.08×10^{21} m^3
Distance from the Sun (average)	1.496×10^8 km (1 AU)
Eccentricity	0.01673
Orbital period (sidereal)	365.256 days
Period of rotation (sidereal)	23.934 hours
Inclination of equator	23.45 degrees
Mean orbital velocity	29.78 km/sec
Acceleration of gravity g (sea level)	9.807 m/sec^2
Solar flux at Earth (above atmosphere)	$1,371 \pm 5$ watts/m^2
Planetary energy fluxes (approximate)	
Solar	10^{17} watts
Geothermal	2.5×10^{13} watts
Tidal friction	3.4×10^{12} watts
Human-made	
Coal-burning	3.06×10^{12} watts
Natural gas-burning	2.00×10^{12} watts
Oil-burning	3.82×10^{12} watts
Nuclear power	0.49×10^{12} watts
Hydroelectric	0.69×10^{12} watts
Total human-made	10.06×10^{12} watts
Number of natural satellites	1 (the Moon)

such satellites help support EARTH SYSTEM SCIENCE. *See also* AQUA SPACECRAFT; LANDSAT, METEOROLOGICAL SATELLITE; *TERRA* SPACECRAFT.

Earth radiation budget (ERB) The EARTH radiation budget is perhaps the most fundamental quantity influencing Earth's climate. The ERB components include the incoming solar radiation; the solar radiation reflected back to space by the clouds, the atmosphere, and Earth's surface; and the long-wavelength thermal radiation emitted by Earth's surface and its atmosphere. The latitudinal variations of Earth's radiation budget are the ultimate driving force for the atmospheric and oceanic circulations and the resulting planetary climate. *See also* GLOBAL CHANGE; MISSION TO PLANET EARTH.

Earth satellite An artificial (human-made) object placed in ORBIT around the PLANET EARTH.

earthshine A SPACECRAFT or SPACE VEHICLE in orbit around EARTH is illuminated by both sunlight and "earthshine." Earthshine consists of sunlight (0.4- to 0.7-micrometer wavelength radiation, or visible light) reflected by Earth and thermal radiation (typically 10.6-micrometer wavelength infrared radiation) emitted by Earth's surface and atmosphere.

Earth's trapped radiation belts The magnetosphere is a region around EARTH through which the solar wind cannot penetrate because of the terrestrial magnetic field. Inside the MAGNETOSPHERE are two belts or zones of very energetic atomic particles (mainly electrons and protons) that are trapped in Earth's magnetic field hundreds of kilometers above the atmosphere. These belts were discovered by Professor James Van Allen of the University of Iowa and his colleagues in 1958. Van Allen made the discovery using simple atomic radiation detectors placed onboard EXPLORER I, the first American SATELLITE.

The two major trapped radiation belts form a doughnut-shaped region around Earth from about 320 to 32,400 kilometers (200 to 20,000 miles) above the equator (depending on solar activity). Energetic PROTONS and ELECTRONS are trapped in these belts. The inner Van Allen belt contains both energetic protons (major constituent) and electrons that were captured from the solar wind or were created in nuclear collision reactions between energetic cosmic ray particles and atoms in Earth's upper atmosphere. The outer Van Allen belt contains mostly energetic electrons that have been captured from the SOLAR WIND.

SPACECRAFT and SPACE STATIONS operating in Earth's trapped radiation belts are subject to the damaging effects of IONIZING RADIATION from charged atomic particles. These particles include protons, electrons, alpha particles (helium nuclei), and heavier atomic nuclei. Their damaging effects include degradation of material properties and component performance, often resulting in reduced capabilities or even failure of spacecraft systems and experiments. For example, solar cells used to provide electric power for spacecraft often are severely damaged by passage through the Van Allen belts. Earth's trapped radiation belts also represent a very hazardous environment for human beings traveling in space.

Radiation damage from Earth's trapped radiation belts can be reduced significantly by designing spacecraft and space stations with proper radiation shielding. Often crew compartments and sensitive equipment can be located in regions shielded by other spacecraft equipment that is less sensitive to the influence of ionizing radiation. Radiation damage also can be limited by selecting mission orbits and trajectories that avoid

long periods of operation where the radiation belts have their highest charged-particle populations. For example, for a spacecraft or space station in low Earth orbit, this would mean avoiding the SOUTH ATLANTIC ANOMALY and, of course, the Van Allen Belts themselves.

eccentricity (symbol: e) A measure of the ovalness of an orbit. For example, when $e = 0$, the orbit is a circle; when $e = 0.9$, the orbit is a long, thin ellipse. (See Figure 30.)

The eccentricity (e) of an ellipse can be computed by the formula:

$$e = \sqrt{[1 - b^2/a^2]}$$

where a is the semimajor axis and b is the semiminor axis. *See also* KEPLERIAN ELEMENTS; ORBITS OF OBJECTS IN SPACE.

Earth system science The modern study of EARTH, facilitated by space-based observations, that treats the PLANET as an interactive, complex system. The four major components of the Earth system are the ATMOSPHERE, the HYDROSPHERE (which includes liquid water and ice), the BIOSPHERE (which includes all living things), and the SOLID Earth (especially the planet's surface and soil).

eccentric orbit An orbit that deviates from a circle, thus forming an ellipse. *See also* ORBITS OF OBJECTS IN SPACE.

Echo satellite A large, inflatable passive communications-relay satellite from which radio signals from one point on Earth could be bounced off and received by another location on Earth. NASA's *Echo 1*, the world's first passive communications-relay satellite, was launched successfully from Cape Canaveral Air Force Station on August 12, 1960.

eclipse 1. The reduction in visibility or the disappearance of a nonluminous body by passing into the shadow cast by another nonluminous body. 2. The apparent cutting off, wholly or partially, of the light from a luminous body by a dark (nonluminous) body coming between it and the observer.

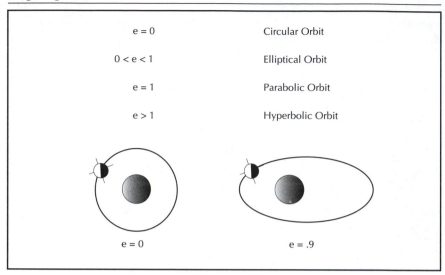

Figure 30. Eccentricity.

The first type of eclipse is exemplified by a *lunar eclipse,* which occurs when the Moon passes through the shadow cast by Earth or when a satellite passes through the shadow cast by its planet; when a satellite passes directly behind its planet, an *occultation* is said to occur. The second type of eclipse is exemplified by a *solar eclipse,* which occurs when the Moon passes between the Sun and Earth.

ecliptic (plane) The apparent annual path of the SUN among the STARS; the intersection of the plane of EARTH'S ORBIT around the Sun with the CELESTIAL SPHERE. Because of the tilt in Earth's AXIS, the ecliptic is a great circle of the celestial sphere inclined at an ANGLE of about 27.4° to the celestial EQUATOR.

ecological system A habitable environment, either created artificially (as in a crewed space vehicle) or occurring naturally (i.e., the environment on the surface of Earth), in which human beings, animals, and/or other organisms can live in mutual relationship with one another and the environment. Under ideal circumstances, the environment furnishes the sustenance for life, and the resulting waste products revert or cycle back into the environment to be used again for the continuous support of life. The interacting system of a biological community and its nonliving environmental surroundings. Also called *ecosystem. See also* CLOSED ECOLOGICAL LIFE SUPPORT SYSTEM.

ecology The study of the relationship of living things to one another and their environment. Remote-sensing data provided by Earth-orbiting satellites are of great value to scientists in their efforts to study and monitor the complex relationships taking place in Earth's ecosphere. *See also* GLOBAL CHANGE; MISSION TO PLANET EARTH; REMOTE SENSING.

ecosphere The CONTINUOUSLY HABITABLE ZONE (CHZ) around a MAIN SEQUENCE STAR of a particular LUMINOSITY in which a PLANET could support environmental conditions favorable to the evolution and continued existence of life. For the chemical evolution of Earthlike, carbon-based living organisms, global TEMPERATURES and atmospheric PRESSURE conditions must allow the retention of liquid water on the planet's surface. A viable ecosphere might lie between about 0.7

and 1.3 ASTRONOMICAL UNITS from a STAR like the SUN. However, if all the surface water has evaporated (the RUNAWAY GREENHOUSE EFFECT) or has completely frozen (the ICE CATASTROPHE), then any Earthlike planet within this ecosphere cannot sustain life.

efficiency (symbol: η) In general, the ratio of useful output of some physical quantity to its total input. In thermodynamics, the ratio of energy output to energy input. For a heat engine, the thermal efficiency ($\eta_{thermal}$) is defined as the ratio of the work (mechanical energy) out to thermal energy (heat) input.

ejecta Any of a variety of rock fragments thrown out by an IMPACT CRATER during its formation and subsequently deposited via ballistic trajectories onto the surrounding terrain. The deposits themselves are called ejecta blankets. It is also material thrown out of a VOLCANO during an explosive eruption.

ejection capsule 1. In an aircraft, aerospace vehicle, or crewed spacecraft, a detachable compartment serving as a cockpit or cabin, which may be ejected as a unit and parachuted to the ground. 2. In a satellite, probe, or uncrewed spacecraft, a box-like unit (usually containing records of observed data or experimental samples) that may be ejected and returned to Earth by a parachute or other deceleration device.

ejector A device consisting of a nozzle, a mixing tube, and a diffuser that uses the kinetic energy of a fluid stream to pump another fluid from a low-pressure region by direct mixing and ejecting both streams.

elasticity The ability of a body that has been deformed by an applied force to return to its original shape when the force is removed.

elasticizer An elastic substance or fuel used in solid rocket propellant to prevent cracking of the propellant grain and to bind it to the combustion-chamber case.

elastic limit The maximum stress that can be applied to a body without producing permanent deformation.

elastomer Rubberlike, polymeric material that can be deformed by a force and then regain its original shape after the force has been removed.

electrical Involving the flow of electricity in a conductor. *Compare* ELECTRONIC.

electric field strength (symbol: E) The electrical force exerted on a unit positive charge at a given point in space. Electric field strength typically is expressed in volts per meter.

electricity Flow of energy due to the motion of electric charges; any physical effect that results from the existence of moving or stationary electric charges (e.g., *static electricity, lightning*).

electric potential (symbol: V) The work done in moving a unit positive charge from infinity to the point in an electric field whose potential is being specified. The unit of electric potential is the volt. For example, if 1 joule is required to transfer a charge of 1 coulomb, then the electric potential is 1 volt.

electric propulsion A ROCKET engine that converts electric POWER into reactive THRUST by accelerating an ionized PROPELLANT (such as mercury, cesium, argon, or xenon) to a very high exhaust VELOCITY. There are three general types of electric rocket engine: electrothermal, electromagnetic, and electrostatic. The electrothermal rocket heats its propellant (such as ammonia) to a high TEMPERATURE. The heated propellant then expands through a NOZZLE, producing thrust. If an electric arc is used to heat the propellant gas, the engine is called an *arc jet engine*. The electromagnetic engine (or PLASMA engine) turns the propellant into plasma and then accelerates that plasma rearward by electric and magnetic fields. The MAGNETOPLASMADYNAMIC (MPD) THRUSTER is an example. The MPD thruster can operate in either a steady state or in a pulsed mode. The electrostatic rocket engine (or ion engine) ionizes propellant ATOMS (typically cesium, mercury, argon, or xenon) and then uses an imposed electric field to accelerate the ionized atoms to very high exhaust velocities. The ELECTRONS removed from the propellant atoms are injected directly into the ion exhaust BEAM. This helps neutralize the positive electric charge in the exhaust beam.

electrode A conductor (terminal) at which electricity passes from one medium into another. The positive electrode is called the *anode;* the negative electrode is called the *cathode.* In semiconductor devices, an element that performs one or more of the functions of emitting or collecting electrons or holes, or of controlling their movements by an electric field. In electron tubes, a conducting element that performs one or more of the functions of emitting, collecting, or controlling the movements of electrons or ions usually by means of an electromagnetic field.

electroexplosive device (EED) A pyrotechnic device in which electrically insulated terminals are in contact with, or adjacent to, a material mixture that reacts chemically (often explosively) when the required electrical energy level is discharged through the terminals. Explosive stage separation devices and missile self-destruct packages (for vehicles that have departed from an acceptable flight trajectory) are examples of EEDs used in aerospace applications.

electrolysis In general, a chemical reaction produced by the passage of an electric current through an electrolyte. Specifically, the process of splitting water into hydrogen and oxygen by means of a direct electric current. A basic electrolysis cell consists of two electrodes immersed in an aqueous conducting solution called an *electrolyte.* A source of direct current (dc) voltage is applied to the electrodes so that an electric current flows through the electrolyte from the anode (positive electrode) to the cathode (negative electrode). As a result, the water in the electrolyte solution is decomposed into

hydrogen gas (H_2), which is released at the cathode, and oxygen gas (O_2), which is released at the anode. Although only the water is split, an electrolyte (i.e., potassium hydroxide) is needed because water itself is a very poor conductor of electricity. *See also* FUEL CELL; HYDROGEN.

electromagnetic Having both electric and magnetic properties; pertaining to magnetism produced or associated with electricity.

electromagnetic (EM) communications In aerospace, the technology involving the development and production of a variety of telecommunication equipment used for electromagnetic transmission of information over any media. The information may be analog or digital, ranging in bandwidth from a single voice or data channel to video or multiplexed channels occupying hundreds of megahertz. Included are onboard satellite communication equipment and laser communication techniques capable of automatically acquiring and tracking signals and maintaining communications through atmospheric and exoatmospheric media.

electromagnetic (EM) gun A gun in which the projectile is accelerated by electromagnetic forces rather than by a chemical explosion as in a conventional gun.

electromagnetic (EM) launcher A device that can accelerate an object to high velocities using the electromotive force produced by a large current in a transverse magnetic field.

electromagnetic pulse (EMP) A large pulse of electromagnetic radiation, effectively reaching out to distances of hundreds of kilometers or more; caused by the interactions of gamma rays from a high-altitude nuclear explosion with atoms in the upper atmosphere. The resulting electromagnetic fields may couple with electrical and electronic systems to produce damaging current and voltage surges. EMP also can be created by nonnuclear means.

electromagnetic radiation (EMR) Radiation made up of oscillating electric and magnetic fields and propagated with the speed of light. Includes (in order of decreasing frequency) gamma radiation, X rays, ultraviolet, visible, and infrared (IR) radiation, and radar and radio waves. *See also* ELECTROMAGNETIC SPECTRUM.

electromagnetic (EM) spectrum When sunlight passes through a prism, it throws a rainbowlike array of colors onto a surface. This display of colors is called the *visible spectrum*. It represents an arrangement in order of wavelength of the narrow band of electromagnetic (EM) radiation to which the human eye is sensitive.

The electromagnetic spectrum comprises the entire range of wavelengths of electromagnetic radiation, from the shortest-wavelength gamma rays to the longest-wavelength radiowaves. (See Figure 31.) The entire EM spectrum includes much more than meets the eye.

As shown in the figure, the names applied to the various regions of the EM spectrum are (going from shortest to longest wavelength) gamma ray, X ray, ultraviolet (UV), visible, infrared (IR) and radio. EM radiation travels at the speed of light (i.e., about 300,000 kilometers per second [~ 186,400 miles per second]) and is the basic mechanism for energy transfer through the vacuum of outer space.

One of the most interesting discoveries of 20th-century physics is the dual nature of electromagnetic radiation. Under some conditions electromagnetic radiation behaves like a wave, while under other conditions it behaves like a stream of particles, called *photons*. The tiny amount of energy carried by a photon is called a *quantum of energy* (plural: quanta). The word "quantum" comes to us from the Latin language and means "little bundle."

The shorter the wavelength, the more energy is carried by a particular form of EM radiation. All things in the universe emit, reflect, and absorb electromagnetic radiation in their own distinctive ways. The way an object does this provides scientists with special characteristics, or a signature, that can be detected by remote sensing instruments. For example, the spectrogram shows bright lines for emission or reflection and dark lines for absorption at selected EM wavelengths. Analyses of the positions and line patterns found in a spectrogram can provide information about the object's composition, surface temperature, density, age, motion, and distance.

For centuries, astronomers have used spectral analyses to learn about distant extraterrestrial phenomena. But up until the space age, they were limited in their view of the universe by the Earth's atmosphere, which filters out most of the EM radiation from the rest of the cosmos. In fact, ground-based astronomers are limited to just the visible portion of the EM spectrum and tiny portions of the infrared, radio, and ultraviolet regions. Space-based observatories now allow us to examine the universe in all portions of the EM

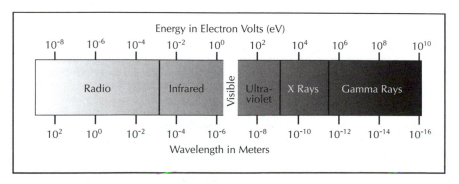

Figure 31. The electromagnetic (EM) spectrum.

spectrum. We now have examined the cosmos in the infrared, ultraviolet, X-ray, and gamma ray portions of the EM spectrum and have made startling discoveries. We also have developed sophisticated remote sensing instruments to look back on Earth in many regions of the electromagnetic spectrum, providing powerful tools for a more careful management of the terrestrial biosphere. *See also* REMOTE SENSING.

electromotive force (emf) The characteristic of an electrical energy source that enables a current to flow in a circuit. It is the sum (algebraic) of the potential differences acting in an electric circuit. The emf (typical unit: volts) is measured by the energy liberated when a unit electric charge passes completely around the circuit.

electron (symbol: e) A stable elementary particle with a unit negative electrical charge (1.602×10^{-19} coulomb) and a rest mass (m_e) of 1/1837 that of a proton (namely, 9.109×10^{-31} kilogram). Electrons surround the positively charged nucleus and determine the chemical properties of the atom. Positively charged electrons, or *positrons*, also exist. Electrons were first discovered in the late 1890s by the British scientist Sir J. J. Thomson (1856–1940).

electronic Of or pertaining to any of the large number and wide variety of devices that involve the generation, transmission, use, or control of electricity.

electronics The branch of physics dealing with the understanding, design, and application of devices based on the conduction of electricity through a vacuum, gas, or semiconductor. Although this term originated with vacuum (electron) tube applications, modern electronics is concerned primarily with semiconductor devices.

electron volt (eV) A unit of energy equivalent to the energy gained by an electron when it experiences a potential difference of one volt. Larger multiple units of the electron volt are encountered frequently—as, for example: *keV* for thousand (or kilo-) electron volts (10^3 eV);

MeV for million (or mega-) electron volts (10^6 eV); and *GeV* for billion (or giga-) electron volts (10^9 eV).

1 electron volt = 1.602×10^{-19} joules

electro-optical countermeasures Countermeasures designed to confuse the sensors of an adversary by jamming, blinding, or dazzling, or by reducing the radiations and reflections produced by one's own assets, or by using decoy targets in conjunction with real targets.

electro-optics The interaction between optics and electronics leading to the transformation of electrical energy into light, or vice versa, with the use of an optical device.

element A chemical substance that cannot be divided or decomposed into simpler substances by chemical means. A substance whose atoms all have the same number of protons (atomic number, Z) and electrons. There are 92 naturally occurring elements, and over 15 human-made or transuranium elements, such as plutonium (atomic number 94).

element set Specific information used to define and locate a particular satellite or orbiting object. Also called *Keplerian elements*. *See also* KEPLERIAN ELEMENTS; ORBITS OF OBJECTS IN SPACE.

ellipse A smooth, oval curve accurately fitted by the orbit of a satellite around a much larger mass. Specifically, a plane curve constituting the locus of all points the sum of whose distances from two fixed points (called *focuses* or *foci*) is constant; an elongated circle. The orbits of planets, satellites, planetoids, and comets are ellipses; the center of attraction (i.e., the primary) is at one focus. *See also* ORBITS OF OBJECTS IN SPACE.

elliptical galaxy A GALAXY with a smooth, elliptical shape without spiral arms and having little or no INTERSTELLAR gas and dust.

elliptical orbit A noncircular Keplerian orbit. *See also* ORBITS OF OBJECTS IN SPACE.

emission line A very small range of wavelengths (or frequencies) in the electromagnetic spectrum within which radiant energy is being emitted by a radiating substance. For example, the small band of wavelengths emitted by a low-density gas when it glows. The pattern of several emission lines is characteristic of the gas and is called the *emission spectrum*. Each radiating substance has a unique, characteristic emission spectrum.

emission spectroscopy Analytical spectroscopic methods that use the characteristic electromagnetic radiation emitted when materials are subjected to thermal or electrical sources for purposes of identification. These thermal or electrical sources excite the molecules or atoms in the sample of material to energy levels above the ground state. As the molecules or atoms return from these higher energy states, electromagnetic radiation is emitted in discrete, characteristic wavelengths or emission lines. The pattern and intensity of these emission lines creates a unique emission spectrum, which enables the analyst to identify the substance.

emission spectrum The distinctive pattern of EMISSION LINES produced by an ELEMENT (or other material substance) when it emits ELECTROMAGNETIC RADIATION in response to heating or other types of energetic activation. Each chemical element has a unique emission spectrum. *See also* SPECTRAL LINE; SPECTRUM.

emissivity (symbol: e or ε) The ratio of the radiant flux per unit area (sometimes called *emittance, E*) emitted by a body's surface at a specified wave-length (λ) and temperature (T) to the radiant flux per unit area emitted by a BLACKBODY radiator at the same temperature and under the same conditions. The greatest value that an emissivity may have is unity (1)—the emissivity value for a black-body radiator—while the least value for an emissivity is zero (0). *See also* STEFAN-BOLTZMANN LAW.

emittance (symbol: E) The radiant flux per unit area emitted by a body. *Spectral emittance* refers to emittance measured at a specified wavelength (λ). *See also* EMISSIVITY.

empirical Derived from observation or experiment.

encounter The close FLYBY or RENDEZVOUS of a SPACECRAFT with a TARGET body. The target of an encounter can be a natural CELESTIAL BODY (such as a PLANET, ASTEROID, or COMET) or a humanmade object (such as another spacecraft).

end effector In robotics, the tool or "hand" at the end of a robotic arm or manipulator. This is the gripper, actuator, or mechanical device by which objects are physically grasped or acted upon.

endoatmospheric Within Earth's atmosphere, generally considered to be at altitudes below 100 kilometers (62 mi).

endoatmospheric interceptor In aerospace defense, an interceptor rocket that attacks incoming reentry vehicles during their terminal flight phase within Earth's atmosphere. *See also* BALLISTIC MISSILE DEFENSE.

endothermic reaction In THERMODYNAMICS, a reaction in which thermal energy (heat) is absorbed from the surroundings; a reaction to which heat must be provided. *Compare* EXOTHERMIC REACTION.

energy (symbol: E) In general, the capacity to do work. Work is done when an object is moved by a force through a distance or when a moving object is accelerated. Energy can be manifested in many different forms, such as mechanical, thermal, chemical, electrical, and nuclear. The First Law of Thermodynamics states that energy is conserved, that is, it can neither be created nor destroyed, but simply changes form (including mass-energy transformations, in which matter is transformed into energy according to Einstein's formula $E = \Delta m\ c^2$). Typical units of energy are the joule (SI unit system) and the British thermal unit (Btu).

Potential energy arises from the position, state, or configuration of a body (or system). *Kinetic energy* is energy associated with a moving object. For example, in classical physics, a body of mass (m) moving at a velocity (V) has a kinetic energy of: $E_{kinetic} = \frac{1}{2}(m V^2)$. In THERMO-DYNAMICS the concept of *internal energy* is used to provide a macroscopic description of the energy within an object due to the microscopic behavior of molecules and atoms. Similarly, in thermodynamic analyses, *heat* or *thermal energy* is treated as disorganized energy in transit, while *work* is treated as organized energy in transit. Thermodynamically, a system never stores work or heat, since these are perceived as transitory energy transfer phenomena, which cease once a particular process has been completed.

Nuclear energy involves liberation of energy as a result of processes in the atomic nucleus, which often involve the transformations of small quantities of matter. *Solar energy* involves the Sun's electromagnetic energy output; while *geothermal energy* involves the thermal energy within Earth as a result of its molten core and the decay of certain radioactive nuclei in its crust.

energy conversion efficiency In rocketry, the efficiency with which a nozzle converts the energy of the working substance (i.e., the propellant) into kinetic energy. This can be expressed as the ratio of the kinetic energy of the jet leaving the nozzle to the kinetic energy of an ideal (hypothetical) jet leaving an ideal nozzle using the same working substance at the same initial state and under the same conditions of velocity and expansion.

energy management In aerospace operations and rocketry, the monitoring of fuel expenditure for the purposes of flight control and navigation.

energy satellite A very large space structure assembled in Earth orbit that takes advantage of the nearly continuous availability of sunlight to provide useful energy to a terrestrial power grid. As proposed most frequently, these large energy satellites would be located in geosynchronous orbit, where they would gather incoming sunlight (solar energy) and then convert it into electric energy by photovoltaic or solar-thermal conversion techniques. The electric energy then would be transmitted to power grids on Earth or to other energy consumption locations in CISLUNAR space (e.g., a space-based manufacturing center) by beams of microwave or laser radiation. *See also* SATELLITE POWER SYSTEM.

engine A machine or thermodynamic device that converts energy, especially thermal energy (heat), into useful work. *See also* ENERGY; HEAT ENGINE; ROCKET ENGINE.

engine bell The conical thrust chamber enclosure, or nozzle, of a rocket engine. It is located behind the combustion chamber. *See also* ROCKET ENGINE.

engine cut-off The specific time when a rocket engine is shut down during a flight. In the case of the space shuttle, for example, this time is referred to as MECO, or "main engine cut-off." Sometimes called *burnout*. *See also* ROCKET.

engine mount A structure used for attaching an engine to a vehicle. *See also* ROCKET.

engine spray The part of a launch pad deluge system that cools a rocket engine and its exhaust during launch, thereby preventing thermal damage to the launch vehicle and the launch pad structure.

enthalpy (symbol: H, h) A property of a thermodynamic system defined as:
$$H = U + pV$$
where H is the total enthalpy (joules), U is the total internal energy (joules), p is pressure (pascals), and V is volume (cubic meters). In many thermodynamic problems, the enthalpy function does not lend itself to a specific physical interpretation although it does have the dimension of energy (joules or British thermal units [BTUs]) or, as an intrinsic property (symbol: h), the dimensions of energy per unit

mass (joules/kg or BTU/lb$_{mass}$). However, since U and the product (pV) appear together in many thermodynamic analyses, the enthalpy function is often an advantageous property to describe a particular process.

entropy (symbol: S, s) 1. A measure of the extent to which the energy of a system is unavailable. A mathematically defined thermodynamic function of state, the increase in which gives a measure of the energy of a system that has ceased to be available for work during a certain process:
$$ds = [dq/T]_{REV}$$
where s is specific entropy (joules/kg-°K), T is absolute temperature (°Kelvin), and q is thermal energy (heat) input per unit mass (joules/kg), for reversible processes. 2. A simplified statistical thermodynamics definition of entropy as a disorder or uncertainty indicator, based on the work of the American scientist Josiah Willard Gibbs (1839–1903), is:
$$s = - k \Sigma p_i \ln p_i$$
where k is the Boltzmann constant and p_i is the probability of the *i*th quantum state of the system. 3. In communication theory, the average information content (nonredundant) of an object. *See also* SECOND LAW OF THERMODYNAMICS; THIRD LAW OF THERMODYNAMICS.

entry Region of the thrust chamber where the contour of the chamber converges to the nozzle throat area. *See also* ROCKET ENGINE.

entry corridor The acceptable range of flight-path angles for an aerospace vehicle or space probe to enter a planetary atmosphere and safely reach the surface or a predetermined altitude in the planet's atmosphere at which parachutes or other soft-landing assistance devices would be employed. If the flight-path angle is too steep (i.e., greater than the entry corridor range), the aerospace vehicle or probe would encounter excessive aerodynamic heating as it plunged into the thickening atmosphere and burn up. If the flight path angle is too shallow (i.e., less than the entry corridor range),

the vehicle or probe would bounce off the atmosphere, much like a smooth pebble skipping across a pond, and go past the target planet back into outer space. For example, a space shuttle orbiter undergoes its gliding atmospheric entry at an initial velocity of about 7.5 kilometers per second (4.66 mps). The flight-path angle is typically –1.2 degrees at the entry interface altitude of 122 kilometers (75.8 mi). *See also* SPACE TRANSPORTATION SYSTEM.

envelope 1. In aerospace engineering, the external boundary defining the limits on the dimensions of the component, subsystem, or system. 2. In aerospace operations, the bounds within which a certain aerospace vehicle, spacecraft, or probe can operate—for example, a *flight envelope.* 3. In mathematics, a curve that bounds the values that a variable can assume.

environment An external condition, or the sum of such conditions, in which a piece of equipment, a living organism, or a system operates; for example, *vibration environment, temperature environment, radiation environment,* or *space environment.* Environments can be natural or human-engineered (artificial) and often are specified by a range of values.

environmental chamber A test chamber in which temperature, pressure, fluid content and composition, humidity, noise, and movement can be controlled so as to simulate the different operational environments that an aerospace component, subsystem, or system might be required to perform in.

environmental satellite *See* EARTH-OBSERVING SPACECRAFT.

eolian Pertaining to, carried by, or caused by the wind. For example, *eolian sand dunes* on the surface of Mars.

ephemeris A collection of data about the predicted positions (or apparent positions) of celestial objects, including artificial satellites, at various times in the future. A satellite ephemeris might contain the orbital elements of satellites and their predicted changes. *See also* ORBITS OF OBJECTS IN SPACE.

epicycle A small circle whose center moves along the circumference of a larger circle, called the deferent. Ancient astronomers, such as PTOLEMY, used the epicycle in an attempt to explain the motions of celestial bodies in their GEOCENTRIC (nonheliocentric) models of the SOLAR SYSTEM.

equation of state (EOS) An equation relating the thermodynamic temperature (T), pressure (p), and volume (V) for an amount of substance (n) in thermodynamic equilibrium. A large number of such equations have been developed. Of these, perhaps the simplest and most widely known is the *ideal* or *perfect gas* equation of state:
$$p V = n \Re T$$
where $\Re$ is the universal gas constant, which has a value of 8314.5 joules per kilogram-mole per degree kelvin [J/(kmol-K)].

equations of motion A set of equations that gives information regarding the motion of a body or of a point in space as a function of time when initial position and initial velocity are known. *See also* NEWTON'S LAWS OF MOTION.

equator An imaginary circle around a celestial body that is everywhere equidistant (90 degrees) from the poles, defining the boundary between the Northern and Southern Hemisphere.

equatorial bulge The excess of a PLANET's equatorial diameter over its polar diameter. The increased size of the equatorial diameter is caused by CENTRIFUGAL FORCE associated with ROTATION about the polar AXIS.

equatorial orbit An orbit with an inclination of zero degrees. The plane of an equatorial orbit contains the equator of the primary body.

equatorial satellite A satellite whose orbital plane coincides, or nearly coincides, with the equatorial plane of the primary body.

equilibrium The state of a thermodynamic system in which there are no net changes—that is, a system in which forces, reactions, energy flows, and transitions balance each other out. For example, the state within a gas in which reactions between components are proceeding equally in the forward and backward direction so that the relative concentrations of the components would not change with time.

equilibrium composition In rocketry, the chemical composition that the exhaust gas would attain, if given sufficient time for the reactants to achieve chemical balance.

equinox One of two points of intersection of the ECLIPTIC and the celestial EQUATOR that the SUN occupies when it appears to cross the celestial equator (that is, has a DECLINATION of 0°). In the Northern Hemisphere, the Sun appears to go from south to north at the vernal equinox, which occurs on or about March 21st. Similarly, the Sun appears to travel from north to south at the autumnal equinox, which occurs on or about September 23 each year. The dates are reversed in the Southern Hemisphere.

erector A vehicle used to support a rocket for transportation and to place the rocket in an upright position within a gantry.

erg The unit of energy or work in the centimeter-gram-second (c.g.s.) unit system. An erg is the work performed by a force of 1 dyne acting through a distance of 1 centimeter.
$$1 \text{ erg} = 10^{-7} \text{ joule}$$

ergometer A bicyclelike instrument used for measuring muscular work and for exercising "in place." Astronauts and cosmonauts use specially designed ergometers to exercise while on extended orbital flights.

ergonomics   That portion of human factors engineering that is concerned with people-machine interfaces and studies the relationship between workers and their working environment.

erosion   1. In general, the progressive loss of original material from a solid due to the mechanical interaction between that surface and a fluid or the impingement of liquid droplets or solid particles. 2. In geology, the collective group of processes whereby rock material is removed and transported. The four major agents of erosion are: running water, wind, glacial ice, and gravity. 3. In environmental science, the wearing away of the land surface by wind and water. Erosion occurs naturally from weather or runoff but can be intensified by land-clearing practices associated with farming, residential or industrial development, road-building, or timber-cutting.

erosive burning   The increased burning of solid propellant that results from the scouring influence of combustion products moving at high velocity over the burning surface. *See also* ROCKET.

escape rocket   A small rocket engine attached to the leading end of an escape tower, which is used to provide additional thrust to the crew capsule to obtain separation of this capsule from the expendable booster vehicle in the event of a launch pad abort or emergency. *See also* MERCURY PROJECT.

escape tower   A trestle tower placed on top of a crew (space) capsule, which during liftoff connects the capsule to the escape rocket. After a successful liftoff and ascent, the escape tower and escape rocket are separated from the capsule. *See also* ESCAPE ROCKET; MERCURY PROJECT.

escape velocity (symbol: V_e)   The minimum velocity needed by a object to climb out of the gravity well (overcome the gravitational attraction) of a celestial body. From classical Newtonian mechanics, the escape velocity from the surface of

a celestial body of mass (M) and radius (R) is given by the equation:
$$V_e = \sqrt{(2\ G\ M)/R}$$
where G is the universal constant of gravitation (6.672×10^{-11} N – m²/kg²), M is the mass of the celestial object (kg), N is Newton (unit of force in the SI system), and R is the radius of the celestial object (m). The table below presents the escape velocity for various objects in the solar system (or an estimated equivalent V_e for those celestial bodies that do not possess a readily identifiable solid surface, such as the giant outer planets and the Sun).

Europa   The smooth, ice-covered MOON of Jupiter, discovered by GALILEO GALILEI in 1610 and currently thought to have a liquid water ocean beneath its frozen surface.

European Space Agency (ESA)   An international organization that promotes the peaceful use of OUTER SPACE and cooperation among the European member states in space research and applications. ESA's 15 member states are Austria, Belgium, Denmark, Finland, France, Germany, Ireland, Italy, the Netherlands, Norway, Portugal, Spain, Sweden, Switzerland, and the United Kingdom. Canada has a special status and

Escape Velocity for Various Objects in the Solar System	
Celestial Body	*Escape Velocity (V_e) (km/sec)*
Earth	11.2
Moon	2.4
Mercury	4.3
Venus	10.4
Mars	5.0
Jupiter	~ 61
Saturn	~ 36
Uranus	~ 21
Neptune	~ 24
Pluto	~ 1
Sun	~ 618

Source: Developed by the author, based on NASA and other astrophysical data.

participates in some projects under a cooperation agreement. ESA has its headquarters in Paris, France. The European Space Research and Technology Center (ESTEC) is situated in Noordwijk, the Netherlands, and serves as the design hub for most ESA spacecraft. The European Space Operations Center (ESOC) is located in Darmstadt, Germany, and is responsible for operating and controlling ESA SATELLITES in ORBIT. The European Space Research Institute (ESRIN) is located in Frascati, Italy (near Rome), and serves as ESA's information technology center. In particular, ESRIN collects, stores, and distributes satellite-derived scientific data among the member states. The European Astronauts Center (EAC) is situated in Cologne, Germany, and trains European astronauts for future missions. ESA has liaison offices in the United States, Russia, and Belgium. In addition, ESA maintains a launch complex in Kourou, French Guiana, and tracking stations in various areas around the world.

evaporation The physical process by which a liquid is transformed to the gaseous state (i.e., the vapor phase) at a temperature below the boiling point of the liquid. *Compare with* SUBLIMATION.

evapotranspiration The loss (discharge) of water from Earth's surface to its atmosphere by evaporation from bodies of water or other surfaces (e.g., the soil) and by transpiration from plants.

event horizon The point of no return for a BLACK HOLE; the distance from a black hole from within which nothing can escape.

evolved star A STAR near the end of its lifetime, when most of its HYDROGEN fuel has been exhausted; a star that has left the main sequence. *See also* MAIN-SEQUENCE STAR.

excited state The state of a molecule, atom, electron, or nucleus when it possesses more than its normal (i.e., ground state) energy. Excess nuclear energy often is released as a gamma ray. Excess molecular energy may appear as thermal energy (heat) or fluorescence. *Compare with* GROUND STATE.

exhaust plume Hot gas ejected from the thrust chamber of a rocket engine. The plume expands as a launch vehicle ascends through Earth's atmosphere, exposing the engine and the vehicle (especially the aft portion) to a greater thermal radiation area.

exhaust stream The stream of gaseous, atomic, or radiant particles that emit from the nozzle of a rocket or other reaction engine.

exhaust velocity The velocity (relative to the rocket) of the gases and particles that exit through the nozzle of a rocket motor. *See also* ROCKET.

exit The aft end of the divergent portion of a rocket nozzle; the plane at which the exhaust gases leave the nozzle. Also called *exit plane*. *See also* NOZZLE; ROCKET.

exit pressure Pressure of the exhaust gas as it leaves a rocket's nozzle; gas pressure at the exit plane of a nozzle. *See also* NOZZLE; ROCKET.

exoatmospheric Outside Earth's atmosphere, generally considered as occurring at altitudes above 100 kilometers (62 mi).

exoatmospheric interceptor In ballistic missile defense (BMD), an interceptor rocket that destroys incoming reentry vehicles above Earth's atmosphere during the late midcourse phase of their flight path. *See also* BALLISTIC MISSILE DEFENSE.

exobiology The multidisciplinary field that involves the study of extraterrestrial environments for living organisms, the recognition of evidence of the possible existence of life in these environments, and the study of any nonterrestrial (i.e., extraterrestrial) life-forms that may be encountered. The challenges of exobiology are being approached from several different directions. First, material samples from alien worlds in our solar system can be obtained for study on Earth, as was accomplished during the Apollo lunar expeditions (1969–1972); or such samples can be studied on the spot by robot explorers, as was accomplished by the Viking Landers (1976). Lunar rock and soil samples have not revealed any traces of life, while the biological results of the Viking Lander experiments involving Martian soils are still unclear and subject to debate.

A second major approach in exobiology involves conducting experiments in terrestrial or space laboratories that attempt either to simulate the primeval conditions that led to the formation of life on Earth and extrapolate these results to other planetary environments or to study the response of terrestrial organisms under environmental conditions found on other worlds.

A third general approach in exobiology involves an attempt to communicate with, or at least listen for signals from, other intelligent life-forms within our galaxy. This effort often is called the search for extraterrestrial intelligence (or SETI, for short). Contemporary SETI activities throughout the world have as their principal aim to listen for evidence of extraterrestrial radio signals generated by intelligent alien civilizations. Also called *astrobiology. See also* APOLLO PROJECT; SEARCH FOR EXTRATERRESTRIAL INTELLIGENCE; VIKING PROJECT.

exoskeleton In robotics, the external supporting structure of a machine, especially an anthropomorphic device. Certain hard-shelled space suits and deep-sea diver suits sometimes are called exoskeletons. The term derives from a biological description of the rigid external covering found in certain animals, such as mollusks. The exoskeleton supports the animal's body and provides attachment points for its muscles.

exosphere The outermost region of Earth's atmosphere. Sometimes called the *region of escape. See also* ATMOSPHERE.

exothermic reaction A chemical or physical reaction in which thermal energy (heat)is released into *Compare with* ENDOTHERMIC REACTION.

expandable space structure A structure that can be packaged in a small volume for launch and then erected to its full size and shape outside Earth's atmosphere.

expanding universe Any model of the UNIVERSE in modern COSMOLOGY that has the distance between widely separated celestial objects (for example, distant galaxies) continuing to grow or expand with time.

expansion geometry The contour of a rocket nozzle, from the throat to the exit plane. *See also* NOZZLE; ROCKET.

expansion ratio (symbol: ε) For a converging-diverging (de Laval-type) rocket nozzle, the expansion ratio (ε) is defined as the ratio of the exit plane area (A_{exit}) to the throat area (A_{throat}). Also called *nozzle area expansion ratio.*

$$\varepsilon = A_{exit}/A_{throat}$$

See also NOZZLE; ROCKET.

expendable launch vehicle (ELV) A ground-launched propulsion vehicle, capable of placing a payload into Earth orbit or Earth-escape trajectory, whose various stages are not designed for or intended for recovery and/or reuse. The U.S. Atlas, Delta, Titan, and Scout rockets are examples of expendable launch vehicles. *See also* LAUNCH VEHICLE.

experimental vehicle (symbol: X) An aircraft, missile, or aerospace vehicle in the research, development, and testing portion of its life cycle and not yet approved for operational use. Often designated by the symbol X. *See also* X-1; X-15.

exploding bridgewire (EBW) An initiator consisting of a small-diameter wire that explodes when high voltage is applied; used to ignite small retrorockets, stage separation rockets, and explosive bolt devices.

exploding galaxies Violent, very energetic explosions centered in certain galactic nuclei where the total MASS of ejected material is comparable to the mass of some 5 million average-size SUNLIKE STARS.

JETS of GAS 1,000 LIGHT-YEARS long are also typical.

Explorer I   The first U.S. EARTH satellite. (See Figure 32). *Explorer I* was launched successfully from CAPE CANAVERAL AIR FORCE STATION on January 31, 1958 by a Juno I four-stage configuration of the Jupiter C launch vehicle. It was a joint project of the Army Ballistic Missile Agency (ABMA) and the Jet Propulsion Laboratory (JPL). ABMA supplied the Jupiter C LAUNCH VEHICLE while JPL supplied the fourth stage rocket and the satellite itself. Dr. James Van Allen of the State University of Iowa provided the instruments, which detected the inner of Earth's two major trapped radiation belts for the first time. *Explorer I* measured three phenomena: COSMIC RAY and radiation levels, the temperature in the spacecraft, and the frequency of collisions with micrometeorites. There was no provision for data storage, and therefore the satellite transmitted information continuously. Earth's trapped radiation belts, discovered by the *Explorer I*, are called the *Van Allen Belts* in honor of Dr. Van Allen.

Explorer satellites   The name "Explorer" has been used by NASA to designate a large series of scientific satellites used to "explore the unknown." Explorer spacecraft were used by NASA since 1958 to study: (1) Earth's atmosphere and ionosphere; (2) the magnetosphere and interplanetary space; (3) astronomical and astrophysical phenomena; and (4) Earth's shape, magnetic field, and surface. Many of the Explorer satellites had project names that were used before they were orbited and then were replaced by Explorer designations once they were placed in orbit. Other Explorer satellites, especially the early ones, were known before launch and after achieving orbit simply by their numerical designations. A brief (but not all inclusive) listing will help illustrate the great variety of important scientific missions performed by these satellites: Aeronomy Explorer, Air Density Explorer, Interplanetary Monitoring Platform (IMP), Ionosphere Explorer, Meteoroid Technol-

Figure 32. Jubilant team leaders (from left to right) Dr. Pickering (JPL), Dr. Van Allen (State University of Iowa), and Dr. von Braun (ABMA) hold aloft a model of the *Explorer I* spacecraft and its solid-rocket final stage. The first U.S. Earth satellite was successfully launched from Cape Canaveral, Florida, on January 31, 1958. *(Courtesy of NASA)*

ogy Satellite (MTS), Radio Astronomy Explorer (RAE), Solar Explorer, and Small Astronomy Satellite (SAS).

Geodetic Satellites (GEOS) also were called *Geodetic Explorer Satellites*. For example, GEOS 1 (*Explorer 29*, launched on November 6, 1965) and GEOS 2 (*Explorer 36*, launched on January 11, 1968) refined our knowledge of Earth's shape and gravity field. SAS-A, an X-ray Astronomy Explorer, became *Explorer 42* when launched on December 12, 1970 by an Italian launch crew from the San Marco platform off the coast of Kenya, Africa. Because it was launched on Kenya's Independence Day, this small spacecraft was also called *Uhuru* (the Swahili word for "Freedom"). It successfully mapped the universe in X-ray wavelengths for four years, discovering X-ray pulsars and providing preliminary evidence of the existence of black holes. *See also* EXPLORER I.

explosion (chemical) The rapid production of a large quantity of hot gas due to very energetic chemical reactions that release thermal energy (heat) and cause a sudden increase in pressure. The volume of this hot gas in much larger than the original volume of the chemical substance that caused the explosion. Trinitrotoluene (TNT) is an example of an explosive chemical substance.

explosion-proof apparatus An apparatus enclosed in a special protective case that is capable of: withstanding an explosion of a specific gas or vapor that may occur within it; or preventing the ignition of a specified gas or vapor surrounding the enclosure by internal sparks, flashes, or explosions of the gas or vapor within. Also an apparatus that operates at such an external temperature that a surrounding flammable atmosphere will not be ignited.

explosive A substance or mixture of substances (i.e., trinitrotoluene (TNT) or gunpowder) that can undergo a rapid chemical change without an outside supply of oxygen and with the release of large quantities of energy—generally accompanied by the evolution of a large volume of hot gases.

explosive-activated device (EAD) An electrically initiated pyrotechnic device, such as an explosive bolt. Used to separate rocket stages or to separate a spacecraft from its rocket vehicle.

explosive bolt A bolt incorporating an explosive charge that can be detonated on command (usually by an electrical signal or pulse), thereby destroying the bolt and releasing pieces of aerospace equipment it was retaining. Explosive bolts and spring-loaded mechanisms can be used to quickly separate launch vehicle stages or a spacecraft from its propulsion system.

explosive decompression A very rapid reduction of air pressure inside the pressurized portion (i.e., crew compartment) of an aircraft, aerospace vehicle, or spacecraft. For example, collision with a large piece of space debris might puncture the wall of one of the pressurized modules on a space station, causing an explosive decompression situation within that module. Air locks would activate, sealing off the stricken portion of the pressurized space habitat.

explosive valve A valve having a small explosive charge that, when detonated, provides high-pressure gas to change the valve position. Also known as a *squib valve.*

exposure suit A garment designed to protect a person from a harmful natural environment, such as frigid water.

external tank (ET) The large tank that contains the propellants for the three space shuttle main engines (SSMEs) and forms the structural backbone of the space shuttle flight system in the launch configuration. At liftoff, the external tank absorbs the total 28,580-kilonewton (6,425,000-pounds-force [lbf]) thrust loads of the three main engines and the two solid rocket boosters (SRBs). When these SRBs separate at an altitude of approximately 44 kilometers (km) (27 miles [mi]), the orbiter vehicle, with the main engines still burning, carries the ET piggyback to near orbital velocity. Then, at an altitude of approximately 113 kilometers (70 mi), some 8.5 minutes into the flight, the now nearly empty tank separates from the orbiter and falls in a preplanned trajectory into a remote area of the Indian Ocean. The ET is the only major expendable element of the space shuttle.

The three main components of the ET are an oxygen tank (located in the forward position), a hydrogen tank (located in the aft position), and a collarlike intertank, which connects the two propellant tanks, houses instrumentation and processing equipment, and provides the attachment structure for the forward end of the solid rocket boosters. In June 1998, a lighter version of the external tank entered service. Made of a special lightweight material called aluminum lithium, the new tank has a mass of approximately 26,300 kilograms (kg) [58,000 pounds-mass (lbm)]—about 3,400 kilogram (7,500 lbm) less mass than the previous version. The reduced tank mass means heavier payloads can be carried into orbit by the space shuttle. *See also* SPACE TRANSPORTATION SYSTEM.

extragalactic Occurring, located, or originating beyond our galaxy (the Milky Way); typically, farther than 100,000 light-years distant.

extragalactic astronomy A branch of astronomy that started around 1930 that studies everything in the universe outside of our galaxy, the Milky Way.

extrasolar Occurring, located, or originating outside of our solar system; as, for example, extrasolar planets.

extrasolar planets Planets that belong to a star other than the Sun.

extraterrestrial (ET) Refers to something that occurs, is located, or originates outside of the planet Earth and its atmosphere.

extraterrestrial catastrophe theory The hypothesis that a large ASTEROID or COMET struck EARTH some 65 million years ago, causing global environmental consequences that annihilated more than 90% of all animal species then living, including the dinosaurs.

extraterrestrial contamination The contamination of one world by life-forms, especially microorganisms, from another world. Taking EARTH'S BIOSPHERE as the reference, planetary contamination is called *forward contamination* when an alien world is contaminated by contact with TERRESTRIAL organisms and *back contamination* when alien organisms are released into Earth's biosphere.

extraterrestrial life Life-forms that may have evolved independent of and now exist outside of the terrestrial biosphere. *See also* EXOBIOLOGY.

extraterrestrial resources The resources to be found in space that could be used to help support an extended human presence and eventually become the physical basis for a thriving solar system–level civilization. These resources include unlimited solar energy, a full range of raw materials (from the Moon, asteroids, comets, Mars, numerous moons of outer planets, etc.), and an environment that is both special (e.g., access to high vacuum, microgravity, physical isolation from terrestrial biosphere) and reasonably predictable.

extravehicular activity (EVA) Extravehicular activity, or EVA, may be defined as the activities conducted in space by an ASTRONAUT or COSMONAUT outside the protective environment of his/her spacecraft, aerospace vehicle, or space station. In the U.S. space program, the first EVA was performed by Astronaut Edward H. White II on June 3, 1965 when he left the protective environment of his *Gemini IV* space capsule and ventured into space (while constrained by an umbilical tether). Since that historic demonstration, EVA has been used successfully during a variety of American and Russian space missions to make critical repairs, perform inspections, help capture and refurbish failed satellites, clean optical surfaces, deploy equipment, and retrieve experiments. The term "EVA," as applied to the SPACE SHUTTLE, includes all activities for which crew members don their spacesuits and life support systems and then exit the orbiter vehicle's crew cabin to perform operations internal or external to the cargo bay.

extreme ultraviolet (EUV) The region of the electromagnetic spectrum corresponding to wavelengths between 10 and 100 nanometers (100 and 1,000 angstroms). *See also* ELECTROMAGNETIC SPECTRUM.

Extreme Ultraviolet Explorer NASA's 70th Explorer-class satellite launched from Cape Canaveral Air Force Station in 1992 into a 525-kilometer (328-mile) Earth orbit. It provided astronomers with a view of the relatively unexplored region of the electromagnetic spectrum—the extreme ultraviolet

(EUV) region (i.e., 10 to 100 nanometers wavelength). It gathered important data about sources of EUV radiation within the "Local Bubble"—a hot, low-density region of the Milky Way galaxy (including our Sun) that is the result of a supernova explosion some 100,000 years ago. Interesting EUV sources include white dwarf stars and binary star systems in which one star is siphoning material from the outer atmosphere of its companion.

extremophile A hardy (TERRESTRIAL) microorganism that can exist under extreme environmental conditions, such as in frigid polar regions or boiling hot springs. Astrobiologists speculate that similar (EXTRATERRESTRIAL) microorganisms might exist elsewhere in this SOLAR SYSTEM, perhaps within subsurface biological niches on MARS or in a suspected liquid water ocean beneath the frozen surface of EUROPA.

eyeballs in, eyeballs out Aerospace jargon, derived from test pilots and used to describe the acceleration experienced by an ASTRONAUT or COSMONAUT at liftoff or when the retrorockets fire. The experience at liftoff is "eyeballs in" (positive g because of vehicle acceleration), while the experience when the retrorockets fire is "eyeballs out" (negative g because of space capsule or aerospace vehicle deceleration).

eyepiece A magnifying LENS that helps an observer view the IMAGE produced by a TELESCOPE.

F

facula A bright region of the SUN's PHOTOSPHERE.

Fahrenheit (symbol: F) A relative temperature scale with the freezing point of water (at atmospheric pressure) given a value of 32°F and the boiling point of water (at atmospheric pressure) assigned the value of 212°F.

fail-operational The ability of an aerospace system to sustain a failure and retain full operational capability for safe mission continuation.

fail-safe 1. Term for philosophy in the design of aerospace system hardware that seeks to avoid the compounding of failures; fail-safe design provisions ensure that the component (e.g., a valve element in the liquid-fuel rocket propulsion system) will move to a predetermined "SAFE" position if electric, pneumatic, or hydraulic power is lost. 2. The ability of an aerospace system to sustain a failure and retain the capability to terminate the mission successfully. For ground support equipment (GSE), the ability to sustain a failure without causing loss of space vehicle systems or loss of personnel capability.

fairing A structural component of a rocket or aerospace vehicle designed to reduce drag or air resistance by smoothing out nonstreamlined objects or sections.

Falcon An air-to-air guided missile (AAGM) that can be carried either internally or externally on interceptor aircraft.

fallaway section A section of a launch vehicle or rocket that is cast off and separates from the vehicle during flight, especially a section that falls back to Earth.

farad (symbol: F) The SI UNIT of electrical capacitance. It is defined as the capacitance of a capacitor whose plates have a potential difference of one volt when charged by a quantity of electricity equal to one coulomb. This unit is named after the 19th-century British scientist Michael Faraday (1791–1867), who was a pioneer in the field of electromagnetism. Since the farad is too large a unit for typical applications, submultiples—such as the *microfarad* (10^{-6}F), the *nanofarad* (10^{-9}F), and the *picofarad* (10^{-12}F)—are encountered frequently.

farside The side of the Moon that never faces Earth. *See also* MOON.

fast-burn booster A ballistic missile that can burn out much more quickly than current versions, possibly before exiting the atmosphere entirely. Such a rapid booster burnout complicates a boost-phase ballistic missile defense (BMD) system. *See also* BALLISTIC MISSILE DEFENSE.

fatigue 1. A weakening or deterioration of metal or other material occurring under load, especially under repeated cyclic or continued loading. Self-explanatory compounds of this term include: *fatigue crack, fatigue failure, fatigue load, fatigue resistance,* and *fatigue test.* 2. The state of the human organism after exposure to any type of physical or psychological stress, as, for example, *pilot fatigue.*

fault tolerance The capability of an aerospace system to function despite one or more critical failures; usually achieved by use of redundant circuits or functions and/or reconfigurable components.

feasibility study A study to determine whether an aerospace mission plan is within the capacity of the resources and/or technologies that can be made available—for example, a NASA feasibility study involving a robotic soil sample return mission to the surface of Mars.

feedback 1. The return of a portion of the output of a device to the input. *Positive feedback* adds to the input; *negative feedback* subtracts from the input. 2. Information concerning results or progress returned to an originating source. 3. In aeronautics, the transmittal of forces initiated by aerodynamic action on control surfaces or rotor blades to the cockpit controls.

femto- (symbol: f) The SI prefix for 10^{-15}. This prefix is used to designate very small quantities, such as a *femtosecond (fs)*, which corresponds to 10^{-15} second—a very brief flash of time.

ferret satellite A military SPACECRAFT designed for the detection, location, recording, and analyzing of ELECTROMAGNETIC RADIATION (for example, enemy RADIO FREQUENCY [RF] transmissions).

ferry flight An (in-the-atmosphere) flight of a space shuttle orbiter mated on top of the Boeing 747 shuttle carrier aircraft. *See also* SPACE TRANSPORTATION SYSTEM (STS).

field 1. In physics, a region in which an object experiences a *force* due to the presence of other objects (i.e., a *gravitational field*) or due to the presence of certain phenomena, such as a *magnetic field* and/or *electric field*. 2. In information and science, the smallest unit of data in a record. 3. In remote sensing, the field of view of a sensor.

field of view (FOV) The area or solid angle that can be viewed through or scanned by an optical instrument.

film badge A "badgelike" packet of ionizing radiation-sensitive materials (including certain types of photographic film) used for measuring and permanently recording the radiation dosage a person has received. An astronaut or cosmonaut often wears a film badge during spaceflight to record his/her exposure to the space radiation environment.

film cooling The cooling of a body or surface, such as the inner surface of a rocket's combustion chamber, by maintaining a thin fluid layer over the affected area.

filter 1. A thin layer of selective material placed in front of a detector that lets through only a selected color (e.g., a blue-light filter for a photometer), group of electromagnetic radiation wavelengths (e.g., thermal infrared band radiation between 10 and 11 micrometers wavelength), group of photon energies (e.g., gamma rays with energies between 1.0 and 1.1 MeV [million electron volts]), or a desired (often narrow) range of particle energies (e.g., solar protons between 2.0 and 2.2 MeV). Narrow-band filters for can be quite selective, passing radiation only in a very narrow region of the electromagnetic spectrum—for example, a 427.8 nanometer (4278 Å) wavelength narrow-band optical filter. Other radiation filters are reasonably broad—for example, a filter that passes an entire "color" in the visible spectrum (i.e., blue-light filter or a red-light filter). 2. A device that removes suspended particulate matter from a liquid or a gas. This removal process is accomplished by forcing the fluid (with the suspended matter) through a permeable material (i.e., one that has many tiny holes or pores). The permeable material retains the solid matter while permitting the passage of the fluid that has now been "filtered" to some level of purity. 3. In electronics, a device that transmits only part of the incident electromagnetic energy and thereby may change its spectral distribution. *High-pass filters* transmit energy above a certain frequency, while *low-pass filters* transmit energy below a certain frequency. *Band-pass filters* transmit energy within a certain frequency band, while *band-stop filters* transmit only energy outside a specific frequency band.

fin 1. In aeronautics, a fixed or adjustable airfoil attached longitudinally to an aircraft, missile, rocket, or similar body to provide a stabilizing effect. 2. In heat transfer, a projecting flat plate or structure that facilitates thermal energy transfer, such as a *cooling fin.*

fire 1. In chemistry, the persistent and rapid reaction that liberates thermal energy (heat) and produces light; the combination of a combustible substance with oxygen that releases thermal energy (i.e., the *exothermic* process involving oxygen and a combustible substance). 2. In aerospace operations, a term meaning to ignite a rocket engine or to launch a rocket vehicle. 3. To detonate the main explosive charge by means of a firing system. 4. In military operations, the command given to discharge a weapon or weapons.

fire arrow An early gunpowder ROCKET attached to a large bamboo stick; developed by the Chinese about 1,000 years ago to confuse and startle enemy troops.

Firebee A remotely controlled target drone that is powered by a turbojet engine. It achieves high subsonic speeds and is designed to be ground-launched or air-launched. The Firebee is used to test, train, and evaluate weapon systems employing surface-to-air and air-to-air missiles. Also called the *BQM-34.*

First Law of Thermodynamics The principle of conservation of energy. For a control mass (i.e., a thermodynamic system of specified matter), the first law can be written in the form of an energy balance:

Energy Output − Energy Input =
Change in Energy Storage
$$W - Q = \Delta E_{storage}$$
where W is the amount of energy transferred out of the control mass as work, Q is the amount of energy transferred into the control mass as heat (thermal energy), and $\Delta E_{storage}$ is the change (increase or decrease) in energy of the control mass (i.e., $E_{final} - E_{initial}$). In thermodynamics, the use of an energy balance often is referred to as a *first-law analysis. See also* THERMODYNAMICS.

first strike The first offensive move of a war. (This term often is associated with nuclear operations.)

fissile material While sometimes used as a synonym for *fissionable material,* this term has a more restricted meaning, namely, any material fissionable (i.e., capable of undergoing *fission*) by neutrons of all energies, including (and especially) thermal (slow) neutrons as well as fast neutrons. Uranium-235 ($^{235}_{92}U$), uranium-233 ($^{233}_{92}U$), and plutonium-239 ($^{239}_{94}Pu$) are fissile. *See also* FISSION (NUCLEAR).

fission (nuclear) The process during which the NUCLEUS of certain heavy RADIOISOTOPES, such as uranium-235, captures a NEUTRON, becomes an unstable compound nucleus, and soon breaks apart. As the compound nucleus splits, or fissions, it forms two lighter nuclei (called *fission products*) and also releases a large amount of ENERGY (about 200 million ELECTRON VOLTS per reaction) plus additional neutrons and GAMMA RAYS.

fission products Medium-mass nuclei (fission fragments) formed by the splitting or fission of heavy nuclei, such as uranium-235 or plutonium-239. Also, the nuclides formed by the radioactive decay of the primary fission products. *See also* FISSION (NUCLEAR).

fixed-area exhaust nozzle An exhaust nozzle exit that remains constant in area.

fixed satellite An Earth satellite that orbits from west to east at such speed as to remain constantly over a given place on Earth's equator. *See also* GEOSTATIONARY EARTH ORBIT.

fixed stars A term used by early astronomers to distinguish between the apparently motionless background STARS and the wandering stars (PLANETS). Modern astronomers now use this term to describe stars that have no detectable PROPER MOTION.

flame bucket A deep, cavelike construction built beneath a launch pad. It is open at the top to receive the hot gases from the rocket or missile positioned above it and is also open on one to three sides below, with a thick metal fourth side bent toward the open side(s) so as to deflect the exhausting gases. *See also* FLAME DEFLECTOR.

flame deflector 1. In a vertical launch pad, any of the obstructions of various designs that intercept the hot exhaust gases of the rocket engine(s), thereby deflecting them away from the launch pad structure or the ground. The flame deflector may be a relatively small device fixed to the top surface of the launch pad, or it may be a heavily constructed piece of metal mounted as a side and bottom of a flame bucket. In the latter case, the flame deflector also may be perforated with numerous holes connected with a source of water. During thrust buildup and the beginning of launch, a deluge of water pours from the holes in such a deflector to keep it from melting. 2. In a captive (hot) test of a rocket engine, an elbow in the exhaust conduit or flame bucket that deflects the flame into the open.

flare A bright eruption from the Sun's chromosphere. Solar flares may appear within minutes and fade within an hour—or else linger for days. They cover a wide range of intensity and size. Flares can cause ionospheric disturbances and telecommunications (radio) fadeouts on Earth. Solar flares also represent a major ionizing radiation hazard to astronauts or cosmonauts walking on the lunar surface or traveling through interplanetary space (i.e., traveling outside the relative protection of Earth's magnetosphere). *See also* SUN.

fleet ballistic missile (FBM) An INTER-CONTINENTAL BALLISTIC MISSILE, usually equipped one or more NUCLEAR WEAPONS, carried by and launched from a submarine. The United States Navy equips a special class of nuclear powered submarines, called FLEET BALLISTIC MISSILE SUBMARINES, with such nuclear WARHEAD–carrying MISSILES. The Trident II (designated D-5) is the sixth generation member of the U.S. Navy's fleet ballistic missile (FBM) program that started in 1956. Since then the Polaris (A1), Polaris (A2), Polaris (A3), Poseidon (C3), and Trident I (C4) have provided a significant deterrent against nuclear aggression. At present, the U.S. Navy deploys Poseidon (C3) and Trident I (C4) missiles, having retired the Polaris family of fleet ballistic missiles. The Trident II (D-5) FBM was first deployed in 1990 on the *USS Tennessee* (SSNB 734). The Trident II (D-5) FBM is a three-STAGE, SOLID PROPELLANT, inertially guided FBN with a range of more than 7,400 kilometers (4,000 NAUTICAL MILES). The new Trident/Ohio class fleet ballistic missile submarines each carry 24 Trident IIs. These missiles can be launched while the submarine is underwater or on the surface. Also called a *submarine-launched ballistic missile* (SLBM).

fleet ballistic missile submarine A nuclear-powered submarine designed to deliver ballistic missile attacks against assigned targets from either a submerged or surface condition. Also designated by the U.S. Navy as *SSBN*.

Fleet Satellite Communications (FLT-SATCOM) System A family of military COMMUNICATIONS SATELLITES operated in GEOSTATIONARY ORBIT. The system was designed to provide a near-global operational communications network capable of supporting the ever-increasing information transfer needs of the United States Navy and Air Force. The U.S. Navy began replacing and upgrading the early SPACECRAFT in its ULTRAHIGH FREQUENCY (UHF) communications satellite network in the 1990s with a CONSTELLATION of customized SATELLITES, collectively referred to as the UHF Follow-On (UFO) spacecraft series. These spacecraft are launched by an evolving family of ATLAS ROCKETs from CAPE CANAVERAL AIR FORCE STATION. For example, the UFO spacecraft, designated F-10 in the series, was successfully placed into ORBIT on November 22, 1999, by an Atlas IIA LAUNCH VEHICLE.

flexible response The capability of aerospace forces for effective reaction to any enemy threat or attack with actions appropriate and adaptable to the existing circumstances.

flexible spacecraft A space vehicle (usually a space structure or rotating satellite) whose surfaces and/or appendages may be subject to and can sustain elastic flexural deformations (vibrations).

flight The movement of an object through the atmosphere or through space, sustained by aerodynamic, aerostatic, or reaction forces or by orbital speed; especially, the movement of a human-operated or human-controlled device, such as a rocket, space probe, aerospace vehicle, or aircraft, as in *space flight, high-altitude flight*, or *high-speed flight*.

flight acceptance test(s) The environmental and other tests that spacecraft, subsystems, components, or experiments scheduled for space flight must pass before launch. These tests are planned to approximate anticipated environmental and operational conditions and have the purpose of detecting flaws in material and workmanship.

flight control room The facility used by the flight control team for direct support of a space shuttle flight from prelaunch countdown through landing rollout. *See also* SPACE TRANSPORTATION SYSTEM.

flight control team Personnel in the Mission Control Center (MCC) on duty to provide support for the duration of each space shuttle flight. The team is composed of the dedicated flight control room teams and the multipurpose support room personnel. *See also* MISSION CONTROL CENTER; SPACE TRANSPORTATION SYSTEM.

flight crew In general, any personnel onboard an aerospace vehicle, space station, or interplanetary spacecraft. With respect to the space shuttle, any personnel engaged in flying the orbiter and/or managing resources onboard (e.g., the commander, pilot, mission specialist).

flight deck A portion of the crew station module of the space shuttle orbiter vehicle. *See also* SPACE TRANSPORTATION SYSTEM.

flight-dependent training With respect to space shuttle operations, the preparation of a mission specialist or payload specialist for a specific flight. Part of the training involves integrated simulations with the rest of the shuttle flight crew and ground support teams. *See also* SPACE TRANSPORTATION SYSTEM.

flight operations Collective term for ground support operations by flight crew or support personnel in preparation for space flight; also, the tasks performed by the astronaut crew during flight.

flight path The line connecting the successive positions occupied, or to be occupied, by an aircraft, missile, or space vehicle as it moves through the atmosphere or outer space.

flight readiness firing A short-duration test operation of a rocket consisting of the complete firing of the liquid-propellant engines while the vehicle is restrained at the launch pad. This test is conducted to verify the readiness of the rocket system for a flight test or operational mission.

flight simulation An aerospace training session in which the flight crew and/or ground operations support personnel practice a portion (or all) of a crewed flight.

flight simulator A training device (often highly computer interactive) that can replicate certain conditions of flight or flight operations. With the arrival of high-resolution graphic displays and virtual reality technologies, a high-fidelity simulator sometimes is referred to as "the real world in a box."

flight test   Test of an aircraft, rocket, missile, or other type of aerospace vehicle by actual flight or launching. Flight tests are planned to achieve specific test objectives and to gain operational information.

flight test vehicle   A rocket, missile, aircraft, or aerospace vehicle used for conducting flight tests. These flight tests can involve either the vehicle's own capabilities or the capabilities of equipment carried onboard the vehicle.

flight unit   A spacecraft that is undergoing or has passed *flight acceptance tests* (i.e., environmental and other tests) that qualify it for launch and space flight. *See also* FLIGHT ACCEPTANCE TEST(S).

flow-to-close valve   Valve in which the flow direction and forces acting on the valving element provide a closing force.

flow-to-open valve   Valve in which the flow direction and forces acting on the valving element provide an opening force.

flow work   In thermodynamics, the amount of work associated with the expansion or compression of a flowing fluid.

fluid   A substance that, when in static equilibrium, cannot sustain a shear stress. A *perfect* or *ideal* fluid is one that has zero viscosity—that is, it offers no resistance to shape change. Actual fluids only approximate this behavior. The term *fluid* is used to refer to both liquids and gases.

fluid-cooled   Term applied to a rocket's thrust chamber or nozzle whose walls are cooled by fluid supplied from an external source, as in *regenerative cooling, transpiration cooling,* or *film cooling.*

fluid interface   Common boundary of two or more surfaces exposed to a fluid (e.g., mating flanges of a duct) or the interface between a fluid and its containing device (e.g., a tube wall).

fluid management   The isolation and separation of liquids from gas in a storage vessel that operates in a reduced- or zero-gravity environment.

fluid mechanics   The major branch of science that deals with the behavior of fluids (both gases and liquids) at rest *(fluid statics)* and in motion *(fluid dynamics).* This scientific field has many subbranches and important applications, including: aerodynamics (the motion of gases, including air), hydrostatics (liquids at rest), and hydrodynamics (the motion of liquids, including water).

fluid-solid interaction   The interaction of a rigid or elastic structure with an incompressible or compressible fluid. Air-blast loading, acoustic interaction, aeroelasticity, and hydroelasticity are examples of such interactions.

fluorescence   Many substances can absorb energy (as for example, from X rays, ultraviolet light, or radioactive particles) and immediately emit this absorbed ("excitation") energy as an electromagnetic photon of visible light. This light-emitting process is called fluorescence; the emitting substances are said to be *fluorescent.*

flux (symbol: φ)   1. In general, the rate of transport or flow of some quantity per unit area; often used in reference to the flow of some form of energy. 2. Specifically, *neutron flux* is defined as the number of neutrons passing through one square centimeter of a given target in one second. For a well-collimated beam of monoenergetic neutrons, the neutron flux can be expressed as $\varphi_n = nv$, where n equals the number of neutrons per cubic centimeter and v equals the neutron velocity (cm/sec). Therefore, the neutron flux (φ_n) has the units neutrons/(cm^2-sec).

flyby   An interplanetary or deep-space mission in which the *flyby spacecraft* passes close to its celestial target (i.e., a distant planet, moon, asteroid, or comet) but does not impact the target or go into orbit around it. Flyby spacecraft follow a continuous trajectory, never to be captured into a planetary orbit. Once the spacecraft has flown past its target (often after years of travel through deep space), it cannot return to recover lost data. So flyby operations often are planned years in advance of the encounter and refined and practiced in the months prior to the encounter date. Flyby operations are conveniently divided into four phases: *observatory phase, far-encounter phase, near-encounter phase,* and *postencounter phase.*

The *observatory phase* of a flyby mission is defined as the period when the celestial target can be better resolved in the spacecraft's optical instruments than it can from Earth-based instruments. The phase generally begins a few months prior to the date of the actual planetary flyby. During this phase the spacecraft is completely involved in making observations of its target, and ground resources become completely operational in support of the upcoming encounter. The *far-encounter phase* includes time when the full disc of the target planet (or other celestial object) can no longer fit within the field of view of the spacecraft's instruments. Observations during this phase are designed to accommodate parts of the planet (e.g., a polar region, an interesting wide-expanse cloud feature such as Jupiter's Red Spot, etc.) rather than the entire planetary disc, and to take advantage of the higher resolution available. The *near-encounter phase* includes the period of closest approach to the target. It is characterized by intensely active observations by all of the spacecraft's science experiments. This phase of the flyby mission provides scientists the opportunity to obtain the highest resolution data about the target. Finally, the *post encounter phase* begins when the near-encounter phase is completed, and the spacecraft is receding from the target. This phase is characterized by day-after-day observations of a diminishing, thin crescent of the planet just encountered. It provides an opportunity to make extensive observations of the night side of the planet. After the postencounter phase is over, the spacecraft stops observing the target and returns to the less intense activities of its interplanetary cruise phase—a phase in which scientific instruments are powered down and navigational corrections made to prepare the spacecraft for an encounter with another celestial object of opportunity or for a final journey of no return into deep space. Some scientific experiments, usually concerning the properties of interplanetary space, can be performed in this cruise phase. *See also* PIONEER 10, 11; VOYAGER.

flying test bed   A rocket, aerospace vehicle, aircraft, or other flying vehicle used to carry objects or devices being flight tested.

flywheel   A massive, rotating wheel that can store energy as kinetic (or motion) energy as its rate of rotation is increased; energy then can be removed from this system by decreasing the rate of rotation of the wheel. An advanced-design flywheel energy storage system has been suggested for lunar surface base applications. During the lunar daytime, solar energy would be converted into electricity using Stirling-cycle heat engines. This electricity would be used for lunar base needs as well as for spinning up several very large flywheels. Excess thermal energy (heat) in the system would be transported away to a radiator by a configuration of heat pipes. During the long lunar nighttime, energy would be extracted from the flywheels and converted into electricity for the surface base.

focal length (common symbol: f)   The distance between the center of a lens or mirror to the focus or focal point.

focal plane   The plane, perpendicular to the optical axis of the lens, in which images of points in the object field of the lens are focused.

focal point   *See* FOCAL LENGTH.

folded optics   Any optical system containing reflecting components for the purpose of reducing the physical length of the system or changing the path of the optical axis.

foot (symbol: ft) A unit of length in the traditional (foot-pound-second [fps]), or English, unit system equal to one-third of a yard. The foot is now defined as equal to 0.3048 meter.

foot-pound (force) (symbol: ft-lb$_{force}$) A unit of work (energy) in the traditional (foot-pound-second [fps]), or English, unit system equal to the work required when a 1-pound weight (i.e., a 1-pound mass at sea level on Earth under the influence of gravity) is raised 1 foot (in height). The foot-pound (force) is now defined as equal to:

1 foot-pound force (lbf) = 1.3558 joules

footprint An area within which a spacecraft or reentry vehicle is intended to land.

force (symbol: F) In physics, the cause of the acceleration of material objects measured by the rate of change of momentum produced on a free body. For a body of constant mass (m) traveling at a velocity (v), the momentum is (m.v). The force (F) is given by:

$$F = d(mv)/dt = m\ dv/dt = ma$$

(Newton's 2nd Law of Motion)

where a is the acceleration. The *newton* (symbol: N) is the SI unit of force. Forces between bodies occur in equal and opposite reaction pairs. *See also* NEWTON'S LAWS OF MOTION.

forced vibration An oscillation of a system in which the response is imposed by the excitation. If the excitation is periodic and continuing, the oscillation is steady state.

fossa A long, narrow, shallow (ditch-like) depression found on the surface of a PLANET or a MOON.

four-dimensional Involving *space-time;* that is, three spatial dimensions (e.g., x, y, and z) plus the dimension of time (t).

Fourier's law (of heat conduction) A fundamental physical principle that states

that when a temperature gradient $\partial T/\partial x$ exists in a body, there is thermal energy transfer through the body from the region of high temperature to the region of low temperature. In its most simple form (i.e., steady-state, one-dimensional heat transfer with uniform thermophysical properties except for the temperature gradient), Fourier's law can be written as:

$$q = -k\ A\ [\partial T/\partial x]$$

where q is the thermal energy transfer rate through the body (e.g., joules/-sec), k is the thermal conductivity e.g., joules per meter per Kelvin per second (J/m-K-s), A is the area in square meters (m^2) through which thermal energy is being transferred, and $\partial T/\partial x$ is the temperature gradient (Kelvin per meter [K/m]). The minus sign (−) is used on the right side of the above expression to fulfill the requirements of the Second Law of Thermodynamics, indicating that heat (thermal energy) flows naturally from a higher-temperature region to a low-temperature region. Note also that the heat transfer rate (q) can be expressed in watts (W), where 1 watt = 1 joule/second. This law is named after the French mathematician Jean Baptiste Joseph Fourier (1768–1830).

fractional orbit bombardment system (FOBS) A potential method of delivering a warhead from a partial satellite orbit and therefore approaching a target on Earth from any direction with very little (if any) warning time.

fratricide The destructive effect of the earlier-detonating weapons in a barrage on those weapons that arrive later.

free electron laser (FEL) A multifrequency (multi-wavelength) laser that achieves optical radiation amplification by a beam of (accelerator-produced) free electrons passing through a vacuum in a transverse periodic magnetic field. This laser technique differs from conventional lasers in which the oscillating electrons are bound to atoms and molecules and therefore produce optical radiation of a specific, characteristic wavelength.

free energy (symbol: ΔG) In thermodynamics, the Gibb's function (G) is defined as:

$$G = H - TS$$

where H is the total ENTHALPY, T is the absolute temperature, and S is the total ENTROPY. The change in the Gibb's function, ΔG, is sometimes called the *free energy* change for a chemical reaction and is useful for indicating the conditions under which a chemical reaction might occur. For example, if ΔG is negative, the chemical reaction will proceed to equilibrium spontaneously, while if ΔG is positive, energy must be supplied if the reaction is to proceed. *See also* THERMODYNAMICS.

free fall The unimpeded fall of an object in a gravitational field. For example, an elevator car whose cable has snapped; the plunge of a skydiver before his/her chute has been opened; or the fall back to Earth of a sounding rocket and its payload (after thrust termination and maximum altitude has been achieved) that is not retarded by a parachute or braking device. *See also* MICROGRAVITY.

free flight Unconstrained or unassisted flight, as: the flight of a rocket after consumption of its propellant or after motor shutoff; the flight of an unguided projectile.

free-flying spacecraft (free-flyer) Any SPACECRAFT or PAYLOAD that can be detached from NASA's SPACE SHUTTLE or the *INTERNATIONAL SPACE STATION* and then operate independently in ORBIT.

free rocket A rocket not subject to guidance or control in flight.

frequency (common symbol: f or ν) In general, the rate of repetition of a recurring or regular event; for example, the number of vibrations of a system per second or the number of cycles of a wave per second. For electromagnetic radiation, the frequency (ν) of a quantum packet of energy (i.e., a photon) is given by: $\nu = E/h$, where E is the photon energy and h is the Planck constant. the SI UNIT of frequency is the *hertz (Hz)*, which is defined as 1 cycle per second.

frequency modulation (FM) An information transfer technique used in telecommunications in which the frequency of the carrier wave is modulated (i.e., increased or decreased) as the signal (to be transferred) increases or decreases in value but the amplitude of the carrier wave remains constant. Specifically, angle modulation of a sine carrier wave in which the instantaneous frequency of the modulated wave differs from the carrier frequency by an amount proportional to the instantaneous value of the modulating wave.

Fresnel lens A thin lens constructed with stepped setbacks so as to have the optical properties of a much thicker lens.

friability The tendency of a crystalline structure to crumble.

friction The force that resists the relative motion of one body (or surface) rolling, sliding, or flowing over another body (or surface) with which it is in contact.

fuel 1. Any substance used to liberate thermal energy, either by chemical or nuclear reaction; a substance that might be used, for example, in a heat engine or a rocket engine. When discussing a liquid-propellant rocket engine, ordinarily the *fuel* (a propellant chemical such as kerosene) is distinguished from the *oxidizer* (e.g., liquid oxygen, or LOX) when these two are stored separately. 2. In nuclear technology, the term "fuel" applies to the fissionable (or fissile) material used to produce energy in a nuclear reactor. Uranium-235 is an example of nuclear reactor fuel.

fuel binder A continuous-phase substance that contributes the principal structural condition to solid propellant but does not contain any oxidizing element, either in solution or chemically bonded.

fuel cell A DIRECT CONVERSION device that transforms chemical ENERGY directly into electrical energy by reacting continuously supplied chemicals. In a modern fuel

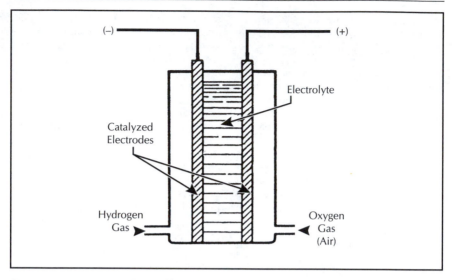

Figure 33. Components of a simple hydrogen-oxygen fuel cell.

cell, an electrochemical CATALYST (such as platinum) promotes a noncombustible reaction between a fuel (such as HYDROGEN) and an oxidant (such as oxygen). (See Figure 33.)

fuel consumption The mass rate of fuel use (typically kg/sec or lb_{mass}/sec) by a rocket engine or power plant.

fuel shutoff The termination of the flow of liquid fuel into a rocket's combustion chamber; or the process of stopping the combustion (burning) of a solid rocket fuel. The event or time marking this particular action. *Compare with* CUTOFF.

functional kill The destruction of a target by disabling vital components in a way not immediately detectable but nevertheless able to prevent the target from functioning properly. For example, the destruction of electronics components in a guidance system of a hostile missile by a space-based neutral particle beam (NPB) weapon system. *See also* BALLISTIC MISSILE DEFENSE.

fuse Igniting device consisting of a detonating or deflagrating train for the propagation of ignition energy.

fuselage The central part of an aircraft or aerospace vehicle that accommodates crew, passengers, payload, or cargo.

fusion (nuclear) The nuclear process by which lighter atomic nuclei join (or fuse) to create a heavier NUCLEUS. The fusion of DEUTERIUM (D) with TRITIUM (T) results in the formation of a HELIUM (He) nucleus and a NEUTRON (n). This D-T reaction also involves the release of 17.6 million ELECTRON VOLTS (MeV) of ENERGY. Thermonuclear reactions are fusion reactions caused by very high temperature (millions of degrees KELVIN). The energy of the SUN and other STARS comes from a variety of thermonuclear fusion reactions including the PROTON-PROTON CHAIN REACTION.

fusion weapon A nuclear weapon that uses the energy of fusion; a thermonuclear weapon or "hydrogen" bomb.

G

g The symbol used for the acceleration due to gravity. For example, at sea level on Earth the acceleration due to gravity is approximately 9.8 meters per second-squared (32.2 feet per second-squared)— that is, "one g." This term is used as a unit of stress for bodies experiencing acceleration. When a spacecraft or aerospace vehicle is accelerated during launch, everything inside it experiences a force that may be as high as several gs.

Gagarin, Yuri A. (1934–1968) The Russian COSMONAUT who became the first human being to travel in OUTER SPACE. He accomplished this feat with his historic one ORBIT of EARTH mission in the VOSTOK 1 SPACECRAFT on April 12, 1961. A popular hero of the former Soviet Union, he died in an aircraft training flight near Moscow on March 27, 1968.

gain 1. An increase or amplification. 2. In electrical engineering, this term is used to describe the ability of an electronic device to increase the magnitude of an electrical input parameter. For example, the gain of a power amplifier is defined as the ratio of the output power to the input power. Gain usually is expressed in decibels. 3. In signal processing, gain is used to denote an increase in signal power in transmission from one point to another. 4. In radar technology, there are two frequently encountered general uses of this term: (a) *antenna gain,* or gain factor, which is defined as the power transmitted along the radar beam axis to that of an isotropic radiator transmitting the same total power; and (b) *receiver gain,* which is the amplification given to an incoming radar signal by the receiver.

gal (derived from: *Galileo*) A unit of acceleration equal to 10^{-2} meter per second-squared (or 1 centimeter per second-squared). The name was chosen to honor GALILEO GALILEI (1564–1642), the famous Italian scientist who conducted early experiments involving the acceleration of gravity. The gal and the "milligal" (1000 milligal = 1 gal) often are encountered in geological survey work, where tiny differences in the acceleration due to gravity at Earth's surface are important.

galactic Of or pertaining to a galaxy, such as the Milky Way galaxy.

galactic cannibalism A postulated model of GALAXY interaction in which a more massive galaxy uses GRAVITATION and tidal FORCEs to pull matter away from a less massive neighboring galaxy.

galactic cluster A diffuse collection of from ten to perhaps several hundred stars, loosely held together by gravitational forces. Also called an *open cluster. See also* GLOBULAR CLUSTER.

galactic cosmic rays (GCRs) Very energetic atomic particles that originate outside the solar system and generally are believed to come from within our Milky Way Galaxy. *See also* COSMIC RAYS.

galactic nucleus The central region of a GALAXY.

galaxy A very large accumulation of from 10^6 to 10^{12} stars. Galaxies—or "island universes," as they are sometimes called—come in a variety of shapes and sizes. They range from dwarf galaxies, such as the Magellanic Clouds, to giant spiral galaxies, such as the Andromeda Galaxy. Astronomers usually classify galaxies as elliptical, spiral (or barred spiral), or irregular. Galaxies are typically tens to hundreds of thousands of light-years across, and the distance between neighboring galaxies is generally a few million light-years. For example, the Andromeda Galaxy is approximately 130,000 light-years in diameter and about 2.2 million light-years away.

Galaxy When capitalized, humans' home galaxy, the MILKY WAY GALAXY.

Galilean satellites The four largest and brightest satellites of the planet, Jupiter. Io, Europa, Ganymede, and Callisto. These major Jovian moons were discovered by the famous Italian scientist Galileo Galilei in 1610. *See also* JUPITER.

Galilean telescope The early REFRACTING TELESCOPE assembled by GALILEO GALILEI in about 1610. It had a CONVERGING LENS as the OBJECTIVE and a DIVERGING LENS as the EYEPIECE.

Galileo Galilei (1564–1642) Italian astronomer, physicist, and mathematician whose innovative use of the telescope to make astronomical observations ignited the scientific revolution of the 17th century. In 1610 he announced some of his early telescopic findings in the publication *Starry Messenger,* including the discovery of the four major MOONS of JUPITER (now called the GALILEAN SATELLITES). Their behavior like a miniature SOLAR SYSTEM stimulated his enthusiastic support for the HELIOCENTRIC COSMOLOGY of NICHOLAS COPERNICUS. Unfortunately, this scientific work led to a direct clash with church authorities, who insisted on retaining the PTOLEMAIC SYSTEM for a number of political and social reasons. By 1632 this conflict earned the fiery Galileo an Inquisition trial at which he was found guilty of heresy (for advocating the COPERNICAN SYSTEM) and confined to house arrest for the remainder of his life.

Galileo Project NASA's highly successful scientific mission to JUPITER launched in October 1989. With electricity supplied by two RADIOISOTOPE-THERMOELECTRIC GENERATOR (RTG) units, the Galileo SPACECRAFT studied the JOVIAN system extensively from December 1995 until February 2003. Upon arrival at the GIANT PLANET, the ORBITER SPACECRAFT, named in honor of GALILEO GALILEI, released a scientific PROBE into the upper portions of Jupiter's ATMOSPHERE. The spacecraft then flew in a highly ELLIPTICAL ORBIT around Jupiter, studying the PLANET and its MOONS—primarily the four GALILEAN SATELLITES: IO, EUROPA, GANYMEDE, and CALLISTO. Right up until the end of its scientific mission, the Galileo spacecraft continued to deliver surprises. In December 2002, its INSTRUMENTS revealed that Jupiter's potato-shaped inner moon, Amalthea, had a very low DENSITY, a fact suggesting, in turn, that the tiny CELESTIAL BODY was probably full of holes. Because of Amalthea's irregular shape and low density, planetary scientists speculate that this moon may have broken into many pieces that now tenuously cling together from the pull of each other's GRAVITY. If true, the moon would possess many empty spaces throughout its interior, where the fragmented pieces do not fit together well. The December 2002 FLYBY of Amalthea also brought the spacecraft closer to Jupiter than at any time since it began orbiting the giant planet on December 7, 1995. After more than 30 close encounters with Jupiter's four largest, moons, the flyby of Amalthea became the spacecraft's last scientific adventure in the Jovian system. On February 28, 2003, the NASA flight team terminated its operation of the Galileo spacecraft. The human directors sent a final set of commands to the far-traveling ROBOT SPACECRAFT, putting it on a course that resulted in its mission-ending plunge into Jupiter's atmosphere in late September 2003.

gamma ray astronomy Branch of ASTRONOMY based on the detection of the energetic GAMMA RAYS associated with SUPERNOVAe, exploding galaxies, QUASARS, PULSARS, and phenomena near suspected BLACK HOLES.

gamma rays (symbol: γ) High-energy, very-short-wavelength packets or quanta of electromagnetic radiation. Gamma ray photons are similar to X rays, except that they are usually more energetic and originate from processes and transitions within the atomic nucleus. Gamma rays typically have energies between 10,000 electron volts and 10 million electron volts (i.e., between 10 keV and 10 MeV) with correspondingly short wavelengths and high frequencies. The processes associated with gamma ray emissions in astrophysical phenomena include: (1) the decay of radioactive nuclei, (2) cosmic ray interactions, (3) curvature radiation in extremely strong magnetic fields, and (4) matter-antimatter annihilation. Gamma rays are very penetrating and are best stopped or shielded against by dense materials, such as lead or tungsten. Sometimes called *gamma radiation*. *See also* GAMMA RAY ASTRONOMY.

gantry A frame that spans over something, such as an elevated platform that runs astride a work area, supported by wheels on each side. Specifically, the term is short for *gantry crane* or *gantry scaffold.*

Ganymede With a diameter of 5,262 kilometers (3,270 mi), the largest MOON of JUPITER and in the SOLAR SYSTEM. Discovered by GALILEO GALILEI in 1610.

gas The state of matter in which the molecules are practically unrestricted by intermolecular forces so that the molecules are free to occupy any space within an enclosure. *See also* IDEAL GAS.

gas cap The gas immediately in front of a body (such as a reentry vehicle or meteor) as it travels through a planetary atmosphere. This region of atmospheric gas is compressed and adiabatically heated, often to incandescence.

gas constant (symbol: R_u) The constant factor in the equation of state for an ideal (perfect) gas. The universal gas constant is

$$R_u = 8.315 \text{ joules per mole per Kelvin}$$
$$[\text{J/(mol-K)}]$$

The gas constant for a particular, nonideal gas is called the *specific gas constant* (symbol: R_{sp}) and is related to the universal gas constant as follows:

$$R_{sp} = R_u/M$$

where M is the molecular weight of the particular gas. *See also* IDEAL GAS.

gas generator 1. An assemblage of parts similar to a small rocket engine in which a propellant is burned to provide hot exhaust gases. These hot gases are then used to: (1) drive the turbine in the turbopump assembly of a rocket vehicle, or (2) pressurize liquid propellants, or (3) provide thrust by exhausting through a nozzle. 2. A device used in a laboratory to generate gases, such as hydrogen and oxygen from water.

gasket A performed deformable part designed to be placed between two adjoining, relatively static parts to prevent the leakage of liquid or gas between the parts.

gas scrubbing The process of contracting a gaseous mixture with a liquid for the purpose of removing gaseous contaminants or entrained liquids or solids.

gas turbine A heat engine that uses high-temperature, high-pressure gas as a working fluid. Its operation approximates the ideal *Brayton cycle,* and a portion of the energy content of the gaseous working fluid is converted directly into work (i.e., rotating shaft action). Often the hot gases for operation of a gas turbine are obtained by the combustion of a fuel in air. These types of gas turbines are also called *combustion turbines.* The simplest combustion gas turbine heat engine is the open cycle system, which has three major components: the compressor, the combustion chamber (or combustor), and the gas turbine itself. The compressor and turbine usually are mounted on the same shaft, so that a portion of the mechanical ("shaft") work produced by the turbine can be used to drive the compressor. *See also* BRAYTON CYCLE.

gauss (symbol: G or Gs) In the centimeter-gram-second (cgs) unit system, the unit of magnetic flux density. Earth's magnetic field at the planet's surface is on the average 0.3 to 0.6 gauss.

$$1 \text{ gauss} = 10^{-4} \text{ tesla}$$

Geiger counter A nuclear radiation detection and measurement instrument. It contains a gas-filled tube containing electrodes between which there is an electrical voltage but no current flowing. When ionizing radiation passes through the tube, a short, intense pulse of current passes from the negative electrode to the positive electrode and is measured or counted. The number of pulses per second measures the intensity of the radiation. This instrument is also called a *Geiger-Müller counter* and is named for Hans Geiger and W. Müller who developed it in the 1920s.

Gemini Project The Gemini Project (1964–1966) was the second U.S. crewed space program and the beginning of sophisticated human space flight. It expanded and refined the scientific and technological endeavors of the Mercury Project and prepared the way for the more sophisticated Apollo Project, which carried U.S. astronauts to the lunar surface. The Gemini Project added a second crew member and a maneuverable spacecraft. New objectives included rendezvous and docking techniques with orbiting spacecraft and extravehicular "walks in space." In all, ten two-person launches occurred, successfully placing 20 astronauts in orbit and returning them safely to Earth. The table summarizes the Gemini Project flights. *See also* APOLLO PROJECT; MERCURY PROJECT.

general relativity Albert Einstein's theory, introduced in 1915, that GRAVITATION arises from the curvature of space and time—the more massive an object, the greater the curvature.

geo- A prefix meaning the planet *Earth,* as in *geology* and *geophysics.*

geocentric Relative to Earth as a center; measured from the center of Earth.

geodesy In general, the science of determining the exact size and shape of bodies in our solar system and of the distribution of mass within these bodies. Specifically, the science that deals mathematically with the size and shape of Earth, Earth's external gravity field, and with surveys of such precision that the overall size and shape of Earth must be taken into consideration.

geodetic Pertaining to GEODESY, the science that deals with the size and shape of Earth.

geographic information system (GIS) A computer-assisted system that acquires, stores, manipulates, compares, and displays geographic data, often including MULTISPECTRAL SENSING data sets from EARTH-OBSERVING SATELLITES.

geoid The figure of Earth as defined by the geopotential surface that most nearly coincides with mean sea level over the entire surface of Earth.

geomagnetic storm Sudden worldwide fluctuations in Earth's magnetic field, associated with solar flare-generated shock waves that propagate from the Sun to Earth. Geomagnetic storms may significantly alter ionospheric current densities over large portions of the middle- and high-latitude regions of Earth, impacting high-frequency (HF) systems. During a solar storm, large numbers of particles are dumped into the high-altitude atmosphere. These particles produce heating effects (especially near the auroral zone), which then alter the near-Earth-space atmospheric densities from about 100 kilometers (62.1 mi) out to about 1,000 kilometers, impacting the operation of spacecraft orbiting Earth at altitudes below 1,000 kilometers. Polar-orbiting spacecraft can be especially impacted, as, for example, by an increase in atmospheric drag.

geophysics The branch of science dealing with Earth and its environment—that is, the solid Earth itself, the hydrosphere (oceans and seas), the atmosphere, and (by extension) near-Earth space. Classi-

Summary of Gemini Project Missions

Mission	Date(s) Recovery Ship	Crew	Ground Elapsed Time (GET) Hr:min	Remarks
Gemini 3 (Molly Brown)	March 23, 1965 Intrepid (A)	USAF Maj, Virgil I. Grissom Navy Lt. Comdr. John W. Young	4:53	3 orbits
Gemini 4	June 3–7, 1965 Wasp (A)	USAF Majors James A. McDivitt and Edward H. White, II	97:56	62 orbits: first U.S. EVA (White)
Gemini 5	Aug. 21–29, 1965 Lake Champlain (A)	USAF Lt. Col. L. Gordon Cooper. Navy Lt. Comdr. Charles Conrad, Jr.	190:55	120 orbits
Gemini 7	Dec. 4–18, 1965 Wasp (A)	USAF Lt. Col. Frank Borman Navy Comdr. James A. Lovell, Jr.	330:35	Longest Gemini flight; rendezvous target for Gemini 6; 206 orbits
Gemini 6	Dec. 15–16, 1965 Wasp (A)	Navy Capt. Walter M. Schirra, Jr. USAF Maj. Thomas P. Stafford	25:51	Rendezvoused within 1 ft of Gemini 7; 16 orbits
Gemini 8	Mar. 16, 1966 L. F. Mason (P)	Civilian Neil A. Armstrong USAF Maj. David R. Scott	10:41	Docked with unmanned Agena 8; 7 orbits
Gemini 9A	June 3–6 1966 Wasp (A)	USAF Lt. Col. Thomas P. Stafford Navy Lt. Comdr. Eugene A. Cernan	72:21	Rendezvous (3) with Agena 9; one EVA; 44 orbits
Gemini 10	July 18–21, 1966 Guadalcanal (A)	Navy Comdr. John W. Young USAF Maj. Michael Collins	70:47	Docked with Agena 10; rendezvoused with Agena 8; two EVAs; 43 orbits
Gemini 11	Sept. 12–15, 1966 Guam (A)	Navy Comdr. Charles Conrad, Jr. Navy Lt. Comdr. Richard F. Gordon, Jr.	71:17	Docked with Agena 11 twice; first tethered flights; two EVAs; highest altitude in Gemini program; 853 miles; 44 orbits
Gemini 12	Nov. 11–15, 1966 Wasp (A)	Navy Capt. James A. Lovell, Jr. USAF Maj. Edwin E. Aldrin, Jr.	94:35	Three EVAs total 5 hrs. 30 min.; 59 orbits

Note: A = Atlantic Ocean
P = Pacific Ocean
EVA = Extravehicular Activity
Source: NASA.

cally, geophysics has been concerned primarily with physical phenomena occurring at and below the surface of Earth. This traditional emphasis gave rise to such companion disciplines as geology, physical oceanography, seismology, hydrology, and geodesy. With the advent of the space age, the trend has been to extend the scope of geophysics to include meteorology, geomagnetism, astrophysics, and other sciences concerned with the physical nature of the universe.

geoprobe A rocket vehicle designed to explore outer space near Earth at a distance of more than 6,440 kilometers (km) (4,000 mi) from Earth's surface. By convention, rocket vehicles operating lower than 6,440 km are called *sounding rockets.*

geosphere The solid *(lithosphere)* and liquid *(hydrosphere)* portions of Earth. Above the geosphere lies the *atmosphere,* at the interface between these two regions is found almost all of the *biosphere,* or zone of life.

Geostationary Operational Environmental Satellite (GOES) GOES weather satellites maintain orbital positions over the same Earth location along the equator at about 35,900 kilometers (22,300 mi) above Earth, giving them the ability to make continuous observations of weather patterns over and near the United States. Operated by the U.S. National Oceanic and Atmospheric Administration (NOAA), these satellites provide both visible light and infrared images of cloud patterns as well as "soundings," or indirect measurements, of the temperature and humidity throughout the atmosphere. NOAA has been operating GOES satellites since 1974. Data from these spacecraft provide input for the forecasting responsibilities of the National Weather Service. Among other applications, the GOES data assist in monitoring storms and provide advance warning of emerging severe weather. The geostationary vantage point of GOES satellites permits the observation of large-scale weather events, which is required for forecasting small-scale events. *See also* WEATHER SATELLITE.

geostationary orbit (GEO) A SATELLITE in a circular ORBIT around EARTH at an ALTITUDE of 35,900 kilometers (22,300 mi) above the EQUATOR that travels around the PLANET at the same rate as Earth spins on its AXIS. COMMUNICATIONS SATELLITES, ENVIRONMENTAL SATELLITES, and SURVEILLANCE SATELLITES use this important orbit. If the SPACECRAFT's orbit is circular and lies in the equatorial plane (to an observer on Earth), the spacecraft appears stationary over a given point on Earth's surface. If the spacecraft's orbit is inclined to the equatorial plane (when observed from Earth), the spacecraft traces out a figure eight path every 24 hours. *See also* GEOSYNCHRONOUS ORBIT; SYNCHRONOUS SATELLITE.

geosynchronous orbit (GEO) An ORBIT in which a SATELLITE completes one REVOLUTION at the same rate as EARTH spins, namely, 23 hours, 56 minutes, and 4.1 seconds. A satellite placed in such an orbit (at approximately 35,900 kilometers [22,300 mi] ALTITUDE above the EQUATOR) revolves around Earth once per day. *See also* GEOSTATIONARY ORBIT; SYNCHRONOUS SATELLITE.

geysering The accumulation of gas in a line and the subsequent expulsion of liquid from the line by a gas bubble.

g-force An inertial force usually expressed in multiples of terrestrial gravity. *See also* G.

giant-impact model The hypothesis that the MOON originated when a MARS-sized object struck a young EARTH with a glancing blow. The giant (oblique) impact released material that formed an ACCRETION DISK around Earth out of which the Moon formed.

giant molecular cloud (GMC) Massive clouds of GAS in INTERSTELLAR space composed primarily of MOLECULES of HYDROGEN (H_2) and dust. GMCs can contain enough MASS to make several million STARS like the SUN and are often the sites of star formation.

giant planets In our solar system, the large, gaseous outer planets: Jupiter, Saturn, Uranus, and Neptune.

giant star A STAR near the end of its life that has swollen in size, such as a BLUE GIANT or a RED GIANT.

gibbous A term used to describe a phase of a moon or planet when more than half, but not all, of the illuminated disk can be viewed by the observer.

giga- (symbol: G) A prefix meaning multiplied by 10^9.

gimbal 1. A device, on which an engine or other object may be mounted, that has two mutually perpendicular and intersecting axes of rotation and therefore provides free angular movement in two directions. 2. In a gyroscope, a support that provides the spin axis with a degree of freedom. 3. To move a reaction engine about on a gimbal so as to obtain pitching and yawing correction moments. 4. To mount something on a gimbal.

gimbaled motor A rocket engine mounted on a gimbal.

Giotto spacecraft Scientific SPACECRAFT launched by the EUROPEAN SPACE AGENCY (ESA) in July 1985 that successfully encountered the NUCLEUS of COMET HALLEY in mid-March 1986 at a distance of about 600 kilometers. The spacecraft was named after the Italian painter Giotto di Bondone (1266–1337), who apparently witnessed the 1301 passage of Comet Halley and then included the first scientific representation of this comet in his famous fresco _Adoration of the Magi_, which can be found in the Scrovegni Chapel in Padua, Italy.

Glenn, John Herschel, Jr. (1921–) The American ASTRONAUT, U.S. Marine Corps officer, and U.S. Senator who was the first American to ORBIT EARTH, a feat that he accomplished on February 20, 1962, as part of NASA's MERCURY PROJECT. Glenn's historic mission aboard the _Friendship_ 7 Mercury SPACE CAPSULE made three

orbits of Earth and lasted about five hours. More than three and a half decades later, he became the oldest human being to travel in OUTER SPACE when he joined the SPACE SHUTTLE _Discovery_ crew on its nine-day STS-95 orbital MISSION (from October 29 to November 7, 1998).

glide 1. A controlled descent by a heavier-than-air aeronautical or aerospace vehicle under little or no engine thrust in which forward motion is maintained by gravity and vertical descent is maintained by lift forces. 2. A descending flight path of a glide, such as a shallow _glide_. 3. To descend in a _glide_.

glide path The flight path (seen from the side) of an aeronautical or aerospace vehicle in a GLIDE.

glider A fixed-wing aircraft specially designed to GLIDE, or to glide and soar. This type of aircraft ordinarily has no powerplant.

global change Earth's environment has been subject to great change over eons. Many of these changes have occurred quite slowly, requiring numerous millennia to achieve their full impact and effect. However, other global changes have occurred relatively rapidly over time periods as short as a few decades or less. These global changes appear in response to such phenomena as the migration of continents, the building and erosion of mountains, changes in the Sun's energy output or variations in Earth's orbital parameters, the reorganization of oceans, and even the catastrophic impact of a large asteroid or comet. Such natural phenomena lead to planetary changes on local, regional, and global scales, including a succession of warm and cool climate epochs, new distributions of tropical rain forests and rich grasslands, the appearance and disappearance of large deserts and marshlands, the advances and retreats of great ice sheets (glaciers), the rise and fall of ocean and lake levels, and even the extinction of vast numbers of species. The last great mass extinction (on a global basis) appears to have occurred some 65 million years ago, possibly due to the impact of a large asteroid. The peak of the most recent period of glaciation generally is considered to have occurred about 18,000 years ago, when average global temperatures were about 5° Celsius (9° Fahrenheit) cooler than today.

Although such global changes are the inevitable results of major natural forces currently beyond human control, it is also apparent to scientists that humans have become a powerful agent for environmental change. For example, the chemistry of Earth's atmosphere has been altered significantly by both the agricultural and industrial revolutions. The erosion of the continents and sedimentation of rivers and shorelines have been influenced dramatically by agricultural and construction practices. The production and release of toxic chemicals have affected the health and natural distributions of biotic populations. The ever-expanding human need for water resources has affected the patterns of natural water exchange that take place in the hydrological cycle (the oceans, surface and ground water, clouds, etc.). One example is the enhanced evaporation rate from large human-engineered reservoirs compared to the smaller natural evaporation rate from wild, unregulated rivers. This increased evaporation rate from large-engineered reservoirs demonstrates how human activities can change natural processes significantly on a local or even regional scale. Such environmental changes may not always be beneficial and often may not have been anticipated at the start of the project or activity. As the world population grows and human civilization undergoes further technological development in the next century, the role of our planet's most influential animal species as an agent of environmental change undoubtedly will expand.

Over the past three decades, scientists have accumulated technical evidence that indicates that ongoing environmental changes are the result of complex interactions among a number of natural and human-related systems. For example, the changes in Earth's climate now are consid-

ered to involve not only wind patterns and atmospheric cloud populations but also the interactive effects of the biosphere and ocean currents, human influences on atmospheric chemistry, Earth's orbital parameters, the reflective properties of our planetary system (Earth's ALBEDO) and the distribution of water among the atmosphere, hydrosphere, and cryosphere (polar ice). The aggregate of these interactive linkages among our planet's major natural and human-made systems that appear to affect the environment has become known as *global change*.

The governments of many nations, including the United States, have begun to address the issues associated with global change. Over the last decade, preliminary results from global observation programs (many involving space-based systems) have stimulated a new set of concerns that the dramatic rise of industrial and agricultural activities during the 19th and 20th centuries may be adversely affecting the overall Earth system. Today the enlightened use of Earth and its resources has become an important contemporary political and scientific issue.

The global changes that may affect both human well-being and the quality of life on this planet include global climate warming, sea-level change, ozone depletion, deforestation, desertification, drought, and a reduction in biodiversity. Although complex phenomena in themselves, these individual global change concerns cannot be fully understood and addressed unless they are studied collectively in an integrated, multidisciplinary fashion. An effective and well-coordinated international research program, which includes use of advanced environmental observation satellite systems, will be required to significantly improve our scientific knowledge of natural and human-induced changes in the global environment and their regional impacts. *See also* MISSION TO PLANET EARTH; REMOTE SENSING.

Global Positioning System (GPS) The NAVSTAR Global Positioning System is a multiservice program within the U.S. Department of Defense. The UNITED STATES

AIR FORCE serves as the designated executive service for management of the system. NAVSTAR GPS is a space-based radio-positioning system nominally consisting of a CONSTELLATION of 24 EARTH-orbiting SATELLITES that provide navigation and timing information to military and civilian users worldwide. GPS satellites orbit Earth every 12 hours, emitting continuous RADIO WAVE SIGNALS. With proper equipment, users can receive these signals and use them to calculate location, time, and VELOCITY. The signals provided by GPS are so accurate that time can be calculated to within a millionth of a second, velocity within a fraction of a kilometer per hour, and location within a few meters or less. GPS receivers have been developed by both military and civilian manufacturers for use in aircraft, ships, and land vehicles. Hand-held systems are also available for use by individuals in the field.

The Global Positioning System has significantly enhanced functions such as mapping, aerial refueling and rendezvous, geodetic surveys, and search and rescue operations. Many military applications were put to the test in the early 1990s during the U.S. involvement in Operations Desert Shield and Storm. Allied troops relied heavily on GPS data to navigate the featureless regions of the Saudi Arabian desert. Forward air controllers, pilots, tank drivers, and support personnel all used the system so successfully that American defense officials cited GPS as a key to the Desert Storm victory.

There are four generations of the GPS satellite, designated as the Block I, the Block II/IIA, the Block IIR, and the Block IIF. Block I satellites were used to test the principles of the Global Positioning System and to demonstrate the efficacy of space-based navigation. Lessons learned from the operation of the 11 Block I satellites were incorporated into later design blocks.

Block II and IIA satellites make up the current constellation. With a MASS of 985 kilograms (2,175 lbm), each GPS Block IIA satellite operates in a specially designated circular 20,350 kilometers (10,988 nautical miles) ALTITUDE ORBIT. The DELTA II LAUNCH VEHICLE lifted the satellites into their characteristic 12-hour orbits from CAPE CANAVERAL AIR FORCE STATION in Florida. GPS Block IIA satellites have a design lifetime of approximately 7.5 years.

Block IIR satellites represent a dramatic improvement over the satellites in the previous design blocks. These satellites have the ability to determine their own position by performing intersatellite ranging with other Block IIR SPACECRAFT. The Block IIR spacecraft have a mass of 1,075 kilograms (2,370 lbm) and a design lifetime of 10 years. The Block IIR satellites are replacing Block II and IIA satellites as the latter reach the end of their service lifetimes.

The more massive 1,700-kilogram (3,760 lbm) Block IIF spacecraft represent the fourth generation of navigation satellite. With a design lifetime of 15 years, these spacecraft will be used as sustainment vehicles, ensuring the vitality of the GPS constellation for the next two decades. Either DELTA IV or ATLAS V EXPENDABLE LAUNCH VEHICLES will place Block IIF satellites into orbit.

Plans to upgrade the current GPS system will reduce vulnerabilities to jamming and respond to national policy that encourages widespread civilian use of this important navigation system without degrading its military utility. For example, upgrades for the satellite and its ground control segment include civil signals and new military signals transmitted at higher power levels. Projected improvements in military end-user equipment will also minimize the impact of adversarial jamming and protect efficient use of the GPS system in time of war by American and friendly forces.

globular cluster   Compact cluster of up to 1 million, generally older, stars. *See also* GALACTIC CLUSTER.

Glonass   Russian global positioning satellites used in a constellation to provide navigation support. *See also* GLOBAL POSITIONING SYSTEM.

Goddard, Robert Hutchings (1882–1945)   American physicist and ROCKET scientist who cofounded ASTRONAUTICS

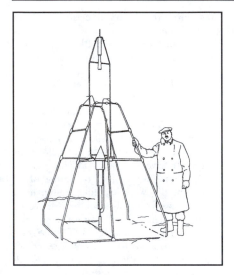

Figure 34. Dr. Robert H. Goddard and the world's first liquid-fuel rocket, which he successfully launched on March 16, 1926.

early in the 20th century along with (but independent of) KONSTANTIN TSIOLKOVSKY and HERMANN J. OBERTH. Regarded as the father of modern rocketry, he successfully launched the world's first LIQUID-PROPEL-LANT ROCKET on March 16, 1926, in a snow-covered field in Auburn, Massachusetts. (See Figure 34.) A brilliant but reclusive inventor, he continued his pioneering rocket experiments at a remote desert site near Roswell, New Mexico. In 1960, recognizing that most modern rockets are really "Goddard rockets," the U.S. government paid his estate $1 million for the use of his numerous patents.

Gorizont A series of geostationary Russian communications satellites.

GOX Gaseous oxygen.

grain 1. In rocketry, the integral piece of molded or extruded solid propellant that comprises both fuel and oxidizer in a solid rocket motor and is shaped to produce, when burned, a specified performance versus time relation. *See also* SOLID-PROPELLANT ROCKET. 2. In photog-

raphy, a small particle of metallic silver remaining in a photographic emulsion after development and fixing. In the agglomerate, these grains form the dark area of a photographic image. Excessive graininess reduces the quality of a photographic image, especially when the image is enlarged or magnified. 3. In materials science, an individual crystal in a polycrystalline metal or alloy.

gram (symbol: g) The fundamental unit of mass in the centimeter-gram-second (CGS) unit system. One gram is equal to one-thousandth (1/1000) of a kilogram.

granules The small bright features found in the photosphere of the Sun. Also called *granulation. See also* SUN.

gravitation The acceleration produced by the mutual attraction of two masses, directed along the line joining their centers of mass, and of magnitude inversely proportional to the square of the distance between the two centers of mass. In classical physics, the force of attraction (F_g) between two masses (m and M), is given by NEWTON'S LAW OF GRAVITATION:
$$F_g = [G\ m\ M] / r^2$$
where G is the gravitational constant (6.6726×10^{-11} N m^{2}kg^{-2}) and r is the distance between the two masses.

gravitational collapse The unimpeded contraction of any MASS caused by its own GRAVITY.

gravitational constant (symbol: G) The coefficient of proportionality in NEWTON'S LAW OF GRAVITATION.
$$G = 6.672 \times 10^{-11}\ \text{newton} - \text{meter}^2 - \text{kilogram}^{-2}\ (\text{N m}^2\ \text{kg}^{-2})$$

graviton The hypothetical QUANTUM (or PARTICLE) of gravitational ENERGY predicted by the German-Swiss-American physicist Albert Einstein (1879–1955) in his GENERAL RELATIVITY theory.

gravity In general, the attraction of a celestial object for any nearby mass.

Specifically, the downward force imparted by Earth on a mass near Earth or on its surface. Since Earth is rotating, the force observed as "gravity" is actually the resultant of the force of gravitation and the centrifugal force arising from its rotation. *See also* GRAVITATION.

gravity anomaly A region on a celestial body where the local force of gravity is lower or higher than expected. If the celestial object is assumed to have a uniform density throughout, then we would expect the gravity on its surface to have the same value everywhere. *See also* MASCON.

gravity assist The change in a spacecraft's direction and velocity achieved by a carefully calculated flyby through a planet's gravitational field. The change in direction and velocity is achieved without the use of supplementary propulsive energy. *See also* ORBITS OF OBJECTS IN SPACE; VOYAGER.

gray (symbol: Gy) In radiation protection, the SI UNIT of absorbed dose of ionizing radiation. One gray (1 Gy) is equal to the absorption of 1 joule of energy per kilogram of matter. This unit is named in honor of L. H. Gray (1905–1965), a British radiologist. The gray is related to the traditional radiation dose unit system as follows:

1 Gy = 100 rad

gray body A body that absorbs some constant fraction, between zero and one, of all electromagnetic radiation incident upon it. The fraction of electromagnetic radiation absorbed is called the *absorptivity* (symbol: α) and is independent of wavelength for a gray body. Many objects fit the gray body model, which represents a surface of absorptive characteristics intermediate between those of a white body and a blackbody. In radiation transport theory, a white body is a hypothetical body whose surface absorbs no electromagnetic radiation at any wavelength, i.e., it is a perfect reflector. Since the white body exhibits zero absorptivity at all wavelengths, it is an idealization that is

exactly opposite to that of the blackbody. The gray body is also called *grey body*. *See also* BLACKBODY.

Great Dark Spot (GDS) A large, dark, oval-shaped feature in the clouds of NEPTUNE, discovered in 1989 by NASA's *VOYAGER* 2 spacecraft.

Great Red Spot (GRS) A distinctive, oval-shaped feature in the southern hemisphere clouds of JUPITER, first noted in 1831 by amateur German astronomer Sammuel Heinrich Schwabe (1789–1875).

Greek alphabet

alpha A, α	iota I, ι	rho P, ρ
beta B, β	kappa K, κ	sigma Σ, σ
gamma Γ, γ	lambda Λ, λ	tau T, τ
delta Δ, δ	mu M, μ	upsilon Y, υ
epsilon E, ε	nu N, ν	phi Φ, φ
zeta Z, ζ	xi Ξ, ξ	chi X, χ
eta H, η	omicron O, o	psi Ψ, ψ
theta Θ, θ	pi Π, π	omega Ω, ω

greenhouse effect The general warming of the lower layers of a planet's atmosphere caused by the presence of "greenhouse gases," such as water vapor (H_2O), carbon dioxide (CO_2), and methane (CH_4). On Earth, the greenhouse effect might occur because our atmosphere is relatively transparent to visible light from the Sun (typically 0.3 to 0.7 micrometer wavelength) but is essentially opaque to the longer-wavelength (typically 10.6 micrometer) thermal infrared radiation emitted by the planet's surface. Because of the presence of greenhouse gases in our atmosphere—such as carbon dioxide, water vapor, methane, nitrous oxide (NO_2) and human-made chlorofluorocarbons (CFCs)—this outgoing thermal radiation from Earth's surface could be blocked from escaping to space, and the absorbed thermal energy causes a rise in the temperature of the lower atmosphere. Therefore, as the presence of greenhouse gases increases in the Earth's atmosphere, more outgoing thermal radiation is trapped, and a global warming trend occurs. (See Figure 35.)

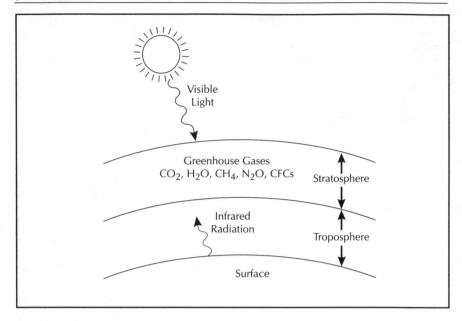

Figure 35. The greenhouse effect in Earth's atmosphere.

Scientists around the world are concerned that human activities, such as increased burning of vast amounts of fossil fuels, are increasing the presence of greenhouse gases in our atmosphere and upsetting the overall planetary energy balance. These scientists are further concerned that we may be creating the conditions of a *runaway greenhouse,* as appears to have occurred in the past on the planet Venus. Such an effect is a planetary climatic extreme in which all of the surface water has evaporated from the surface of a life-bearing or potentially life-bearing planet. Planetary scientists believe that the current Venusian atmosphere allows sunlight to reach the planet's surface, but its thick clouds and rich carbon dioxide content prevent surface heat from being radiated back to space. This condition has led to the evaporation of all surface water on Venus and has produced the present infernolike surface temperatures of approximately 485° Celsius (900° Fahrenheit)—a temperature hot enough to melt lead.

green satellite An Earth-orbiting satellite that collects environmental data. *See also* EARTH OBSERVING SATELLITE; EARTH OBSERVATION SYSTEM (EOS).

Greenwich mean time (GMT) Mean solar time at the meridian of Greenwich, England used as a basis for standard time throughout the world. Normally expressed in four numerals, 0001 through 2400. Also called *universal time, Z-time,* and *zulu time.*

gross liftoff weight (GLOW) The total weight of an aircraft, aerospace vehicle, rocket, as loaded; specifically, the total weight with full crew, full tanks, and payload.

Ground-based Electro-Optical Deep Space Surveillance (GEODSS) There are now more than 8,500 known objects in ORBIT around EARTH. These objects range from active SATELLITEs to pieces of SPACE DEBRIS, or *space junk,* such as expended UPPER STAGE vehicles, fragments from exploded ROCKETs, and even missing tools and cameras from ASTRONAUT-performed EXTRA VEHICULAR ACTIVITY. The Space Control Center (SCC) at CHEYENNE MOUNTAIN AIR FORCE STATION supports the UNITED STATES STRATEGIC COMMAND (USSTRATCOM) missions of space surveillance and the protection of American assets in OUTER SPACE. The SCC maintains a current computerized catalog of all orbiting human-made objects, charts present positions, plots future orbital paths, and forecasts the times and general locations for significant human-made objects reentering Earth's ATMOSPHERE. The center currently tracks more than 8,500 human-made objects in orbit around Earth, of which approximately 20% are functioning PAYLOADS or satellites. The SCC receives data continuously from the Space Surveillance Network (SSN) operated by the UNITED STATES AIR FORCE. These data include information about objects considered to be in DEEP SPACE—that is, at an ALTITUDE of more than 4,800 kilometers (3,000 mi).

Four Ground-based Electro-Optical Deep Space Surveillance (GEODSS) sites around the world play a major role in helping the SCC track human-made objects orbiting Earth. GEODSS is the successor to the Baker-Nunn camera system developed in the mid-1950s. The GEODSS system performs its space-object tracking mission by bringing together TELESCOPEs, low-light-level television, and modern computers. Each operational GEODSS site has at least two main telescopes and one auxiliary telescope. Operational GEODSS sites are located at Socorro, New Mexico (Detachment 1 of the Air Force 18th Space Surveillance Squadron headquartered at Edwards Air Force Base); at Diego Garcia, British Indian Ocean Territories (Detachment 2); at Maui, Hawaii (Detachment 3 collocated on top of Mount Haleakala with the U.S. Air Force Maui Space Surveillance System); and at Morón Air Base, Spain (Detachment 4).

The GEODSS telescopes move across the sky at the same rate as the STARS appear to move. This keeps the distant stars in the same positions in the FIELD OF VIEW. As the telescopes slowly move, the GEODSS cameras take very rapid electronic snapshots of the field of view. Computers then take these snapshot images and overlay them on each other. Star images, which essentially remain fixed, are erased electronically. Human-made space objects circling Earth, however, do not remain fixed, and their movements show up as tiny streaks that can be viewed on a console screen. Computers then measure these streaks and use the data to determine the orbital positions of the human-made objects. The GEODSS system can track objects as small as a basketball orbiting Earth at an altitude of more than 32,200 kilometers (20,000 mi).

ground elapsed time (GET) The time since launch. For example, at two hours GET, the space shuttle crew will open the orbiter's cargo bay doors and prepare to deploy a payload on orbit.

ground receiving station A facility on the surface of Earth that records data transmitted by an Earth observation satellite.

ground state The state of a nucleus, atom, or molecule at its lowest (stable) energy level. *Compare with* EXCITED STATE.

ground support equipment (GSE) Any nonflight (i.e., ground-based) equipment used for launch, checkout, or in-flight support of an aerospace vehicle, expendable rocket, spacecraft, or payload. More specifically, GSE consists of non-flight equipment, devices, and implements that are required to inspect, test, adjust, calibrate, appraise, gauge, measure, repair, overhaul, assemble, transport, safeguard, record, store, or otherwise function in support of a rocket, space vehicle, or the like, either in the research and development phase or in the operational phase. In general, GSE is not considered to include land and buildings but may include equipment needed to support another item of GSE.

ground swath In remote sensing, the width of the terrain strip imaged by a scanner system.

groundtrack The path followed by a spacecraft over Earth's surface.

ground truth In remote sensing, measurements made on the ground to support, confirm, or help calibrate observations made from aerial or space platforms. Typical ground truth data include weather, soil conditions and types, vegetation types and conditions, and surface temperatures. Best results are obtained when these ground truth measurements are performed simultaneously with the airborne or spaceborne sensor measurements. Also referred to as *ground data* and *ground information.*

g-suit (or G-suit) A suit that exerts pressure on the abdomen and lower parts of the body to prevent or retard the collection of blood below the chest under conditions of positive acceleration. *See also* G.

guidance system A system that evaluates flight information; correlates it with target or destination data; determines the desired flight path of the missile, spacecraft, or aerospace vehicle; and communicates the necessary commands to the vehicle's flight control system.

guided missile (GM) An unmanned, self-propelled vehicle moving above the surface of Earth whose trajectory or course is capable of being controlled while in flight. An air-to-air guided missile (AAGM) is an air-launched vehicle for use against aerial targets. An air-to-surface guided missile (ASGM) is an air-launched missile for use against surface targets. A surface-to-air guided missile (SAGM) is a surface-launched guided missile for use against targets in the air. Finally, a surface-to-surface guided missile (SSGM) is a surface-launched missile for use against surface targets. *See also* BALLISTIC MISSILE.

guided missile cruiser (U.S. Navy Designation: CGN) A warship designed to operate with strike and amphibious forces against air, surface, and subsurface threats. Normal armaments consist of 3-inch or 5-inch guns, an advanced area-defense anti-air-warfare missile system, and antisubmarine-warfare weapons.

guided missile destroyer (U.S. Navy Designation: DDG) This U.S. Navy destroyer type is equipped with Terrier/Tartar guided missiles, an improved naval gun battery, long-range sonar, and antisubmarine warfare weapons, including the antisubmarine rocket (ASROC).

guided missile equipment carrier A self-propelled, full-tracked, amphibious, air-transportable, unarmored carrier for various guided missile systems and their equipment.

guided missile frigate (U.S. Navy Designation: FFG) A modern U.S. Navy warship equipped with Tartar or SM-1 missile launchers. This vessel also can be equipped with a 5-inch gun battery or a 54-mm or 70-mm gun battery.

guided missile submarine (nuclear) (U.S. Navy Designation: SSGN) A nuclear-powered submarine designed to have an additional capability to launch guided missile attacks.

guide vane A control surface that may be moved into or against a rocket's jetstream; used to change the direction of the jet flow for thrust vector control.

gun-launch to space (GLTS) An advanced LAUNCH concept involving the use of a long and powerful ELECTROMAGNETIC launcher to hurl small SATELLITES and PAYLOADS into ORBIT. Payloads launched (or perhaps more correctly, "shot") into orbit by an electromagnetic launcher would experience very high peak ACCELERATIONS ranging perhaps from hundreds to thousands of GS (with one g being the normal acceleration due to GRAVITY at EARTH's surface). GLTS has also been suggested for use on the LUNAR surface, where the absence of an

ATMOSPHERE and the MOON's reduced gravitational field (about one-sixth that experienced on Earth's surface) might make this launch approach attractive for hurling raw or refined materials into low lunar orbit.

gyro A device that utilizes the angular momentum of a spinning mass (rotor) to sense angular motion of its base about one or two axes orthogonal (mutually perpendicular) to the spin axis. Also called a *gyroscope.*

H-II launch vehicle An advanced expendable booster sponsored by the National Space Development Agency of Japan (NASDA). The H-II is a two-stage launch vehicle capable of delivering a 2,000-kilogram (kg) (4,400 pound-mass [lbm]) payload into geostationary orbit. The H-II is equipped with high-performance rocket engines using liquid hydrogen and liquid oxygen. The thrust of the first stage is supplemented by two large solid rocket boosters. The first stage uses a large liquid oxygen/liquid hydrogen rocket engine called the LE-7, which employs a staged-combustion cycle. The second stage engine uses a liquid oxygen/liquid hydrogen engine called the LE-5A, which employs a hydrogen bleed cycle. The first launch of the H-II vehicle from the Tanegashima Space Center, Japan, occurred on February 4, 1994; it successfully placed several experimental payloads into their planned orbits. *See also* LAUNCH VEHICLE; NATIONAL SPACE DEVELOPMENT AGENCY OF JAPAN (NASDA).

habitable payload A payload with a pressurized compartment suitable for supporting a crewperson in a shirtsleeve environment.

Hadley Rille A long, ancient lava channel on the MOON that was the landing site for the *Apollo 15* mission during NASA's APOLLO PROJECT.

half-life 1. (radioactive) The time in which half the atoms of a particular radioactive isotope disintegrate to another nuclear form. Measured half-lives vary from millionths of a second to billions of years. The half-life ($T_{1/2}$) is given by the expression:

$$T_{1/2} = (\ln 2)/\lambda = 0.69315/\lambda$$

where λ is the decay constant for the particular radioactive isotope and ln 2 is the natural (Napierian) logarithm of the number 2 with a numerical value of approximately 0.69315. 2. (effective) The time required for a radionuclide contained in a biological system, such as a human being or an animal, to reduce its activity by half as a combined result of radioactive decay and biological elimination. 3. (biological) The time required for the elimination of one-half of the total amount of a toxic substance contained in a biological system by natural processes. 4. (environmental) The time required for a pollutant to lose half its effect on the environment. For example, the half-life of the insecticide DDT in the environment is about 15 years. *See also* LOGARITHM.

halo orbit A circular or ELLIPTICAL ORBIT in which a SPACECRAFT remains in the vicinity of a LAGRANGIAN LIBRATION POINT.

hangfire A faulty condition in the ignition system of a rocket engine.

hard kill The destruction of a target in such a way as to produce unambiguous visible evidence of its neutralization. For example, the explosive fragmentation of a rocket, reentry vehicle, or satellite after a kinetic energy weapon strike would indicate (at a distance) that the target was destroyed.

hard landing A relatively high-velocity impact landing of a "lander" spacecraft or probe on the surface of a planet or moon that destroy all equipment, except perhaps a very rugged instrument package or payload container. This hard landing could be intentional, as occurred during the early U.S. Ranger missions to the lunar surface; or unintentional, as when a retrorocket system fails and the "lander" spacecraft strikes the planetary surface at an unplanned high speed.

hard targets   Ground targets, such as missile silos or deeply buried command centers, that could survive a nuclear blast unless it were to hit within a few hundred meters.

hard vacuum   A very high vacuum; the level of vacuum often associated with deep space. This term often is considered to mean a pressure exerted by any gas of less than [10^{-8} millimeters (mm) of mercury (Hg) (about 1.33×10^{-10} newtons per square meter [N/m^2])].

hardware   1. A generic term dealing with physical items, as distinguished from ideas or "paper designs"; includes aerospace equipment, tools, instruments, devices, components, parts, assemblies, and sub-assemblies. Satellites and rockets often are referred to as *space hardware.* The term also can be used in aerospace engineering to describe the stage of development, as in the passage of a device or component from the design stage into the hardware stage as a finished object. 2. In computer science and data automation, the physical equipment or devices that make up a computer or one of its peripheral components (e.g., a printer), as opposed to the *software,* which operates and controls the computer system.

HARM missile   The AGM-88 High-Speed Antiradiation Missile (HARM) is an air-to-surface tactical missile designed to seek out and destroy enemy radar-equipped air defense systems. An antiradiation missile passively homes in on enemy electronic signal emitters, especially those associated with radar sites used to direct antiaircraft guns and surface-to-air missiles. The AGM-88 can detect, attack, and destroy a target with minimum aircrew input. Used extensively by the U.S. Air Force and Navy.

Harpoon missile   The AGM-84D is an all-weather, over-the-horizon, antiship missile. The Harpoon's active radar guidance, warhead design, and low-level, sea-skimming cruise trajectory assure high survivability and effectiveness. This missile is capable of being launched from surface ships, submarines, or (without its booster component) from aircraft. Originally developed for the U.S. Navy to serve as its basic antiship missile for fleetwide use, the AGM-84D also has been adapted for use on the U.S. Air Force's B-52G bombers.

hatch   A door in the pressure hull of a spacecraft, aerospace vehicle, or space station. The hatch is sealed tightly to prevent the cabin or pressurized module atmosphere from escaping to the vacuum of space outside the pressure hull.

Hawk   A surface-to-air GUIDED MISSILE developed by the United States Army in the late 1950s and continually upgraded since its first field deployment in 1960. Designed to provide air defense protection, the Hawk system is employed as a platoon consisting of an acquisition RADAR, a tracking radar, an identification friend-or-foe (IFF) system, and up to six launchers with three missiles each. The Hawk missile is highly reliable, accurate, and lethal. During the 1991 Persian Gulf War, the Hawk downed several enemy aircraft.

Hawking radiation   A theory proposed in 1974 by the British astrophysicist Stephen W. Hawking (b. 1942) in which he suggests that due to a combination of properties of QUANTUM MECHANICS and GRAVITY, under certain conditions BLACK HOLES can seem to emit radiation.

head (or headrise)   The increase in fluid pressure supplied by a pump; the difference between pressure at the pump inlet and pressure at pump discharge. Fluid pressure sometimes is expressed as an equivalent height of a fluid column, such as 960 millimeters mercury; or else as pascals (newtons per meter-squared [N/m^2] or pounds-force (lbf) per inch-squared [psi]).

head-up display   A display of flight, navigation, attack, or other information superimposed on the pilot's forward field of view.

health physics The branch of science that is concerned with the recognition, evaluation, and control of health hazards from ionizing radiation.

heat Energy transferred by a thermal process. Heat (or *thermal energy*) can be measured in terms of the mechanical units of energy, such as the joule (J), or in terms of the amount of energy required to produce a definite thermal change in some substance, as, for example, the energy required to raise the temperature of a unit mass of water at some initial temperature (e.g., calorie or British thermal unit [BTU]).

1 joule = 0.239 calorie

See also THERMODYNAMICS.

heat balance 1. The equilibrium that exists on the average between the radiation received by a planet and its atmosphere from the Sun and that emitted by the planet and its atmosphere. *See also:* GREENHOUSE EFFECT. 2. The equilibrium that is known to exist when all sources of heat (thermal energy) gain and loss for a given region or object are accounted for. In general, this balance includes conductive, convective, evaporative, and other terms as well as a radiation term.

heat capacity (usual symbol: C) In general, a measure of the increase in energy content of an object or system per degree of temperature rise. This thermodynamic property describes the ability of a substance to store the energy that has been delivered to it as heat (i.e., thermal energy flow). The *specific heat capacity* (C) is related to the *total heat capacity* (C) as follows:

$$c = C/\rho$$

where c has the units (joules/kg-K), C has the units (joules/m^3- K) and ρ is the density of the substance (kg/m^3). The specific heat capacity (c), sometimes shortened to *specific heat,* has very special significance and meaning in classical thermodynamics. For example, the *specific heat at constant volume* (c$_v$) is defined as:

$$c_v = (\partial u/\partial T)_v$$

where u is internal energy (joules/kg) and T is temperature (Kelvin [K]) The expression $(\partial u/\partial T)_v$ means the partial derivative of internal energy (u) with respect to temperature (T) holding volume (v) constant. Similarly, the *specific heat at constant pressure* (c$_p$) is defined as:

$$c_p = (\partial h/\partial T)_p$$

where h is enthaply (joules/kg) and T is temperature (K). The terms c$_v$ and c$_p$ are very important thermodynamic derivative functions.

Heat Capacity Mapping Mission (HCMM) NASA's Heat Capacity Mapping Mission, or HCMM, was launched on April 26, 1978, and operated successfully between April 1978 and September 1980 in a near-polar orbit at 620 kilometers (385 mi) altitude. This Earth-observing spacecraft was the first NASA research effort directed mainly toward observations of the thermal state of Earth's land surface by a satellite. The HCMM sensor measured reflected solar radiation and thermal emission from the surface with a spatial resolution of 600 meters (~1,970 ft). The satellite was placed in an orbit that permitted it to survey thermal conditions during midday and at night. HCMM data have been used to produce temperature difference and apparent thermal inertia images for selected areas within much of North America, Europe, North Africa, and Australia. These data can be used by scientists for rock-type discrimination, soil-moisture detection, assessment of vegetation states, thermal current monitoring in water bodies, urban heat-island assessments, and other environmental studies of Earth.

heat death of the universe A possible ultimate fate of the UNIVERSE suggested in the 19th century by the German theoretical physicist Rudolf Julius Clausius (1822–1888). As he evaluated the consequences of the second law of THERMODYNAMICS on a grand scale, he concluded that the universe would end (die) in a condition of maximum ENTROPY in which there was no ENERGY available for useful work.

heat engine A thermodynamic system that receives energy in the form of heat and that, in the performance of an energy transformation on a working fluid, does work. Heat engines function in cycles. An ideal heat engine works in accordance with the CARNOT CYCLE, while practical heat engines use thermodynamic cycles such as BRAYTON, RANKINE, and STIRLING. The steam engine, which helped create the Industrial Revolution, is a heat engine. Gas turbines and automobile engines are also heat engines.

heater blanket An electrical heater, usually in strip or sheet form, that is wrapped around all or a portion of a cryogenic component (e.g., a valve or actuator) to prevent the temperature within the component from falling below a stated operating minimum.

heat exchanger A device for transferring thermal energy (heat) from one fluid to another without intermixing the fluids. There is a great variety of heat exchanger designs. For example, in a *parallel flow heat exchanger,* both fluids flow in the same direction; while in a *counter-flow heat exchanger,* one fluid flows in the opposite direction of the other. In a *gas-liquid heat exchanger,* one of the fluids is in the gaseous state, while the other is in the liquid state.

heat flux (common symbol: q'') The amount of thermal energy (heat) that is transported across a surface of unit area per unit time; the heat flux (q'') often has the units (joules/meter-squared-second) [$J/(m^2\text{-sec})$] or (British thermal units/foot-squared-second [$BTU/(ft^2\text{-sec})$]).

heat pipe A very-high-efficiency heat (thermal energy) transport device. The basic heat pipe is a closed system that contains a working fluid that transports thermal energy by means of phase change (i.e., through the use of the latent heat of vaporization). An external thermal energy (heat) source is applied to the outside of the evaporator section of the heat pipe. This heat addition causes the working fluid contained in a wick to undergo a phase change from liquid to vapor. The vapor leaves the

wick, flows through the central region of the heat pipe, and eventually arrives at the condenser region. At the condenser section of the heat pipe, the vapor also experiences phase change—this this time, however, going from a gas to a liquid. The condensation process releases heat, which then is removed through the external portion of the evaporator section. The wick (which lines the interior surface of the entire heat pipe) uses capillary action to return the liquid to the evaporator section. Once there, the liquid can again evaporate, continuing the overall heat transport process.

heat shield 1. In general, any device that protects something from unwanted thermal energy (heat). 2. In aerospace engineering, the special structure used to protect a reentry body or aerospace vehicle from aerodynamic heating on flight through a planet's atmosphere; or the protective structure used to block unwanted heat from entering into a spacecraft, its components, or sensing instruments.

heat sink In thermodynamics, a material, region, or environment in which thermal energy (heat) is stored or to which heat is rejected from a heat engine. The term also is used to describe a material capable of absorbing large quantities of heat, such as the special thermal protection materials used on reentry bodies and aerospace vehicles.

heat-sink chamber In rocketry, a combustion chamber in which the heat capacity of the chamber wall limits wall temperature. This approach is effective only for short-duration firings.

heat soak The increase in the temperature of rocket-engine components after firing has ceased. This increase results from heat transfer through contiguous parts of the engine when no active cooling is present.

heat transfer The transfer or exchange of thermal energy (heat) by conduction, convection, or radiation within an object and between an object and its surroundings.

Thermal energy also is transferred when a working fluid experiences phase change, such as evaporation or condensation.

heavy hydrogen An expression for deuterium (D), the nonradioactive isotope of hydrogen with one proton and one neutron in its nucleus.

heavy-lift launch vehicle (HLLV) A conceptual large-capacity, space-lift vehicle capable of carrying tons of cargo into LOW EARTH ORBIT (LEO) at substantially less cost than today's EXPENDABLE LAUNCH VEHICLES (ELVs).

hecto- (symbol: h) A prefix in the SI UNIT system meaning multiplied by 10^2.

hectare A unit of area equal to 10,000 square meters; 1 hectare = 2.4711 acres.

height (symbol: h) Vertical distance; the height above some reference point or plane, as *height above sea level*. The vertical dimension of an object.

height of burst (HOB) The vertical distance from a target on Earth's surface to the point of burst of a nuclear or conventional weapon.

heliocentric Relative to the Sun as a center, as in *heliocentric orbit* or *heliocentric space*.

heliopause The boundary in deep space thought to be roughly circular or perhaps teardrop-shaped, marking the edge of the Sun's influence. It occurs perhaps 100 astronomical units (AU) from the Sun.

heliosphere The space within the boundary of the *heliopause,* containing the Sun and the solar system.

heliostat A mirrorlike device arranged to follow the Sun as it moves through the sky and to reflect the Sun's rays on a stationary collector or receiver

helium (symbol: He) A noble GAS, the second most abundant ELEMENT in the UNIVERSE. Natural helium is mostly the ISOTOPE helium-4, which contains two PROTONS and two NEUTRONS in the NUCLEUS. Helium-3 is a rare isotope of helium containing two protons and one neutron in the nucleus. Helium was initially discovered in the SUN'S SPECTRUM in 1868 by the British physicist Sir Joseph Lockyer (1836–1920) almost four decades before it was found on EARTH.

Hellfire (AGM-114) The Hellfire is a family of AIR-TO-GROUND MISSILES developed by the U.S. Army to provide heavy antiarmor capability for attack helicopters. The first three generations of Hellfire missiles used a laser seeker, while the fourth generation missile, called *Longbow Hellfire,* employs a RADIO FREQUENCY (RF) or RADAR seeker. As part of the war on terrorism, Hellfire missiles have been carried by and successfully fired from the Predator UNMANNED AERIAL VEHICLE (UAV). The U.S. Navy and Marine Corps use versions of the Hellfire missile (namely, the AGM-114B/K/M models) as an air-to-air weapon against hostile helicopters or slow-moving fixed-wing aircraft.

henry (symbol: H) The SI UNIT of inductance (L). Inductance relates the production of an electromotive force (E) in a conductor when there is a change in the magnetic flux (φ) in that conductor. The induced electromotive force (E) is proportional to the time rate of change of the current (dI/dt), namely:

$$E = -L \, (dI/dt)$$

where the inductance (L) serves as a proportionality constant and depends on the geometric design of the circuit.

One henry (H) is defined as the inductance occurring in a closed electric circuit in which an electromotive force (or emf) of 1 volt is produced when the current (I) in the circuit is varied uniformly at the rate of 1 ampere per second. This unit has been named in honor of the American physicist Joseph Henry (1797–1878).

hermetic seal A seal evidencing no detectable leakage or permeation of gas or moisture.

hertz (symbol: Hz) The SI UNIT of frequency. One hertz is equal to 1 cycle per second. Named in honor of the German physicist H. R. Hertz (1857–94).

Hertzsprung-Russell (H-R) diagram A useful graphic depiction of the different types of STARS arranged according to their SPECTRAL CLASSIFICATION and LUMINOSITY. Named in honor of the Danish astronomer Ejnar Hertzsprung (1873–1969) and the American astronomer Henry Norris Russell (1877–1957), who developed the diagram independently of one another.

high-Earth orbit (HEO) An orbit around Earth at an altitude greater than 5,600 kilometers (3,500 mi).

High Energy Astrophysical Observatory (HEAO) A series of three NASA SPACECRAFT placed in EARTH ORBIT (HEAO-1 launched in August 1977; HEAO-2 in November 1978; and HEAO-3 in September 1979) to support X-RAY ASTRONOMY and GAMMA RAY ASTRONOMY. After LAUNCH, NASA renamed HEAO-2 the *Einstein Observatory* in honor of the famous German-Swiss-American physicist Albert Einstein (1879–1955).

high-energy laser (HEL) A space-based laser capable of generating high-energy beams (i.e., 20 kilowatts or greater average power; 1 kilojoule or more per pulse) in the infrared, visible, or ultraviolet portion of the electromagnetic spectrum. When projected to a target this beam would accomplish damage ranging from degradation to destruction. (See Figure 37.)

high explosive (HE) A chemical explosive that when used in its normal manner detonates rather than deflagrates or burns; that is, the rate of advance of the reaction zone into the unreacted material exceeds the speed of sound.

highlands Oldest exposed areas on the surface of the Moon; extensively cratered

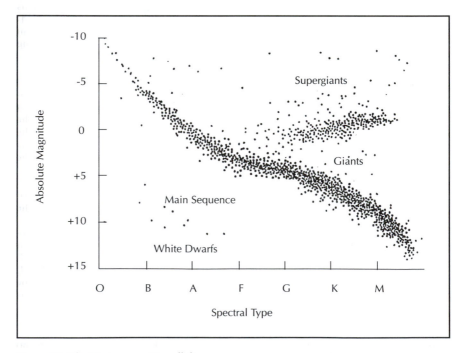

Figure 36. The Hertzsprung-Russell diagram.

Figure 37. Artist's rendering of a space-based, high-energy laser system (circa 2020) that would intercept enemy ballistic missiles in the boost phase, before they can deploy their warheads and decoys. (*Courtesy of U.S. Department of Defense*)

and chemically distinct from the *maria*. *See also* MOON.

high-power microwave (HPM) A proposed space-based weapon system that would be capable of destroying not only targets in space but also airborne and ground targets. This type of weapon system typically would involve peak powers of 100 megawatts or more and a single pulse energy of 100 joules or more. Target damage would result from electronic upset or burnout.

hit-to-kill vehicle A kinetic (hypersonic velocity) vehicle that destroys its target by hitting it and thereby transferring a lethal amount of kinetic energy to it.

hohlraum In radiation heat transfer, a cavity whose walls are in radiative equilibrium with the radiant energy within the cavity. This idealized cavity can be approximated in practice by making a small hole or perforation in the walls of a hollow container of any opaque material. The radiant energy escaping through such a tiny hole will be a reasonable approximation to BLACKBODY radiation at the temperature of the interior of the container.

Hohmann transfer orbit The most efficient ORBIT transfer path between two coplanar circular orbits. The maneuver consists of two impulsive high-THRUST burns (or firings) of a SPACECRAFT's PROPULSION SYSTEM. The first burn is designed to change the original circular orbit to an ELLIPTICAL ORBIT whose PERIGEE is tangent with the lower-ALTITUDE circular orbit and whose APOGEE is tangent with the higher-altitude circular orbit. After coasting for half of the elliptical transfer orbit (that is, the *Hohmann transfer orbit*) and when tangent with the destination circular orbit, the second impulsive high-thrust burn is performed by the spacecraft's onboard propulsion system (sometimes called an APOGEE MOTOR). This circularizes the spacecraft's orbit at the desired new altitude. The technique can also be used to lower the altitude of a SATELLITE from one circular orbit to another circular orbit (of less altitude). In this case, two impulsive retrofirings are required. The first RETROFIRE (that is, a ROCKET firing with the thrust directed opposite to the direction of travel) places the spacecraft on an elliptical transfer orbit. The second retrofire then takes place at perigee of the elliptical transfer orbit. The second spacecraft's orbit at the desired lower altitude. The technique was suggested in 1925 by the German engineer Walter Hohmann (1880–1945).

hold In rocketry, to halt the sequence of events during a countdown until an impediment has been removed so that the countdown to launch can be resumed; for example, "T minus 25 and *holding* because of a sudden change in weather conditions."

holddown test The testing of some system or subsystem in a rocket, while the rocket is firing but restrained in a test stand.

homing guidance A guidance system, by which a missile steers itself toward a target by means of a self-contained mechanism that is activated by some distinguishing characteristics of the target. The process is called: *active homing guidance,* if the target information is received in response to signals transmitted by the missile (e.g., reflected radar beams when the search radar is carried by the missile); *semiactive homing guidance,* if the missile's homing system follows a reflected signal from the target and the source of the reflected signal is other than the missile or the target (e.g., a missile following laser light reflected from the target when the laser beam [also called a target designator beam] is placed or "painted" on the target by the attacking aircraft platform); and *passive guidance,* if the missile follows a signal emitted by the target itself (e.g.,

an infrared signature associated with an exhaust plume).

Honest John An early surface-to-surface missile deployed by the U.S. Army. It has a range of about 50 kilometers (30 mi).

hook Aerospace industry jargon for a design feature to accommodate the addition or upgrade of computer software at some future time; as, for example, the new space platform has been *hooked* and *scarred* to accommodate a telerobotic laboratory capability.

horizon The line marking the apparent junction of Earth and sky.

Horsehead Nebula A DARK NEBULA in the CONSTELLATION Orion that has the shape of a horse's head.

horsepower (symbol: hp) a traditional engineering unit of power defined as:
$$1 \text{ hp} = 550 \text{ foot-pounds force/second}$$
$$= 746 \text{ watts}$$

hot-core injector In rocketry, an injector that produces a central hot-gas combustion region surrounded by a "cold" fuel sheath.

hot fire test A liquid-fuel PROPULSION SYSTEM test conducted by actually firing the ROCKET engine(s) (usually for a short period of time) with the rocket vehicle secured to the LAUNCH PAD by holddown bolts. *Compare with* COLD-FLOW TEST.

Hound Dog An early U.S. Air Force air-to-surface missile with a range of about 800 kilometers (500 miles). It was externally carried under the wing of long-range strategic bombers, such as the B-52.

hour (symbol: h) A unit of time equal to 3,600 seconds or 60 minutes.

"housekeeping" The collection of routine tasks that must be performed to keep an aerospace vehicle or spacecraft functioning properly. For example, the astronaut crew or ground support team performing frequent status checks and making minor adjustments on the life support and waste management systems of a human-occupied space vehicle; or the mission control team performing regular telemetry management checks and trajectory adjustments on an interplanetary spacecraft as it cruises toward its planetary target.

Hubble constant (symbol: H_0) The constant within HUBBLE'S LAW proposed in 1929 by the American astronomer Edwin Powell Hubble (1889–1953) that establishes an EMPIRICAL relationship between the distance to a GALAXY and its VELOCITY of recession due to the expansion of the UNIVERSE. A value of 70 kilometers per second per MEGAPARSEC [km/s/Mpc] ± 7 km/s/Mpc is currently favored by astrophysicists and astronomers.

Hubble's law The hypothesis that the REDSHIFTS of distant galaxies are directly proportional to their distances from EARTH. The American astronomer Edwin Powell Hubble (1889–1953) first proposed this relationship in 1929. It can be expressed as $V = H_0 \times D$, where V is the recessional VELOCITY of a distant GALAXY, H_0 is the HUBBLE CONSTANT, and D is its distance from Earth.

Hubble Space Telescope (HST) A cooperative program between NASA and the EUROPEAN SPACE AGENCY (ESA) to operate a long-lived, space-based optical observatory. Launched on April 25, 1990, by NASA's SPACE SHUTTLE *Discovery* (STS-31 mission), subsequent on-orbit repair and refurbishment missions have allowed this powerful EARTH-orbiting optical observatory to revolutionize knowledge of the size, structure, and makeup of the UNIVERSE. It is named in honor of the American astronomer Edwin Powell Hubble (1889–1953).

human factors engineering The branch of engineering involved in the design, development, testing, and construction of devices, equipment, and artificial living environments to the anthropometric, physiological, and/or psychological requirements of the human beings who will use them. In aeroscope, human factors engineering is involved in such diverse areas as the design of a functional microgravity toilet (for both male and female crew-persons), the creation of crew living and sleeping quarters that are both efficient (from a size perspective) and yet provide a suitable level of aesthetics and privacy, and the development of crew workstations that can accommodate a variety of mission needs and still provide immediate attention to hazardous or emergency situations (i.e., a warning light that "jumps out" immediately and can be recognized easily amid 100 so other lights and indicators).

humidity The amount of water vapor in the air; the "wetness" of the atmosphere. The *relative humidity* represents the ratio of the amount of water vapor present in the air at a given temperature to the greatest amount of water vapor possible for that temperature. For example, a 10% relative humidity is very dry air, while a 90% relative humidity is very moist air.

Huygens probe A scientific PROBE sponsored by the EUROPEAN SPACE AGENCY (ESA) and named after the Dutch astronomer Christiaan Huygens (1629–1695), who discovered TITAN, SATURN's largest MOON, in 1655. The probe will be deployed into the ATMOSPHERE of Titan in 2004 by NASA's CASSINI MISSION SPACECRAFT.

hydraulic Operated, moved, or affected by liquid used to transmit energy.

hydrazine (symbol: N_2H_4) A toxic, colorless liquid that often is used as a rocket propellant because it reacts violently with many oxidizers. It is spontaneously ignitable with concentrated hydrogen peroxide (H_2O_2) and nitric acid. When decomposed by a suitable catalyst, hydrazine is also a good monopropellant and can be used in simple, small rocket engines, such as those for spacecraft attitude control.

hydrodynamics The branch of science that studies the motion of incompressible fluids and fluid-boundary interactions.

hydrogen (symbol: H) A colorless, odorless GAS that is the most abundant chemical ELEMENT in the UNIVERSE. Hydrogen occurs as molecular hydrogen (H_2), atomic hydrogen (H), and ionized hydrogen (that is, broken down into a PROTON and its companion ELECTRON). Hydrogen has three isotopic forms, PROTIUM (ordinary hydrogen), DEUTERIUM (heavy hydrogen), and TRITIUM (radioactive hydrogen). LIQUID HYDROGEN (LH_2) is an excellent high-performance CRYOGENIC chemical PROPELLANT for ROCKET engines, especially those using LIQUID OXYGEN (LO_2) as the OXIDIZER.

hydrogen bomb A nuclear weapon that derives its yield (energy) largely from thermonuclear (fusion) reactions. *See also* FUSION (NUCLEAR).

hydrogen embrittlement Decrease in a metal's tensile strength, notched tensile strength, fatigue strength, resistance-to-crack growth, and especially ductility as a result of absorption by the metal of newly formed gaseous hydrogen.

hydrosphere The water on EARTH's surface (including oceans, seas, rivers, lakes, ice caps, and glaciers) considered as an interactive system. *See also* EARTH SYSTEM SCIENCE.

hydrostatic pressure The pressure exerted by a column of fluid; fluid pressure due to the force of gravity.

hyperbaric Pertaining to breathing atmospheric pressures above sea level normal.

hyperbaric chamber A chamber used to induce an increase in ambient pressure as would occur in descending below sea level, in a water or air environment. It is the only type of chamber suitable for use in the treatment of decompression sickness.

hyperbola An open curve with two branches, all points of which have a constant difference in distance from two fixed points called *focuses* (or foci).

hyperbolic Of or pertaining to a HYPERBOLA.

hyperbolic orbit An ORBIT in the shape of a hyperbola. All INTERPLANETARY FLYBY SPACECRAFT follow hyperbolic orbits, both for EARTH departure and again upon arrival at the target PLANET. *See also* CONIC SECTION.

hyperbolic velocity A velocity sufficient to allow escape from a planetary system or the solar system; for example, all interplanetary spacecraft missions follow hyperbolic orbits, both for Earth departure and then again upon arrival at the target planet. Comets, if not gravitationally captive to the Sun, can have hyperbolic velocities; their trajectories would be escape trajectories or hyperbolas. *See also* ORBITS OF OBJECTS IN SPACE.

hypergolic fuel (hypergol) A rocket fuel that spontaneously ignites when brought into contact with an oxidizing agent (oxidizer); for example, aniline ($C_6H_5NH_2$) mixed with red fuming nitric acid (85% HNO_3 and 15% N_2O_4) produces spontaneous combustion.

hyperoxia An aerospace medicine term describing a condition in which the total oxygen content of the body is increased above that normally existing at sea level.

hypersonic Of or pertaining to speeds equal to or in excess of five times the speed of sound (i.e., $\geq$ Mach 5).

hypersonic glider An unpowered aerospace vehicle designed to fly and maneuver at hypersonic speeds upon reentering a planetary atmosphere.

hypervelocity Extremely high velocity. The term is applied by physicists to speeds approaching the speed of light; but in aerospace applications, the term generally implies speeds of the order of satellite or spacecraft speed and greater (i.e., 5 to 10 kilometers per second and greater).

hypervelocity gun (HVG) A gun that can accelerate projectiles to 5 kilometers per second or more; for example, an electromagnetic gun. *See also* GUN-LAUNCH TO SPACE.

hypervelocity impact A collision between two objects that takes place at a very high relative VELOCITY—typically at a speed in excess of five kilometers per second (3.1 mi/s) A SPACECRAFT colliding with a piece of SPACE DEBRIS or an ASTEROID striking a PLANET are examples.

hypobaric Pertaining to low atmospheric pressure, especially the low pressure of high altitudes.

hypobaric chamber A chamber used to induce a decrease in ambient pressure as would occur in ascending altitude. This type of chamber is used primarily for flight training purposes and for aerospace experiments. Also called an *altitude chamber.*

hypothesis A scientific theory proposed to explain a set of data or observations; can be used as basis for further investigation and testing.

hypoxia An aerospace medicine term describing an oxygen deficiency in the blood, cells, or tissues of the body in such degree as to cause psychological and physiological disturbances.

hysteresis 1. Any of several effects resembling a kind of internal friction, accompanied by the generation of thermal energy (heat) within the substance affected. *Magnetic hysteresis* occurs when a ferromagnetic is subjected to a varying magnetic intensity. *Electric hysteresis* occurs when a dielectric is subjected to a varying electric intensity. *Elastic hysteresis* is the internal friction in an elastic solid subjected to varying stress. 2. The delay of an indicator in registering a change in a parameter being measured.

HZE particles The most potentially damaging cosmic rays, with high atomic number (Z) and high kinetic energy (E). Typically, HZE particles are atomic nuclei with Z greater than 6 and E greater than 100 million electron volts (100 MeV). When these extremely energetic particles pass through a substance, they deposit a large amount of energy along their tracks. This deposited energy ionizes the atoms of the material and disrupts molecular bonds. *See also* COSMIC RAYS.

ice catastrophe   A planetary climatic extreme in which all the liquid water on the surface of a lifebearing or potentially lifebearing planet has become frozen or completely glaciated. *See also* GLOBAL CHANGE.

ideal gas   The pressure (p), volume (V), and temperature (T) behavior of many gases at low pressures and moderate temperatures is approximated quite well by the ideal (or *perfect*) gas equation of state, which is

$$p V = N R_u T$$

where N is the number of moles of gas and R_u is the universal gas constant.

$$R_u = 8314.5 \text{ joules/kg-mole-K} = 1.986$$
$$\text{Btu/lb}_{mass}\text{-mole-}°\text{R}$$

This very useful relationship is based on the experimental work originally conducted by Boyle (Boyle's Law), Charles (Charles Law) and Gay-Lussac (the Gay-Lussac Law). In the ideal gas approximation, we assume that there are no forces exerted between the molecules of the gas and that these molecules occupy negligible space in the containing region. The ideal gas equation above and its many equivalent forms has widespread application in thermodynamics, as, for example, in describing the performance of an IDEAL ROCKET.

ideal nozzle   The nozzle of an ideal rocket, or a nozzle designed according to the ideal gas laws.

ideal rocket   A theoretical rocket postulated for design and performance parameters that are then corrected (i.e., modified) in aerospace engineering practice. The ideal rocket model often assumes one or more of the following conditions: a homogeneous and invariant propellant; exhaust gases that behave in accordance with the ideal (perfect) gas laws; the absence of friction between the exhaust gases and the nozzle walls; the absence of heat transfer across the rocket engine and/or nozzle wall(s); an axially directed velocity of all exhaust gases; a uniform gas velocity across every section normal to the nozzle axis; and chemical equilibrium established in the combustion chamber and maintained in the nozzle. *See also* ROCKET.

igniter   A device used to begin combustion, such as a SQUIB used to ignite the fuel in a solid propellant rocket. The igniter typically contains a specially arranged charge of a ready-burning composition (such as blackpowder) and is used to amplify the initiation of a primer.

ignition   The attainment of self-sustaining combustion of propellants in a rocket engine or motor.

illuminance (symbol: E)   The amount of visible radiation incident on a given surface per unit time per unit area; the luminous flux (φ) per unit area. Illuminance is often measured in *lux* (lumens per square-meter). Also called *illumination*.

image   A graphic representation of a physical object or scene formed by a mirror, lens, or electro-optical recording device. In remote sensing, the term "image" generally refers to a representation acquired by nonphotographic methods. *See also* REMOTE SENSING.

image compensation   Movement intentionally imparted to a photographic film system or electro-optical imagery system at such a rate as to compensate for the forward motion of an air or space vehicle when observing objects on the ground. Without image compensation, the forward motion of the air or space vehicle would cause blurring and degradation of the image produced by the system.

image converter An electro-optical device capable of changing the spectral (i.e., wavelength) characteristics of a radiant image. The conversion of an infrared image to a visible image, an X-ray image to a visible image, and a gamma ray image to a visible image are all examples of image conversion performed by this type of device.

image processing A general term describing the contemporary technology used for acquiring, transferring, analyzing, displaying, and applying digital imagery data. Often this image processing can take place in real time or near real time. It is especially useful in defense applications that involve the use of mobile sensors for real-time target acquisition and guidance, the processing and display of large complex data sets, timely data transmission and compression techniques, and the three-dimensional presentation of battlefield information. The remote sensing of Earth and other planetary objects also has been greatly enhanced by advanced image processing techniques. These techniques (often conducted on today's personal computers) enable remote sensing analysts to compare, interpret, archive, and recall vast quantities of "signature" information emanating from objects on Earth or other planetary bodies. Modern image processing supports applications in such important areas as crop forecasting, rangeland and forest management, land use planning, mineral and petroleum exploration, mapmaking, water quality evaluation, and disaster assessment. *See also* REMOTE SENSING.

imagery Collectively, the representations of objects produced by image-forming devices such as electro-optical sensor systems or photographic systems.

imagery correlation The mutual relationship between the different signatures on imagery from different types of sensors in terms of position and the physical characteristics signified.

imagery exploitation The cycle of processing and printing imagery to the positive or negative state, assembly into imagery packs, identification, interpretation, mensuration, information extraction, preparation of reports, and dissemination of information.

imagery intelligence Intelligence information derived from the exploitation of imagery collected by visual photography, infrared sensors, lasers, electro-optics, and radar sensors (e.g., synthetic aperture radar). Images of objects and target scenes may be reproduced optically or electronically on film, electronic display devices or other media. Also called *IMINT.*

imagery interpretation 1. The process of location, recognition, identification, and description of objects, activities, and terrain represented on imagery. 2. The extraction of information from photographs or other recorded images.

imaging instruments Optical imaging from scientific spacecraft is performed by two families of detectors: *vidicons* and the newer *charge coupled devices* (CCDs). Although the detector technology differs, in each case an image of the target celestial object is focused by a telescope onto the detector, where it is converted to digital data. Color imaging requires three exposures of the same target, through three different color filters selected from a filter wheel. Ground processing combines data from the three black-and-white images, reconstructing the original color by using three values for each picture element (pixel).

A vidicon is a vacuum tube resembling a small cathode ray tube (CRT). An electron beam is swept across a phosphor coating on the glass where the image is focused, and its electrical potential varies slightly in proportion to the levels of light it encounters. This varying potential becomes the basis of the video signal produced. Viking, Voyager, and many earlier spacecraft used vidicon-based imaging systems to send back spectacular images of Mars (Viking) and the outer planets: Jupiter, Saturn, Uranus, and Neptune (Voyager).

The newer charge coupled device (CCD) imaging system is typically a large-scale integrated circuit that has a two-dimensional array of hundreds of thousands of charge-isolated wells, each representing a pixel. Light falling on a well is absorbed by a photoconductive substrate (e.g., silicon) and releases a quantity of electrons proportional to the intensity of the incident light. The CCD then detects and stores an accumulated electrical charges, representing the light level on each well. These charges subsequently are read out for conversion to digital data. CCDs are much more sensitive to light over a wider portion of the electro-magnetic spectrum than vidicon tubes; they are also less massive and require less energy to operate. In addition, they interface more easily with digital circuitry, simplifying (to some extent) onboard data processing and transmission back to Earth. The Galileo spacecraft's solid state imaging (SSI) instrument contained a CCD with an 800 × 800 pixel array. *See also* GALILEO PROJECT; VIKING PROJECT; VOYAGER.

impact 1. A single, forceful collision of one mass in motion with a second mass that may be either in motion or at rest; for example, the *impact* of a meteoroid traveling at high velocity with the surface of the Moon. 2. Specifically in aerospace operations, the action or event of an object, such as a rocket or a space probe, striking the surface of a planet or natural satellite; the time of this event, as in from launch to *impact*. 3. To strike an object or surface, as in the kinetic energy weapon *impacted* its target at a velocity of 5 kilometers per second or the missile *impacted* 20 minutes after launch.

impact area 1. In general, an area having designated boundaries within the limits of which all ordnance is to make contact with the ground. 2. In aerospace operations, the area in which a rocket or missile strikes the surface of Earth or a space probe or spacecraft strikes the surface of a planet. With respect to missile range safety, an area surrounding an approved impact point. The extent and configuration of this area is based on the rocket vehicle or stage dispersion characteristics.

impact crater The CRATER or basin formed on the surface of a planetary body as a result of the high speed IMPACT of a METEOROID, ASTEROID, or COMET. (See Figure 38.)

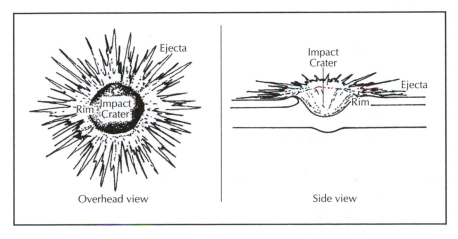

Figure 38. Ideal example of a small fresh-impact crater.

impact line With respect to missile range safety, an imaginary line on the outside of the DESTRUCT LINE and running parallel to it, which defines the outer limits of impact for fragments from a rocket or missile destroyed under command destruct procedures. Also called *impact limit line. See also* ABORT.

impactor 1. The meteoroid, asteroid, or comet that strikes a planetary surface at high velocity creating an impact crater. 2. The hypervelocity projectile fired by a kinetic energy weapon (KEW) system that strikes its target. 3. A piece of space debris that strikes an aerospace vehicle or spacecraft at hypervelocity.

impact pressure 1. Pressure of a moving fluid brought to rest that is in excess of the pressure the fluid has when it does not flow, that is, total pressure less static pressure. Impact pressure is equal to dynamic pressure in incompressible flow; but in compressible flow, impact pressure includes the pressure change owing to the compressibility effect. 2. A measured quantity obtained by placing an open-ended tube, known as an impact tube or PITOT TUBE, in a gas stream and noting the pressure in the tube on a suitable instrument, such as a manometer. Since pressure is exerted at a stagnation point, the impact pressure sometimes is referred to as the stagnation pressure.

impedance (symbol: Z) A quantity describing the total opposition to current flow [both resistance (R) and reactance (X)] in an alternating current (a.c) circuit. For an a.c. circuit, the impedance can be expressed as:
$$Z^2 = R^2 + X^2$$
where Z is the impedance, R is the resistance, and X is the reactance, all expressed in ohms (Ω).

impeller A device that imparts motion to a fluid; for example, in a centrifugal compressor, the impeller is a rotary disk that, faced on one or both sides with radial vanes, accelerates the incoming fluid outward into a diffuser.

impinging-stream injector In a liquid-propellant rocket engine, a device that injects the fuel and oxidizer into the combustion chamber in such a manner that the streams of fluid intersect one another.

implosion The rapid inward collapse of the walls of an evacuated system or device, as a result of the failure of the walls to sustain the ambient pressure. An implosion also can occur when a sudden sharp inward pressure force is exerted uniformly on the surface of an object or composite structure, causing it to crush and collapse inward. A submarine diving below the "crush depth" of its hull design will experience an implosion. The Galileo probe plunging into the depths of the Jovian atmosphere in December 1995 also experienced an implosion at the end of its brief but successful scientific mission. If the application of external pressure is sufficiently rapid and intense, even a solid metal object will experience some degree of "squeezing" and compression as its volume is decreased and its density increased during an implosion.

impulse (symbol: I) In general, a mechanical "jolt" delivered to an object that represents the total change in momentum the object experiences. Physically, the thrust force (F) integrated over the period of time (t_1 to t_2) it is applied. The impulse theorem is expressed mathematically as:
$$\text{Impulse (I)} = {}_1\!\int^2 F \, dt = \text{change of momentum of object,}$$
where F is the time-dependent thrust force applied from t_1 to t_2. If the thrust force is constant (or can be treated as approximately constant) during the period, then the impulse (I) becomes simply:
$$I = F \, \Delta t \text{ (the impulse approximation)}$$
where Δt is the total time increment. Impulse (I) has the units of newton-second (N-s) in the SI system. Also called the total impulse, I_{total}. Compare with but do not confuse with SPECIFIC IMPULSE (I_{sp}).

impulse intensity Mechanical impulse per unit area. The SI UNIT of impulse intensity is the pascal-second (Pa-s). A conventionally used unit of impulse intensity is the *tap*, which is 1 dyne-second per square centimeter.

1 tap = 0.1 Pa-s

impulse kill   The destruction of a target by ablative shock using a directed energy weapon (DEW). The intensity of directed energy (e.g., a high-energy laser pulse) may be so great that the surface of the target violently and rapidly boils off, delivering a mechanical shock to the rest of the target and causing structural failure.

incandescence   Emission of light due to high temperature of the emitting material, as, for example, the very hot filament in an incandescent lamp. Any other emission of light is called *luminescence*.

inch (symbol: in)   A unit of length found in the English (traditional engineering) system. An inch is defined as equal to 1/36 yard. By international agreement, 1 inch = 2.540 centimeters exactly.

inclination (symbol: i)   One of the six KEPLERIAN (ORBITAL) ELEMENTS; inclination describes the ANGLE of an object's ORBITAL PLANE with respect to the central body's EQUATOR. For EARTH-orbiting objects, the orbital plane always goes through the center of Earth, but it can tilt at any angle relative to the equator. By general agreement, inclination is the angle between Earth's equatorial plane and the object's orbital plane measured counterclockwise at the ASCENDING NODE. *See also* ORBITS OF OBJECTS IN SPACE.

incompressible fluid   A fluid for which the density ρ is assumed constant.

increment   An increase in the value of a variable. *Compare with* DECREMENT.

independent system   In aerospace engineering, a system not influenced by other systems. For example, an independent circuit would not require other circuits to be functioning properly in order to perform its task. This "independent" characteristic requires that the circuit to be powered from an independent power supply, be controlled from a single source, and (in terms of redundancy) be tolerant of all credible failure modes in the corresponding redundant circuit or system. To make a system truly independent, the aerospace engineer also must exclude interface items such as mounting brackets and connectors that can create a common failure.

Indian Space Research Organization (ISRO)   The organization that conducts the major portions of the national space program of India, including the development of LAUNCH VEHICLES, SATELLITES and SOUNDING ROCKETS. Under its government-assigned charter, ISRO has produced the Indian National Satellite (INSAT) for communications, television broadcasting, meteorology, and disaster warning and the Indian Remote Sensing (IRS) satellite for resource monitoring and management. ISRO's launch vehicle program started with the development of the SLV-3, first successfully flown on July 18, 1980. This effort was followed by the development and operational use of the Augmented Satellite Launch Vehicle (ASLV) in 1992 and the Polar Satellite Launch Vehicle (PSLV) in 1997. The Geosynchronous Satellite Launch Vehicle (GSLV) is under development.

induction heating   The heating of an electrically conducting material using a varying electromagnetic field to induce eddy currents within the material. This may be an undesirable effect in electric power generation and distribution equipment but a desirable effect in materials processing (e.g., an induction heating furnace).

inert   In aerospace operations, not containing an explosive or other hazardous material such that checkout operations can be conducted in a less restrictive manner and so that anomalies can occur in a safe configuration.

inert gas   Any one of six "noble" gases: helium (He), neon (Ne), argon (Ar), krypton (Kr), xenon (Xe), and radon (Rn), all of which are almost completely chemically inactive. All of these gases are found in Earth's atmosphere in small amounts.

inertia The resistance of a body to a change in its state of motion. Mass (m) is an inherent property of a body that helps us quantify inertia. *Newton's First Law of Motion* sometimes is called the "Law of Inertia"—namely: a body in motion will remain in motion unless acted upon by an external force and a body at rest will remain at rest unless acted upon by an external force. *See also* NEWTON'S LAWS OF MOTION.

inertial guidance system A guidance system designed to project a missile or rocket over a predetermined path, wherein the path of the vehicle is adjusted after launching by (inertial) devices wholly within the missile or rocket and independent of outside information. The inertial guidance system measures and converts accelerations experienced to distance traveled in a certain direction. *See also* GUIDANCE SYSTEM.

inertial orbit An orbit that conforms with Kepler's laws of celestial motion. Kepler's laws describe the behavior of natural celestial bodies and spacecraft that are not under any type of propulsive power. *See also* KEPLER'S LAWS; ORBITS OF OBJECTS IN SPACE.

inertial upper stage (IUS) A versatile ORBITAL TRANSFER VEHICLE (OTV) developed by the UNITED STATES AIR FORCE and manufactured by the Boeing Company. The IUS is a two-STAGE PAYLOAD delivery system that travels from LOW EARTH ORBIT (LEO) to higher ALTITUDE destinations, such as GEOSYNCHRONOUS ORBIT. The versatile UPPER STAGE vehicle is compatible with both the SPACE SHUTTLE and the TITAN IV EXPENDABLE LAUNCH VEHICLE. The IUS consists of two high-performance, SOLID PROPELLANT ROCKET motors, an INTERSTAGE SECTION, and supporting guidance, navigation, and communications equipment. The first stage rocket motor typically contains about 9,700 kilograms (21,400 lbm) of solid propellant and gener-

ates a THRUST of about 185,000 NEWTONS (41,600 lbf). The second stage solid rocket motor contains about 2,720 kilograms (17,600 lbm) of propellant and generates some 78,300 newtons (17,600 lbf) of thrust. The NOZZLE of the second stage has an extendable exit CONE for increased system performance. In July 1999, an IUS system propelled the CHANDRA X-RAY OBSERVATORY from low Earth orbit (LEO) into its operational ELLIPTICAL ORBIT around EARTH that reaches one-third of the distance to the MOON. In addition, the IUS has helped send the GALILEO PROJECT SPACECRAFT on its journey to JUPITER, the MAGELLAN MISSION spacecraft to Venus, and the ULYSSES MISSION spacecraft on its extended journey to explore the polar regions of the SUN. The U.S. Air Force has used IUS rockets to send EARLY WARNING SATELLITES, such as the large DEFENSE SUPPORT PROGRAM spacecraft, from low Earth orbit to their operational locations in geosynchronous orbit.

inferior planet(s) Planets that have orbits that lie inside Earth's orbit around the Sun—namely, Mercury and Venus.

infinity (symbol: ∞) A quantity beyond measurable limits.

in-flight phase The flight of a missile or rocket from launch to detonation or impact; the flight of a spacecraft from launch to the time of planetary flyby, encounter and orbit, or impact.

in-flight start An engine ignition sequence after take-off and during flight. This term includes starts both within and above the sensible atmosphere.

Infrared Astronomical Satellite (IRAS) The *Infrared Astronomical Satellite (IRAS)*, which was launched in January 1983, was the first extensive scientific effort to explore the universe in the infrared portion of the electromagnetic spectrum. *IRAS* was an international effort involving the United States, the United Kingdom, and the Netherlands. By the time *IRAS* ceased operations in November 1983, this successful space-based infrared telescope had completed the first all-sky survey in a wide range of IR wavelengths with a sensitivity 100 to 1,000 times greater than any previous telescope.

infrared (IR) astronomy The branch of astronomy dealing with infrared (IR) radiation from celestial objects. Most celestial objects emit some quantity of infrared radiation. However, when a star is not quite hot enough to shine in the visible portion of the electromagnetic spectrum, it emits the bulk of its energy in the infrared. IR astronomy, consequently, involves the study of relatively cool celestial objects, such as interstellar clouds of dust and gas (typically about 100 Kelvin [K]) and stars with surface temperatures below temperatures below about 6,000 K.

Many interstellar dust and gas molecules emit characteristic infrared signatures that astronomers use to study chemical processes occurring in interstellar space. This same interstellar dust also prevents astronomers from viewing visible light coming from the center of our Milky Way Galaxy. However, IR radiation from the galactic nucleus is absorbed not as severely as radiation in the visible portion of the electromagnetic spectrum, and IR astronomy enables scientists to study the dense core of the Milky Way.

Infrared astronomy also allows astrophysicists to observe stars as they are being formed (called protostars) in giant clouds of dust and gas (called nebula), long before their thermonuclear furnaces have ignited and they have "turned on" their visible emission.

Unfortunately, water and carbon dioxide in Earth's atmosphere absorb most of the interesting IR radiation arriving from celestial objects. Earth-based astronomers can use only a few narrow IR spectral bands or windows in observing the universe; and even these IR windows are distorted by "sky noise" (undesirable IR radiation from atmospheric molecules). With the arrival of the space age, however astronomers have placed sophisticated IR telescopes (such as the INFRARED ASTRO-

NOMICAL SATELLITE) in space, above the limiting and disturbing effects of Earth's atmosphere, and have produced comprehensive catalogs and maps of significant infrared sources in the observable universe.

infrared (IR) imagery That imagery produced as a result of electro-optically sensing electromagnetic radiations emitted or reflected from a given object or target surface in the infrared portion of the electromagnetic spectrum (which extends from approximately 0.72 micrometers [near infrared] to about 1,000 micrometers [far infrared]. *Thermal imagery* is infrared imagery associated with the 8- to 14-micrometer thermal infrared region. Note: Infrared imagery *is not* the same as photographs produced using infrared film.

infrared (IR) photography Photography employing an optical system and direct image recording on "infrared film" that is sensitive to near-infrared wavelengths (about 1.0 micrometer). Note: Infrared photography *is not* the same as infrared imagery.

infrared (IR) radiation That portion of the electromagnetic (EM) spectrum lying between the optical (visible) and radio wavelengths. It is generally considered to span three decades of the EM spectrum, from 1 micrometer to 1,000 micrometers wave-length. The English-German astronomer Sir William Herschel (1738–1822) is credited with the discovery of infrared radiation. *See also* ELECTRO-MAGNETIC SPECTRUM.

infrared (IR) radiometer A telescope-based instrument that measures the intensity of thermal (infrared) energy radiated by targets. This type of instrument often is found as part of the instrument package on a scientific spacecraft. For example, by filling the IR radiometer's field of view completely with the disc of a target planet and measuring its total thermal energy output, the planet's thermal energy balance can be computed, revealing to space scientists the ratio of solar heating to the planet's internal heating.

infrared (IR) sensor A sensor (often placed on a military satellite) that detects the characteristic infrared (IR) radiation from a missile plume, a reentry vehicle, or a "cold body" target in space. Depending on the detector materials and instrument design (e.g., actively cooled versus passively cooled), IR sensors respond to different portions of the IR region of the electromagnetic spectrum and, therefore, support different applications in defense and space science. *See also* BALLISTIC MISSILE DEFENSE; INFRARED ASTRONOMY.

infrasonic frequency A frequency below the audiofrequency range; that is, below the audible limit of about 15 to 20 hertz. The word "infrasonic" may be used as a modifier to indicate a device or system intended to operate at an infrasonic frequency.

inhibit device A electromechanical device that prevents a hazardous event from occurring. This device has direct control—that is, it is not simply a device monitoring a potentially hazardous situation, nor is it in indirect control of some device experiencing the hazardous circumstances. All inhibit devices are independent from each other and are verifiable. A temperature limit switch that shuts down a device or system before a potentially hazardous temperature condition is reached (i.e., at some safe preset temperature limit) is an example of an inhibit device.

inhibitor In general, anything that inhibits; specifically, a substance bonded, taped, or dip-dried onto a solid propellant to restrict the burning surface and to give direction to the burning process.

initiation 1. In general, the process of starting combustion, explosion, or detonation of materials by such means as impact, friction, electrostatic discharge, shock, fragment impact, flame, or heat. 2. The action of a device used as the first element of an explosive train, which, upon receipt of the proper impulse, causes the detonation or burning of an explosive item. 3. With respect to nuclear weapons, the action that sets off a chain reaction in a fissile mass of nuclear material (such as uranium-235 or plutonium-239) that has reached the critical state; generally achieved by the emission of a "burst" or "spurt" of neutrons.

initiator The part of the solid rocket igniter that converts a mechanical, electrical, or chemical input stimulus to an energy output that ignites the energy release system. Also called an *initiation system*.

injection 1. The introduction of fuel and oxidizer in the combustion chamber of a liquid propellant rocket engine. 2. The time following launch when nongravitational forces (e.g., thrust, lift, and drag) become negligible in their effect on the trajectory of a rocket or aerospace vehicle. 3. The process of putting a spacecraft up to escape velocity or a satellite into orbit around Earth.

injection cooling The method of reducing heat transfer to a body by mass transfer cooling, which is accomplished by injecting a fluid into the local flow field through openings in the surface of the body.

injector A device that propels (injects) fuel and/or oxidizer into a combustion chamber of a liquid-propellant rocket engine. The injector atomizes and mixes the propellants so that they can be combusted more easily and completely. Numerous types of injectors have been designed, including splash plate injectors, shower head (nonimpinging) injectors, and spray-type injectors.

inner planets The terrestrial planets: Mercury, Venus, Earth, and Mars. These planets all have orbits around the Sun that lie inside the main asteroid belt.

insertion The process of putting an artificial satellite, aerospace vehicle, or spacecraft into orbit.

instrument 1. (verb) To provide a rocket, aerospace vehicle, spacecraft, or component with instrumentation. 2. (noun) A device that detects, measures, records, or otherwise provides data about certain quantities (e.g., propellant supply available), activities (e.g., vehicle velocity), or environmental conditions (e.g., space radiation environment or surface temperature on a planet).

instrumentation 1. The installation and use of electronic, gyroscopic, and other instruments for the purpose of detecting, measuring, recording, telemetering, processing, or analyzing different values or quantities as encountered in the flight of a rocket, aerospace vehicle, or spacecraft. 2. The collection of such instruments in a rocket, aerospace vehicle, or spacecraft, as, for example, the *instrumentation package* for a scientific spacecraft. 3. The specialized branch of aerospace engineering concerned with the design, composition, arrangement, and operation of such instruments.

intact abort An abort of an aerospace vehicle mission in which the crew, payload, and vehicle are returned to the landing site. *See also* ABORT; ABORT MODES; SPACE TRANSPORTATION SYSTEM.

integral tank A fuel or oxidizer tank built within the normal contours of an aircraft or rocket vehicle and using the skin of the vehicle as a wall of the tank.

integrated circuit (IC) Electronic circuits, including transistors, resistors, capacitors, and their inter-connections, fabricated on a single small piece of semiconductor material (chip). Categories of integrated circuits such as LSI (large-scale integration) and VLSI (very large-scale integration) refer to the level of integration, which denotes the number of transistors on a chip.

integration The collection of activities leading to the compatible assembly of PAYLOAD and LAUNCH VEHICLE into the desired final (flight) configuration.

intelligence The product resulting from the processing of information concerning foreign nations, hostile or poten-

tially hostile elements, or areas of actual or potential operations. The term also is applied to the activity that results in the product and to organizations engaged in such activity. The steps by which information is converted into intelligence and made available to users is called the *intelligence cycle*. There are five major steps in this cycle: (1) planning and direction, (2) collection, (3) processing, (4) production, and (5) dissemination.

intensive property A thermodynamic property that is independent of mass. Temperature (T), pressure (p), specific volume (v), and density (ρ) are examples of intensive properties. The internal energy (u), enthalpy (h), or entropy (s) of a thermodynamic system can be either an extensive property (if it represents the total value of the property) or an intensive property (if it represents the value per unit mass). For example, the total internal energy (U) of a system is expressed in joules and is an extensive thermodynamic property, while the specific internal energy (u) of a system is expressed in joules per kilogram (J/kg) and is an intensive property. Intensive thermodynamic properties are used in many thermodynamic analyses. However, mass (m) and total volume (V) are extensive thermodynamic properties, since these properties vary directly with the mass (or "extent") of the system. *See also* THERMODYNAMICS.

interception The act of destroying a moving target.

intercontinental ballistic missile (ICBM) A ballistic missile with a range capability in excess of 5,500 kilometers (3,000 nautical mi).

interface 1. In general, a common boundary, whether material or nonmaterial, between two parts of a system. 2. In engineering, a mechanical, electrical, or operational common boundary between two elements of an aerospace system, subsystem, or component. 3. In aerospace operations, the point or area where a relationship exists between two or more parts, systems, programs, persons, or procedures wherein physical and functional compatibility is required. 4. In fluid mechanics, a surface separating two fluids across which there is a dissimilarity of some fluid property such as density or velocity, or some derivative of these properties. The equations of motion do not apply at the interface but are replaced by the boundary conditions.

interference fit The retention of a component in a mounting solely by virtue of the friction between the interfaces; the degree of friction (and therefore retention force) is governed by the relative dimensions of the mating parts. Also called *press fit*.

interferometer An instrument that achieves high angular RESOLUTION by combining SIGNALS from at least two widely separated TELESCOPES *(optical interferometer)* or a widely separated ANTENNA ARRAY *(radio interferometer)*. Radio interferometers are one of the basic instruments of RADIO ASTRONOMY. In principle, the interferometer produces and measures interference fringes from two or more coherent WAVE trains from the same source. These instruments are used to measure WAVELENGTHS, to measure the angular width of sources, to determine the angular position of sources (as in SATELLITE tracking), and for many other scientific purposes. *See also* VERY LARGE ARRAY (VLA).

intergalactic Between or among the galaxies Although no place in the universe is truly "empty," the space between clusters of galaxies comes very close. These intergalactic regions are thought to contain less than one atom in every 10 cubic meters (350 cubic ft).

intermediate-range ballistic missile (IRBM) A ballistic missile with a range capability from about 1,000 to 5,500 kilometers (550 to 3,000 nautical mi).

internal energy (symbol: u) A thermodynamic property interpretable through statistical mechanics as a measure of the microscopic energy modes (i.e., molecular activity) of a system. This property appears in a first law of thermodynamics energy balance for a closed system as:

$$du = dq - dw$$

where du is the increment of specific internal energy, dq is the increment of thermal energy (heat) added to system, and dw is the increment of work done by the system per unit mass. *See also* THERMODYNAMICS.

International Space Station (ISS) A major human spaceflight project headed by NASA. Russia, Canada, Europe, Japan, and Brazil are also contributing key elements to this large, modular SPACE STATION in LOW EARTH ORBIT that represents a permanent human outpost in OUTER SPACE for MICROGRAVITY research and advanced space technology demonstrations. On-orbit assembly began in December 1998 with completion originally anticipated by 2004. However, the COLUMBIA accident that took place on February 1, 2003, killing the seven ASTRONAUT crewmembers and destroying the ORBITER vehicle, has exerted a major impact on the *ISS* schedule. Tables 1 and 2 provide a summary of *ISS* construction and crew activities through January 15, 2003.

International System of Units *See* SI UNIT(S).

International Telecommunications Satellite (INTELSAT) The international organization involved in the development of a family of highly successful commercial communications satellites of the same name.

International Ultraviolet Explorer (IUE) A highly successful scientific spacecraft launched in January 1978. Operated jointly by NASA and the European Space Agency (ESA), the *IUE* has helped astronomers from around the world obtain access to the ultraviolet (UV) radiation of celestial objects in unique ways not available by other means. The spacecraft contains a 0.45-meter (1.48-foot) aperture telescope solely for spectroscopy in the wave-length range

Table 1. Vital Statistics for the
International Space Station
(as of 01/15/03)

ISS:	*Major Elements:*
Zarya:	launched Nov. 20, 1998
Unity:	attached Dec. 8, 1998
Zvezda:	attached July 25, 2000
Z1 Truss:	attached Oct. 14, 2000
Soyuz:	docked Oct. 31, 2002
Progress:	docked Sept. 29, 2002
P6 Integrated Truss:	attached Dec. 3, 2000
Destiny:	attached Feb. 10, 2001
Canadarm2:	attached April 22, 2001
Joint Airlock:	attached July 15, 2001
Pirs:	attached Sept. 16, 2001
SO Truss:	attached April 11, 2002
S1 Truss:	attached Oct. 10, 2002
P1 Truss:	attached Nov. 26, 2002
Weight:	178,594 kg (393,733 lbs.)
Habitable Volume:	425 cubic meters (15,000 cubic feet)
Surface Area (solar arrays):	892 square meters (9,600 square feet)
Dimensions:	
Width:	73 meters (240 feet) across solar arrays
Length:	44.5 meters (146 feet) from Destiny Lab to Zvezda 52 meters (170 feet) with a Progress resupply vessel docked
Height:	27.5 meters (90 feet)

Source: NASA.

Table 2. Expedition Crews to the *International Space Station*
(as of 01/15/03)

Expedition One:	Launch: 10/31/00
	Land: 03/18/01
	Time: 140 days, 23 hours, 28 minutes
	Crew: Commander William Shepherd, Soyuz Commander Yuri Gidzenko, Flight Engineer Sergei Krikalev
Expedition Two:	Launch: 03/08/01
	Land: 08/22/01
	Time: 167 days, 6 hours, 41 minutes
	Crew: Commander Yury Usachev, Flight Engineer Susan Helms, Flight Engineer James Voss

(continues)

Table 2. Expedition Crews to the *International Space Station*
(as of 01/15/03) *(continued)*

Expedition Three:	Launch: 08/10/01 Land: 12/17/01 Time*: 128 days, 20 hours, 45 minutes Crew: Commander Frank Culbertson, Soyuz Commander Vladimir Dezhurov, Flight Engineer Mikhail Tyurin
Expedition Four:	Launch: 12/05/01 Land: 06/19/02 Time*: 195 days, 19 hours, 39 minutes Crew: Commander Yury Onufrienko, Flight Engineer Dan Bursch, Flight Engineer Carl Walz
Expedition Five:	Launch: 06/06/02 Land: 12/07/02 Time*: 184 days, 22 hours, 14 minutes Crew: Commander Valery Korzun, NASA ISS Science Officer Peggy Whitson, Flight Engineer Sergei Treschev
Expedition Six:	Launch: 11/23/02 Land: 05/03/03 Time: 161 days, 19 hours, 17 minutes Crew: Commander Ken Bowersox, Flight Engineer Nikolai Budarin, NASA ISS Science Officer Don Pettit

Source: NASA.

from 115 to 325 nanometers (nm) [1,150 to 3,250 angstroms (Å)]. *IUE* data helps support fundamental studies of comets and their evaporation rate when they approach the Sun and of the mechanisms driving the stellar winds that make many stars lose a significant fraction of their mass (before they die slowly as white dwarfs or suddenly in supernova explosions). This long-lived international spacecraft also has assisted astrophysicists in their search to understand the ways by which BLACK HOLES possibly power the turbulent and violent nuclei of active galaxies.

Internet An enormous global computer network that links many government agencies, research laboratories, universities, private companies, and individuals. This worldwide computer network has its origins in the "ARPA-net"—a small experimental computer network estab-lished by the Advanced Research Project Agency (ARPA) of the U.S. Department of Defense in the 1970s to permit rapid com-munication among universities, laborato-ries, and military project offices.

interplanetary Between the planets; within the solar system.

interplanetary dust (IPD) Tiny parti-cles of matter (typically less than 100 micrometers in diameter) that exist in space within the confines of our solar sys-tem. By convention, the term applies to all solid bodies ranging in size from submi-crometer diameter to tens of centimeters in diameter, with corresponding masses ranging from 10^{-17} gram to approximately 10 kilograms. Near Earth the IPD flux is taken as approximately 10^{-13} to 10^{-12} gram per square meter per second [g/m²-s)]. Space scientists have made rough esti-mates that Earth collects about 10,000 metric tons of IPD per year. They also estimate that the entire IPD "cloud" in the solar system has a total mass of between 10^{+16} and 10^{+17} kilograms.

interstage section A section of a MISSILE or ROCKET that lies between stages. *See also* STAGING.

interstellar Between or among the stars.

interstellar communication and contact Several methods of achieving contact with (postulated) intelligent extraterrestrial life-forms have been suggested. These methods include: (1) interstellar travel by means of starships, leading to physical contact between different civilizations; (2) indirect contact through the use of robot (i.e., uncrewed) interstellar probes; (3) serendipitous contact, such as finding a derelict alien starship or probe drifting in the main asteroid belt; (4) interstellar communication involving the transmission and reception of electromagnetic signals; and (5) very "exotic" techniques involving information transfer through the modulation of gravitons, neutrinos, or streams of tachyons; the use of some form of telepathy; and perhaps even matter transport by means of distortions in the space-time continuum that help "beat" the speed-of-light barrier that now underlies our understanding of the physical universe. *See also* SEARCH FOR EXTRATERRESTRIAL INTELLIGENCE (SETI).

interstellar medium (ISM) The gas and dust particles that are found between the stars in our galaxy, the Milky Way. Up until about three decades ago, the interstellar medium was considered to be an uninteresting void. Today, through advances in astronomy (especially infrared and radioastronomy), we now know that the interstellar medium contains a rich and interesting variety of atoms and molecules as well as a population of fine-grained dust particles. Over 100 interstellar molecules have been discovered to date, including many organic molecules considered essential in the development of life. Interstellar dust (which "reddens" the visible light from the stars behind it because of its preferential scattering of shorter-wavelength photons) is considered to consist of very fine silicate particles, typically 0.1 micrometers in diameter. These interstellar "sands" sometimes may have an irregularly shaped coating of water ice, ammonia ice, or solidified carbon dioxide.

interstellar probe A highly automated interstellar spacecraft sent from our solar system to explore other star systems. Most likely this type of probe would make use of very smart machine systems capable of operating autonomously for decades or centuries. Once the robot probe arrives at a new star system, it would begin a detailed exploration procedure. The target star system is scanned for possible life-bearing planets, and if any are detected, they become the object of more intense scientific investigations. Data collected by the "mother" interstellar probe and any miniprobes (deployed to explore individual objects of interest within the new star system) are transmitted back to Earth. There, after light-years of travel, the signals are intercepted and analyzed by scientists, and interesting discoveries and information are used to enrich our understanding of the universe.

Robot interstellar probes also might be designed to carry specially engineered microorganisms, spores, and bacteria. If a probe encounters ecologically suitable planets on which life has not yet evolved, then it could "seed" such barren but potentially fertile worlds with primitive life-forms or at least life precursors. In that way, human beings (through their robot probes) would not only be exploring neighboring star systems but would be participating in the spreading of life itself through some portion of the Milky Way Galaxy.

interstellar travel In general, matter transport between star systems in a galaxy; specifically, travel beyond our own solar system. The "matter" transported may be: (1) a robot interstellar probe on an exploration mission; (2) an automated spacecraft that carries a summary of the cultural and technical heritage of a civilization (e.g., the plaques on the *Pioneer 10* and *11* spacecraft and the "recorded message" on the *Voyager 1* and 2 spacecraft); (3) a "crewed" starship on a long-term round-trip voyage of scientific exploration; or even (4) a giant "interstellar ark" that is designed to transport a portion of the human race on a one-way mission beyond the solar system in search of new, suitable planetary systems to explore and inhabit.

intravehicular activity (IVA) ASTRONAUT or COSMONAUT activities performed inside an orbiting SPACECRAFT or AEROSPACE VEHICLE. *Compare with* EXTRA VEHICULAR ACTIVITY.

inverse-square law A relation between physical quantities of the form: x proportional to $1/y^2$, where y is usually a distance; and x terms are of two kinds, forces and fluxes. NEWTON'S LAW OF GRAVITATION is an example of the inverse-square law. This relationship is:
$$F = [Gm_1 m_2]/r^2$$
where F is the gravitational force between two objects having masses, m_1 and m_2, respectively, G is the gravitational constant, and r is the distance between these two masses. Note that the force of gravitational attraction decreases as the inverse of the distance squared (i.e., as $1/r^2$).

inverter A device for changing direct current (d.c.) into alternating current (a.c.)

inviscid fluid A hypothesized "perfect" fluid that has zero coefficient of viscosity. Physically this means that shear stresses are absent despite the occurrence of shearing deformations in the fluid. The inviscid fluid also glides past solid boundaries without sticking. No real fluids are inviscid. In fact, since real fluids are viscous, they stick to solid boundaries during flow processes creating thin boundary layers where shear forces are significant. However, the inviscid (perfect) fluid approximation provides a useful model that approximates the behavior of real fluids in many flow situations.

in vitro Literally, "in glass." A life sciences term describing biological tests using cells from an organism but occurring outside that living organism in a laboratory apparatus, such as a glass test tube.

in vivo Literally, "in a living organism (body)." A life sciences term describing biological experiments that take place within (or using) a living organism, such as a test animal or human volunteer. Astronauts often participate in an interesting variety of in vivo life science experiments while on extended flights in microgravity.

Io The pizza-colored, volcanic GALILEAN SATELLITE of JUPITER with a diameter of 3,630 kilometers (2,255 mi).

ion An atom or molecule that has lost or (more rarely) gained one or more electrons. By this *ionization process,* it becomes electrically charged.

ion engine An electrostatic rocket engine in which a propellant (e.g., cesium, mercury, argon, or xenon) is ionized and the propellant ions are accelerated by an imposed electric field to very high exhaust velocity. *See also* ELECTRIC PROPULSION.

ionization The process of producing ions by the removal of electrons from, or the addition of electrons to, atoms or molecules. High temperatures, electrical discharges, or nuclear radiations can cause ionization.

ionizing radiation Any type of NUCLEAR RADIATION that displaces ELECTRONS from ATOMS or MOLECULES, thereby producing IONS within the irradiated material. Examples include alpha (α) radiation, beta (β) radiation, gamma (γ) radiation, PROTONS, NEUTRONS, and X RAYS. *See also* ALPHA PARTICLE; BETA PARTICLE; GAMMA RAYS.

ionosphere That portion of EARTH's upper ATMOSPHERE extending from about 50 to 1,000 kilometers (30 to 620 mi), in which IONS and free ELECTRONS exist in sufficient quantity to reflect RADIO WAVES.

ionospheric storm Disturbances of Earth's ionosphere (caused by solar flare activity), resulting in anomalous variations in its characteristics and effects on radio wave propagation. In general, ionospheric storms can persist for several days and cover large portions of the globe. Two types of storms of particular interest are: the polar-cap absorption (PCA) event and the geomagnetically induced storm. As the name implies, the polar-cap absorption (PCA) event occurs at high latitudes and is associated with the arrival of very energetic solar flare protons. Radio communications blackouts usually occur as a result of a PCA event. The geomagnetically induced ionospheric storm occurs along with periods of special auroral activity. Unlike PCA events, the geomagnetic storms are more intense at night. Finally, the sudden ionospheric disturbance (SID) appears within a few minutes of the occurrence of strong solar flares and causes a fade-out of long-distance radio communications on the sunlit part of Earth.

irradiance (symbol: E) The total radiant flux (electromagnetic radiation of all wavelengths) received on a unit area of a given surface; usually measured in watts per meter-squared (W/m^2). *Compare with* ILLUMINANCE.

irregular galaxy A GALAXY with a poorly defined structure or shape.

irreversible process In thermodynamics, a process that cannot return both the system and the surroundings to their original conditions. Four common causes of irreversibility in a thermodynamic process are: friction, heat transfer through a finite temperature difference, the mixing of two different substances, and unrestrained expansion of a fluid. The exhausting of combustion gases through a rocket engine's nozzle is an irreversible process.

isentropic Of equal or constant entropy. *See also* ENTROPY; THERMODYNAMICS.

Ishtar Terra A very large highland plateau in the northern hemisphere of VENUS, about 5,000 kilometers (3,100 mi) long and 600 kilometers (375 mi) wide.

island universe(s) Term introduced in 18th century by the German philosopher Immanuel Kant (1724–1804) to describe other galaxies.

isobaric Of equal or constant pressure.

isochoric Of equal or constant volume; usually applied to a thermodynamic process during which the volume of the system remains unchanged.

isomer 1. In nuclear physics, one of two or more nuclides with the same numbers of neutrons and protons in their nuclei but with different energies; a nuclide in an excited state and a similar nuclide in the ground state are isomers. 2. In chemistry, one of two or more molecules having the same atomic composition and molecular weight but differing in geometric configuration.

isomeric transition A radioactive transition from one nuclear isomer (in an excited energy state) to another of lower energy. This transition often is accompanied by the emission of a gamma ray or by internal conversion (IC) followed by the emission of X rays and/or Auger electrons.

isothermal process In thermodynamics, any process or change of state of a system that takes place at constant temperature. *See also* THERMODYNAMICS.

isotope One of two or more atoms with the same *atomic number* (Z) (i.e., the same chemical element) but with different atomic weights. An equivalent statement is that the nuclei of isotopes have the same number of protons but different numbers of neutrons. Therefore, carbon-12 ($^{12}_6C$), carbon-13 ($^{13}_6C$), and carbon-14 ($^{14}_6C$), are all isotopes of the element carbon. The subscripts denote their common atomic number (i.e., Z = 6), while the superscripts denote their differing *atomic mass numbers* (i.e., A = 12, 13, and 14, respectively), or approximate atomic weights. Isotopes usually have very nearly the same chemical properties but different physical and nuclear properties. For example, the isotope carbon-14 is radioactive, while the isotopes carbon-12 and carbon-13 are both stable (nonradioactive).

isotope power system A device that uses the thermal energy deposited by the absorption of alpha or beta particles from a radioisotope (radioactive material) as the heat source to produce electric power by direct energy conversion or dynamic conversion (thermodynamic cycle) processes. (Note: Radioisotopes that emit gamma rays are not normally considered suitable for this application because of the shielding requirements.) A variety of radioisotope thermoelectric generators (RTGs), powered by the alpha-emitting radioisotope plutonium-238 ($^{238}_{94}Pu$), has been used very successfully to provide continuous, long-term electric power to spacecraft such as *Pioneer 10* and *11, Voyager 1* and *2, Ulysses, Galileo,* and the *Viking* Mars Landers. *See also* GALILEO PROJECT; PIONEER 10, 11; SPACE NUCLEAR POWER; ULYSSES MISSION; VIKING PROJECT; VOYAGER.

isotropic Having uniform properties in all directions.

isotropic radiator An energy source that radiates (emits) particles or photons uniformly in all directions.

James Webb Space Telescope (JWST)

NASA's *James Webb Space Telescope,* previously called the *Next Generation Space Telescope (NGST),* is scheduled for launch in 2010. An EXPENDABLE LAUNCH VEHICLE will send the SPACECRAFT on a three-month journey to its operational location, about 1.5 million kilometers (940,000 mi) from EARTH in ORBIT around the second LAGRANGE LIBRATION POINT (or L2) of the Earth-SUN system. This distant location will provide an important advantage for the spacecraft's INFRARED RADIATION (IR) imaging and SPECTROSCOPY INSTRUMENTS. At that location, a single shield on just one side of the observatory will protect the sensitive TELESCOPE from unwanted THERMAL RADIATION from both Earth and the Sun. As a result, the spacecraft's INFRARED SENSORS will be able to function at a TEMPERATURE of 50 KELVINS (K) without the need for complicated refrigeration equipment.

The telescope is being designed to detect ELECTROMAGNETIC RADIATION whose WAVELENGTH lies in the range from 0.6 to 28 MICROMETERS but will have an optimum performance in the 1 to 5 micrometer region. The *JWST* will be able to collect infrared SIGNALS from celestial objects that are much fainter than are those now being studied with very large ground-based infrared telescopes (such as the KECK OBSERVATORY) or the current generation of space-based infrared telescopes (such as the SPACE INFRARED TELESCOPE FACILITY). With a primary MIRROR diameter of at least 6 meters (19.7 ft), the *JWST* will make infrared measurements that are comparable in spatial RESOLUTION (that is, image sharpness) to images currently collected in the visible LIGHT portion of the spectrum by the *HUBBLE SPACE TELESCOPE.*

In September 2002, NASA officials renamed the NGST to honor James E. Webb (1906–1992), the civilian space agency's administrator, from February 1961 to October 1968 during the development of the APOLLO PROJECT. Webb also initiated many other important space science programs within the fledgling agency. NASA is developing the *JWST* to observe the faint infrared signals from the first STARS and galaxies in the UNIVERSE. Data from the orbiting observatory should help scientists better respond to lingering fundamental questions about the universe's origin and ultimate fate. One important astrophysical mystery involves the nature and role of DARK MATTER.

jansky (symbol: Jy) A unit used to describe the strength of an incoming electromagnetic wave signal. The jansky frequently is used in radio and infrared astronomy. It is named after the American radio engineer Karl G. Jansky (1905–1950), who discovered extraterrestrial radiowave sources in the 1930s, a discovery generally regarded as the birth of radio astronomy.

1 jansky (Jy) = 10^{-26} watts per meter-squared per hertz [W/(m²-Hz)]

See also RADIO ASTRONOMY.

jet 1. A strong, well-defined stream of fluid either issuing from an orifice or moving in a contracted duct, such as the jet of combustion gases coming out of a reaction engine or the jet in the test section in a wind tunnel. 2. A tube or nozzle through which fluid passes or emerges in a "jet," such as the jet in a liquid propellant rocket motor's injector. 3. A jet engine. 4. A jet aircraft.

jet propulsion Reaction propulsion in which the propulsion unit obtains oxygen from the air, as distinguished from rocket propulsion, in which the unit carries its own oxidizer (i.e., oxygen-producing material). In connection with aircraft propulsion, the term refers to a turbine jet unit (usually burning hydrocarbon fuel) that discharges hot gas through a tail pipe and a nozzle, producing a thrust that propels the aircraft. *See also* ROCKET PROPULSION.

Jet Propulsion Laboratory (JPL) NASA's Jet Propulsion Laboratory (JPL) is located near Pasadena, California, approximately 32 kilometers (20 miles) northeast of Los Angeles. JPL is a government-owned facility operated by the California Institute of Technology under a NASA contract. In addition to the Pasadena site, JPL operates the worldwide Deep Space Network (DSN), including a DSN station, at Goldstone, California.

This laboratory is engaged in activities associated with deep-space automated scientific missions, such as subsystem engineering, instrument development, and data reduction and analysis required by deep space flight. The Cassini spacecraft's exploration of SATURN is a current JPL mission. *See also* NATIONAL AERONAUTICS AND SPACE ADMINISTRATION (NASA).

jetstream A *jet* issuing from an orifice into a much more slowly moving medium, such as the stream of combustion products ejected from a reaction engine. This term also is used in meteorology, but appears as two words, namely *jet stream*. *See also* JET STREAM.

jet stream A strong band of wind or winds in the upper troposphere or in the stratosphere, moving in the general direction from west to east and often reaching velocities of hundreds of kilometers an hour. *See also* JETSTREAM.

jettison To discard. For example, when the propellant in the booster stage of a multistage rocket vehicle is used up, the now-useless booster stage is *jettisoned*— that is, it is separated from the upper stage(s) of the launch vehicle and allowed

to fall back to Earth, usually impacting in a remote ocean area.

Johnson Space Center (JSC) NASA's Lyndon B. Johnson Space Center (JSC) is located about 32 kilometers (20 miles) southeast of downtown Houston, Texas. This center was established in September 1961 as NASA's primary center for: (1) the design, development, and testing of spacecraft and associated systems for human space flight; (2) the selection and training of astronauts; (3) the planning and supervision of crewed space missions; and (4) extensive participation in the medical, engineering, and scientific experiments aboard human crew space flights. The Mission Control Center (MCC) at JSC is the central facility from which all U.S. crewed space flights are monitored.

joule (symbol: J) The SI UNIT of energy or work in the International System of Units. It is defined as the work done (or its energy equivalent) when the point of application of a force of one newton moves a distance of one meter in the direction of the force.

1 joule (J) = 1 newton-meter (N-m) = 1 watt-second (W-s) = 10^7 erg = 0.2388 calories

This unit is named in honor of the British physicist James Prescott Joule (1818–1889).

Jovian Of or relating to the planet Jupiter.

Jovian planet One of the major, or giant, planets that is characterized by a great total mass, low average density, and an abundance of the lighter elements (especially hydrogen and helium). The Jovian planets are Jupiter, Saturn, Uranus, and Neptune. *See also* JUPITER; NEPTUNE; SATURN; URANUS.

JP-4 A liquid, hydrocarbon fuel for jet and rocket engines, the chief ingredient of which is kerosene.

Juno The Juno I and Juno II were early NASA expendable launch vehicles adapted from existing U.S. Army missiles. These rocket vehicles were named after the ancient Roman goddess, Juno, queen of the gods, and wife of Jupiter, king of the Gods. The Juno I, a four-stage configuration of the Jupiter C vehicle, orbited the first U.S. satellite, *Explorer 1,* on January 31, 1958. The Juno V was the early designation of the launch vehicle that eventually became the Saturn I rocket. *See also* EXPLORER I; LAUNCH VEHICLE.

Jupiter The largest planet in our solar system with more than twice the mass of all the other planets and their moons combined and a diameter of approximately 143,000 kilometers (km). It is the fifth planet from the Sun and is separated from the four terrestrial planets by the main asteroid belt. The giant planet, named after Jupiter, king of the gods in Roman mythology (Zeus in Greek mythology), rotates at a dizzying pace—about once every 9 hours and 55 minutes. It takes Jupiter almost 12 Earth-years to complete a journey around the Sun. Its mean distance from the Sun is about 5.2 astronomical units (AU), or 7.78 × 10⁸km. Table 1 provides a summary of the physical and dynamic characteristics of this giant planet.

The planet is distinguished by bands of colored clouds that change their appearance over time. One distinctive feature, the Great Red Spot, is a huge oval-shaped atmospheric storm that has persisted for at least three centuries. In July 1994, 20 large fragments from the Comet Shoemaker-Levy crashed into the clouds of Jupiter, producing gigantic "Earth-size" disturbances in the Jovian cloud system. Scientists estimated that the largest cometary fragments (about 3 to 5 km in diameter) produced a blast in the Jovian atmosphere equivalent perhaps to as much as 6 million megatons per fragment.

Jupiter's atmosphere, explored in situ by the Galileo atmospheric probe (on

---

Table 1. Physical and Dynamic Properties of Jupiter

Diameter (equatorial)	142,982 km
Mass	1.9×10^{27} kg
Density (mean)	1.32 g/cm^3
Surface gravity (Equatorial)	23.1 m/s^2
Escape velocity	59.5 km/s
Albedo (visual geometric)	0.52
Atmosphere	Hydrogen (~89%), helium (~11%), also ammonia, methane, and water.
Natural satellites	16
Rings	3 (1 main, 2 minor)
Period of rotation (a "Jovian day")	0.413 day
Average distance from Sun	7.78×10^8 km (5.20 AU) [43.25 light-min]
Eccentricity	0.048
Period of revolution around Sun (a "Jovian year")	11.86 years
Mean orbital velocity	13.1 km/s
Magnetosphere	Yes (intense)
Radiation belts	Yes (intense)
Mean atmospheric temperature (at cloud tops)	~129 K
Solar flux at planet (at top of atmosphere)	50.6 W/m^2 (at 5.2 AU)

Source: NASA.

Table 2. Properties of the Moons of Jupiter

Moon	Diameter (km)	Semimajor Axis of Orbit (km)	Period of Rotation (days)
Sinope	30 (approx)	23,700,000	758 (retrograde)
Pasiphae	40 (approx)	23,500,000	735 (retrograde)
Carme	30 (approx)	22,600,000	692 (retrograde)
Ananke	20 (approx)	21,200,000	631 (retrograde)
Elara	80 (approx)	11,740,000	260.1
Lysithea	20 (approx)	11,710,000	260
Himalia	180 (approx)	11,470,000	251
Leda	16 (approx)	11,110,000	240
Callisto[a]	4,820	1,880,000	16.70
Ganymede[a]	5,262	1,070,000	7.16
Europa[a]	3,130	670,900	3.55
Io[a]	3,640	422,000	1.77
Thebe (1979J2)	100 (approx)	222,000	0.675
Amalthea	270 (approx)	181,300	0.498
Adrastea (1979J1)	40 (approx)	129,000	0.298
Metis (1979J3)	40 (approx)	127,900	0.295

[a] Galilean satellite.
Source: NASA.

December 7, 1995) is composed primarily of hydrogen and helium in approximately "stellar composition" abundances —that is, about 89% hydrogen (H) and 11% helium (He). In fact, Jupiter is sometimes called a near star. If it had been only about 100 times larger, nuclear burning could have started in its core, and Jupiter would have become a star to rival the Sun itself. Its interesting complement of natural satellites even resembles a miniature solar system.

Jupiter has 16 known natural satellites. The four largest of Jupiter's moons are often called the Galilean satellites. They are: Io, Europa, Ganymede, and Callisto. These moons were discovered in 1610 by the Italian astronomer Galileo Galilei. Very little actually was known about the Jovian moons until the *Pioneer 10* and *11* and *Voyager 1* and *2* spacecraft encountered Jupiter between 1973 and 1979. These flybys provided a great deal of valuable information and initial imagery, including the discovery of three new moons (Metis, Adrastea, and Thebe), active volcanism on Io, and a possible liquid-water ocean beneath the smooth icy surface of Europa. The highly elliptical orbit of the Galileo spacecraft provided the opportunity for several closeup inspections of Jupiter's major moons and greatly improved the current knowledge base for the Jovian system. Table 2 provides selected physical and dynamic data for moons of Jupiter. *See also* GALILEO PROJECT; *PIONEER 10, 11;* VOYAGER.

Jupiter C A modified version of the REDSTONE ballistic missile developed by the U.S. Army and a direct descendant of the V-2 ROCKET developed in Germany during World War II. The Juno I, a four-stage configuration of the Jupiter C rocket, orbited the first American satellite *(Explorer I)* on January 31, 1958. *See also* EXPLORER I; JUNO.

K

Kapustin Yar A minor Russian launch complex that is located on the banks of the Volga River near Volgograd at approximately 48.4 degrees north latitude and 45.8 degrees east longitude. This complex originally was built to support the early Soviet ballistic missile test program; its first missile launch occurred in 1947. Since the launch of *Kosmos I* in 1962, the Kapustin Yar site has conducted relatively few (less than 100) space launches in comparison to the other two Russian launch complexes: Baikonur Cosmodrome (now in Kazakhstan) and Plesetsk Cosmodrome (south of Archangel in the northwest corner of Russia), which is sometimes called the world's busiest spaceport. However, Kapustin Yar has been used for small and intermediate-size payloads, including those launched for or in cooperation with other nations. For example, on April 19, 1975, with 50 scientists and technicians from India in attendance, the Indian satellite *Aryabhata* was sent into orbit from this complex.

Keck Observatory Located near the 4,200-meter (13,780 ft) high summit of Mauna Kea, Hawaii, the W. M. Keck Observatory possess two of the world's largest optical/infrared reflector TELESCOPES, each with a 10-meter (32.8-ft) diameter primary MIRROR. Keck I started astronomical observations in May 1993 and its twin, Keck II, in October 1996. The pair of telescopes operate as an optical INTERFEROMETER.

keep-out zone A volume of space around a space asset (e.g., a military surveillance satellite) that is declared to be forbidden (i.e., "off-limits") to parties who are not owners/operators of the asset. Enforcement of such a zone is intended to protect space assets against attack, especially by space mines. Keep-out zones can be negotiated or unilaterally declared. The right to defend such a zone by force and the legality of unilaterally declared zones under the terms of the current Outer Space Treaty remain to be determined in space law.

kelvin (symbol: K) The International System (SI) unit of thermodynamic temperature. One kelvin is defined as the fraction 1/273.16 of the thermodynamic temperature of the triple point of water. The absolute zero of temperature has a temperature of 0K. This unit is named after the British physicist Lord Kelvin (1824–1907). *See also* SI UNITS.

Kennedy Space Center (KSC) NASA's John F. Kennedy Space Center (KSC) is located on the east coast of Florida 241 kilometers (km) (150 miles [mi]) south of Jacksonville and approximately 80 km (50 mi) east of Orlando. KSC is immediately north and west of Cape Canaveral Air Force Station. The center is about 55 km (34 mi) long and varies in width from 8 km (5 mi) to 16 km (10 mi). The total land and water occupied by the installation is 56,817 hectares (140,393 acres). Of this total area, 34,007 hectares (84,031 acres) are NASA-owned. The remainder is owned by the State of Florida. This large area, with adjoining bodies of water, provides the buffer space necessary to protect the nearby civilian communities during space vehicle launches. Agreements have been made with U.S. Department of the Interior supporting the use of the nonoperational (buffer) area as a wildlife refuge and national seashore.

KSC was established in the early 1960s to serve as the launch site for the Apollo-Saturn V lunar landing missions. After the Apollo program ended in 1972, Launch Complex 39 was used to support both the Skylab program (early nonpermanent U.S. space station) and the Apollo-Soyuz test project (an international rendezvous and docking demonstration involving spacecraft from the United States and Russia [former Soviet Union]).

The Kennedy Space Center now serves as the primary center within NASA for the test, checkout, and launch of space vehicles. KSC responsibility includes the launching of crewed (space shuttle) vehicles at Launch Complex (LC) 39 and NASA uncrewed, expendable launch vehicles (such as a Delta II rocket) at both nearby Cape Canaveral Air Force Station and at Vandenberg Air Force Base in California.

The assembly, checkout, and launch of the space shuttle vehicles and their payloads takes place at the center. Weather conditions permitting, the orbiter vehicle lands at the Kennedy Space Center (after an orbital mission) and undergoes "turnaround," or processing between flights. The Vehicle Assembly Building (VAB) and the two shuttle launch pads at LC 39 may be the best-known structures at KSC, but other facilities also play critical roles in prelaunch processing of payloads and elements of the space shuttle system. Some buildings, such as the 160-meter (525-foot) tall VAB, were originally designed for the Apollo Project in the 1960s and then altered to accommodate the space shuttle. Other facilities, such as the Orbiter Processing Facility (OPF) high bays, were designed and built exclusively for the space shuttle program. *See also* NATIONAL AERONAUTICS AND SPACE ADMINISTRATION (NASA); SPACE TRANSPORTATION SYSTEM (STS).

Kepler, Johannes (Johann) (1571–1630)

German astronomer and mathematician who developed three laws of planetary motion that described the ELLIPTICAL ORBITS of the PLANETS around the SUN and provided the empirical basis for the acceptance of NICHOLAS COPERNICUS's HELIO-CENTRIC hypothesis. KEPLER'S LAWS gave ASTRONOMY its modern, mathematical foundation. His publication *The New Star (De Stella Nova)* described the SUPERNOVA in the CONSTELLATION Ophiuchus that he first observed (with the NAKED EYE) on October 9, 1604.

Keplerian elements

The Keplerian elements, or *orbital elements,* are the six parameters that uniquely specify the position and path of a satellite (natural or human made) in its orbit as a function of time. These elements and their characteristics are described in Figure 39. *See also* ORBITS OF OBJECTS IN SPACE.

Kepler Mission

Scheduled for launch in 2006, NASA's Kepler Mission SPACECRAFT will carry a unique space-based TELESCOPE specifically designed to search for EARTH-LIKE PLANETS around STARS beyond the SOLAR SYSTEM. As of 2003, astronomers had discovered more than 80 EXTRA SOLAR PLANETS. However, these discoveries, made using ground-based telescopes, all involved GIANT PLANETS similar to JUPITER. Such giant extrasolar planets are most likely to be gaseous worlds composed primarily of HYDROGEN and HELIUM and unlikely to harbor life. None of the detection methods used to date have had the capability of finding Earthlike planets—tiny objects that are between 30 and 600 times less massive than Jupiter. The Kepler Mission's specialized one-meter-diameter telescope will look for the TRANSIT signature of extrasolar planets. A transit occurs each time a planet crosses the LINE OF SIGHT (LOS) between that planet's parent star and a distant observer. As the orbiting extrasolar planet blocks some of the LIGHT from its parent star, the Kepler spacecraft's sensitive PHOTOMETER will record a periodic dimming of the incident starlight. Astronomers will then use such periodic signature data to detect the presence of the planet and to determine its size and ORBIT. Three transits of a star, all with consistent period, brightness changes, and duration, will provide a robust method for detecting and confirming the presence of extrasolar planets around neighboring stars. Scientists can

Keplerian Elements

a   semi-major axis, gives the size of the orbit

e   eccentricity, gives the shape of the orbit

i   inclination angle, gives the angle of the orbit plane to the central body's equator

Ω   right ascension of the ascending node, gives the rotation of the orbit plane from reference axis

ω   argument of perigee, gives the rotation of the orbit in its plane

θ   true anomaly, gives the location of the satellite on the orbit

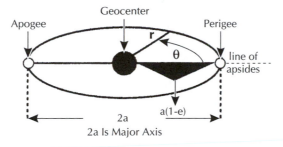

Figure 39. The six Keplerian, or classical orbital elements, that uniquely define the position of a satellite.

use the measured orbit of the planet and the known properties of the parent star to determine if a newly discovered extrasolar planet lies in the CONTINUOUSLY HABITABLE ZONE (CHZ), that is, at a distance from the parent star where liquid water could exist on the planet's surface. NASA named this cornerstone mission in the search for EXTRATERRESTRIAL LIFE after the famous German astronomer and mathematician JOHANNES KEPLER.

Kepler's laws Three empirical laws describing the motion of the planets in their orbits around the Sun, first formulated by the German astronomer JOHANNES KEPLER (1571–1630), on the basis of the detailed observations of the Danish astronomer Tycho Brahe (1546–1601). These laws are: (1) the orbits of the planets are ellipses, with the Sun at a common focus; (2) as a planet moves in its orbit, the line joining the planet and the Sun sweeps over equal areas in equal intervals of time (sometimes called the *law of equal areas*); and (3) the square of the orbital period (P) of any planet is proportional to the cube of its mean distance (a) from the Sun (i.e., the semimajor axis "a" for the elliptical orbit). An empirical statement of Kepler's third law is that for any planet, P^2/a^3 = a constant. About a century after Kepler's initial formulation of these laws of planetary motion, NEWTON'S LAW OF GRAVITATION and LAWS OF MOTION provided the mathematical basis for a more complete physical understanding the motion of planets and satellites.

kerosene A liquid hydrocarbon fuel used in certain rocket and jet engines. Also spelled kerosine.

keV A unit of energy, corresponding to 1,000 electron volts; keV means kilo-electron volts. *See also* ELECTRON VOLT; SI UNITS.

kill assessment The detection and assimilation of information indicating the destruction of an object under attack. Kill assessment is one of the many functions to be performed by a battle management system.

kill probability A measure of the probability of destroying a target.

kilo- (symbol: k) An SI UNIT system prefix meaning that a basic space/time/mass unit is multiplied by (1,000); as, for example, a *kilogram* (kg) or a *kilometer (km)*. Note, however, that in computer technology and digital data processing, kilo designated with a capital "K" refers to a precise value of 1,024 (which corresponds in binary notation to 2^{10}). Therefore, a kilobyte (abbreviated KB or K-byte) is actually 1,024 bytes; similarly, a kilobit (Kb or K-bit) stands for 1,024 bits.

kilogram (symbol: kg) The fundamental unit of mass in the International System (SI).
   1 kilogram (kg) = 1,000 grams = 2.205
   pounds-mass (lbm) (approximately)
See also SI UNITS.

kilometer (km) An SI UNIT of distance corresponding to 1,000 meters or approximately 0.6214 statute miles (0.54 nautical miles).
   1 kilometer (km) = 1,000 meters = 3280.8
   = 0.62137 statute miles

kiloparsec (kpc) A distance of 1,000 parsecs or approximately 3,261.6 light-years.

kiloton (symbol: kT) A unit of energy defined as 10^{12} calories (4.2×10^{12} joules). This unit is used to describe the energy released in a nuclear detonation. It is approximately the amount of energy that would be released by the explosion of 1,000 metric tons (1 kiloton) of TNT (trinitrotoluene).

kinetic energy (symbol: KE or E_{KE}) The energy that a body possess as a result of its motion, defined as one-half the product of its mass (m) and the square of its speed (v), that is: $E_{KE} = 1/2 \, m \, v^2$.

kinetic energy interceptor An interceptor that uses a nonexplosive projectile moving at very high speed (e.g., several kilometers per second) to destroy a target on impact. The projectile may include homing sensors and onboard rockets to improve its accuracy, or it may follow a preset trajectory (as with a projectile launched from a high-velocity gas or electromagnetic gun).

kinetic energy weapon (KEW) A weapon that uses kinetic energy, or the energy of motion, to destroy or "kill" an object. Kinetic energy weapon projectiles strike their targets at very high velocities (typically more than 5 kilometers per second) and achieve a great deal of impact damage, which destroys the target. Typical KEW projectiles include spherical or shaped plastic pellets (fired by high-velocity gas guns) or tiny metal rods, pellets, or shaped shells (fired by electromagnetic rail-gun systems.) Even tiny grains of sand, explosively dispersed into a cloud directly in the path of a high-velocity reentry vehicle, can achieve a "kinetic energy" kill, as the hostile reentry vehicle (and its warhead) is quite literally "sand-blasted" out of existence. *See also* BALLISTIC MISSILE DEFENSE.

kinetic kill vehicle (KKV) In space defense, a rocket that homes in on its target and then kills it either by striking it directly or by showering it with fragments of material from a specially designed exploding device.

kinetic theory A physical theory, initially developed in the 19th century, that describes the "macroscopic" physical properties of matter (e.g., pressure and temperature) in terms of the "microscopic" behavior (i.e., motions and interactions) of the constituent atoms and molecules. For example, the simple kinetic theory of a gas explains pressure in terms of a statistical treatment of the continuous impacts of gas molecules (or atoms) on the walls of the container—that is, the pressure is described as being proportional to the average translational kinetic energy of the system of particles per unit volume.

Kirchhoff's law (of radiation) The physical law of radiation heat transport states that at a given temperature, the ratio of the emissivity (ε) to the absorptivity (α) for a given wavelength is the same for all bodies and is equal to the emissivity of an ideal BLACKBODY at that temperature and wavelength.

Kirkwood gaps Gaps or "holes" in the main asteroid belt between Mars and Jupiter where essentially no asteroids are located. These gaps initially were explained in 1857 by Daniel Kirkwood, an American astronomer (1814–1895), as being the result of complex orbital resonances with Jupiter. Specifically, the gaps correspond to the absence of asteroids with orbital periods that are simple fractions (e.g., 1/2, 2/5, 1/3, 1/4, etc.) of Jupiter's orbital period. Consider, for example, an asteroid that orbits the Sun in exactly half the time it takes Jupiter to orbit the Sun. (This example corresponds to a 2:1 orbital resonance.) The minor planet in this particular orbital resonance would then experience a regular, periodic pull (i.e., gravitational tug) from Jupiter at the same point in every other orbit. The cumulative effect of these recurring gravitational pulls is to deflect the asteroid into a chaotic (elongated) orbit that could then cross the orbit of Earth or Mars. *See also* ASTEROID.

klystron A high-powered electron tube that converts direct current (d.c.) electric energy into radio-frequency waves (microwaves) by alternately speeding up and slowing down the electrons (i.e., by electron beam velocity oscillation).

knot (symbol: kn) A unit of speed originating in the late 16th-century sailing industry, which represents a speed equal to 1 nautical mile per hour.

1 knot (kn) = 1 nautical mile/hr = 1.1508 statute miles/hr = 1.852 kilometers/hr

Korolev, Sergei (1907–1966) The Russian (Ukraine-born) ROCKET engineer who was the driving technical force behind the initial INTERCONTINENTAL BALLISTIC MISSILE (ICBM) program and the early OUTER SPACE exploration projects of the former Soviet Union. In 1954 he started work on the first Soviet ICBM, the R-7. This powerful rocket system was capable of carrying a massive PAYLOAD across continental distances. As part of COLD WAR politics Soviet premier Nikita Khrushchev (1894–1971) allowed Korolev to use this military rocket to place the first ARTIFICIAL SATELLITE *(Sputnik 1)* into orbit around EARTH on October 4th, 1957, an event now generally regarded as the beginning of the Space Age.

Kuiper Belt A region in the outer SOLAR SYSTEM beyond the ORBIT of NEPTUNE that contains millions of icy PLANETESIMALS (small solid objects). These icy objects range in size from tiny PARTICLES to Plutonian-sized planetary bodies. The Dutch-American astronomer Gerard Peter Kuiper (1905–1973) first suggested the existence of this disk-shaped reservoir of icy objects in 1951. *See also* OORT CLOUD.

L

Lagrangian libration point(s) In CELES-TIAL MECHANICS, one of five points in OUTER SPACE (called L_1, L_2, L_3, L_4, and L_5) where a small object can experience a stable ORBIT in spite of the FORCE of GRAVITY exerted by two much more massive celestial bodies when they orbit about a common CENTER OF MASS. Joseph Louis Lagrange (1736–1818), a French mathematician, calculated the existence and location of these points in 1772. Three of the points (called L_1, L_2, and L_3) lie on the line joining the center of mass of the two large bodies. A small object placed at any of these points is actually in unstable equilibrium. Any slight displacement of its position results in the object's rapid departure from the point due to the gravitational influence of the more massive bodies. However, the fourth and fifth libration points (called L_4 and L_5) are locations of stable equilibrium. The TROJAN GROUP of ASTEROIDS can be found in such stable Lagrangian points 60 degrees ahead of (L_4) and 60 degrees behind (L_5) JUPITER's orbit around the SUN. Space visionaries have suggested placing large SPACE SETTLEMENTS in the L_4 and L_5 libration points of the EARTH–MOON system. (See Figure 40.)

lambert (symbol: L) A unit of luminance defined as equal to 1 LUMEN of (light) flux emitted per square centimeter of a perfectly diffuse surface. This unit is named in honor of the German mathematician Johann H. Lambert (1728–1777).

laminar boundary layer In fluid flow, the layer next to a fixed boundary layer. The fluid velocity is zero at the boundary but the molecular viscous stress is large because the velocity gradient normal to the wall is large. The equations describing the flow in the laminar boundary layer are the Navier-Stokes equations containing only inertia and molecular viscous terms.

laminar flow In fluid flow, a smooth flow in which there is no crossflow of fluid particles occurring between adjacent stream lines. The flow velocity profile for laminar flow in circular pipes is parabolic in shape with a maximum flow in the center of the pipe and a minimum flow at the pipe walls. *Compare with* TURBULENT FLOW.

lander (spacecraft) A spacecraft designed to reach the surface of a planet and capable of surviving long enough to telemeter back data to Earth. The U.S. *Viking 1* and 2 landers on Mars and Surveyor landers on the Moon are examples of highly successful lander missions. The Russian Venera landers were able to briefly survive the extremely harsh surface environment of Venus and carried out chemical composition analyses of Venusian rocks and relayed images of the surface.

Landsat A family of versatile NASA-developed EARTH-OBSERVING SATELLITES that have pioneered numerous applications of MULTISPECTRAL SENSING. The first SPACECRAFT in this series, *Landsat-1* (originally called the Earth Resources Technology Satellite, or ERTS-1), was launched successfully in July 1972 and quite literally changed the way scientists and nonscientists alike could study EARTH from the vantage point of OUTER SPACE. *Landsat-1* was the first civilian spacecraft to collect relatively high-RESOLUTION images of the PLANET's land surfaces and did so simultaneously in several important WAVELENGTH BANDS of the ELECTROMAGNETIC

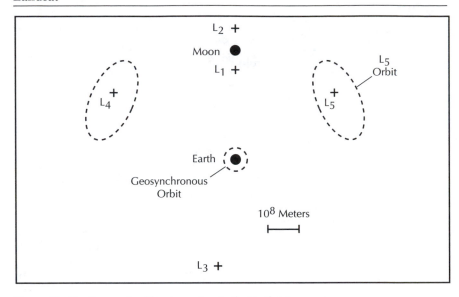

Figure 40. Five Lagrangian libration points in the Earth-Moon system.

SPECTRUM. NASA made these information-packed visible to near-infrared images available to researchers around the globe, who quickly applied the new data to a wide variety of important disciplines, including agriculture, water resource evaluation, forestry, urban planning, and pollution monitoring. *Landsat-2* (launched in January 1975) and *Landsat-3* (launched in March 1978) were similar in design and capability to *Landsat-1.*

In July 1982, NASA introduced its second-generation civilian REMOTE SENSING spacecraft with the successful launch of *Landsat-4.* That satellite had both an improved multispectral scanner (MSS) and a new thematic mapper (TM) INSTRUMENT. *Landsat-5,* carrying a similar complement of advanced instruments, was placed successfully into POLAR ORBIT in March 1984. Unfortunately, an improved *Landsat-6* spacecraft failed to achieve ORBIT in October 1993.

On April 15, 1999, a Boeing DELTA II EXPENDABLE LAUNCH VEHICLE lifted off from VANDENBERG AIR FORCE BASE in California and successfully placed the *Landsat-7* spacecraft into a 705-kilometer (438-mi) ALTITUDE, SUN-SYNCHRONOUS (POLAR) ORBIT. This operational orbit allows the spacecraft to completely image Earth every 16 days. *Landsat-7* carries an enhanced thematic mapper plus (ETM+)—an improved REMOTE SENSING instrument with eight bands sensitive to different wavelengths of visible and INFRARED RADIATION. *Landsat-7* sustains the overall objectives of the Landsat program by providing continuous, comprehensive coverage of Earth's land surfaces, thereby extending the unique collection of environmental and GLOBAL CHANGE data that began in 1972. Because of the program's continuous global coverage at the 30-meter (98-feet) spatial resolution, scientists can use Landsat data in global change research, in regional environmental change studies, and for many other civil and commercial purposes. The 1992 Land Remote Sensing Act identified data continuity as the fundamental goal of the Landsat program. *Landsat-7* was developed as a triagency program among NASA, the National Oceanic and Atmospheric Administration (NOAA), and the Department of the Interior's U.S. Geological Survey (USGS)—the agency responsible for receiving, processing, archiving, and distributing the data. Landsat data are now being applied in many interesting areas, including precision agriculture, cartography, GEOGRAPHIC INFORMATION SYSTEMS, water management, flood and hurricane damage assessment, environmental monitoring and protection, global change research, rangeland management, urban planning, and geology.

Large Magellanic Cloud (LMC) An IRREGULAR GALAXY about 20,000 LIGHT-YEARS in diameter and approximately 160,000 light-years from EARTH. *See also* MAGELLANIC CLOUDS.

large space structures The building of very large and complicated structures in space will be one of the hallmarks of space activities in the 21st century. The capability to supervise deployment and construction operations on orbit is a crucial factor in the effective use of space. Space planners and visionaries already have identified three major categories of large space structures: (1) very large surfaces (i.e., 100s square meters to perhaps a few square kilometers in area) to provide power (solar energy collection), to reject thermal energy (from a multimegawatt power-level space nuclear reactor), or to reflect sunlight (illumination of Earth's surface from space); (2) very large antennas (50-meter to 100-meter diameter and more) to receive and transmit energy over the radio-frequency portion of the electromagnetic spectrum; and (3) large space platforms (perhaps 200 meters or more in dimension) that provide the docking, assembly, checkout, and resupply facilities to service spacecraft and orbital transfer vehicles or that accommodate the highly automated "factory" facilities for space-based manufacturing operations. In the beginning, these large space structures will be deployed on-orbit—that is, erected or inflated from some stowed launch configuration. However, there is a limit to the size of such deployable and erectable systems. Larger, more sophisticated space structures (i.e., those with dimensions exceeding 25 meters or so) will require assembly on orbit—most likely by teams of space workers supported by fleets of

robotic spacecraft that can intelligently assist in their fabrication, assembly, and maintenance.

laser A device capable of producing an intense beam of coherent light. The term is an acronym derived from the expression *l*ight *a*mplification by the stimulated *e*mission of *r*adiation. In a laser, the beam of light is amplified when photons (quanta of light) strike excited atoms or molecules. These atoms or molecules are thereby stimulated to emit new photons (in a cascade or chain reaction) that have the same wavelength and are moving in phase and in the same direction as the original photon. Modern lasers come in a wide variety of wave-lengths and power levels. Very-high-energy lasers, for example, can destroy a target by heating, melting, or vaporizing the object. Other types of lasers (operating at much lower power levels) are used in such diverse areas as medicine, communications, environmental monitoring, and geophysics research (i.e., to study movement of Earth's crust). *See also* BALLISTIC MISSILE DEFENSE.

laser designator A device that emits a low-power beam of laser light that is used to mark a specific place or object. A weapon system equipped with a special tracking sensor that detects this reflected laser light can then home in on the designated target with great accuracy.

Laser Geodynamics Satellite (LAGEOS) A series of PASSIVE, spherical SATELLITES launched by NASA and the Italian Space Agency (Agenzia Spaziale Italiana [ASI]) and dedicated exclusively to SATELLITE LASER RANGING (SLR) technologies and demonstrations. These SLR activities include measurements of global tectonic plate motion, regional crustal deformation near plate boundaries, EARTH'S GRAVITY field, and the orientation of Earth's polar AXIS and spin rate. The *LAGEOS-1* satellite was launched in 1976 and the *LAGEOS-2* in 1992. Each satellite was placed into a nearly circular, 5,800-kilometer (3,600 mi) ORBIT but with a different INCLINATIONS (110 DEGREES for *LAGEOS-1* and 52 degrees for *LAGEOS-2*.)

The small, spherical satellites are just 60 centimeters (24 in) in diameter but have a mass of 405 kilograms (900 lbm). This compact, dense design makes their orbits as stable as possible, thereby accommodating satellite laser ranging measurements of submillimeter precision. The massive, compact SPACECRAFT have a design life of 50 years.

The LAGEOS satellite looks like a dimpled golf ball with its surface covered by 426 nearly equally spaced RETRORE-FLECTORS. Each retroreflector has a flat, circular front face with a prism-shaped back. These retroreflectors are three-dimensional prisms that reflect LASER LIGHT directly back to its source. A timing signal starts when the laser BEAM leaves the ground station and continues until the PULSE, reflected from one of the spacecraft's retroreflectors, returns to the ground station. Since the SPEED OF LIGHT is constant, the distance between the ground station and the satellite can be determined precisely. This process is called *satellite laser ranging* (SLR). Scientists use this technique to accurately measure movements of Earth's surface—centimeters per year in certain locations. For example, *LAGEOS-1* data have shown that the island of Maui (in Hawaii) is moving toward Japan at a rate of approximately 7 centimeters (2.8 in) per year and away from South America at 8 centimeters (3.1 in) per year.

laser guided weapon A weapon that utilizes a seeker to detect laser energy reflected from a laser-marked/designated target and through signal processing provides guidance commands to a control system that guides the weapon to the point from which the laser energy is being reflected.

laser propulsion system A proposed form of advanced propulsion in which a high-energy laser beam is used to heat a propellant (working fluid) to extremely high temperatures and then the heated propellant is expanded through a nozzle to produce thrust. If the high-energy laser is located away from the space vehicle, as, for example, on Earth or the Moon or at a

different orbital position in space, the laser energy used to heat the propellant would be beamed into the space vehicle's thrust chamber. This concept offers the advantage that the space vehicle being propelled need not carry a heavy onboard power supply. A fleet of laser-propelled orbital transfer vehicles operating between low-Earth orbit and lunar orbit has been suggested as part of a lunar base transportation system. This propulsion concept also has been suggested for deep space probes on one-way missions to the edges of the solar system and beyond. In this case, the laser energy ("light energy") might first be gathered by a large optical collector and then converted into electric energy, which would be used to operate some type of advanced electric propulsion system. *See also* BEAMED ENERGY PROPULSION.

laser radar An active remote sensing technique analogous to radar but that uses laser light rather than radio- or microwave radiation to illuminate a target. Also called *ladar* or *lidar*. *See also* LIDAR.

laser rangefinder A device that uses laser energy for determining the distance from the device to a place or object. A short pulse of laser light usually is placed on the target and the time required for the reflected laser signal to return to the device is used to determine the range.

latch A device that fastens one object or part to another but is subject to ready release on demand so the objects or parts can be separated. For example, a sounding rocket can be held on its launcher by a latch or several latches and then quickly released after ignition and proper thrust development.

latent heat The amount of thermal energy (heat) added to or removed from a substance to produce an isothermal (constant temperature) change in phase. There are three basic types of latent heat. The *latent heat of fusion* is the amount of thermal energy added or removed from a unit mass of substance to cause an isothermal change in phase between liquid and solid (e.g., freezing water or melting an ice cube). The *latent heat of vaporization* is the amount of thermal energy added to or removed from a unit mass of substance to cause an isothermal change of phase between liquid and gas (e.g., evaporating water or condensing steam). The latent heat of vaporization is sometimes called the *latent heat of condensation,* when the change in phase is from gas to liquid. Finally, the *latent heat of sublimation* is the amount of thermal energy added to or removed from a unit mass of substance to cause an isothermal change in phase from solid to vapor (e.g., a block of "dry ice" disappearing in a cloud of carbon dioxide vapor). In the thermodynamic literature, the latent heat is often treated as the ENTHALPY of: vaporization, condensation, melting, solidification (fusion), sublimation, as appropriate. Note: The term "latent heat" is used to describe only the amount of thermal energy (heat) added to or removed from a substance that causes a change in phase but no change in temperature; term "sensible heat" is used to describe energy transfer as heat that produces a temperature change. This somewhat confusing nomenclature has its roots in the historic evolution of the science of thermodynamics.

lateral 1. Of or pertaining to the side; directed or moving toward the side. 2. Of or pertaining to the lateral axis; directed, moving, or located along, or parallel to, the lateral axis.

Latin space designations The language of the Roman Empire (from over two millennia ago) is used today by space scientists to identify places and features found on other worlds. This is done by international agreement to avoid confusion or contemporary language favoritism in the naming of newly discovered features on the planets and their moons. As in biology and botany, Latin is treated as a "neutral" language in space science. Two of the more common Latin terms that are encountered in space science and planetary geology are: *mare maria* (pl:) meaning "sea" e.g., *Mare*

Tranquillitatis—the Sea of Tranquility—site of the first human landing (*Apollo 11* mission) on the Moon; and *mons* monte (pl:) meaning "mountain" e.g., *Olympus Mons*—the largest volcano on Mars). As an interesting historical footnote, when Galileo first explored the Moon with his newly invented telescope (circa 1610), he thought the dark (lava-flow) regions he saw where bodies of water and mistakenly called them *maria* (or seas).

latitude The angle between a perpendicular at a location and the equatorial plane of Earth. Latitude is measured in degrees north or south of the equator with the equator being defined as 0 degrees, the North Pole as 90 degrees north and the South Pole as 90 degrees south latitude, respectively.

launch 1. (noun) a. The action taken in launching a rocket or aerospace vehicle from a planetary surface. b. The transition from static repose to dynamic flight of a missile or rocket. c. The action of sending forth a rocket, probe, or other object from a moving vehicle such as an aircraft, aerospace vehicle, or spacecraft. 2. (verb) a. To send off a rocket or missile under its own propulsive power. b. To send off a missile, rocket, or aircraft by means of a catapult, as in the case of the German V-1 cruise missile (World War II era), or by means of inertial force, as in the release of a bomb from a flying aircraft. c. To give a payload or space probe an added boost for flight into space just before separation from its launch vehicle.

launch aircraft The aircraft used to air launch a missile, rocket, human-crewed experimental aircraft, air-launched space booster or remotely piloted vehicle (RPV). Launch aircraft, such as the NASA/Air Force B-52s used during the X-15 hypersonic flight program, often are called "motherships." See also PEGASUS; X-15.

launch azimuth The initial compass heading of a powered vehicle at launch. The term often is applied to space launch vehicles.

launch complex In general, the site, facilities, and equipment used to launch a rocket, missile, or aerospace vehicle. *See also* LAUNCH SITE.

launch configuration The assembled combination of strap-on booster rockets, primary launch vehicle, spacecraft/payload, and upper stages (if appropriate) that must be lifted off the ground at launch.

launch crew The group of engineers, technicians, managers, and safety personnel that prepare and launch a rocket, missile, or aerospace vehicle.

launching angle The angle between a horizontal plane and the longitudinal axis of a rocket, missile, or aerospace vehicle being launched.

launching rail A rail that gives initial support and guidance to a missile or rocket launched in a nonvertical position.

launchpad In general, the load-bearing base of platform from which a rocket, missile, or aerospace vehicle is launched. Often simply called "the pad."

launch site A defined area from which an aerospace or rocket vehicle is launched, either operationally or for test purposes. Major launch sites in the United States include the Eastern Range at Cape Canaveral Air Force Station/Kennedy Space Center on the central east coast of Florida, the Western Range at Vandenberg Air Force Base on the west central coast of California, and NASA's Wallops Island facility on the east coast of Virginia. Major Russian launch sites include Plesetsk, Kapustin Yar, and Tyuratam (also called the Baikonur Cosmodrome), which is located in Kazakhstan. The European Space Agency (ESA) has a major launch site in Kourou, French Guiana, on the northeast coast of South America.

launch time The time at which an aircraft, missile, rocket, or aerospace vehicle is scheduled for flight.

launch under attack (LUA) Execution by National Command Authorities (e.g., the president) of Single Integrated Operational Plan (SIOP) forces subsequent to tactical warning of strategic nuclear attack against the United States and prior to first impact of an enemy's nuclear warheads.

launch vehicle (LV) An expendable (ELV) or reusable (RLV) ROCKET-propelled vehicle that provides sufficient THRUST to place a SPACECRAFT into ORBIT around EARTH or to send a PAYLOAD on an INTERPLANETARY TRAJECTORY to another CELESTIAL BODY. Sometimes called BOOSTER or space lift vehicle. (See the table and Figure 41.)

launch window An interval of time during which a launch may be made to satisfy some mission objective. Usually a short period of time each day for a certain number of days. An interplanetary launch window generally is constrained within a number of weeks each year by the location of Earth in its orbit around the Sun, in order to permit the launch vehicle to use Earth's orbital motion for its trajectory as well as to time the spacecraft to arrive at its destination when the target planet is in the correct position. A launch that involves a rendezvous in Earth orbit with another spacecraft or platform must be timed precisely to account for the orbital motion of the "target" space object, creating launch windows of just a few minutes each day. Similarly, an interplanetary launch window is further constrained to just a few hours each appropriate day in order to take full advantage of Earth's rotational motion.

layered defense An approach to ballistic missile defense (BMD) consisting of several relatively independent layers of missile defense technologies/systems designed to operate against different portions of the trajectory of an enemy's ballistic missile. For example, there could be a boost-phase (first layer) defense with any remaining targets being passed on to succeeding layers (i.e., mid-course, terminal) of the defense system. *See also* BALLISTIC MISSILE DEFENSE.

Characteristics of Some of the World's Launch Vehicles

Country	Launch Vehicle	Stages	First Launch	Performance
China	Long March 2 (CZ-2C)	2 hypergolic, optional solid upper stage	1975	3,175 kg to LEO
	Long March 2F	2 hypergolic, 4 hypergolic strap-on rockets	1992	8,800 kg to LEO
	Long March 3	2 hypergolic, 1 cryogenic	1984	5,000 kg to LEO
	Long March 3A	2 hypergolic, 1 cryogenic	1994	8,500 kg to LEO
	Long March 4	3 hypergolic	1988	4,000 kg to LEO
Europe (ESA/ France)	Ariane 40	2 hypergolic, 1 cryogenic	1990	4,625 kg to LEO
	Ariane 42P	2 hypergolic, 1 cryogenic, 2 strap-on solid rockets	1990	6,025 kg to LEO
	Ariane 42L	2 hypergolic, 1 cryogenic 2 hypergolic strap-on rockets	1993	3,550 kg to GTO
	Ariane 5	2 large solid boosters, cryogenic core, hypergolic upper stage	1996	18,000 kg to LEO, 6,800 kg to GTO
India	Polar Space Launch Vehicle (PSLV)	2 solid stages, 2 hypergolic, 6 strap-on solid rockets	1993	3,000 kg to LEO
Israel	Shavit	3 solid-rocket stages	1988	160 kg to LEO
Japan	M-3SII	3 solid-rocket stages, 2 strap-on solid rockets	1985	770 kg to LEO
	H-2	2 cryogenic, 2 strap-on solid rockets	1994	10,000 kg to LEO, 4,000 kg to GTO
Russia	Soyuz	2 cryogenic, 4 cryogenic strap-on rockets	1963	6,900 kg to LEO
	Rokot	3 hypergolic	1994	1,850 kg to LEO
	Tsyklon	3 hypergolic	1977	3,625 kg to LEO
	Proton (D-I)	3 hypergolic	1968	20,950 kg to LEO
	Energia	cryogenic core, 4 cryogenic strap-on rockets, optional cryogenic upper stages	1987	105,200 kg to LEO
USA	Atlas I	1-1/2 cryogenic lower stage, 1 cryogenic upper stage	1990	5,580 kg to LEO, 2,250 GTO
	Atlas II	1-1/2 cryogenic lower stage, 1 cryogenic upper stage	1991	6,530 kg to LEO, 2,800 kg to GTO
	Atlas IIAS	1-1/2 cryogenic lower stage, 1 cryogenic upper stage, 4 strap-on solid rockets	1993	8,640 kg to LEO, 4,000 kg to GTO
	Delta II	1 cryogenic, 1 hypergolic, 1 solid stage, 9 strap-on solid	1990	5,050 kg to LEO, 1,820 kg to GTO
	Athena I	2 solid stages	1995	815 kg to LEO
	Pegasus (aircraft-launched)	3 solid stages	1990	290 kg to LEO
	Space shuttle	2 large solid-rocket boosters, cryogenic core	1981	25,000 kg to LEO
	Taurus	4 solid stages	1994	1,300 to LEO
	Titan 4	2 hypergolic stages, 2 large strap-on solid rockets, variety of upper stages	1989	18,100 to LEO

LEO, low Earth orbit; GTO, geostationary transfer orbit.
Source: NASA, DoD, OTA (U.S. Congress), and others.

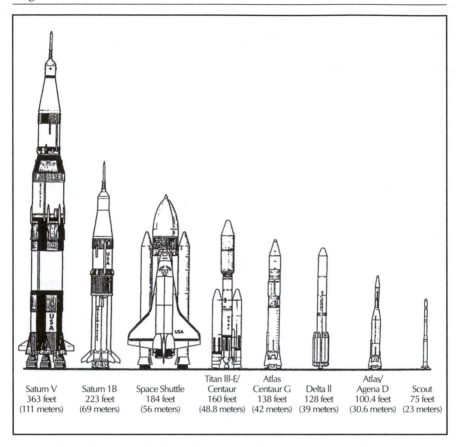

Saturn V	Saturn 1B	Space Shuttle	Titan III-E/ Centaur	Atlas Centaur G	Delta II	Atlas/ Agena D	Scout
363 feet	223 feet	184 feet	160 feet	138 feet	128 feet	100.4 feet	75 feet
(111 meters)	(69 meters)	(56 meters)	(48.8 meters)	(42 meters)	(39 meters)	(30.6 meters)	(23 meters)

Figure 41. Major U.S. launch vehicles that supported space exploration in the 20th century.

length (symbol: L, l, or *l*) The dimension of an aircraft, rocket, or aerospace vehicle from nose to tail; the measurement of this dimension.

lens A curve piece of glass, quartz, plastic, or other transparent material that has been ground, polished, and shaped accurately to focus light from a distant object so as to form an image of that object.

libration A real or apparent oscillatory motion, particularly the apparent oscillation of the Moon. Because of libration, more than half (about 59%) of the Moon's surface actually can be seen over time by an observer on Earth despite the fact that it is in synchronous motion around our planet (i.e., keeping the same side always facing Earth).

libration points The unique positions of gravitational balance measured from a celestial body and its satellite at which points smaller objects can orbit in stable equilibrium with the other two bodies. *See also* LAGRANGIAN LIBRATION POINTS.

lidar An active remote-sensing technique analogous to radar but that uses laser light rather than radio- or microwave radiation to illuminate a target. The laser light is reflected off a target and then detected. Using time-of-flight techniques to analyze the returned beam, information on both the distance to and velocity of the target can be obtained. The term is an acronym for *light detection and ranging*. Sometimes referred to as *ladar* (*laser detection and ranging*).

life cycle   All the phases through which an item, component, or system passes from the time engineers envision and initially develop it until the time it is either consumed in use (for example, expended during a MISSION) or disposed of as excess to known requirements (for example, a flight spare that is not needed because of the success of the original system). NASA engineers generally use the following life cycle phases: Pre-Phase A (conceptual study), Phase A (preliminary analysis), Phase B (definition), Phase C/D (design and development), and the operations phase.

life support system (LSS)   The system that maintains life throughout the AEROSPACE flight environment, including (as appropriate) travel in OUTER SPACE, activities on the surface of another world (for example, the LUNAR surface), and ascent and descent through EARTH'S ATMOSPHERE. The LSS must reliably satisfy the human crew's daily needs for clean AIR, potable water, food, and effective waste removal.

lift   1. That component (commonly used symbol, L) of the total aerodynamic force acting upon a body perpendicular to the undisturbed airflow relative to the body. *See also* AERODYNAMIC FORCES. 2. To lift off, or to take off in a vertical ascent. Said of a rocket or an aerostat.

liftoff   The action of a ROCKET or AEROSPACE VEHICLE as it separates from the LAUNCH PAD in a vertical ascent. In AEROSPACE operations, this term only applies to vertical ascent, while TAKEOFF applies to ascent at any ANGLE.

light   Electromagnetic radiation, ranging from about 0.4 to 0.7 micrometers (microns) in wave-length, to which the human eye is sensitive. Light helps us form our visual awareness of the universe and its contents.

light-gathering power (LGP)   The ability of a TELESCOPE or other optical INSTRUMENT to collect LIGHT.

light-minute (lm)   A unit of length equal to the distance traveled by a beam of light (or any electromagnetic wave) in the vacuum of outer space in one minute. Since the speed of light (c) is 299,792.5 kilometers per second (km/s) in free space, a light-minute corresponds to a distance of approximately 18 million kilometers.

lightning protection   Aerospace launch complexes and the launch vehicle themselves must be protected from lightning strikes. At any given instant, more than 2,000 thunderstorms are taking place throughout the world. All of these combine to produce about 100 lightning flashes per second.

On November 14, 1969, an out-of-the-blue lightning bolt struck and temporarily disabled the electrical systems on the *Apollo 12* spacecraft onboard a Saturn V launch vehicle about 1.6 kilometers (1 mile) above NASA's Kennedy Space Center (KSC) on the way to the Moon. On March 26, 1987 a NASA Atlas/Centaur vehicle and its payload were lost when the vehicle was hit by lightning while climbing up through the atmosphere after launch from Cape Canaveral Air Force Station (CCAFS). The lightning strike caused the rocket's computer to upset and issue an extreme yaw command that led to the vehicle's breakup in flight.

Lightning tends to strike the highest points in any given area, and aerospace safety engineers take special care to protect tall structures at a launch complex from these high-voltage bolts. The U.S. Lightning Protection Code for structures requires a pathway, or conductor, that will lead a lightning bolt safely to the ground. Additional protection is provided by circuit breakers, fuses, and electrical surge arrestors.

KSC operates extensive lightning protection and detection systems in order to keep workers, the space shuttle vehicle, launchpads, and processing facilities safe from harm. While the NASA lightning protection system is exclusively on KSC property, the lightning detection system incorporates equipment and personnel both at the Kennedy Space Center and CCAFS, which is located just east of the NASA complex. The Launch Pad Lightning Warning System (LPLWS) and the Lightning Surveillance System (LSS) provide the data necessary for evaluating weather conditions that can lead to the issuance of lightning warnings for the entire KSC/CCAFS launch complex. When the threat of lightning is present (typically within 8 kilometers [5 miles] of critical facilities), a lightning policy is put into effect. This lightning policy restricts certain outdoor activities and hazardous operations, such as the start of launch vehicle fueling.

light-second (ls) A unit of length equal to the distance traveled by a beam of light (or any electromagnetic wave) in the vacuum of outer space in one second. Since the speed of light (c) in free space is 299,792.5 kilometers per second (km/s), a light-second corresponds to a distance of approximately 300,000 kilometers.

light water Ordinary water (H_2O), as distinguished from heavy water (D_2O), where D is deuterium.

light-year (ly) The distance LIGHT (or other forms of ELECTROMAGNETIC RADIATION) can travel in one YEAR. One light-year is equal to a distance of approximately 9.46×10^{12} KILOMETERS, or 63,240 ASTRONOMICAL UNITS (AU).

limb The visible outer edge or observable rim of the DISK of a CELESTIAL BODY—for example, the "limb" of EARTH, the SUN, or the MOON.

limit switch A mechanical device that can be used to determine the physical position of equipment. For example, an extension on a valve shaft mechanically trips a limit switch as it moves from open to shut or shut to open. The limit switch gives "ON/OFF" output that corresponds to valve position. Normally, limit switches are used to provide full open or full shut indications. Many limit switches are the push-button variety. When the valve extension comes in contact with the limit switch, the switch depresses to complete, or turn on, the electrical circuit. As the valve extension moves away from the limit switch, spring pressure opens the switch, turning off the circuit.

line (or duct) Enclosed passageway (usually circular in cross section with relatively thin walls) that conveys fluid under pressure.

linear 1. Of or pertaining to a line. 2. Having a relation such that a change in one quantity is accompanied by an exactly proportional change in a related quantity, such as input and output of electronic equipment.

linear acceleration The time rate of change of linear velocity, namely: $a = dv/dt$.

linear accelerator A long straight tube (or series of tubes) in which charged particles, such as electrons or protons, gain in energy by the action of oscillating electromagnetic fields. *See also* ACCELERATOR.

linear energy transfer (LET) The energy lost (ΔE) per unit path length (Δx) by ionizing radiation as it passes through matter. Often it is used as a measure of the relative ability of different types of ionizing radiation to produce a potentially harmful effect on living matter; that is, the higher the LET value, the greater the potential for biological effect. LET = $\Delta E / \Delta x$ = energy deposited/unit path length; often expressed in units of: keV/micrometer.

line of apsides The line connecting the two points of an orbit that are nearest and farthest from the center of attraction, as the *perigee* and *apogee* of a satellite in orbit around Earth or the *perihelion* and *aphelion* of a planet around the Sun; the major axis of any elliptical orbit and extending indefinitely in both directions. *See also* ORBITS OF OBJECTS IN SPACE.

line of flight The line in air or space along which an aircraft, rocket, aerospace vehicle, or spacecraft travels. This line can be either the intended (i.e., planned) line of movement or the actual line of movement of the vehicle or craft.

line of force A line indicating the direction that a force acts.

line of nodes The straight line created by the inter-section of a satellite's orbital plane and a reference plane (usually the equatorial plane) of the planet or object being orbited.

line of sight (LOS) The straight line between a SENSOR or the eye of an observer and the object or point being observed. Sometimes called the optical path.

liner 1. With respect to solid-propellant rocket motors, the thin layer of material used to bond the solid propellant to the motor case; this material also can provide thermal insulation. 2. With respect to a rocket nozzle, the thin layer of ablative material placed on the nozzle wall (especially in the throat area) to help reduce heat transfer from the hot exhaust gases to the nozzle wall structure.

line spectra The discontinuous lines produced when bound electrons go from a higher energy level to a lower energy level in an excited atom, spontaneously emitting electromagnetic radiation. This radiation is essentially emitted at a single frequency, which is determined by the transition or "jump" in energy. Each different "jump" in energy level, therefore, has its own characteristic frequency. The overall collection radiation emitted by the excited atom is called the line spectra. Since these line spectra are characteristic of the atom, they can be used for identification purposes.

link In telecommunications, a general term used to indicate the existence of communications path-ways and/or facilities between two points. In referring to communications between a ground sta-tion and a spacecraft or satellite, the term "uplink" describes communications from the ground site to the spacecraft, while the term "downlink" describes communications from the spacecraft to the ground site.

liquid A phase of matter between a solid and a gas. A substance in a state in which the individual particles move freely with relation to each other and take the shape of the container but do not expand to fill the container. *Compare with* FLUID.

liquid fuel A rocket fuel that is liquid under the conditions in which it is stored and used in a rocket engine. *See also* LIQUID PROPELLANT.

liquid-fuel rocket A ROCKET vehicle that uses chemical PROPELLANTs in LIQUID form for both the FUEL and the OXIDIZER.

liquid hydrogen (LH$_2$) A liquid propellant used as the fuel with liquid oxygen serving as the oxidizer in high-performance cryogenic rocket engines. Hydrogen remains a liquid only at very low (cryogenic) temperatures, typically about 20 kelvins (K) (–253° Celsius or –423° Fahrenheit) or less. This cryogenic temperature requirement creates significant propellant storage and handling problems. However, the performance advantage experienced with the use of liquid hydrogen/liquid oxygen in high-energy space boosters often offsets these cryogenic handling problems. *See also* LIQUID PROPELLANT.

liquid oxygen (LOX or LO$_2$) A cryogenic liquid propellant requiring storage at temperatures below 90 kelvins (K) (i.e., –183° Celsius or –298° Fahrenheit). Often used as the oxidizer with RP-1 (a kerosene-type) fuel in many expendable boosters or with liquid hydrogen (LH$_2$) fuel in high-performance cryogenic propellant rocket engines. *See also* LIQUID PROPELLANT.

liquid propellant Any liquid combustible fed to the combustion chamber of a rocket engine. The petroleum used as a rocket fuel is a type of highly refined kerosene, called RP-1 (for *refined petroleum-1*). Usually it is burned with liquid oxygen (LOX) in rocket engines to provide thrust.

Cryogentic propellants are very cold liquid propellants. The most commonly used are LOX, which serves as an oxidizer, and liquid hydrogen (LH_2), which serves as the fuel. Cooling and compressing gaseous hydrogen and oxygen into liquids vastly increases their density, making it possible to store them (as cryogenic liquids) in smaller tanks. Unfortunately, the tendency of cryogenic propellants to return to their gaseous form unless kept supercool makes them difficult to store for long periods of time. Therefore, cryogenic propellants are less satisfactory for use in strategic nuclear missiles or tactical military rockets, which must be kept launch-ready for months to years at a time. However, cryogenic propellants are favored for use in space boosters, because of the relatively high thrust achieved per unit mass of chemical propellant consumed. For example, the space shuttle's main engines burn LH_2 and LOX.

Hypergolic propellants are liquid fuels and oxidizers that ignite on contact with each other and need no ignition source. This easy start and restart capability makes these liquid propellants attractive for both crewed and uncrewed spacecraft maneuvering systems. Another positive feature of hypergolic propellants is the fact that, unlike cryogentic propellants, these liquids do not have to be stored at extremely low temperatures. One favored hypergolic combination is monomethyl hydrazine (MMH) as the fuel and nitrogen tetroxide (N_2O_4) as the oxidizer. However, both of these fluids are highly toxic and must be handled under the most stringent safety conditions. Hypergolic propellants are used in the core liquid propellant stages of the Titan family of launch vehicles.

Some of the desirable properties of a liquid rocket propellant are: high specific gravity (to minimize tank space); low freezing point (to permit cold-weather operation); stability (to minimize chemical deterioration during storage); favorable thermophysical properties (e.g., a high boiling points and high thermal conductivity—especially valuable in support of regenerative cooling of the combustion chamber and nozzle throat); and low vapor pressure (to support more efficient propellant pumping operations). Some of the physical hazards encountered in the use of liquid propellants include: toxicity, corrosion, ignitability (when exposed to air or heat), and explosive potential (when the propellant deteriorates, is shocked, or excessively heated). *See also* ROCKET.

liquid-propellant rocket engine A ROCKET engine that uses chemical PROPELLANTS in LIQUID form for both the FUEL and the OXIDIZER.

liter (symbol: l or L) A unit of volume in the metric (SI) system. Defined as the volume of 1 kilogram of pure water at standard (atmospheric) pressure and a temperature of 4° Celsius. Also spelled *litre*. 1 liter = 0.2642 gallons = 1.000028 cubic decimeters (dm^3).

lithium hydroxide (LiOH) A white crystalline compound used for removing carbon dioxide from a closed atmosphere, such as found on a crewed spacecraft, aerospace vehicle, or space station. Spacesuit life support systems also use lithium hydroxide canisters to purge the suit's closed atmosphere of the carbon dioxide exhaled by the astronaut occupant.

little Green men (LGM) A popular expression (originating in science fiction literature) for extraterrestrial beings, presumably intelligent.

live testing The testing of a rocket, missile, or aerospace vehicle by actually launching it. *Compare* STATIC TESTING.

lobe An element of a beam of focused electromagnetic (e.g., radio frequency, or RF) energy. Lobes define surfaces of equal power density at varying distances and directions from the radiating antenna. Their configuration is governed by two factors: (1) the geometrical properties of

Uranus must use a high-gain antenna (HGA) because the distances involved. An LGA sometimes is mounted on top of a HGA's subreflector.

LOX Liquid oxygen at a temperature of 90 kelvins (K) (−183° Celsius) or lower. A common propellant (the oxidizer) for liquid rocket engines. Encountered in a variety of aerospace uses, such as *LOX tank, LOX valve,* or *loxing* (e.g., to fill a rocket vehicle's propellant tank with liquid oxygen).

lox-hydrogen engine A rocket engine that uses liquid hydrogen as the fuel and liquid oxygen (LOX) as the oxidizer, such as the space shuttle main engines (SSMEs).

lubrication (in space) Because of the very low PRESSURE conditions encountered in OUTER SPACE, conventional lubricating oils and greases evaporate very rapidly. Even soft metals (such as copper, lead, tin, and cadmium) that are often used in bearing materials on EARTH will evaporate at significant rates in space. Responding to the harsh demands on materials imposed by the space environment, AEROSPACE engineers use two general lubrication techniques in SPACE VEHICLE design and operation: thick-film lubrication and thin-film lubrication. In thick-film lubrication (also known as hydrodynamic or hydrostatic lubrication), the lubricant remains sufficiently viscous during operation so that the moving surfaces do not come into physical contact with each other. Thin-film lubrication takes place whenever the film of lubricant between two moving surfaces is squeezed out so that surfaces actually come into physical contact with each other (on a microscopic scale). Aerospace engineers lubricate the moving components of a SPACECRAFT or SATELLITE with dry films, liquids, metallic coatings, special greases, or combinations of these materials. Liquid lubricants are often used on space vehicles with missions of a year or more duration. *See also* SPACE TRIBOLOGY.

lumen (symbol: lm) The SI UNIT of luminous flux. It is defined as the luminous flux emitted by a uniform point source with an intensity of one candela in a solid angle of one steradian. *See also* CANDELA.

luminosity (symbol: *L*) The rate at which a STAR or other luminous object emits ENERGY, usually in the form of ELECTROMAGNETIC RADIATION. The luminosity of the SUN is approximately 4×10^{26} WATTS. *See also* STEFAN-BOLTZMANN LAW.

luminous intensity Luminous energy per unit time per unit solid angle; the intensity (i.e., flux per unit solid angle) of visible radiation weighted to take into account to variable response of the human eye as a function of the wavelength of light; usually expressed in candelas. Sometimes called *luminous flux density*.

lumped mass Concept in engineering analysis wherein a mass is treated as if it were concentrated at a point.

Luna A series of Russian spacecraft sent to the Moon in the 1960s and 1970s. For example, *Luna 9*, a 1,581-kilogram (kg) (3,485 (pounds-mass [lbm]) spacecraft, was launched on January 31, 1966 and successfully soft-landed on the Ocean of Storms on February 3, 1966. This spacecraft transmitted several medium-resolution photographs of the lunar surface before its batteries failed four days after landing. The lander spacecraft also returned data on radiation levels at the landing site. *Luna 12* was an orbiter spacecraft that was launched on October 22, 1966 successfully orbited the Moon and returned television pictures of the surface. *Luna 16*, launched on September 12, 1970, was the first successful automated (robotic) sample-return mission to the lunar surface. (Of course, the U.S. *Apollo 11* and *12* lunar landing missions in July 1969 and November 1969 respectively, already had returned lunar samples collected by astronauts who walked on the Moon's surface.) After landing on the Sea of Fertility, this robot spacecraft deployed a drill that bore 35 centimeters (about 1 foot) into the surface. The lunar soil sample, which had a mass of about 0.1 kg (0.22 lbm), was transferred automatically to a return vehicle, which left the lunar surface and landed in the former Soviet Union on September 24, 1970, *Luna 17* introduced the first robot roving vehicle, called Lunokhod I, on the lunar surface. The spacecraft successfully touched down on the Sea of Rains and deployed the sophisticated Lunokhod I rover. This eight-wheel vehicle was radio-controlled from Earth. The rover covered 10.5 kilometers (6.5 miles) during a surface exploration mission that lasted 10.5 months. The rover's cameras transmitted more than 20,000 images of the Moon's surface and instruments of the vehicle analyzed properties of the lunar soil at many hundreds of locations. *Luna 20* (launched February 14, 1972) and *Luna 24* (launched August 9, 1976) were also successful robot soil sample return missions. *Luna 21,* launched in January 1973, successfully deployed another robot rover, Lunokhod 2, in the Le Monnier crater in the Sea of Tranquility. This 840-kg (1,852 lbm) rover vehicle traveled about 37 kilometers (23 miles) during its four-month-long surface exploration mission. Numerous photographs were taken and surface experiments conducted by this robot rover under radio-control by Russian scientists and technicians on Earth. *See also* MOON.

lunar   Of or pertaining to the Moon.

lunar base   A permanently inhabited complex on the surface of the Moon. In the first permanent lunar base camp, a team of from 10, up to perhaps 100, lunar workers will set about the task of fully investigating the Moon. (The word "permanent" here means that the facility will always be occupied by human beings, but individuals probably will serve tours of from one to three years before returning to Earth.) These lunar base inhabitants will take advantage of the Moon as a science-in-space platform and perform the fundamental engineering studies needed to confirm and define the specific roles the Moon will play in the full development of space in the 21st century. For example, the discovery of frozen volatiles (including water) in the perpetually frozen recesses of the Moon's polar regions could change

lunar base logistics strategies and accelerate development of a large lunar settlement of up to 10,000 or more inhabitants. Many lunar base applications have been proposed. Some of these concepts include: (1) a lunar scientific laboratory complex; (2) a lunar industrial complex to support space-based manufacturing; (3) an astrophysical observatory for solar system and deep space surveillance; (4) a fueling station for orbital transfer vehicles that travel through cislunar-space; and (5) a training site and assembly point for the first human expedition to Mars.

lunar crater   A depression, usually circular, on the surface of the Moon. It frequently occurs with a raised rim called a *ringwall.* Lunar craters range in size up to 250 kilometers (155 miles) in diameter. The largest lunar craters are sometimes called "walled plains." The smaller crater—say, 15 to 30 kilometers (9.3 to 18.6 miles) across—often are called "craterlets"; while the very smallest, just a few hundred meters across, are called "beads." Many lunar craters have been named after famous people, usually astronomers. *See also* IMPACT CRATER; MOON.

lunar day   The period of time associated with one complete orbit of the Moon about Earth. It is equal to 27.322 "Earth" days. The lunar day is also equal in length to the sidereal month.

lunar eclipse   The phenomenon observed when the Moon enters the shadow of Earth. A lunar eclipse is called *penumbral,* if the Moon enters only the penumbra of Earth; *partial,* if the Moon enters the umbra without being totally immersed; and *total,* if the Moon is entirely immersed in the umbra.

lunar excursion module (LEM)   *See* LUNAR MODULE.

lunar gravity   The force imparted by the Moon to a mass that is at rest relative to the lunar surface. It is approximately one-sixth the acceleration experienced by a mass at rest on the surface (sea level) of

Earth. On the Moon's surface, the acceleration due to gravity is about 1.62 meters per second per second (m/s^2) [5.32 feet per second per second (ft/s^2)].

lunar highlands The light-colored, heavily cratered mountainous part of the lunar landscape. *See also* MOON.

lunar maria Vast expanses of low-ALBEDO (that is, comparatively dark) ancient volcanic lava flows found mostly on the NEARSIDE surface of the MOON.

lunar module (LM) The lander spacecraft used to bring the Apollo astronauts to the surface of the Moon. The lunar module (LM) lifted off from Earth enclosed in a compartment of the *Saturn V* launch vehicle, below the command-service module (CSM) that housed the astronauts. Once the craft was on its way to the Moon, the CSM pulled the LM from its storage area. In lunar orbit, the two Apollo astronauts who were to explore the lunar surface climbed into the LM, undocked, and "flew" the lander spacecraft to the Moon's surface. The descent stage of the LM contained a main, centrally located retrorocket engine that accomplished the "soft-landing." Upon completion of the surface expedition, the crew lifted off from the lunar surface using the top segment of the LM, which was powered by another, smaller rocket. The "four-legged" bottom portion of the LM remained on the Moon. After reaching lunar orbit with the now-smaller LM spacecraft, the crew rendezvoused with and transferred back into the Apollo CSM. The ascent stage of the LM was then jettisoned and placed in a decaying orbit that sent it crashing into the lunar surface to avoid possible interference with the Apollo CSM or a future mission. (See Figure 42.)

During the aborted *Apollo 13* mission, the LM served as a lifeboat for all three Apollo astronauts, providing them the emergency life support that helped get them safely home after the CSM suffered a catastrophic explosion on the way to the Moon.

Figure 42. *Apollo 11* astronaut Edwin "Buzz" Aldrin, Jr. descends the ladder of the lunar module and becomes the second human to walk on the Moon (July 20, 1969). *(Courtesy of NASA)*

The LM is also referred to as the *lunar excursion module,* or simply "LEM." *See also* APOLLO PROJECT.

lunar orbit Orbit of a spacecraft around the Moon.

Lunar Orbiter Five Lunar Orbiter missions were launched by NASA in 1966 through 1967 with the purpose of mapping the Moon's surface prior to the landings by the Apollo astronauts (which occurred from 1969 to 1972). All five missions were highly successful; 99% of the lunar surface was photographed with a 60-meter (197-feet) spatial resolution or better. The first three Lunar Orbiter missions were dedicated to imaging 20 potential Apollo landing sites, which had been preselected based on observations from Earth. The fourth and fifth missions were committed to broader scientific objectives and were flown in high-altitude polar orbits around the Moon. *Lunar Orbiter 4* photographed the entire nearside and 95% of the farside, and *Lunar Orbiter 5* completed the farside coverage and acquired medium- (20-meter) and high- (2-meter) resolution images of 36 preselected areas.

These probes were sent into orbit around the Moon to gather information and then purposely crashed at the end of each mission to prevent possible interference with future projects. See also APOLLO PROJECT; MOON.

lunar probe A probe for exploring and reporting conditions on or about the Moon.

Lunar-Prospector A NASA Discovery Program SPACECRAFT designed for a low-ALTITUDE, POLAR ORBIT investigation of the MOON. The spacecraft was launched successfully from CAPE CANAVERAL AIR FORCE STATION, Florida, by a Lockheed Athena II vehicle (formerly called the LOCKHEED MARTIN LAUNCH VEHICLE) on January 6, 1998. After swinging into ORBIT around the Moon on January 11, the *Lunar Prospector* used its complement of INSTRUMENTS to perform a detailed study of surface composition. The spacecraft also searched for resources—especially suspected deposits of water ice in the permanently shadowed regions of the lunar poles. The 126-KILOGRAM (278 lbm) spacecraft carried a GAMMA RAY SPECTROMETER, a NEUTRON spectrometer, a MAGNETOMETER, an ELECTRON reflectometer, an ALPHA PARTICLE spectrometer, and a Doppler GRAVITY experiment.

The data from this mission complemented the detailed imagery data from the CLEMENTINE mission. Data from the *Lunar Prospector's* neutron spectrometer hinted at the presence of significant amounts of water ice at the LUNAR poles. While still subject to confirmation, the presence of large quantities of water ice at the lunar poles would make the Moon a valuable supply depot for any future human settlement of the Moon and regions beyond.

After a highly successful 19-month scientific mapping mission, flight controllers decided to turn the spacecraft's originally planned end-of-life crash into the lunar surface into an impact experiment that might possibly confirm the presence of water ice on the Moon. Therefore, as its supply of attitude control fuel neared exhaustion, the spacecraft was directed to crash into a

CRATER near the Moon's south pole on July 31, 1999. Observers from EARTH attempted to detect signs of water in the impact plume, but no such signal was found. However, this impromptu impact experiment should be regarded only as a long shot opportunity and not a carefully designed scientific procedure. In contrast, the *Lunar Prospector's* scientific data have allowed scientists to construct a detailed map of the Moon's surface composition. These data have also greatly improved knowledge of the origin, evolution, and current inventory of lunar resources.

lunar rover(s) Crewed or automated (robot) vehicles used to help explore the Moon's surface. The Lunar Rover Vehicle (LRV), shown in Figure 43, also was called a space buggy and the "Moon car." It was used by American astronauts during the *Apollo 15, 16,* and *17* expeditions to the Moon. This vehicle was designed to climb over steep slopes, go over rocks, and move easily over sandlike lunar surfaces. It was able to carry more than twice its own mass (about 210 kilograms [463 pounds-mass]) in passengers, scientific instruments, and lunar material samples. This electric-powered (battery) vehicle could travel about 16 kilometers (10 miles) per hour on level ground. The vehicle's power came from two 36-volt silver zinc batteries that drove independent 1/4-horsepower electric motors in each wheel. Apollo astronauts used their space buggies to explore well beyond their initial lunar-landing sites. With these vehicles, they were able to gather Moon rocks and travel much farther and quicker across the lunar surface than if they had to explore on foot. For example, during the *Apollo 17* expedition, the lunar rover traveled 19 kilometers (12 miles) on just one of its three excursions.

Automated or robot rovers also can be used to explore the lunar surface. For example, during the Russian *Luna 17* mission to the Moon in 1970, the "mother" spacecraft soft-landed on the lunar surface in the Sea of Rains and deployed the Lunokhod 1, robot rover vehicle. Controlled from Earth by radio signals, this

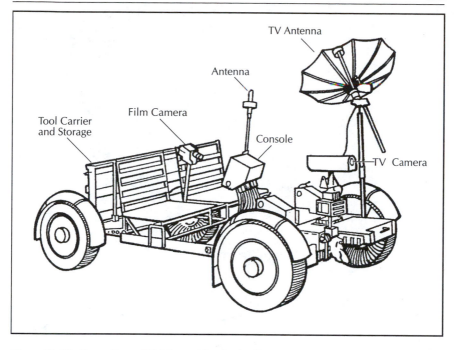

Figure 43. The Lunar Rover Vehicle used by the Apollo astronauts to explore the Moon's surface far from their landing site.

eight-wheeled lunar rover vehicle traveled for months across the lunar surface, transmitting more than 20,000 television images of the surface and performing more than 500 lunar soil tests at various locations. The Russian *Luna 21* mission to the Moon in January 1973 successfully deployed another robot rover, called Lunokhod 2. *See also* APOLLO PROJECT; LUNA; MOON.

lunar satellite A human-made object that is placed in orbit (one or more revolutions) around the Moon. For example, the American Lunar Orbiter spacecraft was a lunar satellite. *See also* LUNAR ORBITER.

Lunokhod A Russian eight-wheeled robot vehicle, controlled by radio signals from Earth and used to conduct surface exploration of the lunar surface. *See also* LUNA; LUNAR ROVER(S).

lux (symbol: lx) The SI unit of illuminance. It is defined as 1 lumen per square meter. *See also* LUMEN.

M

Mace An improved version of the air force's Matador missile (MGM-1C), differing primarily in its improved guidance system, longer-range, low-level attack capability, and higher-yield warhead. The Mace was designated as the MGM-13. This early "winged missile" was guided by a self-contained radar guidance system (MGM-13A) or by an inertial guidance system (MG-13B).

Mach number (symbol: M) A number expressing the ratio of the speed of a body or of a point on a body with respect to the surrounding air or other fluid, or the speed of a flow, to the speed of sound in the medium. It is named after the Austrian scientist Ernst Mach (1838–1916).

If the Mach number is less than one (i.e., M < 1), the flow is called *subsonic,* and local disturbances can propagate ahead of the flow. If the Mach number is equal to unity (i.e., M = 1), the flow is called *sonic.* If the Mach number is greater than one (i.e., M > 1), the flow is called *supersonic,* and disturbances cannot propagate ahead of the flow, with the general result that oblique shock waves form (i.e., shock waves inclined in the direction of flow). The formation and characteristics of such oblique shock waves involve complicated compressible flow phenomena. Sometimes the oblique shock wave is attached to the surface of the body, but this usually occurs only when the supersonic stream encounters a sharp, narrow-angled object. Most times, however, the oblique shock wave takes the form of a rounded, detached shock that appears in front of a more blunt-shaped object moving at supersonic speeds through a compressible fluid medium.

MACHO The *ma*ssive *c*ompact *h*alo *o*bject hypothesized to populate the outer regions (or halo) of galaxies, accounting for most of the DARK MATTER. Current theories suggest that MACHOs might be low-LUMINOSITY STARS, JOVIAN PLANETS, or possibly BLACK HOLES. *See also* WIMP.

Magellanic Clouds The two dwarf, irregularly shaped neighboring galaxies that are closest to our MILKY WAY GALAXY. The LARGE MAGELLANIC CLOUD (LMC) is about 160,000 LIGHT-YEARS away, and the SMALL MAGELLANIC CLOUD (SMC) is approximately 180,000 light-years away. Both can be seen with the NAKED EYE in the Southern Hemisphere. Their presence was first recorded in 1519 by the Portuguese explorer Ferdinand Magellan (1480–1521), after whom they are named.

Magellan **mission** A NASA solar system exploration mission to the planet Venus. On May 4, 1989, the 3,550 kilogram (7,825 pound-mass) *Magellan* spacecraft was delivered to Earth orbit by the space shuttle *Atlantis* during the STS 30 mission and then sent on an interplanetary trajectory to the cloud-shrouded planet by a solid-fueled inertial upper stage (IUS) rocket system. *Magellan* was the first interplanetary spacecraft to be launched by the space shuttle. On August 10, 1990, the *Magellan* (named for the famous 16th-century Portuguese explorer Ferdinand Magellan) was inserted into orbit around Venus and began initial operations of its very successful radar mapping mission.

Magellan used a sophisticated imaging radar system to make the most detailed map of Venus ever captured during its four years in orbit around Earth's "sister planet" from 1990 to 1994. After concluding its radar mapping mission, *Magellan* made global maps of the Venusian gravity field. During this phase of the mission, the spacecraft did not use its radar mapper but instead transmitted a constant radio signal back to Earth. When it passed over an area on Venus with higher than normal gravity, the spacecraft would speed up slightly in its orbit. This movement then would cause the frequency of *Magellan*'s radio signal to change very slightly due to the DOPPLER EFFECT. Because of the ability of the radio receivers in the NASA Deep Space Network to measure radio frequencies extremely accurately, scientists were able to construct a very detailed gravity map of Venus. In fact, during this phase of its mission, the spacecraft provided high-resolution gravity data for about 95% of the planet's surface. Flight controllers also tested a new maneuvering technique called *aerobraking*, a technique that uses a planet's atmosphere to slow or steer a spacecraft.

The craters revealed by *Magellan*'s detailed radar images suggested to planetary scientists that the Venusian surface is relatively young—perhaps "recently" resurfaced or modified about 500 million years ago by widespread volcanic eruptions. The planet's current harsh environment has persisted at least since then. No surface features were detected that suggest the presence of oceans or lakes at any time in the planet's past. Furthermore, scientists found no evidence of plate tectonics, that is, the movements of huge crustal masses.

Magellan's mission ended with a dramatic plunge through the dense atmosphere to the planet's surface. This was the first time an operating planetary spacecraft has ever been crashed intentionally. Contact was lost with the spacecraft on October 12, 1994, at 10:02 Universal Time (3:02 A.M. Pacific Daylight Time). The purpose of this last maneuver was to gather data on the Venusian atmosphere before the spacecraft ceased functioning during its fiery descent. Although much of the *Magellan* is believed to have been vaporized by atmospheric heating during this final plunge, some sections may have survived and hit the planet's surface intact. *See also* VENUS.

magma Molten rock material beneath the surface of a planet or moon. It may be ejected to the surface by volcanic activity.

magnet A substance with the property of attracting certain other substances. Some metals, such as cobalt, iron, and nickel, can be magnetized to attract other magnetic metals. A magnet has a MAGNETIC FIELD, which is a region around it where there are forces exerted on other magnetic materials. Earth has a magnetic field, as do several other planets. An electric current, flowing through a conductor, also creates a magnetic field.

magnetic field A region of space with a detectable magnetic force at every location within the region. Often it is represented as an array of imaginary lines of force that exist in relation to a magnetic pole.

magnetic field strength (symbol: H) The strength of the magnetic force on a unit magnetic pole in a region of space affected by other magnets or electric currents. The magnetic field strength is a vector quantity that is measured in amperes per meter.

magnetic flux (symbol: φ_M) A measure of the total size of a magnetic field. The flux through any area in the medium surrounding a magnet is equal to the integral of the magnetic flux density *(B)* over the area *(A)*. That is,

$$\varphi_M = \int B \cdot d A$$

The weber (Wb) is the SI UNIT of magnetic flux.

magnetic flux density (symbol: B) The magnetic flux that passes through a unit area of a MAGNETIC FIELD in a direction at right angles to the magnetic force. It is a vector quantity. The SI UNIT of magnetic flux density is the tesla (T), which is equal to one weber of magnetic flux per square meter.

magnetic storm A sudden, worldwide disturbance of Earth's magnetic field, believed caused by solar disturbances, such as a solar flare. The onset of a magnetic storm may occur in an hour, while the gradual return to normal conditions may take several days.

magnetohydrodynamics (MHD) converter A device that directly converts the kinetic energy of a hot working fluid into electrical energy. This direct energy conversion (DEC) process is achieved by flowing a high-temperature, electrically conducting gas or gas-liquid metal combination through a stationary magnetic field. MHD conversion systems can either be closed-cycle or open-cycle systems. In a closed-cycle MHD system, the working fluid is continuously recirculated; in an open-cycle MHD system, the working fluid passes just once through the converter region and is then discharged.

magnetometer An instrument for measuring the strength and sometimes the direction of a magnetic field. Magnetometers often are used to detect and measure the interplanetary and solar magnetic fields in the vicinity of a scientific spacecraft. These direct-sensing instruments typically detect the strength of magnetic fields in three dimensions (i.e., in the x-, y-, and z-planes). As a magnetometer sweeps an arc through a magnetic field when the spacecraft rotates, an electrical signal is produced proportional to the magnitude and structure of the field.

magnetoplasmadynamic (MPD) thruster An advanced electric propulsion device capable of operating with a wide range of propellants in both pulsed and steady-state modes. MPD thrusters are well suited for orbit transfer and spacecraft maneuvering applications. In an MPD thruster device, the current flowing from the cathode to the anode sets up a ring-shaped magnetic field. This magnetic field then pushes against the plasma in the arc. As propellant, such as argon, flows through the arc plasma, it is ionized and forced away by the magnetic field. A thrusting force therefore is created by the interaction of an electrical current and a magnetic field. *See also* ELECTRIC PROPULSION.

magnetosphere The region around a planet in which charged atomic particles are influenced by the planet's own MAGNETIC FIELD rather than by the magnetic field of the Sun, as projected by the solar wind. Because Earth has its own magnetic field, the interaction of Earth's magnetic field and the solar wind results in this very dynamic and complicated region surrounding Earth. As shown in Figure 44 studies by spacecraft and probes have now mapped much of the region of magnetic field structures and streams of trapped particles around Earth. The *solar wind,* a plasma of electrically charged particles (mostly protons and electrons) that flows at speeds of 1 million kilometers per hour or more from the Sun, shapes Earth's magnetosphere into a teardrop, with a long magnetic tail (called the *magnetotail*) stretching out opposite the Sun.

Earth and other planets of the solar system exist in the *heliosphere*—the region of space dominated by the magnetic influence of the Sun. Interplanetary space is not empty but filled with the solar wind. The geomagnetic field of Earth presents an obstacle to the solar wind, behaving much like a rock in a swiftly flowing stream of water. A shock wave, called the BOW SHOCK, forms on the sunward side of Earth and deflects the flow of the solar wind. The bow shock slows down, heats, and compresses the solar wind, which then flows around the geomagnetic field, creating Earth's magnetosphere. The steady pressure of the solar wind compresses the otherwise spherical field lines of Earth's magnetic field on the sunward side at about 15 Earth radii, or some 100,000 kilometers, a distance still inside the Moon's orbit around Earth. On the night side of Earth away from the Sun, the solar wind pulls the geomagnetic field lines out to form a long magnetic tail (i.e., the *magnetotail*). The magnetotail is believed to extend for hundreds of Earth radii, although it is not known precisely

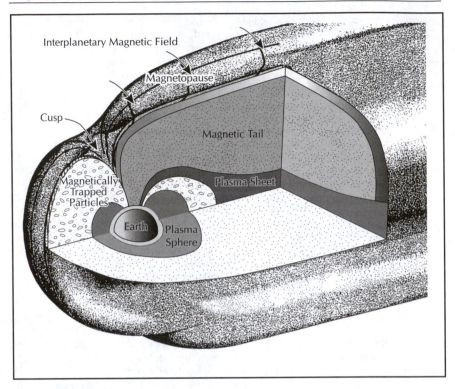

Figure 44. A synoptic view of Earth's magnetosphere.

how far it actually extends into space away from Earth.

The outermost boundary of Earth's magnetosphere is called the *magnetopause*. Some solar wind particles do pass through the magnetopause and become trapped in the inner magnetosphere. Some of those trapped particles then travel down through the *polar cusps* at the North and South poles and into the uppermost portions of the Earth's atmosphere. These trapped solar wind particles then have enough energy to trigger the *aurora,* which are also called the *Northern (aurora borealis)* and *Southern Lights (aurora australis)* because they occur in circles around the North and South poles. These spectacular aurora are just one dramatic manifestation of the many connections among the Sun, the solar wind, and Earth's magnetosphere and atmosphere.

magnitude A number, now measured on a logarithmic scale, used to indicate the relative brightness of a celestial body. The smaller the magnitude number, the greater the brightness. When ancient astronomers studied the heavens, they observed objects of varying brightness and color and decided to group them according to their relative brightness. The 20 brightest stars of the night sky were called *stars of the first magnitude.* Other stars were called 2nd, 3rd, 4th, 5th, and 6th magnitude stars according to their relative brightness. The 6th magnitude stars are the faintest stars visible to the unaided (i.e., "naked") human eye under the most favorable observing conditions. By convention the star Vega (Alpha Lyra) was defined as having a magnitude of zero. Even brighter objects, such as the star Sirius or the planets Mars and Jupiter thus acquired *negative magnitude* values.

In 1856 the British astronomer N.R. Pogson, proposed a more precise logarithmic magnitude in which a difference of five magnitudes represented a relative brightness ratio of 100 to 1. Consequently, with Pogson's proposed scale, two stars differing by five magnitudes would differ in brightness by a factor of 100. Today the Pogson magnitude scale is used almost universally in astronomy. (See Figure 45.) *See also* STAR(S).

main-belt asteroid An asteroid located in the main asteroid belt, which occurs between Mars and Jupiter. This term sometimes is limited to asteroids found in the most populous portion of about 2.2 to 3.3 astronomical units (AUs) from the Sun. *See also* ASTEROID.

main-sequence star A STAR in the prime of its life that shines with a constant LUMINOSITY achieved by steadily converting HYDROGEN into HELIUM through THERMONUCLEAR FUSION in its CORE.

main stage 1. In a multistage rocket vehicle, the stage that develops the greatest amount of thrust, with or without booster engines. 2. In a single-stage rocket vehicle powered by one or more engines, the period when "full thrust" (i.e., at or above 90% of the rated thrust) is attained. 3. A sustainer engine that is considered as a stage after booster engines have fallen away, as in the main stage of the Atlas launch vehicle. *See also* ROCKET.

main valve The valve, usually located just upstream of the thrust chamber injector, that controls the propellant to the injector in a liquid rocket engine. *See also* ROCKET.

make safe One or more actions necessary to prevent or interrupt complete function of the system. Among the necessary actions are: (1) install safety devices such as pins or locks; (2) disconnect hoses, linkages, or batteries; (3) bleed (i.e., drain) fluids from accumulators and reservoirs; (4) remove explosive devices such as initiators, fuzes, and detonators; and (5) intervene by welding components in place or fixing their movement with lockwires. Often used to describe the "disarming" or "disabling" of a weapon or rocket system.

mandrel The tool that forms the geometry of the central cavity in the casting of a solid-propellant grain.

maneuvering reentry vehicle (MaRV) A reentry vehicle that can maneuver in the late-midcourse or terminal phase of its flight, either to enhance its accuracy or to avoid antiballistic missiles. Maneuvers within the atmosphere usually can be accomplished by aerodynamic means, while maneuvers in space could be accomplished by small rockets. *See also* BALLISTIC MISSILE DEFENSE.

maneuver pad Data or information on spacecraft attitude, thrust values, event times, and the like that is transmitted in advance of a maneuver.

manipulators Mechanical devices used for handling objects; frequently involving

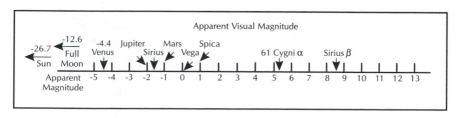

Figure 45. The apparent visual magnitude of several celestial objects.

remote operations (i.e., teleoperation) and/or hazardous substances or environmental conditions. That portion of a robot system which is capable of grasping or handling. A manipulator often has a versatile end effector (i.e., the special tool or "grasping element" installed at the end of the manipulator) that can respond to a variety of different handling requirements. For example, the U.S. space shuttle has a very useful manipulator called the *remote manipulator system (RMS)*.

man-machine interface The boundary where human and machine characteristics and capabilities are joined in order to obtain optimum operating conditions and maximum efficiency of the combined *man-machine system*. A joystick and a control panel are examples of man-machine interfaces. Also called *human-machine interface*.

manned An aerospace vehicle or system that is occupied by one or more persons, male or female. The terms "crewed," "human," or "personed" are preferred today in the aerospace literature. For example, a "manned mission to Mars" should be called a "human mission to Mars."

manned vehicle An older aerospace term describing a rocket or spacecraft that carried one or more human beings, male or female. Used to distinguish that craft from a robot (i.e., pilotless) aircraft, a ballistic missile, or an automated (and uncrewed) satellite or planetary probe. The expression "crewed vehicle" or "personed vehicle" is now preferred.

manometer An instrument for measuring the pressure of fluids (either gases or liquids) both above and below atmospheric pressure.

man-rated A launch vehicle, spacecraft, aerospace system, or component considered safe and reliable enough to be used by human crew members. The term "human-rated" is now preferred.

man-tended capability A space facility or orbiting platform designed to operate without a permanent human crew on board. However, astronauts will visit this type of orbiting facility periodically to conduct servicing operations and maintenance. The HUBBLE SPACE TELESCOPE (HST) is an example. The term "crew-tended" or "human-tended" is now preferred.

maria (singular: mare) Latin word for "seas." It was originally used by GALILEO GALILEI to describe the large, dark, ancient lava flows on the LUNAR surface, thought by early astronomers to be bodies of water on the MOON. Despite the physical inaccuracy, this term is still used by astronomers. *See* LATIN SPACE DESIGNATIONS.

Mariner spacecraft A series of NASA planetary exploration spacecraft that performed reconnaissance flyby and orbital missions to Mercury, Mars, and Venus in the 1960s and 1970s. *Mariner 2* encountered Venus at a distance of about 41,000 kilometers (km) (25,480 miles [mi]) on December 14, 1962. It provided important scientific data about conditions in interplanetary space (e.g., solar wind, interplanetary magnetic field, cosmic rays, etc.) and on the planet Venus. *Mariner 4* flew by Mars on July 14, 1965, and returned images of the planet's surface. The closest encounter distance for its mission was 9,846 km (6,118 mi) from the Martian surface. The *Mariner 5* spacecraft passed within 4,000 km (2,486 mi) of Venus on October 19, 1967. Its instruments successfully measured both interplanetary and Venusian magnetic fields, charged particles, plasmas, as well as certain properties of the Venusian atmosphere. *Mariner 6* passed within 3,431 km (2,132 mi) of the Martian surface on July 31, 1969. The spacecraft's instruments took images of the planet's surface and measured ultraviolet and infrared emissions of the Martian atmosphere. The *Mariner 7* spacecraft was identical to the *Mariner 6* spacecraft. *Mariner 7* passed within 3,430 km (2,131 mi) of Mars on August 5, 1969, and acquired images of the planet's surface. Emissions of the Martian atmosphere also were measured. *Mariner 9* arrived at and orbited Mars on November 14, 1971. This spacecraft gathered data on the atmospheric composition, density, pressure, and temperature of the Martian atmosphere as well as performing studies of the Martian surface. After depleting its supply of attitude control gas, the spacecraft was turned off on October 27, 1972. *Mariner 10* was the seventh successful launch in the Mariner series. (*Mariner 1* and *Mariner 8* experienced launch failures, while *Mariner 3* ceased transmissions nine hours after launch and entered solar orbit.) The *Mariner 10* spacecraft was the first to use the gravitational pull of one planet (Venus) to reach another planet (Mercury). It passed Venus on February 5, 1974, at a distance of 4,200 km (2,610 mi). The spacecraft then crossed the orbit of Mercury at a distance of 704 km (437 mi) from the surface on March 29, 1974. A second encounter with Mercury occurred on September 21, 1974, at an altitude of about 47,000 km (29,210 mi). A third and final Mercury encounter occurred on March 16, 1975, when the spacecraft passed the planet at an altitude of 327 km (203 mi). Many images of the planet's surface were acquired during these flybys, and magnetic field measurements were performed. When the supply of attitude control gas became depleted on March 24, 1975, this highly successful mission was terminated. *See also* MARS; MERCURY; VENUS.

marmon clamp A ring-shaped clamp, consisting of three equal-length segments held together by explosive bolts, used to couple the main subsections of a rocket vehicle.

Mars The fourth planet in the solar system with an equatorial diameter of 6,794 kilometers. Throughout human history Mars, the Red Planet, has been at the center of astronomical thought. The ancient Babylonians, for example, followed the motions of this wandering red light across the night sky and named it after Nergal, their god of war. In time the Romans, also honoring their own god of war, gave the planet its present name. The

presence of an atmosphere, polar caps, and changing patterns of light and dark on the surface caused many pre-space age astronomers and scientists to consider Mars an "Earthlike planet"—the possible abode of extraterrestrial life. In fact, when actor Orson Welles broadcast a radio drama in 1938 based on H. G. Wells's science-fiction classic *War of the Worlds,* enough people believed the report of invading Martians to create a near panic in some areas.

Over the past three decades, however, sophisticated robot spacecraft—flybys, orbiters, and landers—have shattered these romantic myths of a race of ancient Martians struggling to bring water to the more productive regions of a dying world. Spacecraft-derived data have shown instead that the Red Planet is actually a "halfway" world. Part of the Martian surface is ancient, like the surfaces of the Moon and Mercury, while part is more evolved and Earthlike. Contemporary information about Mars is presented in Table 1.

In August and September 1975, two Viking spacecraft were launched on a mission to help answer the question: Is there life on Mars? Each Viking spacecraft consisted of an orbiter and lander. While scientists did not expect these spacecraft to discover Martian cities bustling with intelligent life, the exobiology experiments on the lander were designed to find evidence of primitive life-forms, past or present. Unfortunately, the results sent back by the two robot landers were teasingly inconclusive.

The Viking Project was the first mission to successfully soft-land a robot spacecraft on another planet (excluding the Earth's Moon.) All four Viking spacecraft (two orbiters and two landers) exceeded by considerable margins their design goal lifetime of 90 days. The spacecraft were launched in 1975 and began to operate around or on the Red Planet in 1976. When the *Viking 1* lander touched down on the Plain of Chryse on July 20, 1976, it found a bleak landscape. Several weeks later its twin, the *Viking 2* lander, set down on the Plain of Utopia and discovered a more gentle, rolling landscape.

Table 1. Physical and Dynamic Data for Mars

Diameter (equatorial)	6,794 km
Mass	6.42×10^{23} kg
Density (mean)	3.9 g/cm^3
Surface gravity	3.73 m/sec^2
Escape velocity	5.0 km/sec
Albedo Atmosphere	0.15
(main components by volume)	
Carbon dioxide (CO_2)	95.32%
Nitrogen (N_2)	2.7%
Argon (Ar)	1.6%
Oxygen (O_2)	0.13%
Carbon monoxide (CO)	0.07%
Water vapor (H_2O)	0.03%[a]
Natural satellites	2 (Phobos and Delmos)
Period of rotation (a Martian day)	1.026 days
Average distance from Sun	2.28×10^8 km (1.523 AU)
Eccentricity	0.093
Period of revolution around Sun (a Martian year)	687 days
Mean orbital velocity	24.1 km/sec
Solar flux at planet (at top of atmosphere)	590 W/m^2 (at 1.52 AU)

[a] variable
Source: Based on NASA data.

One by one these robot explorers finished their highly successful visits its to Mars. The *Viking Orbiter 2* spacecraft ceased operation in July 1978; the Lander 2 fell silent in April 1980; *Viking Orbiter 1* managed at least partial operation until August 1980; the *Viking Lander 1* made its final transmission on November 11, 1982. NASA officially ended the Viding mission to Mars on May 21, 1983.

As a result of these interplanetary missions, we now know that Martian weather changes very little. For example, the highest atmosphere temperature recorded by either Viking lander was minus 21° Celsius (mid-summer at the *Viking 1* site) while the lowest recorded temperature was minus 124° Celsius (at the more northerly *Viking 2* site during winter).

The atmosphere of Mars was found to be primarily carbon dioxide (CO_2). Nitrogen, argon, and oxygen are present in small percentages, along with trace amounts of neon, xenon, and krypton. The Martian atmosphere contains only a wisp of water (about 1/1000th as much as found in Earth's atmosphere). But even this tiny amount can condense out and form clouds that ride high in the Martian atmosphere or form patches of morning fog in valleys. There is also evidence that Mars had a much denser atmosphere in the past—one capable of permitting liquid water to flow on the planet's surface. Physical features resembling riverbeds, canyons and gorges, shorelines, and even islands hint that large rivers and maybe even small seas once existed on the Red Planet.

Mars has to small, irregularly shaped moons called Phobos ("fear") and Deimos ("terror"). These natural satellites were discovered in 1877 by the American astronomer Asaph Hall (1829–1907). They both have ancient, cratered surfaces with some indication of REGOLITHS to depths of possibly 5 meters or more. The physical properties of these two moons are presented in Table 2. It is hypothesized that these moons actually may be asteroids "captured" by Mars.

Scientists believe that at least 12 unusual meteorites found on Earth actually are pieces of Mars that were blasted off the Red Planet by ancient meteoroid impact collisions. One particular Martian meteorite, called ALH84001, has stimulated a great deal of interest in the possibility of life on Mars. In the summer of 1996, a NASA research team at the Johnson Space Center (JSC) announced that they had found evidence in ALH84001 that "strongly suggests primitive life may have existed on Mars more than 3.6 billion years ago." The NASA team found the first organic molecules thought to be of Martian origin; several mineral features characteristic of biological activity; and possibly microfossils (i.e., very tiny fossils) of primitive, bacterialike organisms inside this ancient Martian rock that fell to Earth as a meteorite.

Stimulated by the exciting possibility of life on Mars, NASA has launched or will soon launch several new ROBOT SPACECRAFT intended to accomplish more focused scientific investigations of the Red Planet. Begin-

ning in 1996, some of these missions have proven highly successful, while others have ended in disappointing failures.

This new wave of exploration started on November 7, 1996, when NASA launched the MARS GLOBAL SURVEYOR (MGS) from CAPE CANAVERAL AIR FORCE STATION, Florida. The spacecraft arrived at Mars on September 12, 1997, an event representing the first successful mission to the Red Planet in two decades. After spending a year and a half carefully trimming its orbit from a looping ELLIPSE to a more useful circular track around the planet, the MGS began its mapping mission in March 1999. Using its high-resolution camera, the MGS observed the planet from its low-altitude, nearly polar orbit over the course of an entire Martian year—the equivalent of nearly two Earth years. At the conclusion of its primary scientific mission on January 31, 2001, the spacecraft entered an extended mission phase. The MGS has successfully studied the entire Martian surface, atmosphere, and interior, returning an enormous amount of valuable scientific data. Among its most significant scientific contributions so far are high-resolution images of gullies and debris flow features that invitingly suggest there may be current

Table 2. Physical and Dynamic Properties of the
Martian Moons, Phobos and Deimos

Property	Phobos	Deimos
Characteristic dimensions (both are irregularly shaped)		
Longest dimension	27 km	15 km
Intermediate dimension	21 km	12 km
Shortest dimension	19 km	11 km
Mass	10.8×10^{15} kg	1.8×10^{15} kg
Density	1.9 g/cm^3	1.8 g/cm^3
Albedo	0.06	0.07
Surface gravity	1 cm/sec^2	0.5 cm/sec^2
Rotation	Synchronous	Synchronous
Semimajor axis of orbit	9,378 km	23,436
Eccentricity	0.018	0.0008
Sidereal period (approx.)	0.319 days	1.262 days

Source: Based on NASA data.

sources of liquid water, similar to an aquifer, at or near the surface of the planet. These findings are shaping and guiding upcoming robot missions to Mars.

NASA launched the MARS PATHFINDER mission to the Red Planet on December 4, 1996, using a Delta II expendable launch vehicle. The mission, formerly called the Mars Environmental Survey (or MESUR) Pathfinder, had as its primary objective the demonstration of innovative, low-cost technology for delivering an instrumented LANDER and free-ranging robotic ROVER to the Martian surface. The *Mars Pathfinder* not only accomplished that important objective, but it also returned an unprecedented amount of data and operated well beyond its anticipated design life. From *Mars Pathfinder*'s innovative airbag bounce and roll landing on July 4, 1997, until its final data transmission on September 27, the lander/minirover team returned numerous images of the Ares Vallis landing site and useful chemical analyses of proximate rocks and soil deposits. Data from this successful mission have suggested that ancient Mars was once warm and wet, further stimulating the intriguing question of whether life could have emerged on that planet when liquid water flowed on its surface and its atmosphere was significantly thicker.

However, the exhilaration generated by these two successful missions was quickly dampened by two glaring failures. On December 11, 1998, NASA launched the MARS CLIMATE ORBITER (MCO) . This spacecraft was to serve as both an INTERPLANETARY WEATHER SATELLITE and a data relay satellite for another mission, called the MARS POLAR LANDER (MPL). The MCO also carried two science INSTRUMENTS— an atmospheric sounder and a color imager. However, just as the spacecraft arrived at the Red Planet on September 23, 1999, all contact was lost with it. NASA engineers have concluded that because of human error in programming the final TRAJECTORY, the spacecraft most probably attempted to enter orbit too deep in the planet's atmosphere and consequently burned up.

NASA used another Delta II expendable launch vehicle to send the *MPL* to the Red Planet on January 3, 1999. The *MPL* was an ambitious mission to land a robot spacecraft on the frigid Martian terrain near the edge of the planet's southern polar cap. Two small PENETRATOR PROBES (called Deep Space 2) piggybacked with the lander spacecraft on the trip to Mars. After an uneventful interplanetary journey, the *MPL* and its companion Deep Space 2 experiments were mysteriously lost when the spacecraft arrived at the planet on December 3, 1999.

Undaunted by the disappointing sequential failures, NASA officials sent the 2001 *Mars Odyssey* mission to the Red Planet on April 7, 2001. The scientific instruments onboard the orbiter spacecraft were designed to determine the composition of the planet's surface, to detect water and shallow buried ice, and to study the IONIZING RADIATION environment in the vicinity of Mars. The spacecraft arrived at the planet on October 24, 2001, and successfully entered orbit around it. After executing a series of AEROBRAKE maneuvers that properly trimmed it into a near-circular polar orbit around Mars, the spacecraft began to make scientific measurements in January 2002. The orbiting spacecraft will continue to collect scientific data until the end of its primary scientific mission (projected for July 2004). Beyond then, it will function primarily as a communications relay satellite until about October 2005, supporting information transfer from the 2003 Mars Exploration Rover Mission back to scientists on Earth.

NASA's 2003 Mars Exploration Rover Mission was launched in 2003 (between June 10 and July 17). Separate Delta II expendable launch vehicles will send each of the twin Mars ROVERS from Cape Canaveral Air Force Station to different locations on Mars. With much greater mobility than the *Mars Pathfinder* minirover, the new robot explorers can travel up to 100 meters (about 320 ft) per Martian day across the surface of the planet. Each rover carries a complement of instruments that enable it to search for evidence that liquid water was previously present on the planet's surface. The 2003 Mars Exploration Rovers (referred to as

Spirit and Opportunity) are identical twins. Their landing sequence at two different locations on Mars will resemble the airbag bounce and roll technique successfully demonstrated during the *Mars Pathfinder* mission.

In 2003 NASA also participated in a mission called MARS EXPRESS being planned by the EUROPEAN SPACE AGENCY (ESA) and the Italian Space Agency. The spacecraft's scientific instruments will study the atmosphere and surface of Mars from polar orbit. The mission's main objective will be to search from orbit for suspected subsurface water sites and then to deliver a small lander to a suitable location for closer scientific scrutiny. The small lander is called Beagle 2, after the famous ship in which the British naturalist Charles Darwin (1809–1882) made his great voyage of scientific discovery. After coming to rest on the surface of Mars, Beagle 2 will perform EXOBIOLOGY and geochemistry research.

In 2005 NASA plans to launch a powerful new scientific spacecraft called the *Mars Reconnaissance Orbiter (MRO)*. This mission to the Red Planet will focus on scrutinizing those candidate water-bearing locations previously identified by the *Mars Global Surveyor* and 2001 *Mars Odyssey* missions. The MRO will be capable of measuring thousands of Martian landscapes at a resolution of between 0.2 and 0.3 meters (0.7 to 1.0 ft). By way of comparison, this spacecraft's imaging capability will be good enough to detect and identify rocks on the surface of Mars that are as small as beach balls. Such high-resolution imagery data will help bridge the gap between detailed localized surface observations accomplished by landers and rovers and the synoptic global measurements made by orbiting spacecraft.

Possibly as early as 2009, NASA plans to develop and launch a long-duration, long-range mobile science laboratory. This effort will demonstrate the efficacy of developing and deploying truly "smart landers"—advanced robotic systems that are capable of autonomous operation, including hazard avoidance and navigation around obstacles to reach very

promising but difficult-to-reach scientific sites. NASA also proposes to create a new family of small scout missions. These missions would involve airborne robotic vehicles (such as a Mars airplane) and special miniature surface landers or penetrator probes, possibly delivered to interesting locations around Mars by the airborne robotic vehicles. These scout missions, beginning in about 2007, would greatly increase the number of interesting sites studied and set the stage for even more sophisticated robotic explorations in the second decade of the 21st century.

As presently planned, NASA intends to launch its first MARS SAMPLE RETURN MISSION (MSRM) in about 2014. Future advances in robot spacecraft technology will enhance and accelerate the search for possible deposits of subsurface water and life (existent or extinct) on the Red Planet. Two decades of intensely focused scientific missions by robot spacecraft will not only greatly increase scientific knowledge about Mars, but the effort will set the stage for the first MARS EXPEDITION by human explorers.

Mars base For automated Mars missions, the spacecraft and robotic surface rovers generally will be small and self-contained. For human expeditions to the surface of the Red Planet, however, two major requirements must be satisfied: life support (habitation) and surface transportation (mobility). Habitats, power supplies, and life support systems will tend to be more complex in a permanent Martian surface base that must sustain human beings for years at a time. Surface mobility systems will also grow in complexity and sophistication as early Martian explorers and settlers travel tens to hundreds of kilometers from their base camp. (See Figure 46.) At a relatively early time in any Martian surface base program, the use of Martian resources to support the base must be tested vigorously and then quickly integrated in the development of an eventually self-sustaining surface infrastructure.

In one possible scenario, the initial Martian habitats will resemble standardized lunar base (or space station) pressurized

Figure 46. An artist's rendering of a 21st-century Mars base near Pavonis Mons, a large shield volcano on the Martian equator that overlooks the ancient water-eroded canyon in which the base is located. The base infrastructure depicted includes a habitation module, a power module, central base work facilities, a green house, a launch and landing complex, and even a Martian airplane (ultralight construction). In the foreground, human explorers have taken their surface rover vehicle to an interesting spot where one of the explorers has just made the discovery of the century—a large fossil of an ancient Martian creature! *(Courtesy of NASA/JSC. Artist: Pat Rawlings)*

modules and will be transported from cislunar space to Mars in prefabricated condition by interplanetary nuclear-electric propulsion (NEP) cargo ships. These modules will be configured and connected as needed on the surface of Mars and then covered with a meter or so thickness of Martian soil for protection against the lethal effects of solar flare radiation or continuous exposure to cosmic rays on the planet's surface. (Unlike Earth's atmos-

phere, the very thin Martian atmosphere does not shield very well against ionizing radiations from space.) *See also* MARS.

Mars Climate Orbiter (MCO) Originally called part of the *Mars Surveyor '98* mission, NASA launched the *Mars Climate Orbiter* on December 11, 1998, from CAPE CANAVERAL AIR FORCE STATION, Florida, using a DELTA II EXPENDABLE LAUNCH VEHICLE. The mission of this ORBITER SPACECRAFT was to circle MARS and serve as both an INTERPLANETARY WEATHER SATELLITE and a COMMUNICATIONS SATELLITE, relaying data back to EARTH from the other part of the *Mars Surveyor '98* mission, a lander called the MARS POLAR LANDER. The *MCO* also carried two science INSTRUMENTS: an atmospheric sounder and a color imager. Unfortunately, when it arrived at the Red Planet, all contact with the spacecraft was lost on September 23, 1999. NASA engineers believe that the *MCO* burned up in the MARTIAN ATMOSPHERE due to a fatal error in its arrival TRAJECTORY. This human-induced computational error caused the spacecraft to enter too deeply into the planet's atmosphere and encounter destructive AERODYNAMIC HEATING.

Mars expedition The first human-crewed MISSION to visit MARS in the 21st century. Current concepts suggest a 600- to 1,000-day duration mission that most likely will start from EARTH ORBIT, a total crew size of up to 15 ASTRONAUTS, and about 30 days allocated for surface excursion activities on the RED PLANET.

Mars Express A mission to MARS launched in June 2003 and developed by the EUROPEAN SPACE AGENCY (ESA) and the Italian Space Agency. After the 1,042-kilogram (2,300-lbm) spacecraft arrives at the RED PLANET in December 2003, its scientific INSTRUMENTS will begin to study the ATMOSPHERE and surface of Mars from a POLAR ORBIT. The main objective of the *Mars Express* mission is to search from orbit for suspected subsurface water locations and then to deliver a small LANDER spacecraft to more closely investigate the most suitable candidate site. The small lander is named Beagle 2 in honor of the famous ship in which the British naturalist Charles Darwin (1809–1882) made his great voyage of scientific discovery. After coming to rest on the surface of Mars, Beagle 2 will perform EXOBIOLOGY and geochemistry research.

Mars Global Surveyor (MGS) NASA launched the *Mars Global Surveyor* MISSION from CAPE CANAVERAL AIR FORCE STATION, Florida, on November 7, 1996, using a DELTA II EXPENDABLE LAUNCH VEHICLE. The safe arrival of this ROBOT SPACECRAFT at MARS on September 12, 1997, represented the first successful mission to the RED PLANET in two decades. *MGS* was designed as a rapid, low-cost recovery of the *MARS OBSERVER* mission objectives. After a year and a half trimming its ORBIT from a looping ELLIPSE to a circular track around the PLANET, the spacecraft began its primary mapping mission in March 1999. Using a high-RESOLUTION camera, the *MGS* spacecraft observed the planet from a low-ALTITUDE, nearly polar orbit over the course of one complete Martian year, the equivalent of nearly two EARTH years. Completing its primary mission on January 31, 2001, the spacecraft entered an extended mission phase.

The *MGS* science INSTRUMENTs include a high-resolution camera, a thermal emission SPECTROMETER, a LASER ALTIMETER, and a MAGNETOMETER/ELECTRON reflectometer. With these instruments, the spacecraft successfully studied the entire Martian surface, ATMOSPHERE, and interior, returning an enormous amount of valuable scientific data in the process. Among the key scientific findings of this mission so far are high resolution images of gullies and debris flow features that suggest there may be current sources of liquid water, similar to an aquifer, at or near the surface of the planet. Magnetometer readings indicate the Martian MAGNETIC FIELD is not globally generated in the planet's CORE but appears to be localized in particular areas of the CRUST. Data from the spacecraft's laser altimeter have provided the first three-dimensional

views of the northern ice cap on Mars. Finally, new TEMPERATURE data and close-up images of the Martian MOON, PHOBOS, suggest that its surface consists of a powdery material at least one meter (3 ft) thick—most likely the result of millions of years of METEOROID IMPACTS.

Marshall Space Flight Center (MSFC) A major NASA complex in Huntsville, Alabama. It was the center where Dr. Wernher von Braun (1912–1977) and his team of German rocket scientists developed a series of highly successful rockets, including the giant Saturn V vehicle that sent human expeditions to the lunar surface. *See also* NATIONAL AERONAUTICS AND SPACE ADMINISTRATION.

Mars Observer mission   NASA's *Mars Observer,* the first of the Observer series of planetary missions, was designed to study the geoscience of Mars. The primary science objectives for the mission were to: (1) determine the global elemental and mineralogical character of the surface; (2) define globally the topography and gravitational field of the planet; (3) establish the nature of the Martian magnetic field; (4) determine the temporal and spatial distribution, abundance, sources, and sinks of volatiles (substances that readily evaporate) and dust over a seasonal cycle; and (5) explore the structure and circulation of the Martian atmosphere. The 1,018-kilogram (2,245 pound-mass) spacecraft was launched successfully on September 25, 1992. Unfortunately, for unknown reasons, contact with the *Mars Observer* was lost on August 22, 1993, just three days before scheduled orbit insertion around Mars. Contact with the spacecraft was not reestablished, and it is not known whether this spacecraft was able to follow its automatic programming and go into Mars orbit or if it flew by Mars and is now in a heliocentric orbit. Although none of the primary objectives of the mission was achieved, cruise mode (i.e., interplanetary) data were collected up to loss of contact. *See also* MARS; *MARS GLOBAL SURVEYOR.*

Mars Pathfinder NASA launched the *Mars Pathfinder* MISSION to the RED PLANET using a DELTA II EXPENDABLE LAUNCH VEHICLE on December 4, 1996. This mission, previously called the Mars Environmental Survey (or MESUR) Pathfinder, had the primary objective of demonstrating innovative technology for delivering an instrumented LANDER and free-ranging robotic ROVER to the MARTIAN surface. The *Mars Pathfinder* not only accomplished this primary mission but also returned an unprecedented amount of data, operating well beyond the anticipated design life.

Mars Pathfinder used an innovative landing method that involved a direct enter into the Martian ATMOSPHERE assisted by a PARACHUTE to slow its descent through the PLANET's atmosphere and then a system of large airbags to cushion the IMPACT of landing. From its airbag-protected bounce and roll landing on July 4, 1997, until the final data transmission on September 27, the robotic lander/rover team returned numerous close-up images of MARS and chemical analyses of various rocks and soil found in the vicinity of the landing site.

The landing site was at 19.33 N, 33.55 W, in the Ares Vallis region of Mars, a large outwash plain near CHRYSE PLANITIA (the Plains of Gold), where the VIKING 1 LANDER had successfully touched down on July 20, 1976. Planetary geologists speculate that this region is one of the largest outflow channels on Mars—the result of a huge ancient flood that occurred over a short period of time and flowed into the Martian northern lowlands.

The lander, renamed by NASA the Carl Sagan Memorial Station, first transmitted engineering and science data collected during atmospheric entry and landing. The American astronomer Carl Sagan (1934–1996) popularized ASTRONOMY and ASTROPHYSICS and wrote extensively about the possibility of EXTRATERRESTRIAL LIFE. Then, the lander's imaging system (which was on a pop-up mast) obtained views of the rover and the immediate surroundings. These images were transmitted back to EARTH to assist the human flight team in planning the robot rover's operations on the surface of Mars. After some initial maneuvers to clear an airbag out of the way, the lander deployed the ramps for the rover. The 10.6-KILOGRAM (23-lbm) minirover had been stowed against one of the lander's petals. Once commanded from Earth, the tiny robot explorer came to life and rolled onto the Martian surface. Following rover deployment, the bulk of the lander's remaining tasks were to support the rover by imaging rover operations and relaying data from the rover back to Earth. SOLAR CELLS on the lander's three petals, in combination with rechargeable batteries, powered the lander, which also was equipped with a meteorology station.

The rover, renamed Sojourner (after the American civil rights crusader Sojourner Truth), was a six-wheeled vehicle that was teleoperated (that is, driven over great distances by remote control) by personnel at the JET PROPULSION LABORATORY (JPL) in Pasadena, California. The rover's human controllers used images obtained by both the rover and the lander systems. TELEOPERATION at INTERPLANETARY distances required that the rover be capable of some semiautonomous operation, since the time delay of the SIGNALS averaged between 10 and 15 minutes depending on the relative positions of Earth and Mars.

For example, the rover had a hazard avoidance system, and surface movement was performed very slowly. The small rover was 280 millimeters high (about 11 in), 630 millimeters (24.8 in) long, and 480 millimeters (18.9 in) wide with a ground clearance of 130 millimeters (5.1 in). While stowed in the lander, the rover had a height of just 180 millimeters (about 7.1 in). However, after deployment on the Martian surface, the rover extended to its full height and rolled down a deployment ramp. The relatively far-traveling little rover received its supply of electrical ENERGY from its 0.2-square meter [2.16-sq ft] array of solar cells. Several nonrechargeable batteries provided backup power.

The rover was equipped with a black-and-white imaging system. This system provided views of the lander, the surrounding Martian terrain, and even the rover's own wheel tracks that helped scientists estimate soil properties. An ALPHA PARTICLE X-RAY SPECTROMETER (APXS) onboard the rover was used to assess the composition of Martian rocks and soil.

Both the lander and the rover outlived their design lives—the lander by nearly three times and the rover by 12 times. Data from this very successful lander/rover surface mission suggest that ancient Mars was once warm and wet, stimulating further scientific and popular interest in the intriguing question of whether life could have emerged on the planet when it had liquid water on the surface and a thicker atmosphere.

Mars Polar Lander (MPL) Originally designated as the LANDER portion of the *Mars Surveyor '98* mission, NASA launched the *Mars Polar Lander* SPACECRAFT from CAPE CANAVERAL AIR FORCE STATION, Florida, on January 3, 1999, using a DELTA II EXPENDABLE LAUNCH VEHICLE. *MPL* was an ambitious MISSION to land a ROBOT SPACECRAFT on the frigid surface of MARS near the edge of the PLANET's southern polar cap. Two small PENETRATOR PROBES (called Deep Space 2) piggybacked with the lander spacecraft on the trip to Mars. After an uneventful INTERPLANETARY journey, all contact with the *MPL* and the Deep Space 2 experiments was lost as the spacecraft arrived at the planet on December 3, 1999. The missing lander was equipped with cameras, a robotic arm, and instruments to measure the composition of the Martian soil. The two tiny penetrators were to be released as the lander spacecraft approached Mars and then follow independent ballistic trajectories, impacting on the surface and plunging below it in search of water ice.

The exact fate of the lander and its two tiny microprobes remains a mystery. Some NASA engineers believe that the *MPL* might have tumbled down into a steep canyon, while others speculate the *MPL* may have experienced too rough a landing and become disassembled. A third hypothesis suggests the *MPL* may have suffered a fatal failure during its descent through the Martian atmosphere. No firm conclusions could be drawn because the NASA mission controllers were completely unable to communicate with the missing lander or either of its hitchhiking planetary penetrators.

Mars Sample Return Mission (MSRM)
The purpose of a Mars Sample Return Mission is, as the name implies, to use a combination of robot spacecraft and lander systems to collect soil and rock samples from Mars and then return them to Earth for detailed laboratory analysis. A wide variety of options for this type of mission are being explored. For example, one or several small robot rover vehicles could be carried and deployed by the lander vehicle. These rovers (under the control of operators on Earth) would travel away from the original landing site and collect a wider range of rock and soil samples for return to Earth. Another option is to design a nonstationary, or mobile, lander that could travel (again guided by controllers on Earth) to various surface locations and collect interesting specimens. After the soil collection mission was completed, the upper portion of the lander vehicle would lift off from the Martian surface and rendezvous in orbit with a special "carrier" spacecraft. This automated rendezvous/return "carrier" spacecraft would remove the soil sample canisters from the ascent portion of the lander vehicle and then depart Mars orbit on a trajectory that would bring the samples back to Earth. After an interplanetary journey of about one year, this automated "carrier" spacecraft, with its precious cargo of Martian soil and rocks, would achieve orbit around Earth.

To avoid any potential problems of extraterrestrial contamination of Earth's biosphere by alien microorganisms that might possibly be contained in the Martian soil or rocks, the sample canisters might first be analyzed in a special human-tended orbiting quarantine facility. An alternate return mission scenario would be to bypass an Earth-orbiting quarantine process altogether and use a direct reentry vehicle operation to bring the encapsulated Martian soil samples to Earth.

Whatever sample return mission profile ultimately is selected, contemporary analysis of Martian meteorites (that have fallen to Earth) has stimulated a great scientific interest in obtaining well-documented and well-controlled "virgin" samples of Martian soil and rocks. Carefully analyzed in laboratories on Earth, these samples will provide a wealth of important and unique information about the Red Planet. These samples might even provide further clarification of the most intriguing question of all: Is there (or, at least, has there been) life on Mars? A successful Mars Sample Return Mission is also considered a significant and necessary step toward eventual human expeditions to Mars in the 21st century. *See also* EXTRATERRESTRIAL CONTAMINATION; MARS; MARS PATHFINDER; MARS SURFACE ROVER(S); MARTIAN METEORITES.

Mars surface rover(s) Automated robot rovers and human-crewed mobility systems used to satisfy a number of surface exploration objectives on MARS in the 21st century. *See also MARS PATHFINDER; 2003 MARS EXPLORATION ROVER MISSION.*

Martian Of or relating to the planet Mars.

2001 Mars Odyssey NASA launched the *2001 Mars Odyssey* MISSION to the RED PLANET from CAPE CANAVERAL AIR FORCE STATION on April 7, 2001. The ROBOT SPACECRAFT, previously called the *Mars Surveyor 2001 Orbiter,* is designed to determine the composition of the PLANET's surface, to detect water and shallow buried ice, and to study the IONIZING RADIATION environment in the vicinity of MARS. The spacecraft arrived at the planet on October 24, 2001, successfully entered ORBIT, and then performed a series of AEROBRAKE maneuvers to trim itself into the POLAR ORBIT around Mars for scientific data collection. The scientific mission began in January 2002.

Mars Odyssey has three primary science instruments: the thermal emission imaging system (THEMIS), a gamma ray spectrometer (GRS), and the Mars radiation environment experiment (MARIE).

THEMIS is examining the surface distribution of minerals on Mars, especially those that can form only in the presence of water. The GRS is determining the presence of 20 chemical ELEMENTs on the surface of Mars, including shallow, subsurface pockets of hydrogen that act as a proxy for determining the amount and distribution of possible water ice on the planet. Finally, MARIE is analyzing the Martian radiation environment in a preliminary effort to determine the potential hazard to future human explorers. The spacecraft will continue to collect scientific data until the nominal end of its primary scientific mission in July 2004. At that point, it will function as an orbiting communications relay until about October 2005, supporting information transfer between the *2003 MARS EXPLORATION ROVER* MISSION and scientists on Earth.

NASA selected the name *2001 Mars Odyssey* for this important spacecraft as a tribute to the vision and spirit of space exploration embodied in the science fact and science fiction works of the famous British writer Sir Arthur C. Clarke (b. 1917).

2003 Mars Exploration Rover Mission NASA launched identical twin MARS ROVERS to operate on the surface of the RED PLANET in the summer of 2003. With much greater mobility than the *MARS PATHFINDER* minirover, each of these powerful new robot explorers will be able to travel up to 100 meters (about 320 ft) per Martian day across the surface of the planet. Each rover carries a complement of sophisticated INSTRUMENTS that allows it to search for evidence that liquid water was present on the surface of MARS in ancient times. The landing will resemble the successful airbag bounce and roll arrival demonstrated during the *Mars Pathfinder* mission. Each rover will visit a different region of the planet. Immediately after landing the rover will perform reconnaissance of the particular landing site by taking panoramic (360°) visible (color) and infrared images. Then each rover will drive off to begin its surface exploration MISSION.

Using images and spectra taken daily by the rovers, scientists on EARTH will use TELECOMMUNICATIONS and TELEOPERATIONS to supervise the overall scientific program. With intermittent human guidance, the pair of mechanical explorers will function like "robot prospectors" capable of examining particular rock or soil targets and evaluating composition and texture at the microscopic level. Each rover has a set of five instruments with which to analyze rocks and soil samples. The instruments carried by each rover include a panoramic camera (Pancam), a miniature thermal emission SPECTROMETER (Mini-TES), a Mössbauer spectrometer (MB), an alpha particle X-ray spectrometer (APXS), magnets, and a microscopic imager (MI). There is also a special rock abrasion tool (or RAT) that will allow each rover to expose fresh rock surfaces for additional study of interesting targets.

Each rover has a mass of 180 kilograms (about 400 lbm) and a range of up to 100 meters (320 ft) per sol (Martian day). Surface operations should last for at least 90 sols and extend into late May 2004. If either or both rovers remain healthy, surface science activities could continue beyond that time. Communications back to Earth will be accomplished primarily by means of Mars-orbiting spacecraft, such as the *2001 MARS ODYSSEY*, serving as data relays.

Martian meteorites Scientists now believe that at least 12 unusual meteorites are pieces of Mars that were blasted off the Red Planet by meteoroid impact collisions. These interesting meteorites impact collisions. These interesting meteorites were previously called *SNC meteorites,* after the first three type samples discovered (namely: Shergotty, Nakhla, and Chassigny) but are now generally referred to as *Martian meteorites*. The *Chassigny meteorite* was discovered in Chassigny, France on October 3, 1815. It establishes the name of the *chassignite* type subgroup of the SNC meteorites. Similarly, the *Shergotty meteorite* fell on Shergotty, India on August 25, 1865 and provides the name of the *shergottite* type subgroup of SNC meteorites. Finally, the *Nakhla meteorite* was found in Nakhla, Egypt on June 28, 1911, and establishes the name for the *nakhlite* type subgroup of SNC meteorites.

Martian Meteorites				
Name	*Classification*	*Mass (kg)*	*Find/Fall*	*Year*
Shergotty	S-basalt (pyx-plag)	4.00	fall	1865
Zagami	S-basalt	18.00	fall	1962
EETA 79001	S-basalt	7.90	find-A	1980
QUE94201	S-basalt	0.012	find-A	1995
ALHA77005	S-lherzolite (ol-pyx)	0.48	find-A	1978
LEW88516	S-lherzolite	0.013	find-A	1991
Y793605	S-lherzolite	0.018	find-A	1995
Nakhla	N-clinopyroxenite	40.00	fall	1911
Lafayette	N-clinopyroxenite	0.80	find	1931
Gov. Valadares	N-clinopyroxenite	0.16	find	1958
Chassigny	C-dunite(olivine)	4.00	fall	1815
ALH84001	orthopyroxenite	1.90	find-A	1993

Classification: S = shergottite, N = nakhlte, C = chassignite. ALH84001 is none of these. *find-A* designates Antarctic meteorites (all recent finds). *Year* is recovery date for non-Antarctic meteorite, and date of Martian classification for Antarctic meteorites.

Source: NASA.

All 12 known SNC meteorites are igneous rocks crystallized from molten lava in the crust of the parent planetary body. The Martian meteorites discovered so far on Earth represent five different types of igneous rocks, ranging from simple plagioclase-pyroxene basalts to almost monomineralic cumulates of pyroxene or olivine. These Martian meteorites are summarized in the table.

The only natural process capable of launching Martian rocks to Earth is meteoroid impact. To be ejected from Mars, a rock must reach a velocity of 5 kilometers (3.1 miles [mi]) per second or more. (The escape velocity for Mars is 5 km/sec [3.1 mi/sec].) During a large meteoroid impact on the surface of Mars, the kinetic energy of the incoming cosmic "projectile" causes shock deformation, heating, melting, and vaporization as well as crater excavation and ejection of target material. The impact and shock environment of such a collision provide scientists with an explanation as to why the Martian meteorites are all igneous rocks. Martian sedimentary rocks and soil would not be consolidated sufficiently to survive the impact as intact rocks and then wander through space for millions of years and eventually land on Earth as meteorites.

One particular Martian meteorite, called ALH84001, has stimulated a great deal of interest in the possibility of life on Mars. In the summer of 1996, a NASA research team at the Johnson Space Center (JSC) announced that they had found evidence in ALH84001 that "strongly suggests primitive life may have existed on Mars more than 3.6 billion years ago." The NASA research team found the first organic molecules thought to be of Martian origin; several mineral features characteristic of biological activity; and possibly microscopic fossils of primitive, bacterialike organisms inside of an ancient Martian rock that fell to Earth as a meteorite. While the NASA research team did not claim that they had conclusively proved life existed on Mars some 3.6 billion years ago, they did believe that "they have found quite reasonable evidence of past life on Mars."

Martian meteorite ALH84001 is a 1.9-kilogram (4.2 pound-mass (lbm)), potato-sized igneous rock that has been age-dated to about 4.5 billion years, the period when the planet Mars formed. This rock is believed to have originated underneath the Martian surface and to have been extensively fractured by impacts as meteorites bombarded the planet during the early history of the solar system. Between 3.6 and 4.0 billion years ago, Mars is believed to have been a warmer and wetter world. "Martian" water is thought to have penetrated fractures in the subsurface rock, possibly forming an underground water system. Since the water was saturated with carbon dioxide from the Martian atmosphere, carbonate materials were deposited in the fractures. The NASA research team estimates that this rock from Mars entered Earth's atmosphere about 13,000 years ago and fell in Antarctica as a meteorite. ALH84001 was discovered in 1984 in the Allan Hills ice field of Antarctica by an annual expedition of the National Science Foundation's Antarctic Meteorite Program. It was preserved for study at the NASA JSC Meteorite Processing Laboratory, but its possible Martian origin was not fully recognized until 1993. It is the oldest of the Martian meteorites yet discovered. *See also* MARS; METEORITE; METEOROID.

mascon A term meaning "mass concentration." An area of mass concentration or high density within a celestial body, usually near the surface. In 1968 data from five U.S. lunar orbiter spacecraft indicated that regions of high density, or "mascons," existed under circular maria (i.e., the extensive dark areas) on the Moon. The Moon's gravitational attraction is somewhat higher over such mass concentrations, and their presence perturbs (causes variations in) the orbits of spacecraft around the Moon. *See also* MOON.

mass (symbol: m) The amount of material present in an object. This fundamental unit describes "how much" material makes up an object. The SI UNIT for mass is the kilogram (kg), while it is the pound-mass (lbm) in the traditional or engineering unit system. The term "mass" and "weight" are very often confused in the traditional or engineering unit system, because they have similar names: the pound-mass (lbm) and the pound-force (lbf). However, these terms are *not* the same, since "weight" is a derived unit that describes the action of the local force of gravity on the "mass" of an object. An object of 1 kilogram mass on Earth will also have a 1 kilogram mass on the surface of Mars. However, the "weight" of this 1-kilogram-mass object will be different on the surface of each planet, since the acceleration of gravity is different on each planet (i.e., about 9.8 meters per second per second on Earth's surface versus 3.7 meters per second per second on Mars).

mass driver An electromagnetic device (currently under study and development) that can accelerate payloads (nonliving and nonfragile) to a very high terminal velocity. Small, magnetically levitated vehicles, sometimes called "buckets," would be used to carry the payloads. These buckets would then contain superconducting coils and be accelerated by pulsed magnetic fields along a linear trace or guideway. When these buckets reach an appropriate terminal velocity (e.g., several kilometers per second), they release their payloads and are then decelerated for reuse. Mass drivers have been suggested as a way "shooting" lunar ores into orbit around the Moon for collection and subsequent use by space-based manufacturing facilities. See also GUN-LAUNCH TO SPACE.

mass-energy equation (mass-energy equivalence) The statement developed by Albert Einstein (1879–1955), German-born American physicist, that "the mass of a body is a measure of its energy content," as an extension of his 1905 *Special Theory of Relativity*. The statement subsequently was verified experimentally by measurements of mass and energy in nuclear reactions. This equation is written as: $E = m\,c^2$, and illustrates that when the

energy of a body changes by an amount, E, the mass, m, of the body will change by an amount equal to E/c^2. (Note: The factor c^2, the square of the speed of light in a vacuum, may be regarded as the conversion factor relating units of mass and energy.) This famous equation predicted the possibility of releasing enormous amounts of energy in the nuclear chain reaction (e.g., as found in a nuclear explosion) by the conversion of some of the mass in the atomic nucleus into energy. This equation is sometimes called the *Einstein equation*. See also RELATIVITY.

mass flow rate (symbol: ṁ) The mass of fluid flowing through or past a reference point per unit time. Typical units are kilograms per second (kg/s) and pounds-mass per second (lbm/s).

mass fraction The fraction of a rocket's (or of a rocket stage's) mass that is taken up by propellant. The remaining mass is structure and payload.

mass number (symbol: A) The number of nucleons (i.e., the number of protons and neutrons) in an atomic nucleus. It is the nearest whole number to an atom's atomic weight. For example, the mass number of the isotope uranium-235 is 235.

mass spectrometer An instrument used to measure the relative atomic masses and relative abundances of isotopes. A sample (usually gaseous) is ionized, and the resultant stream of charged particles is accelerated into a high vacuum region where electric and magnetic fields deflect the particles and focus them on a detector. A *mass spectrum* (i.e., a series of lines related to mass/charge values) then is created. This characteristic pattern of lines helps scientists identify different molecules.

mass transfer cooling Heat transfer or thermal control accomplished primarily by the use of an amount of mass to cool a surface or region. There are three general categories of mass transfer cooling: (1) *transpiration cooling*, (2) *film cooling*, or (3) *ablation cooling*. Aerospace designers have long recognized mass transfer to the boundary layer as an effective technique for significantly altering the adverse aerodynamic heating phenomena associated with a body in hypersonic flight through a planetary atmosphere. Similarly, rocket engineers often use some form of mass transfer cooling to prevent the overheating or destruction of thermally sensitive rocket nozzle components.

Matador An early medium-range, surface-to-surface winged missile developed by the U.S. Air Force in the 1950s.

materials processing in space (MPS) *Materials processing* is the science by which ordinary and comparatively inexpensive raw materials are made into useful crystals, chemicals, metals, ceramics, and countless other manufactured products. Modern materials processing on Earth has taken us into the space age and opened up the microgravity environment of Earth orbit. The benefits of extended periods of "weightlessness" promise to open up new and unique opportunities for the science of materials processing. In the microgravity environment of an orbiting spacecraft, scientists can use materials processing procedures that are all but impossible on Earth.

In orbit, materials processing can be accomplished without the effects of gravity, which on Earth causes materials of different densities and temperatures to separate and deform under the influences of their own masses. However, when scientists refer to an orbiting object as being "weightless," they do not literally mean there is an absence of gravity. Rather, they are referring to the microgravity conditions, or the absence of relative motion between objects in a free-falling environment, as experienced in an Earth-orbiting spacecraft. These useful "free-fall" conditions can be obtained only briefly on Earth using drop towers or "zero-gravity" aircraft. Extended periods of microgravity can be achieved only on an orbiting spacecraft, such as the space shuttle orbiter, a space station, or a crew-tended free-flying platform.

Hydrostatic pressure places a strain on materials during solidification processes on Earth. Certain crystals are sufficiently dense and delicate that they are subject to strain under the influence of their own weight during growth. Such strain-induced deformations in crystals degrade their overall performance. In microgravity, heat-treated, melted, and resolidified crystals and alloys can be developed free of such deformations.

Containerless processing in microgravity eliminates the problems of container contamination and wall effects. Often these are the greatest source of impurities and imperfections when a molten material is formed on Earth. But in space a material can be melted, manipulated, and shaped free of contact with a container wall or crucible, by acoustic, electromagnetic, or electrostatic fields. In microgravity, the surface tension of the molten material helps hold it together, while on Earth this cohesive force is overpowered by gravity.

Over the next few decades, space-based materials processing research will emphasize both scientific and commercial goals. Potential space-manufactured products include special crystals, metals, ceramics, glasses, and biological materials. Processes will include containerless processing and fluid and chemical transport. As research in these areas progresses, specialized new materials and manufactured products could become available in the 21st century for use in space as well as on Earth. *See also* DROP TOWER; INTERNATIONAL SPACE STATION; MICROGRAVITY; ZERO-GRAVITY AIRCRAFT.

mating The act of fitting together two major components of a system, such as the mating of a launch vehicle and a spacecraft. Also, the physical joining of two orbiting spacecraft either through a docking or a berthing process.

matter-antimatter propulsion Spacecraft propulsion by use of matter-antimatter annihilation reactions.

Maverick (AGM-65) A tactical, air-to-surface guided missile designed for close air support, interdiction, and defense suppression missions. It provides standoff capability and high probability of strike against a wide range of tactical targets, including armor, air defenses, ships, transportation equipment, and fuel storage facilities. Maverick A and B models have an electro-optical television guidance system; the Maverick D and G models have an imaging infrared guidance system. The U.S. Air Force accepted the first AGM-65A Maverick missile in August 1972. AGM-65 missiles were used successfully by F-16 and A-10 aircraft in 1991 to attack armored targets in the Persian Gulf during Operation Desert Storm.

max-Q An aerospace term describing the condition of maximum dynamic pressure—the point in the flight of a launch vehicle when it experiences the most severe aerodynamic forces, as it rises up through Earth's atmosphere toward outer space.

Maxwell Montes A mountain range on VENUS located in ISHTAR TERRA containing the highest peak (11 km altitude) on the PLANET. It is named after the Scottish theoretical physicist James Clerk Maxwell (1831–1879).

"mayday" A distress call.

mean solar day The duration of one rotation of Earth on its axis, with respect to the uniform motion of the mean Sun. The length of the mean solar day is 24 hours of *mean solar time* or 24 hours 3 minutes 56.555 seconds if *mean sidereal time*. A mean solar day beginning at midnight is called a *civil day;* one beginning at noon is called an *astronomical day.*

mean solar time Time based on Earth's rate of rotation, which originally was assumed to be constant. *See also* MEAN SOLAR DAY.

mean time between failures (MTBF) The average time between the failure of elements in a system composed of many elements.

mean time to repair (MTTR) In a multielement system, the average time required to repair the system in the event of a failure.

medium-range ballistic missile (MRBM) A ballistic missile with a range capability from about 1,100 kilometers (km) (680 miles [mi]) to 2,780 km (1,725 mi).

mega- (symbol: M) A prefix in the SI UNIT system meaning multiplied by 1 million (10^6), as, for example, megahertz (MHz), meaning 1 million hertz.

megaparsec (Mpc) One million parsecs; a distance of approximately 3,260,000 light-years. *See also* LIGHT-YEAR; PARSEC.

megaton (MT) A measure of the explosive yield of a nuclear weapon that is equivalent to the explosion of 1 million tons of trinitrotoluene (TNT), a chemical high explosive. By definition, 1 MT = 10^{15} calories or 4.2×10^{15} joules. (Here, a ton represents 1,000 kilograms or approximately 2,000 pounds-mass of TNT.)

megawatt (MW) One million watts. *See also* WATT.

melting In thermodynamics, the transition of a material from the solid phase to the liquid phase, generally as a result of heating.

Mercury The innermost planet in the solar system, orbiting the Sun at approximately 0.4 astronomical unit (AU). This planet, named for the messenger god of Roman mythology, is a scorched, primordial world that is only 40% larger in diameter than Earth's Moon.

NASA's *Mariner 10* spacecraft provided the first close-up views of Mercury. This spacecraft was launched from Cape Canaveral in November 1973. After traveling almost five months (including a flyby of Venus), this spacecraft passed within 805 kilometers (km) (483 miles [mi]) of Mercury on March 29, 1974.

Mariner 10 then looped around the Sun and made another rendezvous with Mercury on September 21, 1974. This encounter process was repeated a third time on March 16, 1975, before the control gas used to orient the spacecraft was exhausted. This triple flyby of the planet Mercury by *Mariner 10* is sometimes referred to as Mercury I, II, and III in the technical literature.

The images of Mercury transmitted back to Earth by *Mariner 10* revealed an ancient, heavily cratered world that closely resembled Earth's Moon. Unlike the Moon, however, huge cliffs (called *lobate scarps*) crisscross Mercury. These great cliffs apparently were formed when Mercury's interior cooled and shrank, compressing the planet's crust. The cliffs are as high as 2 km (1.2 mi) and as long as 1,500 km (900 mi).

To the surprise of scientists, instruments on board *Mariner 10* discovered that Mercury has a weak magnetic field. It also has a wisp of an atmosphere—a trillionth of the density of Earth's atmosphere and made up mainly of traces of helium, hydrogen, sodium, and potassium.

Temperatures on the suplit side of Mercury reach approximately 700 kelvins (K) (427° Celsius)—a temperature that exceeds the melting point of lead; on the dark side, temperatures plunge to a frigid 100 kelvins (K) (minus 173° Celsius). Quite literally, Mercury is a world seared with intolerable heat in the daytime and frozen at night.

In the late 1960s, scientists on Earth bounced radar signals off the surface of Mercury. Analysis of the scattered radar signals indicated that the planet actually rotated slowly on its axis with a period of about 59 days. Consequently, the "days" and "nights" on this planet are quite long by terrestrial standards, 1 Mercurian day equals 59 Earth-days. It takes the planet approximately 88 days to orbit around the Sun.

Mercury's surface features include large regions of gently rolling hills and numerous impact craters like those found on the Moon. Many of these craters are

Physical and Dynamic Properties of the Planet Mercury	
Radius (mean equatorial)	2,439 km
Mass	3.30×10^{23} kg
Mean density	5.44 g/cm^3
Acceleration of gravity (at the surface)	3.70 m/sec^2
Escape velocity	4.25 km/sec
Normal albedo (averaged over visible spectrum)	0.125
Surface temperature extremes	100 K to 700 K
Atmosphere	negligible
Number of natural satellites	none
Flux of solar radiation	
Aphelion	6,290 W/m^2
Perihelion	14,490 W/m^2
Semimajor axis	5.79×10^7 km (0.387 AU)
Perihelion distance	4.60×10^7 km (0.308 AU)
Aphelion distance	6.98×10^7 km (0.467 AU)
Eccentricity	0.20563
Orbital inclination	7.004 degrees
Mean orbital velocity	47.87 km/sec
Sidereal day (a Mercurean "day")	58.646 Earth days
Sidereal year (a Mercurean "year")	87.969 Earth days

Source: NASA.

surrounded by blankets of ejecta (material thrown out at the time of a meteoroid impact) and secondary craters that were created when chunks of ejected material fell back down to the planet's surface. Because Mercury has a higher gravitational attraction than the Moon, these secondary craters are not spread as widely from each primary craters as occurs on the Moon. One major surface feature discovered by *Mariner 10* is a large impact basin called Caloris, which is about 1,300 km (780 mi) in diameter. Scientists now believe that Mercury has a large iron-rich core—the source of its weak, but detectable, magnetic field. The table presents some contemporary physical—and dynamic-property data about the Sun's closest planetary companion.

Because Mercury lies deep in the Sun's gravity well, detailed exploration with sophisticated orbiters and landers will require the development of advanced planetary spacecraft that take advantage of intricate "gravity-assist" maneuvers involving both the Earth and Venus. *See also* MARINER SPACECRAFT.

Mercury Project America's pioneering project to put a human being into orbit. The series of six suborbital and orbital flights was designed to demonstrate that human beings could withstand the high acceleration of a rocket launching, a prolonged period of weightlessness, and then a period of high deceleration during reentry.

The Mercury Project became an official program of NASA on October 7, 1958. Seven astronauts were chosen in April 1959, after a nationwide call for jet pilot volunteers. The one-person Mercury spacecraft was designed and built with a maximum orbiting mass of about 1,452 kilograms (3,200 pounds-mass [lbm]). Shaped somewhat like a bell, this small spacecraft was about 189 centimeters (74.5 inches) wide across the bottom and about 2.7 meters (9 feet) tall. The astronaut escape tower added another 5.2 meters (17 feet) for an overall length of approximately 8 meters (26 feet) at launch.

Two boosters were chosen: the U.S. Army Redstone with its 346,944 newtons (78,000 pounds-force [lbf]) thrust for the suborbital flights and the U.S. Air Force Atlas with its 1,601,280 newtons (360,000 lbf) thrust for the orbital missions.

On May 5, 1961, Astronaut Alan B. Shepard, Jr., was launched from Complex 5 at Cape Canaveral Air Force Station (CCAFS) by a Redstone booster on the first U.S. crewed space flight. His brief suborbital mission lasted just 15 minutes, and his *Freedom 7* Mercury capsule traveled 186.7 kilometers (116 miles) high into space. On July 21, 1961 another Redstone booster hurled Astronaut Virgil I. "Gus" Grissom through the second and last suborbital flight in the *Liberty Bell 7* Mercury capsule.

Following these two successful suborbital missions, NASA then advanced the project to the Mercury-Atlas series of orbital missions. Another space milestone was reached on February 20, 1962 when Astronaut John H. Glenn, Jr., became the first American in orbit, circling the Earth three times in the *Friendship 7* spacecraft.

On May 24, 1962, Astronaut M. Scott Carpenter completed another three-orbit flight in the *Aurora 7* spacecraft. Astronaut Walter M. Schirra, Jr., doubled the flight time in space and orbited six times, landing the *Sigma 7* Mercury capsule in a Pacific Ocean recovery area. All previous Mercury Project landings had been in the Atlantic Ocean. Finally, on May 15–16, 1963, Astronaut L. Gordon Cooper, Jr., completed a 22 orbit mission in the *Faith 7* spacecraft, triumphantly concluding the program and paving the way for the Gemini Project and ultimately the Apollo Project, which took American astronauts to the lunar surface. The table on the following page summarizes the Mercury Project missions. *See also* APOLLO PROJECT; GEMINI PROJECT.

meridians Great circles that pass through both the North and South poles; also called *lines of longitude.*

mesosphere The region of Earth's atmosphere above the stratosphere that is characterized by temperature decreasing with height. The top of this layer, called

Mission	Date (s)/Recovery Ship, Ocean	Astronaut	Mission Duration	Remarks
		Mercury Project Missions (1961–1963)		
Mercury Redstone 3 (Freedom 7)	May 5, 1961 Lake Champlain, Atlantic	Navy Comdr. Alan B. Shepard, Jr.	0:15:22	suborbital
Mercury Redstone 4 (Liberty Bell 7)	July 21, 1961 Randolph, Atlantic	USAF Maj. Virgil I. Grissom	0:15:37	suborbital
Mercury Atlas 6 (Friendship 7)	Feb. 20, 1962 Noa, Atlantic	Marine Lt. Col. John H. Glenn	4:55:23	3 orbits
Mercury Atlas 7 (Aurora 7)	May 24, 1962 Pierce, Atlantic	Navy Lt. Comdr. Scott Carpenter	4:56:05	3 orbits
Mercury Atlas 8 (Sigma 7)	Oct 3, 1962 Kearsarge, Pacific	Navy Comdr. Walter M. Schirra, Jr.	9:13:11	6 orbits
Mercury Atlas 9 (Faith 7)	May 15–16, 1963 Kearsarge, Pacific	USAF Maj. L. Gordon Copper	34:19:49	22 orbits

Note: Names in parentheses in the first column are those given to Mercury spacecraft.
Source: NASA.

the *mesopause,* occurs between 80 and 85 kilometers. (48–50 miles) altitude. The mesopause is the coldest region of the entire atmosphere and has a temperature of approximately 180 kelvins (K) (–93° Celsius). *See also* ATMOSPHERE.

meteor The luminous phenomena that occurs when a meteoroid enters Earth's atmosphere. Sometimes called a *shooting star. See also* METEOROIDS.

meteorite Metallic or stony material that has passed through the atmosphere and reached Earth's surface. Meteorites are of extraterrestrial origin. Known sources of meteorites include asteroids and comets; some even originated on the Moon or on Mars. Presumably meteorites from the Moon and Mars were ejected by impact cratering events. The term "meteorite" also is applied to meteoroids that land on planetary bodies other than Earth. *See also* MARTIAN METEORITES; METEOROIDS.

meteoroids An all encompassing term that refers to solid objects found in space, ranging in diameter from micrometers to kilometers and in mass from less than 10^{-12} gram to more than 10^{+16} grams. If

these pieces of extraterrestrial material are less than 1 gram, they often are called *micrometeoroids.* When objects of more than approximately 10^{-6} gram reach Earth's atmosphere, they are heated to incandescence (i.e., they glow with heat) and produce the visible effect popularly called a *meteor.* If some of the original meteoroid survives its glowing plunge into Earth's atmosphere, the remaining unvaporized chunk of space matter is then called a *meteorite.*

Scientists currently think that meteoroids originate primarily from asteroids and comets that have perihelia (portions of their orbits nearest the Sun) near or inside Earth's orbit around the Sun. The parent celestial objects are assumed to have been broken down into a collection of smaller bodies by numerous collisions. Recently formed meteoroids tend to remain concentrated along the orbital path of their parent body. These "stream meteoroids" produce the well-known meteor showers that can be seen at certain dates from Earth.

Meteoroids generally are classified by composition as stony meteorites (chondrites), irons, and stony-irons. Of the meteorites that fall on Earth, stony meteorites make up about 93%, irons about

Time Between Meteoroid Collisions for a Space Shuttle Orbiter in Low Earth Orbit (300 km Altitude)	
Minimum Meteoroid Mass (g)	*Estimated Time Between Collisions (yr)*
10	350,000
1	25,000
0.1	1,800
0.01	130

Source: Based on NASA data.

5.5%, and stony-irons about 1.5%. Space scientists estimate that about 10^{+7} kilograms (or 10,000 metric tons) of "cosmic rocks" now fall on our planet annually.

Some of the meteorites found on Earth are thought to have their origins on the Moon or on Mars. It is now believed that these lunar or Martian meteorites were ejected by ancient impact-cratering events. If an asteroid or comet of sufficient mass and velocity hits the surface of the Moon, a small fraction of the material ejected by the impact collision could depart from the Moon's surface with velocities greater than its escape velocity (2.4 kilometers per second [km/sec]). A fraction of that ejected material eventually would reach Earth's surface, with Moon-to-Earth transit times ranging from under 1 million years to upward of 100 million years. Similarly, a very energetic asteroid or comet impact collision on Mars (which has an escape velocity of 5.0 km/sec) could be the source of the interesting "Martian" meteorites found recently in Antarctica.

Micrometeoroids and small meteoroids large enough to damage an Earth-orbiting spacecraft are considered "somewhat rare" by space scientists. For example, the table presents a contemporary estimate for the time between collisions between an object the size of a space shuttle orbiter in LOW-EARTH ORBIT and a meteoroid of mass greater than a given meteoroid mass.

On a much larger "collision scale," meteoroid impacts now are considered by planetary scientists to have played a basic role in the evolution of planetary surfaces in the early history of the solar system. Although still dramatically evident in the cratered surfaces found on other planets and moons, here on Earth this stage of surface evolution has essentially been lost due to later crustal recycling and weathering processes. *See also* ASTEROID; COMET; MARTIAN METEORITES.

meteorological rocket A rocket designed for routine upper-air observation (as opposed to scientific research), especially in that portion of Earth's atmosphere inaccessible to weather balloons, namely above 30,500 meters (100,000 feet) altitude.

meteorological satellite An Earth-observing spacecraft that senses some or most of the atmospheric phenomena (e.g., wind and clouds) related to weather conditions on a local, regional, or hemispheric scale. *See also* WEATHER SATELLITE.

meteorology The branch of science dealing with phenomena of the atmosphere. This includes not only the physics, chemistry, and dynamics of the atmosphere but also many of the direct effects of the atmosphere upon Earth's surface, the oceans, and life in general. Meteorology is concerned with actual weather conditions, while *climatology* deals with average weather conditions and long-term weather patterns. *See also* WEATHER SATELLITE.

meter (symbol: m) The fundamental SI UNIT of length. 1 meter = 3.281 feet. Also spelled *metre* (British spelling).

metric system The international system (SI) of weights and measures based on the *meter* as the fundamental unit of length, the *kilogram* as the fundamental unit of mass, and the *second* as the fundamental unit of time. Also called the *mks system*. *See also* SI UNITS.

metrology The science of dimensional measurement; sometimes includes the science of weighing.

MeV   An abbreviation for 1 million electron volts, a common energy unit encountered in the study of nuclear reactions. 1 MeV = 10^6 eV.

micro- (symbol: μ)   A prefix in the SI UNIT system meaning divided by 1 million; for example, a micrometer (mm) is 10^{-6} meter. The term also is used as a prefix to indicate something is very small, as in *micrometeoroid* or *micromachine*.

microbar (symbol: μb)   A unit of pressure in the centimeter-gram-second (cgs) system equal to 1 dyne per square centimeter. 1 μb = 1 dyne/cm^2. *See also* BAR.

microgravity   Because the inertial trajectory of a spacecraft (e.g., the space shuttle orbiter) compensates for the force of Earth's gravity, an orbiting spacecraft and all its contents approach a state of free-fall. In this state of free-fall, all objects inside the spacecraft appear "weightless."

It is important to understand how this condition of weightlessness, or the apparent lack of gravity, develops. NEWTON'S LAW OF GRAVITATION states that any two objects have a gravitational attraction for each other that is proportional to their masses and inversely proportional to the square of the distance between their centers of mass. It is also interesting to recognize that a spacecraft orbiting Earth at an altitude of 400 kilometers is only 6% farther away from the center of Earth than it would be if it were on Earth's surface. Using Newton's law, we find that the gravitational attraction at this particular altitude is only 12% less than the attraction of gravity at the surface of Earth. In other words, an Earth-orbiting spacecraft and all its contents are very much under the influence of Earth's gravity! The phenomenon of weightlessness occurs because the orbiting spacecraft and its contents are in a continual state of free-fall.

Figure 47 describes the different orbital paths a falling object may take when "dropped" from a point above Earth's sensible atmosphere. With no tangential-velocity component, an object would fall straight down (trajectory 1) in this simplified demonstration. As the object receives an increasing, tangential-velocity component, it still "falls" toward Earth under the influence of terrestrial gravitational attraction, but the tangential-velocity component now gives the object a trajectory that is a segment of an ellipse. As shown in trajecto-

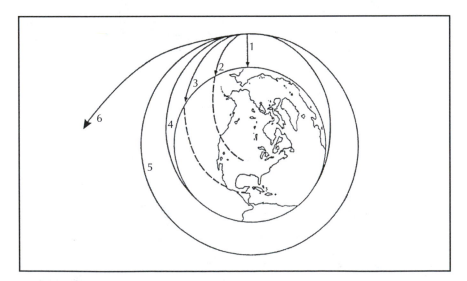

Figure 47. Various orbital paths of a falling body around Earth.

ries 2 and 3 in the figure, as the object receives a larger tangential velocity, the point where it finally hits Earth moves farther and farther away from the release point. If we keep increasing this velocity component, the object eventually "misses Earth" completely (trajectory 4). As the tangential velocity is increased further, the object's trajectory takes the form of a circle (trajectory 5) and then a larger ellipse, with the release point representing the point of closest approach to Earth (or "perigee"). Finally, when the initial tangential-velocity component is about 41% greater than that needed to achieve a circular orbit, the object follows a parabolic, or escape, trajectory and will never return (trajectory 6).

Einstein's principle of equivalence states that the physical behavior inside a system in free-fall is identical to that inside a system for removed from other matter that could exert a gravitational influence. Therefore, the term "zero gravity" (also called "zero g") or "weightlessness" often is used to describe a free-falling system in orbit.

Sometimes people ask what is the difference between mass and weight. Why do we say, for example, "weightlessness" and not "masslessness"? *Mass* is the physical substance of an object—it has the same value everywhere. *Weight*, on the other hand, is the product of an object's mass and the local acceleration of gravity (in accordance with Newton's second low of motion, $F = ma$). For example, you would weigh about one-sixth as much on the Moon as on Earth, but your mass remains the same in both places.

A "zero-gravity" environment is really an ideal situation that can never be totally achieved in an orbiting spacecraft. The venting of gases from the space vehicle, the minute drag exerted by a very thin terrestrial atmosphere at orbital altitude, and even crew motions create nearly imperceptible forces on people and object alike. These tiny forces are collectively called "microgravity." In a microgravity environment, astronauts and their equipment are almost, but not entirely, weightless.

Microgravity represents an intriguing experience for space travelers. However, life in microgravity is not necessarily eas-ier than life on Earth. For example, the caloric (food-intake) requirements for people living in microgravity are the same as those on Earth. Living in microgravity also calls for special design technology. A beverage in an open container, for instance will cling to the inner or outer walls and, if shaken, will leave the container as free-floating droplets or fluid globs. Such free-floating droplets are not merely an inconvenience. They can annoy crew members, and they represent a definite hazard to equipment, especially sensitive electronic devices and computers.

Therefore, water usually is served in microgravity through a specially designed dispenser unit that can be turned on or off by squeezing and releasing a trigger. Other beverages, such as orange juice, typically are served in sealed containers through which a plastic straw can be inserted. When the beverage is not being sipped, the straw is simply clamped shut.

Microgravity living also calls for special considerations in handling solid foods. Crumbly foods are provided only in bit-size pieces to avoid having crumbs floating around the space cabin. Gravies, sauces, and dressings have a viscosity (stickiness) that generally prevents them from simply lifting off food trays and floating away. Typical space food trays are equipped with magnets, clamps, and double-adhesive tape to hold metal, plastic, and other utensils. Astronauts are provided with forks and spoons. However, they must learn to eat without sudden starts and stops if they expect the solid food to stay on their eating utensils.

Personal hygiene is also a bit challenging in microgravity. For example, space shuttle astronauts must take sponge baths rather than showers or regular baths. Because water adheres to the skin in microgravity, perspiration also can be annoying, especially during strenuous activities. Waste elimination in microgravity represents another challenging design problem. Special toilet facilities are needed that help keep an astronaut in place (i.e., prevent drifting). The waste products themselves are flushed away by a flow of air and a mechanical "chopper-type" device.

Sleeping in microgravity is another interesting experience. For example, shuttle and space station astronauts can sleep either horizontally or vertically while in orbit. Their fireproof sleeping bags attach to rigid padded boards for support. But the astronauts themselves quite literally sleep "floating in air."

Working in microgravity requires the use of special tools (e.g., torqueless wrenches), handholds, and foot restraints. These devices are needed to balance or neutralize reaction forces. If these devices were not available, an astronaut might find him-/herself helplessly rotating around a "work piece."

Exposure to microgravity also causes a variety of physiological (bodily) changes. For example, space travelers appear to have "smaller eyes," because their faces have become puffy. They also get rosy cheeks and distended veins in their foreheads and necks. They may even be a little bit taller than they are on Earth, because their body masses no longer "weigh down" their spines. Leg muscles shrink, and anthropometric (measurable postural) changes also occur. Astronauts tend to move a slight crouch, with head and arms forward.

Some space travelers suffer from a temporary condition resembling motion sickness. This condition is called space sickness or *space adaptation syndrome*. In addition, sinuses become congested, leading to a condition similar to a cold.

Many of these microgravity-induced physiological effects appear to be caused by fluid shifts from the lower to the upper portions of the body. So much fluid goes to the head that the brain may be fooled into thinking that the body has too much water. This can result in an increased production of urine.

Extended stays in microgravity tend to shrink the heart, decrease production of red blood cells, and increase production of white blood cells. A process called resorption occurs. This is the leaching of vital minerals and other chemicals (e.g., calcium, phosphorous, potassium, and nitrogen) from the bones and muscles into the body fluids that are then expelled as urine. Such mineral and chemical losses can have adverse physiological and psychological effects. In addition, prolonged exposure to a microgravity environment might cause bone loss and a reduced rate of bone-tissue formation.

While a relatively brief stay (say from 7 to 70 days) in microgravity may prove a nondetrimental experience for most space travelers, long-duration (i.e., one to several years) missions such as a human expedition to Mars, could require the use of artificial gravity (created through the slow rotation of the living modules of the spacecraft) to avoid any serious health effects that might arise from such prolonged exposure to a microgravity environment. While cruising to Mars, this artificial gravity environment also would help condition the astronauts for activities on the Martian surface, where they will once again experience the "tug" of a planet's gravity.

Besides providing an interesting new dimension for human experience, the microgravity environment of an orbiting space system offers the ability to create new and improved materials that cannot be made on Earth. Although microgravity can be simulated from here on Earth using drop towers, special airplane trajectories, and sounding rocket flights, these techniques are only short-duration simulations (lasting only seconds to minutes) that are frequently "contaminated" by vibrations and other undesirable effects. However, the long-term microgravity environment found in orbit provides an entirely new dimension for materials science research, life science research, and even manufacturing of specialized products. *See also* DROP TOWER; MATERIALS PROCESSING IN SPACE; ZERO-GRAVITY AIRCRAFT.

micrometeoroids Tiny particles of meteoritic dust in space ranging in size from about 1 to 200 or more micrometers (microns) in diameter.

micrometer 1. An SI UNIT of length equal to one-millionth (10^{-6}) of a meter; also called a *micron*. 1 μm = 10^{-6} m. 2. An instrument or gauge for making very precise linear measurements (e.g., thicknesses and small diameters) in which the displacements measured correspond to the travel of a screw of accurately known pitch.

micron (symbol: μm) An SI UNIT of length equal to one-millionth (10^{-6}) of a meter. Also called a *micrometer.*

microsecond (symbol: μs) A unit of time equal to one-millionth (10^{-6}) of a second.

microwave A comparatively short-wavelength electromagnetic (EM) wave in the radio-frequency portion of the EM spectrum. The term "micro-wave" usually is applied to those EM wavelengths that are measured in centimeters, approximately 30 centimeters to 1 millimeter (with corresponding frequencies of 1 gigahertz [GHz] to 300 gigahertz [GHz]). *See also* ELECTROMAGNETIC SPECTRUM.

Mie scattering Any scattering produced by spherical particles without special regard to the comparative size if radiation wavelength and particle diameter.

mil 1. As a unit of length, a mil is 1/1000 (10^{-3}) of an inch. 2. As a unit of angular measurement, a mil is 1/6400 of a circle or 1/1000 (10^{-3}) of a radian. (A radian is approximately equal to 57.296 degrees.)

mile (symbol: mi) 1. (statute or land mile) A unit of distance in the English unit system equal to 1,760 yards or 5,280 feet. 1 statute mile = 1,609.344 meters. Note: In this book, "mile (mi)" indicates statute mile, unless otherwise specified. 2. (nautical mile) The nautical mile is defined as the length of 1 minute of latitude. By international agreement, 1 nautical mile = 1,852 meters (i.e., 1 minute of arc at 45 degrees latitude). Note: When a unit of distance is expressed in nautical miles in this book, it is specifically identified as "nautical miles (n.mi.)."

milestone An important event or decision point in a program or plan. The term originates from the use of stone markers set up on roadsides to indicate the distance in miles to a given point. Milestone charts are used extensively in aerospace programs and planning activities.

military satellite (MILSAT) A satellite used for military or defense purposes such as missile surveillance, navigation, and intelligence gathering.

Milky Way Galaxy Our home galaxy. The immense band of stars stretching across the night sky represents out "inside view" of the Milky Way. Classified as a spiral galaxy, the Milky Way is characterized by the following general features: a spherical central bulge at the galactic nucleus; a thin disk of stars, dust, and gas formed in a beautiful, extensive pattern of spiral arms; and a halo defined by an essentially spherical distribution of globular clusters. This disk is between 2,000 and 3,000 light-years thick and is some 100,000 light-years in diameter. It contains primarily younger, very luminous, metal-rich stars (called Population I stars) as well as gas and dust. Most of the stars found in the halo are older, metal-poor stars (called Population II stars), while the galactic nucleus appears to contain a mixed population of older and younger stars. Some astrophysicists now speculate that a massive black hole, containing millions of "devoured" solar masses, may lie at the center of many galaxies, including our own. Current estimates suggest that our galaxy contains between 200 and 600 billion solar masses. (A solar mass is a unit used for comparing stellar masses, with one solar mass being equal to the mass of our Sun.) Our solar system is located about 30,000 light-years from the center of the galaxy. (See Figure 48.) *See also* BLACK HOLE(S); GALAXY; STARS.

milli- (symbol: m) The SI UNIT system prefix meaning multiplied by 1/1000 (10^{-3}). For example a *millivolt* (mV) is 0.001 volt; a *millimeter* (mm) is 0.001 meter; and a *millisecond* (msec) is 0.001 second.

millibar (symbol: mbar or mb) A unit of pressure equal to 0.001 bar (i.e., 10^{-3} bar) or 1,000 dynes per square centimeter.

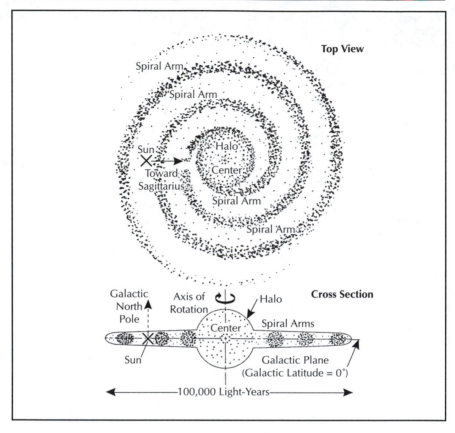

Top View

Spiral Arm

Spiral Arm

Sun

Halo

Toward
Sagittarius

Center

Spiral Arm

Spiral Arm

Galactic
North
Pole

Axis of
Rotation

Halo

Cross Section

Center

Spiral Arms

Sun

Galactic Plane
(Galactic Latitude = 0°)

◄————————100,000 Light-Years————————►

Figure 48. Top view and cross section of the Milky Way Galaxy, a gigantic rotating disk of stars, gas, and dust about 100,000 light-years in diameter and about 3,000 light-years thick, with a bulge, or halo, around the center of mass.

The millibar is used as a unit of measure of atmospheric pressure, with a *standard atmosphere* being equal to about 1,013 millibars or 29.92 inches (760 mm) of mercury 1 mbar = 100 newtons/m^2 = 1,000 dynes/cm^2. *See also* BAR; DYNE.

millimeter (symbol: mm) One-thousandth (1/1000;10^{-3}) of a second. msec 1 mm = 0.001 m = 0.1 cm = 0.03937 in.

millisecond (symbol: msec or ms) One-thousandth (1/1000, 10^{-3}) of a second. 1 msec = 0.001 sec = 1,000 μsec.

Milstar The Milstar (*Mi*litary *S*trategic and *T*actical *R*elay) System is an advanced

military COMMUNICATIONS SATELLITE that provides the Department of Defense (DOD), the National Command Authorities (NCA), and the armed forces of the United States worldwide assured, survivable communications. Designed to penetrate enemy jammers and overcome the disruptive effects of NUCLEAR DETONATIONS, the Milstar is the most robust and reliable satellite communications system ever deployed the DOD. The program originated near the end of the COLD WAR with the primary objective of creating a secure, nuclear survivable, space-based communication system for the National Command Authorities. Today, in the age of information warfare, the system will support global high-priority defense communications needs with a CONSTELLATION of improved satellites (Milstar II) that provide an exceptionally low probability of interception or detection by hostile forces.

The Milstar satellite has a MASS of approximately 4,540-kilograms (10,000 lbm) and a design life of approximately 10 years. The operational Milstar satellite constellation consists of four spacecraft positioned around EARTH in GEOSYNCHRONOUS ORBITS. Each Milstar satellite serves as a smart switchboard in space directing message traffic from terminal to terminal anywhere on Earth. Since the satellite actually processes the communications signal and can link with other Milstar satellites through crosslinks, the requirement for ground-controlled switching is significantly reduced. In fact, the satellite establishes, maintains, reconfigures, and disassembles required communications circuits as directed by the users. Milstar terminals provide encrypted voice, data, teletype, or facsimile communications. A key goal of the contemporary Milstar system is to provide interoperable communications among users of U.S. Army, Navy, and Air Force Milstar terminals. Timely and protected communications among different units of the American armed forces is essential if such combat forces are to bring about the rapid and successful conclusion of a high-intensity modern conflict.

The first Milstar satellite was launched on February 7, 1994, by a TITAN IV ROCKET from CAPE CANAVERAL AIR FORCE STATION (CCAFS). On January 15, 2001, another Titan IV EXPENDABLE LAUNCH VEHICLE successfully lifted off from CCAFS with a Milstar II communications satellite as its PAYLOAD. There are now three operational Milstar satellites in orbit; another satellite was lost due to BOOSTER failure, and two additional satellites are in production.

mini- An abbreviation for "miniature."

miniature homing vehicle (MHV) An air-launched, direct-ascent (i.e., "pop-up") kinetic-energy kill, antisatellite (ASAT) weapon.

minor planet An ASTEROID.

minute 1. A unit of time equal to the 60th part of an hour; that is, 60 minutes = 1 hour. 2. A unit of angular measurement such that 60 minutes (60′) equal 1 degree (1°) of arc.

Minuteman (LGM-30) A three-stage solid-propellant ballistic missile that is guided to its target by an all-inertial guidance and control system. These strategic missiles are equipped with nuclear warheads and designed for deployment in hardened and dispersed underground silos. The LGM-30 Minuteman intercontinental ballistic missile (ICBM) is an element of the U.S. strategic deterrent force. The "L" in LGM stands for silo-configuration; "G" means surface attack; and "M" means guided missile.

The Minuteman weapon system was conceived in the late 1950s and deployed in the mid-1960s. Minuteman was a revolutionary concept and an extraordinary technical achievement. Both the missile and basing components incorporated significant advances beyond the relatively slow-reacting, liquid-fueled, remotely controlled ICBMs of the previous generation of missiles (such as the Atlas and the Titan). From the beginning, Minuteman missiles have provided a quick-reacting, inertially guided, highly survivable component of America's nuclear Triad. Minuteman's maintenance concept capitalizes on high reliability and a "remove-and-replace" approach to achieve a near 100% alert rate.

Through state-of-the-art improvements, the Minuteman system has evolved over three decades to meet new challenges and assume new missions. Modernization programs have resulted in new versions of the missile, expanded targeting options, and significantly improved accuracy. For example, when the Minuteman I became operational in October 1962, it had a single-target capability. The Minuteman II became operational in October 1965. While looking similar to the Minuteman I, the Minuteman II had greater range and targeting capability. Finally, the Minuteman III became operational in June 1970. This missile, with its improved third stage and the postboost vehicle, can deliver multiple independently targetable reentry vehicles and their penetration aids onto multiple targets. Over 500 Minuteman III's are currently deployed at bases in the United States.

Mir ("peace" in Russian) A third-generation Russian SPACE STATION of modular design that was assembled in ORBIT around a core MODULE launched in February 1986. Although used extensively by many COSMONAUTS and guest researchers (including American ASTRONAUTS), the massive station was eventually abandoned because of economics and the construction of the *INTERNATIONAL SPACE STATION* that started in 1998. *Mir* was safely deorbited into a remote area of the Pacific Ocean in March 2001.

mirror A surface capable of reflecting most of the LIGHT (that is, ELECTROMAGNETIC RADIATION in the visible portion of the SPECTRUM) falling upon it.

mirror matter A popular name for antimatter, which is the "mirror image" of ordinary matter. For example, an antielectron (also called a positron) has a positive charge, while an ordinary electron has a negative charge. *See also* ANTIMATTER.

missile Any object thrown, dropped, fired, launched, or otherwise projected with the purpose of striking a target. Short for *ballistic missile* or *guided missile*. "Missile" should *not* be used loosely as a synonym for rocket or launch vehicle.

missile assembly-checkout facility A building, van, or other type of structure located near the operational missile launching location and designed for final assembly and checkout of the missile system.

Missile Defense Agency (MDA) An organization within the U.S. Department of Defense with the assigned mission to develop, test, and prepare for deployment a

MISSILE defense system. MDA is integrating advanced interceptors, land-, sea-, air- and space-based SENSORS, and BATTLE MANAGEMENT command and control systems in order to design and demonstrate a layered missile defense system that can respond to and engage all classes and ranges of ballistic missile threats. The contemporary missile defense systems being developed and tested by MDA are primarily based on HIT-TO-KILL VEHICLE technology and sophisticated, data-fusing battle management systems. *See also* BALLISTIC MISSILE DEFENSE.

missile intercept zone That geographical division of the destruction area where surface-to-air missiles have primary responsibility for the destruction of incoming airborne objects.

missile release line The line at which an attacking aircraft could launch an air-to-surface missile against a specific target.

missile silo A hardened protective container, buried in the ground, in which land-based long-range ballistic missiles are placed for launching.

missilry The art and science of designing, developing, building, launching, directing, and sometimes guiding a rocket-propelled missile; any phase or aspect or aspect of this art or science. Also called *missilery*.

missing mass *See* DARK MATTER.

mission 1. In aerospace operations, the performance of a set of investigations or operations in space to achieve program goals. For example, the *Voyager 2* "Grand Tour" *mission* to the outer planets. 2. In military operations, a duty assigned to an individual or unit; a task. Also, the dispatching of one or more aircraft to accomplish one particular task. For example, a search-and-destroy *mission*.

mission control center (MCC) In general, the operational headquarters of a space mission. Specifically, the NASA facility responsible for providing total support for all phases of a crewed space flight (e.g., space shuttle flight)—prelaunch, ascent, on-orbit activities, reentry, and landing. It provides systems monitoring and contingency support, maintains communications with the crew and onboard systems, performs flight data collection, and coordinates flight operations. NASA's MCC is located at the Johnson Space Center in Houston, Texas.

mission specialist The space shuttle crew member (and career astronaut) responsible for coordinating payload/Space Transportation System (STS) interaction and, during the payload operation phase of a shuttle flight, directing the allocation of STS and crew resources to accomplish the combined pay-load objectives.

Mission To Planet Earth (MTPE) *See* EARTH-OBSERVING SATELLITE; EARTH SYSTEM SCIENCE.

mixture ratio In liquid-propellant rockets, the mass flow rate of oxidizer to the combustion chamber divided by the mass flow rate of fuel to the combustion chamber.

MKS system The International System of Units based on the *meter* (length), the *kilogram* (mass), and the second (time) as fundamental units of measure. *See also* SI UNITS.

mobile launch platform (MLP) The structure on which the elements of the space shuttle are stacked in the Vehicle Assembly Building (VAB) and then moved to the launch pad (Complex 39) at the NASA Kennedy Space Center in Florida. *See also* SPACE TRANSPORTATION SYSTEM.

mobility aids Handrails and footrails that help crew members move about an orbiting spacecraft or space station.

mock test An operational test of a complete launch vehicle or rocket system performed without actually firing the rocket engine(s).

mock-up A full-size replica or dummy of something, such as a spacecraft, often made of some substitute material, such as wood, and sometimes incorporating actual functioning pieces of equipment, such as engines or power supplies. Mock-ups are used to study construction procedures, to examine equipment interfaces, or to train personnel.

modeling A scientific investigative technique that uses a mathematical or physical representation of a system or theory. This representation, or "model," accounts for all, or at least some, of the known properties of the system or the characteristics of the theory. Models are used frequently to test the effects of changes of system components on the overall performance of the system or the effects of variation of critical parameters on the behavior of the theory.

modulation The process of modifying a radio frequency (RF) signal by shifting its phase, frequency, or amplitude to carry information. The respective processes are called *phase modulation* (PM), *frequency modulation* (FM), and *amplitude modulation* (AM).

module 1. A self-contained unit of a launch vehicle or spacecraft that serves as a building block for the overall structure. It is common in aerospace practice to refer to the module by its primary function, for example, the Apollo spacecraft "command module" that was used in the Apollo Project to the Moon. 2. A pressurized, crewed laboratory suitable for conducting science, applications, and technology activities; for example, the spacelab module in the Space Transportation System. 3. A one-package assembly of functionally related parts, usually a "plug-in" unit arranged to function as a system or subsystem; a "black box."

mole (symbol: mol) The SI UNIT of the amount of substance. It is defined as the amount of substance that contains as many *elementary units* as there are atoms in 0.012 kilograms of carbon-12, a quantity known as *Avogadro's number (N_A)*. (The Avogadro Number, N_A, has a value of about 6.022×10^{23} molecules/mole.) These elementary units may be atoms, molecules, ions, or radicals, and are specified. For example, often it is convenient to express the amount of an "ideal" gas in a given volume in terms of the number of moles (n). As previously stated, a mole of any substance is that mass of the substance that contains a specific number of atoms or molecules. Therefore, the number of moles of a substance can be related to its mass (*m*) (in grams) by the equation:

$$n = m/M_W$$

where M_W is the *molecular weight* of the substance (expressed as grams per mole or g/mol.) For the case of oxygen (O_2) which has a molecular weight (M_W) of 32 g/mol, the mass of *1 mole* of oxygen is 32.0 grams; *a 1/2 mole* of oxygen would have a mass of 16.0 grams, and so on.

molecule A group of atoms held together by chemical forces. The atoms in the molecule may be identical, as in hydrogen (H_2), or different, as in water (H_2O) and carbon dioxide (CO_2). A molecule is the smallest unit of matter that can exist by itself and retain all its chemical properties.

Molniya launch vehicle A Russian launch vehicle that is descended from the first Russian intercontinental ballistic missile (ICBM). This vehicle, first used in 1961, consists of three cryogenic stages and four cryogenic strap-on motors. It can place a 1,590-kilogram (3,500 pound-mass [Ibm]) payload into a highly elliptical *(Molniya)* orbit. This vehicle was used to launch communication satellites (of the same name, i.e., "Molniya" satellites), as well as many of the Russian interplanetary missions in the 1960s.

Molniya orbit A highly elliptical 12-hour orbit that places the apogee (about 40,000 kilometers [km] (24,855 miles [mi])) of a spacecraft over the Northern Hemisphere and the perigee (about 500 km [311 mi]) of the spacecraft over the Southern Hemisphere. Developed and used by the Russians for their communications satel-

lites (called *Molniya satellites*). A satellite in a Molniya orbit spends the bulk of its time (i.e., apogee) above the horizon in view of the high northern latitudes and very little of its time (i.e., perigee) over southern latitudes.

Molniya satellite One of a family of Russian communications spacecraft that operate in highly elliptical orbits, called *Molniya orbits,* so that the spacecraft spends most of its time above the horizon in the Northern Hemisphere.

moment (symbol: M) A tendency to cause rotation about a point or axis, as of a control surface about its hinge or of an airplane about its center of gravity; the measure of this tendency, equal to the product of the force and the perpendicular distance between the point or axis of rotation and the line of action of the force.

moment of inertia (symbol: I) For a massive body made up of many particles or "point masses" (m_i), the moment of inertia (I) about an axis is defined as the sum (Σ) of all the products formed by multiplying each point mass or particle (m_i) by the square of its distance $(r_i)^2$. from the line or axis of rotation; that is: $I = \Sigma_i m_i (r_i)^2$. The moment of inertia can be considered as the analog in rotational dynamics of mass in linear dynamics.

momentum (linear) The linear momentum (p) of a particle of mass (m) is given by the equation: $p = m \cdot v$, where v is the particle's velocity. Newton's Second Law of Motion states that the time rate of change of momentum of a particle is equal to the resultant force (F) on the particle, namely: $F = dp/dt$.

monopropellant A liquid rocket propellant consisting of a single chemical substance (such as hydrazine) that decomposes exothermally and produces a heated exhaust jet without the use of a second chemical substance. Often used in attitude control systems on spacecraft and aerospace vehicles.

Moon The Moon (the term is capitalized when used in this sense) is Earth's only natural satellite and closest celestial neighbor. (See Figure 49.) While life on Earth is made possible by the Sun, it is also regulated by the periodic motions of the Moon. For example, the months of our year are measured by the regular motions of the Moon around Earth, and the tides rise and fall because of the gravitational tug-of-war between Earth and the Moon. Throughout history, the Moon has had a significant influence on human culture, art, and literature. Even in the space age, it has proved to be a major technical stimulus. It was just far enough away to represent a real technical challenge to reach it; yet it was close enough to allow us to be successful on the first concentrated effort. Starting in 1959 with the U.S. *Pioneer 4* and the Russian *Luna 1* lunar flyby missions, a variety of American and Russian missions have been sent to and around the Moon. The most exciting of these missions were the APOLLO PROJECT's human expeditions to the Moon from 1968 to 1972.

In 1994 the *Clementine* spacecraft, which was developed and flown by the Ballistic Missile Defense Organization of the U.S. Department of Defense as a demonstration of certain advanced space technologies, spent 70 days in lunar orbit mapping the Moon's surface. Subsequent analysis of the *Clementine* data offered tantalizing hints that water ice might be present in some of the permanently shadowed regions at the Moon's poles. NASA launched the LUNAR PROSPECTOR mission in January 1998 to perform a detailed study of the Moon's surface composition and to hunt for signs of the suspected deposits of water ice. Data from the Lunar Prospector mission strongly suggested the presence of water ice in the lunar polar regions, although the results require additional confirmation. Water on the Moon (as trapped surface ice in the permanently shadowed regions of the Moon's poles) would be an extremely valuable resource that would open up many exciting possibilities for future LUNAR BASE development.

From evidence gathered by the early uncrewed lunar missions (such as Ranger, Surveyor, and the lunar orbiter spacecraft), and by the Apollo missions, lunar scientists have learned a great deal more about the Moon and have been able to construct a geologic history dating back to its infancy. The table on the following page provides selected physical and dynamic properties of the Moon.

Because the Moon does not have any oceans or other free-flowing water and lacks a sensible atmosphere, appreciable erosion, or "weathering," has not occurred there. In fact, the Moon is actually a "museum world." The primitive materials that lay on its surface for billions of years are still in an excellent state of preservation. Scientists believe that the Moon was formed over 4 billion years ago and then differentiated quite early, perhaps only 100 million years later. Tectonic activity ceased eons ago on the Moon. The lunar crust and mantle are quite thick, extending inward to more than 800 kilometers (480 miles). However, the deep interior of the Moon is still unknown. It may contain a small iron core at its center, and there is some evidence that the lunar interior may be hot and even partially molten. Moonquakes have been measured within the lithosphere and interior, most being the

Figure 49. An excellent view of the near-full moon, as photographed by the *Apollo 16* crew (June 1972). Most of the lunar area in this picture is on the far side of the Moon and not visible from Earth. *(Courtesy of NASA)*

Physical and Astrophysical Properties of the Moon	
Diameter (equatorial)	3,476 km
Mass	$7,350 \ 10^{22}$ kg
Mass (Earth's mass = 1.0)	0.0123
Average density	3.34 g/cm^3
Mean distance from Earth (center-to-center)	384,400 km
Surface gravity (equatorial)	1.62 m/sec^2
Escape velocity	2.38 km/sec
Orbital eccentricity (mean)	0.0549
Inclination of orbital plane (to ecliptic)	5° 09′
Sidereal month (rotation period)	27.322 days
Albedo (mean)	0.07
Mean visual magnitude (at full)	–12.7
Surface area	37.9×10^6 km^2
Volume	2.20×10^{10} km^3
Atmospheric density (at night on surface)	2×10^5 molecules/cm^3
Surface temperature	102 K–384 K

Source: NASA.

result of gravitational stresses. Chemically, Earth and the Moon are quite similar, although compared to Earth, the Moon is depleted in more easily vaporized materials. The lunar surface consists of highlands composed of alumina-rich rocks that formed from a globe-encircling molten sea and maria made up of volcanic melts that surfaced about 3.5 billion years ago. However, despite all we have learned in the past three decades about our nearest celestial neighbor, lunar exploration really has only just started. Several puzzling mysteries still remain, including the origin of the Moon itself.

Recently, a new lunar origin theory has been suggested: a cataclysmic birth of the Moon. Scientists supporting this theory suggest that near the end of Earth's accretion from the primordial solar nebula materials (i.e., after its core was formed, but while Earth was still in a molten state), a Mars-size celestial object (called an "impactor") hit Earth at an oblique angle. This ancient explosive collision sent vaporized-impactor and molten-Earth material into Earth orbit, and the Moon then formed from these materials.

As previously mentioned, the surface of the Moon has two major regions with dis-

tinctive geologic features and evolutionary histories. First is the relatively smooth, dark areas that Galileo originally called "maria" (because he thought they were seas or oceans). Second is the densely cratered, rugged highlands (uplands), which Galileo called "terrae." The highlands occupy about 83% of the Moon's surface and generally have a higher elevation (as much as 5 kilometers above the Moon's mean radius.) In other places the maria lie about 5 kilometers below the mean radius and are concentrated on the "near side" of the Moon (i.e., on the side of the Moon always facing Earth).

The main external geologic process modifying the surface of the Moon is meteoroid impact. Craters range in size from very tiny pits only micrometers in diameter to gigantic basins hundreds of kilometers across.

The surface of the Moon is strongly brecciated, or fragmented. This mantle of weakly coherent debris is called regolith. It consists of shocked fragments of rocks, minerals, and special pieces of glass formed by meteoroid impact. Regolith thickness is quite variable and depends on the age of the bedrock beneath and on the proximity of craters and their ejecta blankets. Generally, the maria are covered by 3 to 16 meters of regolith, while the older highlands have developed a "lunar soil" at least 10 meters thick. *See also* LUNAR BASE; LUNAR PROSPECTOR.

moon A small natural body that orbits a larger one; a natural satellite.

mother spacecraft Exploration SPACE-CRAFT that carries and deploys one or several ATMOSPHERIC PROBES and ROVER or LANDER SPACECRAFT when arriving at a target PLANET. The mother spacecraft then relays its data back to EARTH and may orbit the planet to perform its own scientific mission. NASA's GALILEO PROJECT spacecraft to JUPITER and the CASSINI MISSION spacecraft to SATURN are examples.

multi- More than one; for example, *multimedia*.

multiple-degree-of-freedom system A mechanical system (e.g., a robot arm) for which two or more coordinates are required to define completely the position of the system at any instant.

multiple independently-targetable reentry vehicle (MIRV) A package of two or more reentry vehicles that can be carried by a single ballistic missile and guided to separate targets. MIRVed ballistic missiles use a warhead-dispensing mechanism called a postboost vehicle to target and release the reentry vehicles.

multiple scattering Any scattering process in which radiation being observed or detected is scattered more than once before reaching the sensor system.

multipropellant A rocket propellant consisting of two or more substances fed separately to the combustion chamber.

multispectral sensing A method of using many different bands of the electromagnetic spectrum (e.g., the visible, near-infrared, and thermal infrared bands) to sense a target. When several bands are used to sense a target, deceptive measures become much less effective.

multistage launch A launch that uses several stages to boost the payload into orbit. After the first-stage rocket booster has consumed its propellant, it is jettisoned and the second-stage rocket is fired. When the second-stage rocket engine's propellant is depleted, the second-stage rocket also is discarded, and so on. In the early aerospace literature also called a *step-rocket launch*.

multistage rocket A vehicle having two or more rocket units, each firing after the one in back of it has exhausted its propellant. Normally, each unit, or stage, is jettisoned after completing its firing. Also called a *multiple-stage rocket* or, infrequently, a *step rocket*.

mutual assured destruction (MAD) The strategic situation that existed during the cold war in which either superpower (i.e., either the United States or the former Soviet Union) could inflict massive nuclear destruction on the other, no matter which side struck first.

nadir 1. The direction from a spacecraft directly down toward the center of the planet. Opposite of the *zenith*. 2. That point on the celestial sphere directly beneath the observer and directly opposite the zenith.

naked eye The normal human eye unaided by any optical instrument, such as a TELESCOPE. The use of corrective lenses (glasses) or contact lenses that restore an individual's normal vision are included in the concept of naked eye observing.

nano- (symbol: n) A prefix in the SI UNIT system meaning multiplied by 10^{-9}.

nanometer (nm) A billionth of a meter (i.e., 10^{-9} meter)

nanosecond (ns) A billionth of a second (i.e., 10^{-9} second).

nanotechnology A general term that describes the manufacture and application of microminiature machines, electronic devices, and chemical and biological sensors all of which have characteristic dimensions on the order of a micron (10^{-6} meter) or less.

NASA *See* NATIONAL AERONAUTICS AND SPACE ADMINISTRATION.

NASDA *See* NATIONAL SPACE DEVELOPMENT AGENCY OF JAPAN.

National Aeronautics and Space Administration (NASA) The civilian space agency of the United States, created in 1958 by an act of Congress (i.e., the National Aeronautics and Space Act of 1958). NASA belongs to the executive branch of the federal government. Its overall mission is to plan, direct, and conduct civilian (including scientific) aeronautical and space activities for peaceful purposes. This mission is carried out at NASA headquarters in Washington, D.C., supported by centers and field facilities throughout the United States. Current NASA centers and facilities include the Ames Research Center (Moffett Field, California), the Dryden Flight Research Center (Edwards AFB, California), the Glenn Research Center (at Lewis Field in Cleveland, Ohio), the Goddard Institute for Space Studies (located in New York City, New York), the Goddard Space Flight Center (Greenbelt, Maryland), the Independent Verification and Validation Facility (Fairmont, West Virginia), the contractor-operated JET PROPULSION LABORATORY (Pasadena, California), the JOHNSON SPACE CENTER (Houston, Texas), the KENNEDY SPACE CENTER (Cape Canaveral, Florida), the Langley Research Center (Hampton, Virginia), the MARSHALL SPACE FLIGHT CENTER (Huntsville, Alabama), the Stennis Space Center (Mississippi), the Wallops Flight Facility (Wallops Island, Virginia), and the White Sands Test Facility (White Sands, New Mexico).

NASA headquarters in Washington, D.C., exercises management over the spaceflight centers, research centers, and other installations that make up the civilian space agency. Responsibilities include general determination of programs and projects, establishment of policies and procedures, and evaluation and review of all phases of the NASA AEROSPACE program. NASA's human spaceflight program placed ASTRONAUTS on the MOON during the APOLLO PROJECT and currently involves the SPACE TRANSPORTATION SYSTEM (SPACE SHUTTLE) and the INTERNATIONAL SPACE STATION. Contemporary NASA programs dedicated to the exploration of the universe include the HUBBLE SPACE TELESCOPE, the CHANDRA X-RAY OBSERVATORY, and the SPACE INFRARED TELESCOPE FACILITY. NASA scientific investigation of the SOLAR SYSTEM continues with such projects as the CASSINI MISSION to SATURN, the 2001 MARS ODYSSEY SPACECRAFT, and the 2003 MARS EXPLORATION ROVERS MISSION. For more than four decades, NASA has also sponsored monitoring EARTH from space with the AQUA and TERRA spacecraft now providing leading-edge collections of environmental data in support of EARTH SYSTEM SCIENCE activities. Since its inception, NASA has also supported many important space technology applications including the development of WEATHER SATELLITES, COMMUNICATIONS SATELLITES, and the civilian application of MULTISPECTRAL SENSING from satellites such as LANDSAT.

National Oceanic and Atmospheric Administration (NOAA)

In 1970 the National Oceanic and Atmospheric Administration (NOAA) was established as an agency within the U.S. Department of Commerce (DOC) to ensure the safety of the general public from atmospheric phenomena and to provide the public with an understanding of EARTH's environment and resources. NOAA conducts research and gathers data about the global oceans, the ATMOSPHERE, OUTER SPACE, and the SUN through five major organizations: the National Weather Service, the National Ocean Service, the National Marine Fisheries Service, the National Environmental Satellite, Data and Information Service, and NOAA Research. NOAA Research. NOAA research and operational activities are supported by the seventh uniformed service of the U.S. government. The NOAA Corps contains the commissioned personnel who operate NOAA's ships and fly its aircraft.

The National Environmental Satellite, Data, and Information Service (NESDIS) is responsible for the daily operation of American environmental satellites, such as the GEOSYNCHRONOUS OPERATIONAL ENVIRONMENTAL SATELLITE (GOES). The prime customer for environmental satellite data is the National Weather Service, but NOAA also distributes these data to many other users within and outside the government. NOAA's operational environmental satellite system is composed of geostationary operational environmental satellites for short-range warning and "nowcasting" and polar orbiting environmental satellites (POES) for long-term forecasting. Both types of satellites are necessary for providing a complete global weather monitoring system. The satellites also carry search and rescue (SAR) capabilities for locating people lost in remote regions on land (including victims of aircraft crashes) or stranded at sea as a result of maritime disasters and accidents.

As a result of a presidential directive in May 1994, NESDIS now operates the SPACECRAFT in the Department of Defense's DEFENSE METEOROLOGICAL SATELLITE PROGRAM (DMSP). The executive order combined the U.S. military and civilian operational METEOROLOGICAL SATELLITE systems into a single national system capable of satisfying both civil and national security requirements for space-based remotely sensed environmental data. As part of this merger, a triagency (NOAA, NASA, and DOD) effort is underway to develop and deploy the NATIONAL POLAR-ORBITING OPERATIONAL SATELLITE SYSTEM (NPOESS) starting in about 2010. In addition to operating satellites, NESDIS also manages global databases for meteorology, oceanography, SOLID EARTH geophysics, and SOLAR-terrestrial sciences. *See also* GLOBAL CHANGE.

National Polar-orbiting Operational Environmental Satellite System (NPOESS)

The planned U.S. system of advanced polar-orbiting environmental satellites that converges existing American polar-orbiting METEOROLOGICAL SATELLITE systems—namely the DEFENSE METEOROLOGICAL SATELLITE PROGRAM and the Department of Commerce's POLAR-ORBITING ENVIRONMEN-

TAL SATELLITE (POES)—into a single national program. Responding to a 1994 presidential directive, a triagency team of Department of Defense, Department of Commerce (NOAA), and NASA personnel are currently developing this new system of polar-orbiting environmental satellites. When deployed (starting in about 2010), NPOESS will continue to monitor global environmental conditions and to collect and disseminate data related to weather, the atmosphere, the oceans, various land masses, and the near-Earth space environment. The global and regional environmental imagery and specialized environmental data from NPOESS will support the peacetime and wartime missions of the Department of Defense as well as civil mission requirements of organizations such the National Weather Service within the NATIONAL OCEANIC AND ATMOSPHERIC ADMINISTRATION (NOAA). In particular, NPOESS will use INSTRUMENTS that sense surface and atmospheric RADIATION in the visible, infrared, and MICROWAVE bands of the ELECTROMAGNETIC SPECTRUM, monitor important parameters of the space environment, and measure distinct environmental parameters such as soil moisture, cloud levels, sea ice, and ionospheric scintillation.

National Radio Astronomy Observatory (NRAO)

A collection of government-owned RADIO ASTRONOMY facilities throughout the United States, including the RADIO TELESCOPE in Green Bank, West Virginia.

National Reconnaissance Office (NRO)

The National Reconnaissance Office (NRO) is the national program to meet the needs of the U.S. government through spaceborne reconnaissance. The NRO is an agency of the Department of Defense (DoD) and receives its budget through that portion of the National Foreign Intelligence Program (NFIP) known as the National Reconnaissance Program (NRP), which is approved both by the secretary of defense and the director of Central Intelligence (DCI). The existence of the NRO was declassified by the deputy secretary of defense, as recommended by the DCI, on September 18, 1992.

The mission of the NRO is to ensure that the United States has the technology and spaceborne assets needed to acquire intelligence worldwide. This mission is accomplished through research, development, acquisition, and operation of the nation's intelligence satellites. The NRO's assets collect intelligence to support such functions as intelligence and warning, monitoring of arms control agreements, military operations and exercises, and monitoring of natural disasters and environmental issues.

The director of the NRO is appointed by the president and confirmed by the Congress as the assistant secretary of the air force for space. The secretary of defense has the responsibility, which is exercised in concert with the DCI, for the management and operation of the NRO. The DCI establishes the collection priorities and requirements for the collection of satellite data. The NRO is staffed by personnel from the Central Intelligence Agency (CIA), the military services, and civilian DoD personnel.

The *Corona* program was the first satellite photo reconnaissance program developed by the United States. Corona was a product of the cold war. Soviet nuclear weapons coupled with the long-range bomber and missile delivery systems created the possibility of a devastating surprise attack—a "nuclear Pearl Harbor"—and President Dwight D. Eisenhower demanded better intelligence. But the Iron Curtain and the sheer size of the Soviet land-mass made traditional intelligence collection methods marginally useful, at best. The U-2 reconnaissance aircraft served admirably for four years collecting data over the Soviet Union, but its effectiveness was known to be limited by advancements in Soviet air defenses. That, coupled with the urgency for a national space program stimulated by *Sputnik* (the first satellite launched by the former Soviet Union on October 4, 1957), gave birth to the Corona satellite program.

The air force, under a program with the innocuous designation Weapon System 117L, Advanced Reconnaissance System, had been developing a family of satellites since 1956 in connection with development of the ballistic missile, which would serve as the satellite launch vehicle. But priority for the satellite program had been low, and funding was limited.

Following *Sputnik,* all that changed. The most promising near-term project was split out from Weapon System-117L in early 1958 and named Corona. The program was provided increased priority and funding and designated for streamlined management under a joint air force/CIA team. That team was told to take a space booster, a spacecraft, a reentry vehicle, a camera, photographic film, and a control network, all untested, and to make them work together as a system quickly. However, strong leadership, coupled with unwavering White House support, would be needed (and provided) to take this pioneering reconnaissance satellite program through many early setbacks. And there were numerous setbacks! All but one of the first 13 Corona launches (numbered 0–12) resulted in failure of one type or another. But each failure was investigated, identified, and corrected as the team moved closer to an operational photo reconnaissance capability.

Then, on August 10, 1960, *Corona XIII* was launched with diagnostic equipment rather than an actual cameral payload. A day later the film capsule reentered and was recovered in the Pacific Ocean. On August 18 *Corona XIV,* with a camera/film payload, was launched in a perfect mission ending with a successful midair capsule recovery. The film retrieved *Corona XIV* was processed quickly and exploited by the intelligence community. This film capsule provided more coverage than all previous U-2 flights over the Soviet Union combined. The United States now had the space-based "eyes" so necessary to protect its people and keep the peace throughout the cold war. The 145th and final Corona launch took place on May 15, 1972.

From a technical perspective, Corona was the first space program to recover an object from orbit and the first to deliver photo reconnaissance information from a satellite. It would go on to be the first pro-

gram to use multiple reentry vehicles, pass the 100 mission mark, and produce stereoscopic space imagery. Its most remarkable technological advance, however, was the improvement in its ground resolution from an initial 25- to 40-foot (about 7.6 to 12.2-meter) capability to an ultimate 6-foot (1.82-meter) resolution.

National Space Development Agency of Japan (NASDA)

Japan's space development activities are primarily implemented by the National Space Development Agency and the Institute of Space and Astronautical Science (ISAS), in cooperation with other related organizations and in accordance with the space development program created and approved by the Space Activities Commission, an advisory committee to the prime minister. NASDA was established on October 1, 1969, and the organization's activities are limited to the peaceful uses of OUTER SPACE. The agency is primarily engaged in research and development involving SATELLITES and LAUNCH VEHICLES for practical uses, launch and tracking operations for Japanese satellites, and promoting the development of civilian REMOTE SENSING technologies and MATERIALS PROCESSING IN SPACE (MPS).

NASDA's H-II ROCKET is the centerpiece of Japan's space program. It is the first fully domestically manufactured Japanese ROCKET vehicle and has been designed to send 2,000-kilogram-class missions to GEOSTATIONARY ORBIT. The first STAGE of this modern EXPENDABLE LAUNCH VEHICLE is a high-performance LIQUID OXYGEN/LIQUID HYDROGEN engine called the LE-7. The vehicle's second stage uses the LE-5A liquid hydrogen/liquid oxygen engine. It is a reignitable engine that offers higher performance and reliability. H-II vehicle Number 1 was successfully launched on February 4, 1994, but vehicle Number 8 suffered a launch failure on November 15, 1999, providing a checkered pattern of success and failure that testifies to the difficulties of developing truly advanced high-performance rocket systems using CRYOGENIC PROPELLANTS. The H-II also uses two strap-on SOLID-PROPELLANT BOOSTER ROCKETS.

The main NASDA facilities include the Tsukuba Space Center (Ibaraki), Kakuda Propulsion Center, and Tanegashima Space Center. The Tanegashima Space Center is located on the southeastern portion of Tanegashima Island, Kagoshima. This center includes the Takesake Range for small rockets, the Osaki Range for the H-II launch vehicle, and the Yoshinobu Launch Complex for the H-II expendable vehicle.

NASDA is participating in the INTERNATIONAL SPACE STATION (ISS) and has also sponsored Japanese astronauts on flights onboard the U.S. SPACE SHUTTLE. For example, astronaut Takao Doi served as a MISSION SPECIALIST during the STS-87 mission (November 20, 1997, through December 5, 1997) and successfully performed two extra vehicular activities (EVAs).

NATO III A constellation of military communications satellites launched from Cape Canaveral, Air Force Station, Florida between April 1976 and November 1984. These satellites, designated A, B, C, and D, have provided communications for North Atlantic Treaty Organization (NATO) officials in Belgium, Canada, Denmark, the United Kingdom (England), Germany, as well as Greece, Iceland, and Italy. The United States, Netherlands, Norway, Portugal and Turkey also have used the satellite communications system. NATO III simultaneously accommodated hundreds of NATO users and provided voice and facsimile services. The 2.13-meter (7-foot) diameter, 2.74-meter- (9-foot-) high cylinder-shaped, spinning spacecraft are located in geostationary orbit. Each satellite has three "horn" antennas mounted on a platform that spines in the opposite direction of the body of the spacecraft, enabling the antennas to point at the same place on Earth constantly.

nautical mile (nm) A unit of distance used mainly in navigation. Defined as the distance spanned by one minute of arc of any great circle on Earth. Also called the *sea mile*.

1 nautical mile = 1.852 kilometers = 1.15 statute (land) miles

navigation satellite A SPACECRAFT placed into a well-known, stable ORBIT around EARTH that transmits precisely timed RADIO WAVE SIGNALS useful in determining locations on land, at sea, or in the air. Such SATELLITES are deployed as part of an interactive CONSTELLATION. *See also* GLOBAL POSITIONING SYSTEM.

Navstar The Navstar Global Positioning System is a constellation of advanced radio navigation satellites developed for the U.S. Department of Defense and operated by the U.S. Air Force. In addition to serving defense navigation needs this system, also supports many civilian applications. *See also* GLOBAL POSITIONING SYSTEM (GPS).

near-Earth asteroid (NEA) An inner SOLAR SYSTEM ASTEROID whose ORBIT around the SUN brings it close to EARTH, perhaps even posing a collision threat in the future. *See also* ASTEROID DETECTION SYSTEM; EARTH-CROSSING ASTEROID.

Near Earth Asteroid Rendezvous (NEAR) mission NASA's Near Earth Asteroid Rendezvous (NEAR) mission was launched on February 17, 1996, from CAPE CANAVERAL AIR FORCE STATION by a DELTA II EXPENDABLE LAUNCH VEHICLE. It is the first of NASA's Discovery missions, a series of small-scale SPACECRAFT designed to proceed from development to flight in less than three years for a cost of less than $150 million. The NEAR spacecraft was equipped with an X-RAY/GAMMA RAY SPECTROMETER, a NEAR-INFRARED imaging spectrograph, a multispectral camera fitted with a CHARGE-COUPLED DEVICE (CCD) imaging DETECTOR, a LASER ALTIMETER, and a MAGNETOMETER. The primary goal of this mission is to rendezvous with and achieve ORBIT around the NEAR-EARTH ASTEROID, Eros-433 (also referred to as 433 Eros and sometimes just simply Eros).

Eros-N33 is an irregularly shaped S-class asteroid about $13 \times 13 \times 33$ KILOMETERS in size. This asteroid, the first near Earth asteroid to be found, was discovered on August 13, 1898, by the German astronomer Gustav Witt (1866–1946). In

Greek mythology, Eros (Roman name: Cupid) was the son of Hermes (Roman name: Mercury) and Aphrodite (Roman name: Venus) and served as the god of love.

As a member of the AMOR GROUP of asteroids, Eros has an orbit that crosses the orbital path of MARS but does not intersect the orbital path of EARTH around the SUN. The asteroid follows a slightly ELLIPTICAL ORBIT, circling the Sun in 1.76 years at an INCLINATION of 10.8 DEGREES to the ECLIPTIC. Eros-433 has a PERIHELION of 1.13 ASTRONOMICAL UNITS (AU) and an APHELION of 1.78 AU. The closest approach of Eros to Earth in the 20th century occurred on January 23, 1975, when the asteroid came within 0.15 AU (about 22 million kilometers) of our home planet.

After launch and departure from Earth orbit, NEAR entered the first part of its CRUISE PHASE. It spent most of this phase in a minimal activity (hibernation) state that ended a few days before the successful flyby of the asteroid Mathilde-253 on June 27, 1997. During that encounter, the spacecraft flew within 1,200 kilometers of Mathilde-253 at a relative VELOCITY of 9.93 kilometers per second (km/s). Imagery and other scientific data were collected.

On July 3, 1997, the NEAR spacecraft executed its first major DEEP SPACE maneuver, a two-part propulsive BURN of its main 450-NEWTON (HYDRAZINE/nitrogen tetroxide-fueled) thruster. This maneuver successfully decreased the spacecraft's velocity by 279 meters per second (m/s) and lowered perihelion from 0.99 AU to 0.95 AU. Then, on January 23, 1998, the spacecraft performed an Earth-GRAVITY ASSIST FLYBY—a critical maneuver that altered its orbital inclination from 0.5 to 10.2 degrees and its aphelion distance from 2.17 AU to 1.77 AU. This gravity assist maneuver gave the NEAR spacecraft orbital parameters that nearly matched those of the TARGET asteroid, Eros-433.

The original mission plan was to RENDEZVOUS with and achieve orbit around Eros-433 in January 1999 and then to study the asteroid for approximately one year. However, a SOFTWARE problem caused an ABORT of the first encounter burn, and NASA revised the mission plan

to include a flyby of Eros-433 on December 23, 1998, and then an encounter and orbit on February 14, 2000. The radius of the spacecraft's orbit around Eros-433 was brought down in stages to a 50 × 50 kilometers orbit on April 30, 2000, and decreased to a 35 × 35 kilometers orbit on July 14, 2000. The orbit was then raised over the succeeding months to 200 kilometers × 200 kilometers and next slowly decreased and altered to a 35 × 35 kilometers RETROGRADE ORBIT around the asteroid on December 13, 2000. The mission ended with a touchdown in the "saddle" region of Eros on February 12, 2001. NASA renamed the spacecraft NEAR-Shoemaker in honor of American astronomer and geologist Eugene M. Shoemaker (1928–1997) following his untimely death in an automobile accident on July 18, 1997.

near infrared That portion of the electromagnetic spectrum involving the shorter wavelengths in the infrared region; generally considered to extend from just beyond the visible red portion of the spectrum (about 0.7 micrometer wavelength) out to about 3.0 micrometers wavelength. *See also* ELECTROMAGNETIC SPECTRUM; INFRARED (IR) RADIATION.

nearside The side of the Moon that always faces Earth. *See also* MOON.

nebula (plural: nebulas or nebulae) A cloud of INTERSTELLAR GAS or dust. It can be seen as either a dark hole against a brighter background (called a DARK NEBULA) or as a luminous patch of LIGHT (called a bright nebula).

Nemesis A postulated dark stellar companion to the Sun whose close passage every 26 million years is thought to be responsible for the cycle of mass extinctions that seem also to have occurred on Earth at 26-million-year intervals. This "death star" companion has been named for the Greek goddess of retributive justice or vengeance. If it really does exist, it might be a white dwarf, a rogue star that was captured by the Sun, or possibly a tiny but gravitationally influential neutron star. The passage of such a death star through the Oort Cloud (a postulated swarm of comets surrounding the solar system) could lead to a massive shower of comets into the solar system. One or several of these "perturbed" comets impacting on Earth then would trigger massive extinctions and catastrophic environmental changes within a very short period of time.

neper (symbol: N or N_p) A natural logarithmic unit (x) used to express the ratio of two power levels, P_1(input) and P_2(output), such that x (nepers) = 1/2 ln (P_1/P_2). The unit is named after John Napier (1550–1617), the Scottish mathematician who developed natural logarithms (symbol: In). This unit is often encountered in telecommunications engineering. 1 neper = 8.686 decibels. *See also* BEL.

Neptune The outermost of the Jovian planets and the first planet to be discovered using theoretical predictions. Neptune's discovery was made by J. G. Galle (1812–1910) at the Berlin Observatory in 1846. This discovery was based on independent orbital perturbation (disturbance) analyses by the French astronomer Jean Joseph Urbain Le Verrier (1811–1877) and the British scientist J. Adams (1819–1892). It is considered to be one of the triumphs of 19th-century theoretical astronomy.

Because of its great distance from Earth, little was known about this majestic blue giant planet until the *Voyager 2* spacecraft swept through the Neptunian system on August 25, 1989. Neptune's characteristic blue color comes from the selective absorption of red light by the methane (CH_4) found in its atmosphere—an atmosphere consisting primarily of hydrogen (over 89%) and helium (about 11%) with minor amounts of methane, ammonia ice, and water ice. At the time of the *Voyager 2* encounter, Neptune's most prominent "surface" feature was called the Great Dark Spot

Physical and Dynamic Properties of Neptune	
Diameter (equatorial)	49,532 km
Mass	1.02×10^{26} kg
Density (mean)	1.64 g/cm^3
Surface gravity	11 m/sec^2 (approx.)
Escape velocity	23.5 km/sec (approx.)
Albedo (visual geometric)	0.4 (approx.)
Atmosphere	Hydrogen (~80%), helium (~18.5%) methane (~1.5%)
Temperature (blackbody)	33.3 K
Natural satellites	8
Rings	6 (Galle, LeVerrier, Lassell, Arago, unnamed, Adams [arcs])
Period of rotation (a Neptunian day)	0.6715 day (16 hr 7 min)
Average distance from Sun	4.5×10^9 km (30.06 AU)
Eccentricity	0.0086
Period of revolution around Sun (a Neptunian year)	165 yr
Mean orbital velocity	5.48 km/sec
Magnetic field	Yes (strong, complex; tilted 50° to planet's axis of rotation)
Radiation belts	Yes (complex structure)
Solar flux at planet (at top of atmosphere)	1.5 W/m^2 (at 30 AU)

Source: NASA.

(GDS), which was somewhat analogous in relative size and scale to Jupiter's Red Spot. However, unlike Jupiter's Red Spot, which has been observed for at least 300 years, Neptune's GDS, which was located in the Southern Hemisphere in 1989, had disappeared by June of 1994, when the Hubble Space Telescope looked for it. Then, a few months later, a nearly identical spot appeared in Neptune's northern hemisphere. Neptune is an extraordinarily dynamic planet that continues to surprise space scientists. The *Voyager 2* encounter also revealed the existence of six additional satellites and an interesting ring system. The table on the preceding provides contemporary physical and dynamic data for Neptune.

Triton, Neptune's largest moon, is one of the most interesting and coldest objects (about 35 kelvins (K) surface temperature) yet discovered in the solar system. Because of its inclined retrograde orbit, density (2.0 grams per centimeter cubed). rock and ice composition, and frost-covered (frozen nitrogen) surface, space scientists consider Triton to be a "first cousin" to the planet Pluto. Triton shows remarkable geologic history; *Voyager 2* images have revealed active geyserlike eruptions spewing invisible nitrogen gas and dark dust several kilometers into space. *See also* PLUTO; URANUS; VOYAGER.

NERVA Nuclear Engine for Rocket Vehicle Application. The acronym applied to a series of nuclear reactors developed as part of the overall U.S. nuclear rocket development program (i.e., the ROVER Program) from 1959 to 1973. *See also* NUCLEAR ROCKET; SPACE NUCLEAR PROPULSION.

neutral Without an electrical charge; neither positive nor negative.

neutral atmosphere That portion of a planetary atmosphere consisting of neutral (uncharged) atoms or molecules, as opposed to electrically charged ions.

neutral buoyancy simulator Orbiting objects in space, being "weightless,"

behave a little like neutral buoyancy objects in water here on Earth. Astronauts have trained for certain on-orbit tasks by working with and moving space hardware while underwater in a large tank or submersion facility. These activities help simulate the actual mission activities to be performed during weightlessness.

neutral burning With respect to a solid-propellant rocket, a condition of propellant burning such that the burning surface or area remains constant, thereby producing a constant pressure and thrust.

neutrino (symbol: v) An uncharged fundamental PARTICLE with no (or possibly very little) MASS that interacts only weakly with matter.

neutron (symbol: n) An uncharged elementary particle with a mass slightly greater than that of the proton. It is found in the nucleus of every atom heavier than hydrogen. A free neutron is unstable, with a half-life of about 12 minutes, and decays into an electron, a proton, and a neutrino. Neutrons sustain the fission chain reaction in a nuclear reactor.

neutron star A very small (typically 20 to 30 KILOMETERS in diameter), superdense stellar object—the gravitationally collapsed CORE of a massive STAR that has undergone a SUPERNOVA explosion. Astrophysicists hypothesize that PULSARS are rapidly spinning neutron stars that possess intense MAGNETIC FIELDS.

New Horizons Mission *See* PLUTO-KUIPER BELT MISSION.

new moon The Moon at conjunction, when little or none of it is visible to an observer on Earth because the illuminated side is away from him or her. *See also* MOON.

Newton, Sir Isaac (1642–1727) The brilliant though introverted English physicist and mathematician whose law of GRAVITATION, three laws of motion, development of the calculus, and design of a new type of REFLECTING TELESCOPE make him one of the greatest scientific minds in human history. Through the patient encouragement and financial support of the English mathematician Edmond Halley (1656–1742), Newton published his great work, *Mathematical Principles of Natural Philosophy* (or *The Principia*), in 1687. This monumental book transformed the practice of physical science and completed the scientific revolution started by NICHOLAS COPERNICUS, JOHANNES KEPLER, and GALILEO GALILEI. *See also* NEWTONIAN TELESCOPE; NEWTON'S LAW OF GRAVITATION; NEWTON'S LAWS OF MOTION.

newton (symbol: N) The unit of force in the International System of Units. It is defined as the force that provides a 1-kilogram mass with an acceleration of 1 meter per second per second.

$$1 \text{ N} = 1 \text{ kg-m/s}^2$$

Newtonian mechanics The system of mechanics based on Newton's laws of motion in which mass and energy are considered as separate, conservative mechanical properties, in contrast to their treatment in relativistic mechanics, in which mass and energy are treated as equivalent (from the Einstein mass-energy equivalency formula: $E = m\ c^2$). *See also* NEWTON'S LAWS OF MOTION.

newtonian telescope A reflecting telescope in which a small plane mirror reflects the convergence beam from the objective (primary mirror) to an eyepiece at the side of the telescope. After the second reflection the rays travel approximately perpendicular to the longitudinal axis of the telescope. This type of telescope, the first reflecting telescope to be constructed was developed by Sir Isaac Newton (1642–1727) around 1670.

Newton's law of gravitation Every particle of matter in the universe attracts every other particle with a force (F), acting along the line joining the two particles, proportional to the product of the masses (m_1 and m_2) of the particles and inversely proportional to the square of the distance (r) between the particles, or

$$F = [G\ m_1\ m_2]/r^2$$

where F is the force of gravity (newtons), m_1, m_2 are the masses of the (attracting) particles (kg); r is the distance between the particles (m); and G is the universal gravitational constant.

$$G = 6.6732\ (\pm 0.003) \times 10^{-11} \text{ N m}^2/\text{kg}^2$$
$$\text{(in SI units)}$$

Newton's laws of motion A set of three fundamental postulates that form the basis of the mechanics of rigid bodies. These laws were formulated in about 1685 by the brilliant English scientist and mathematician Sir Isaac Newton (1642–1727), as he was studying the motion of the planets around the Sun. Newton described this work in the book *Mathematical Principles of Natural Philosophy,* which is often referred to simply as Newton's *Principia.*

Newton's first law is concerned with the principle of inertia and states that if a body in motion is not acted upon by an external force, its momentum remains constant. This law also can be called the *law of conservation of momentum.*

The second law states that the rate of change of momentum of a body is proportional to the force acting upon the body and is in the direction of the applied force. A familiar statement of this law is the equation: $F = m\ a$, where F is the vector sum of applied forces, m is the mass, and a is the vector acceleration of the body.

Newton's third law is the principle of action and reaction. It states that for every force acting upon a body, there is a corresponding force of the same magnitude exerted by the body in the opposite direction.

nickel-cadmium (nicad) batteries Long-lived, rechargeable batteries frequently used in space-craft applications as a secondary source of power. For example, in a typical Earth-orbiting spacecraft, solar cell arrays capture sunlight and provide electric power to operate the spacecraft's electrical systems and also to charge the nickel-cadmium batteries. Then, when the spacecraft is shadowed by Earth and is not illuminated by the Sun, spacecraft electric power is provided by the nickel-cadmium batteries. This cycle continues many times during the lifetime of the spacecraft. While the nicad batteries have a low energy density (typically 15 to 30 watt-hours per kilogram), they have a good deep-discharge tolerance and long cycle life, making their use nearly standard in contemporary spacecraft operations.

Nike Hercules A U.S. Army air defense surface-to-air guided missile system that provides nuclear or conventional, medium- to high-altitude air defense coverage against manned bombers and air-breathing missiles. The system is designed to operate in either a mobile or fixed-site configuration and has a capability of performing surface-to-surface missions. Also designated as the MIM-14.

The army developed two other surface-to-air missiles in the Nike missile family: the Nike Ajax and the Nike Zeus.

Nimbus satellites A series of second-generation weather satellites created by NASA and the National Oceanographic and Atmospheric Administration (NOAA) as a follow-up to the Tiros satellites, which were the first polar orbiting meteorological satellites. The Nimbus satellites were more complex than the Tiros satellites. For example, they carried advanced television (TV) cloud-mapping cameras and an infrared radiometer that permitted the collection of cloud images at night for the first time. From 1964 to 1978 seven Nimbus spacecraft were placed in polar orbit from Vandenberg Air Force Base, California, creating a 24-hour-per-day capability to observe weather conditions on the planet. *See also* WEATHER SATELLITE.

NOAA *See* NATIONAL OCEANIC AND ATMOSPHERIC ADMINISTRATION.

node 1. In orbital mechanics, one of the two points of intersection of the orbit of a planet, planetoid, or comet with the ecliptic, or of the orbit of a satellite with the plane of the orbit of its primary. By convention, that point at which the body crosses to the north side of the reference plane is called the *ascending node;* the other node is called the *descending node.* The line connecting these two nodes is called the *line of nodes.* 2. In physics, a point, line, or surface in a standing wave where some characteristic of the wave field has essentially zero amplitude. By comparison, a position of maximum amplitude in a standing wave is referred to as an *antinode.* 3. In communications, a network junction or connection point (e.g., a terminal).

"nominal" In the context of aerospace operations and activities, a word meaning "within prescribed or acceptable limits." For example, the Mars mission is now on a "nominal" interplanetary trajectory; or the pressure in the combustion chamber is "nominal."

noncoherent communication Communications mode wherein a spacecraft generates its downlink frequency independent of any uplink frequency. *See also* TELECOMMUNICATIONS.

nondestructive testing Testing to detect internal and concealed defects in materials and components using techniques that do not damage or destroy the items being tested. X rays, gamma rays, and neutron irradiation, as well as ultrasonics, frequently are used to accomplish nondestructive testing.

nonequilibrium In thermodynamics, a state within a gas mixture in which reactions have not reached equilibrium. If left in this condition, the composition of the gas mixture would change in time.

nonequilibrium composition In rocketry, the exhaust gas chemical composition resulting from incomplete chemical reaction of the products of combustion.

nonflight unit An aerospace unit that is not intended for flight usage. Also called *test article* or a *hangar queen.*

nonimpinging injector An injector used in liquid rocket engines that employs parallel streams of propellant usually emerging normal to the face of the injector. In this type injector, propellant mixing usually is obtained by turbulence and diffusion. The World War II era German V-2 rocket used a nonimpinging injector. *See also* INJECTOR.

nonionizing electromagnetic radiation Radiation that does not change the structure of atoms but does heat tissue and may cause harmful biological effects. Microwaves, radio waves, and low-frequency electromagnetic fields from high-voltage transmission lines are considered to be nonionizing radiation.

normal 1. Equivalent to usual, regular, rational, or standard conditions. 2. Perpendicular. A line is *normal* to another line or a plane when it is perpendicular to it. A line is *normal* to a curve or curved surface when it is perpendicular to the tangent line or plane at the point of tangency.

normally closed valve Powered valve that returns to a closed position on shutoff or on failure of the actuating energy or signal.

normally open valve Powered valve that returns to an open position on shutoff or on failure of the actuating energy or signal.

North American Aerospace Defense Command (NORAD) The North American Aerospace Defense Command is a binational (U.S. and Canadian) organization charged with the missions of AEROSPACE warning and aerospace control of North America. Aerospace warning includes the monitoring of human-made objects in space and the detection, validation, and warning of an attack against North America whether by AIRCRAFT, MISSILES, or SPACE VEHICLES. Aerospace control includes ensuring air sovereignty and AIR DEFENSE of the airspace of Canada and the United States. To achieve these mission objectives NORAD relies on mutual support arrangements with other commands, including UNITED STATES STRATEGIC COMMAND (USSTRATCOM). The commander of NORAD is responsible to both the president of the United States and the prime minister of Canada.

The headquarters for the NORAD commander is located at Peterson AFB, Colorado, and NORAD's command and control center is a short distance away at CHEYENNE MOUNTAIN AIR STATION (CMAS). Deep within the underground complex at Cheyenne Mountain are all the facilities needed by NORAD personnel to collect and coordinate data from a worldwide system of SENSORS. Data from these sensors provide an accurate picture of any emerging aerospace threat to North America. NORAD has three subordinate regional headquarters located at Elmendorf AFB, Alaska, Canadian Forces Base, Winnipeg, Manitoba, and Tyndall AFB, Florida. These regional headquarters receive direction from the NORAD commander and then take appropriate actions to control air operations within their respective geographic areas of responsibility.

As part of the aerospace warning mission, NORAD's commander must provide an integrated tactical warning and attack assessment (called an ITW/AA) concerning a particular aerospace attack on North America to the governments of both Canada and the United States. A portion of the necessary threat information comes from warning systems controlled directly by NORAD, while other attack assessment information comes to NORAD from other commands under various mission support agreements.

NORAD's aerospace control mission includes detecting and responding to any air-breathing threat to North America. NORAD uses a network of ground-based RADAR systems and fighter aircraft to detect, intercept, and (if necessary) engage any air-breathing threat to the continent. The SUPERSONIC aircraft presently used by NORAD consist of U.S. F-15 and F-16 fighters and Canadian CF-18 fighters. NORAD also assists in the detection and monitoring of aircraft suspected of illegal drug trafficking. This information is passed on to civilian law enforcement agencies to help combat the flow of illegal drugs by air into the United States and Canada. Responding to the threat of terrorism, NORAD provides its military response capabilities to civil authorities and homeland defense organizations to counter domestic airspace threats.

nose cone The cone-shaped leading edge of a rocket vehicle. Designed to protect and contain a warhead or payloads, such as satellites, instruments, animals, or auxiliary equipment. The outer surface and structure of a nose cone is built to withstand the high temperatures associated with the aerodynamic heating and vibrations (buffeting) that results from flight through the atmosphere. *See also* REENTRY VEHICLE; ROCKET.

nova (plural: novae) A star that exhibits a sudden and exceptional brightness, usually of a temporary nature, and then returns to its former luminosity. A *supernova* is a much brighter explosion of a giant star during which a large fraction of its mass is blown outward into space. The term derives from the Latin word *nova*, which means "new." *See also* STAR(S).

nozzle In a rocket engine, the carefully shaped aft portion of the thrust chamber that controls the expansion of the exhaust gas so that the thermal energy of combustion is converted effectively into kinetic energy of the combustion product gases, thereby propelling the rocket vehicle. The nozzle is a major component of a rocket engine, having a significant influence on the overall engine performance and representing a large fraction of the engine structure. The design of the nozzle consists of solving simultaneously two different problems: defining the shape of the wall that forms the expansion surface and delineating the nozzle structure and hydraulic system.

In general, the nozzle shape is selected to maximize performance within the constraints placed on the rocket system. The nozzle structure of a large rocket must provide strength and rigidity to a system in which mass is at a high premium and the loads are not readily predictable. The maxi-

mum loads on the nozzle structure often occur during the start transients (i.e., initial rapid rise in pressure and fluid flow rates) before full flow is established.

The nozzles used on liquid rocket engines now are inherently very efficient components, since they are the product of many years of engineering development. (See Figure 50.) With increasing area ratio (i.e., exit area to throat area), the nozzle becomes a proportionally larger part of the rocket engine. Because there is relatively little to be gained in trying to increase the efficiencies of conventional bell and conical nozzles, recent aerospace engineering activities in nozzle development have focused on obtaining the same nozzle efficiency from a shorter package through the use of short bell nozzles and annular nozzles, such as expansion-deflection (E-D) and plug (aerospike) nozzles.

An *aerospike nozzle* is an annular nozzle that allows the combustion gas to expand from one surface—a centerbody spike—to ambient pressure. A *bell nozzle* is a nozzle with a circular opening for a throat and an axisymmetric contoured wall downstream of the throat that gives the nozzle a characteristic bell shape. The *expansion-deflection, or E-D, nozzle* is a nozzle that has an annular throat that discharges exhaust gas with a radial outward component. A *plug nozzle* is an annular nozzle that discharges exhaust gas with a radial inward component—a truncated aerospike.

The nozzle configuration best suited to a particular application depends on a variety of factors, including the altitude regime in which the nozzle will be used, the diversity of stages in which the same nozzle will be employed, and limitations imposed by available development time and funding. The propellant combination, chamber pressure, mixture ratio, and thrust level for a rocket engine generally are determined on the basis of factors other than the nozzle configuration. These parameters will have only a minor effect on the selection of a nozzle configuration but will have a significant effect on the selection of nozzle cooling method.

The throat of the nozzle is the region of transition from subsonic to supersonic

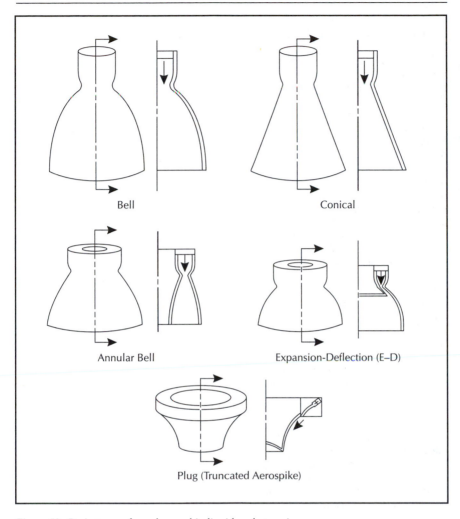

Figure 50. Basic types of nozzles used in liquid rocket engines.

flow. For typical rocket nozzles, local mass flux and, consequently, the rate of heat transfer to the wall are highest in this area of the nozzle. The shape of the nozzle wall in the vicinity of the throat dictates the distribution of exhaust-gas flow across the nozzle at the throat. The nozzle wall geometry immediately upstream of the throat determines the distribution of gas properties at the throat. The expansion geometry extends from the throat to the nozzle exit. The function of this part of the nozzle is to accelerate the exhaust gases to a high velocity in a short distance, while providing near-ideal performance. Both engine length and propellant specific impulse strongly influence the payload capability of a rocket vehicle. Therefore, aerospace engineers give considerable attention to the design of the nozzle expansion geometry in order to obtain the maximum performance from a nozzle length consistent with optimum vehicle payload.

The *area ratio* of a nozzle is defined as the ratio of the geometric flow area of the nozzle exit to the geometric flow area of the nozzle throat. This important nozzle parameter is also called the *expansion area ratio*. An *ideal nozzle* is a nozzle that provides theoretically perfect performance for the given area ratio when analyzed on the basis of one-dimensional point-source flow. The *aerodynamic throat area* of a nozzle is the effective flow area of the throat, which is less than the geometric flow area because the flow is not uniform. A *nozzle extension* is a nozzle structure that has been added to the main nozzle in order to increase the expansion area ratio or to provide a change in nozzle construction. Finally, *overexpansion* is a condition that occurs when a nozzle expands the exhaust gas to an ambient pressure that is higher than that for which the nozzle was originally designed. *See also* ROCKET.

nozzle extension Nozzle structure that is added to the main nozzle in order to increase the expansion area ratio or to provide a change in nozzle construction. *See also* NOZZLE.

nuclear detonation A NUCLEAR EXPLOSION resulting from FISSION or FUSION reactions in the nuclear materials contained in a NUCLEAR WEAPON.

nuclear detonation detection and warning system A system employed to provide surveillance coverage of critical friendly target areas and indicate location, height of burst, yield, and ground zero (actual or projected) of nuclear detonations.

nuclear-electric propulsion (NEP) system A propulsion system in which a nuclear reactor is used to produce the electricity needed to operate the electric propulsion engine(s). Unlike the solar-electric propulsion (SEP) system, the nuclear electric propulsion (NEP) system can operate anywhere in the solar system and even beyond, since the NEP system's performance is independent of its position relative to the Sun. Furthermore, it can provide shorter trip times and greater payload capacity than any of the advanced chemical propulsion technologies that might be used in the next two decades or

so for detailed exploration of the outer planets, especially Saturn, Uranus, Neptune, Pluto, and their respective moons. Closer to Earth, a NEP orbital transfer vehicle (OTV) can serve effectively in cislunar space, gently transporting large cargoes and structures from LOW EARTH ORBIT to geosynchronous orbit or to a logistics depot in lunar orbit that supports the needs of an expanding lunar settlement. Finally, advanced nuclear electric propulsion systems can become the main space-based transport vehicles for both human explorers and their equipment as permanent surface bases are established on Mars in the 21st century. *See also* ELECTRIC PROPULSION; SPACE NUCLEAR POWER; SPACE NUCLEAR PROPULSION.

nuclear energy Energy released as a result of interactions involving atomic nuclei; that is, reactions in which there is a rearrangement of the constituents (i.e., protons and neutrons) of the nucleus. Any nuclear process in which the total mass of the products is less than the mass before interaction is accompanied by the release of energy. The amount of energy release is related to the decrease in (nuclear) mass by the famous Einstein mass-energy equivalence formula:
$$E = mc^2$$
where E is the energy release, m is the *decrease* in mass, and c is the speed of light. The disappearance of 1 atomic mass unit (i.e., the mass of a proton or a neutron) results in the release of about 931 million electron volts (MeV) of energy. *See also* FISSION (NUCLEAR); FUSION (NUCLEAR).

nuclear explosion A catastrophic, destructive event in which the primary cause of damage to human society and/or the environment is a direct result of the rapid (explosive) release of large amounts of ENERGY from NUCLEAR REACTIONS (FISSION and/or FUSION).

nuclear fission The splitting of a heavy atomic nucleus into two smaller masses accompanied by the release of neutrons, gamma rays, and about 200 million electron volts (MeV) of energy. *See also* FISSION (NUCLEAR).

nuclear fusion The nuclear reaction (process) whereby several smaller nuclei combine to form a more massive nucleus, accompanied by the release of a large amount of energy (which represents the difference in nuclear masses between the sum of the smaller particles and the resultant combined, or "fused," nucleus). Nuclear fusion is the source of energy in stars, including the Sun. For example, the nuclear fusion processes now taking place within the Sun convert approximately 5 million metric tons of mass into energy every second. *See also* FUSION (NUCLEAR).

nuclear radiation Particles (e.g., alpha particles, beta particles, and neutrons) and very energetic electromagnetic radiation (i.e., gamma ray photons) emitted from atomic nuclei during various nuclear reaction processes, including radioactive decay, fission, and fusion. Nuclear radiations are ionizing radiations.

nuclear reaction A reaction involving a change in an atomic nucleus, such as neutron capture, radioactive decay, fission, or fusion. A nuclear reaction is distinct from and far more energetic than a chemical reaction, which is limited to changes in the electron structure surrounding the nucleus. *See also* FISSION (NUCLEAR); FUSION (NUCLEAR).

nuclear reactor A device in which a fission chain reaction can be initiated, maintained, and controlled. Its essential component is the core, which contains the fissile fuel. Depending on the type of reactor, its purpose and design, the reactor also can contain a moderator, a reflector, shielding, coolant, and control mechanisms. *See also* FISSION (NUCLEAR); SPACE NUCLEAR POWER; SPACE NUCLEAR PROPULSION.

nuclear rocket In general, a rocket vehicle that derives its propulsive thrust from nuclear energy sources, primarily fission. There are two general classes of nuclear rockets: the *nuclear-thermal rocket* and the *nuclear-electric propulsion (NEP) system*. The nuclear thermal rocket uses a nuclear reactor to heat the propellant (generally hydrogen) to extremely high temperatures before it is expelled through a rocket nozzle. The nuclear electric propulsion system uses the nuclear reactor to produce electric power, which in turn is used to operate an electric propulsion system. (See Figure 51.) *See also* NUCLEAR-ELECTRIC PROPULSION (NEP) SYSTEM; SPACE NUCLEAR PROPULSION.

Nuclear Rocket Development Station (NRDS) Major test facilities for the U.S. nuclear rocket program (from 1959 to 1973) were located at the Nuclear Rocket Development Station (NRDS) at the Nevada Test Site (NTS) in southern Nevada. Both nuclear reactors and complete nuclear rocket engine assemblies were tested (in place) at the NRDS, which had three major test areas: Test cells A and C and Engine Test Stand Number One (ETS-1). The reactor test facilities were designed to test a reactor in an upward-firing position, while the engine test facility tested nuclear rocket engines in a downward-firing mode. During a nuclear rocket test at the NRDS, the surrounding desert basin (called Jackass Flats after some of the indigenous wildlife) literally became an inferno as the very hot hydrogen gas spontaneously ignited upon contact with the air and burned (with atmospheric oxygen) to form water. *See also* NUCLEAR ROCKET; SPACE NUCLEAR PROPULSION.

nuclear weapon A device in which the explosion results from the energy released by reactions involving atomic nuclei, either fission or fusion, or both. The basic fission weapon can either be a gun-assembly type weapon or an implosion-type weapon, while a thermonuclear weapon involves the use of the energy of a fission weapon to ignite the desired thermonuclear reactions.

nuclear yields The energy released in the detonation of a nuclear weapon, measured in terms of the kilotons or megatons

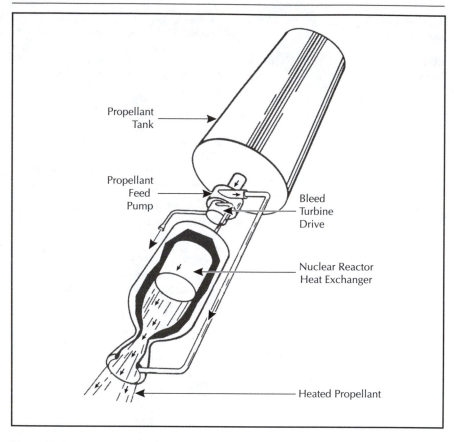

Figure 51. Basic components of a nuclear thermal rocket. The propellant tank carries liquid hydrogen (at cryogenic temperatures). This single propellant is heated to extremely high temperatures as it passes through the nuclear reactor, and then very hot gaseous hydrogen is expelled out the rocket's nozzle to produce thrust.

of TNT (trinitrotoluene) required to produce the same energy release. Yields are categorized as: *very low,* less than 1 kiloton; *low,* 1–10 kilotons; *medium,* over 10–50 kilotons; *high,* 50–500 kilotons; and *very high,* over 500 kilotons.

nucleon The common name for a major constituent particle of the atomic nucleus. It is applied to protons and neutrons. The number of nucleons in a nucleus represents the atomic mass number (A) of that nucleus. For example, the isotope uranium-235 has 235 nucleons in its nucleus (92 protons and 143 neutrons). This iso-

tope also is said to have an atomic mass number of 235 (i.e., A = 235).

nucleosynthesis The production of heavier chemical elements from the fusion or joining of lighter chemical elements (e.g., hydrogen or helium nuclei) in thermonuclear reactions in stellar interiors. *See also* FUSION (NUCLEAR); STAR(S).

nucleus (plural: nuclei) 1. (atomic) The small, positively charged central region of an ATOM that contains essentially all of its MASS. All nuclei contain both PROTONS and NEUTRONS except the nucleus of ordinary (light) HYDROGEN, which consists of a single proton. The number of protons in the nucleus determines what ELEMENT an atom is, while the total number of protons and neutrons determines a particular atom's nuclear properties (such as RADIOACTIVITY, the ability to FISSION, etc.). ISOTOPES of a given chemical element have the same number of protons in their nuclei but different numbers of neutrons, resulting in different ATOMIC MASSes—as, for example, uranium-235 and uranium-238. 2. (cometary) The small (few KILOMETER diameter), permanent, SOLID ice-rock central body of a COMET. 3. (galactic) The central region of a GALAXY a few LIGHT-YEARS in diameter characterized by a dense cluster of STARS or possibly even the hidden presence of a massive BLACK HOLE. *See also* ACTIVE GALACTIC NUCLEUS.

nuclide A general term applicable to all atomic (isotopic) forms of all the ELEMENTs; nuclides are distinguished by their ATOMIC NUMBER, relative mass number (ATOMIC MASS), and ENERGY state. *Compare with* ISOTOPE.

objective The main LIGHT-gathering LENS or MIRROR of a TELESCOPE. It is sometimes called the *primary lens* or *primary mirror* of a telescope.

Oberth, Hermann J. (1894–1989) Transylvanian-born German ROCKET scientist who, like KONSTANTIN TSIOLKOVSKY and ROBERT GODDARD, helped establish the field of ASTRONAUTICS and vigorously promoted the concept of space travel throughout his life. His inspirational 1923 publication, *The Rocket Into Planetary Space,* provided a comprehensive discussion of all the major aspects of space travel, and his 1929 award-winning book, *Roads to Space Travel,* popularized the concept of space travel for technical and nontechnical readers alike.

oblate spheroid A sphere flattened such that its polar diameter is smaller than its equatorial diameter.

obscurant A substance, such as chaff or smoke, used to conceal an object from observation by a radio-frequency (i.e., radar) or optical sensor. Chaff helps conceal an object from radar detection by creating numerous false images and returns, while smoke simply helps hide an object from direct observation by an optical sensor.

observable universe The portions of the UNIVERSE that can be detected and studied by the LIGHT they emit. *Compare with* DARK MATTER.

observatory The place (or facility) from which astronomical observations are made. For example, the KECK OBSERVATORY is a ground-based observatory, while the *HUBBLE SPACE TELESCOPE* is a space-based (or EARTH-orbiting) observatory.

observed In astronomy and navigation, pertaining to a value that has been measured, in contrast to one that has been computed.

occult, occulting The disappearance of one celestial object behind another. For example, a solar eclipse is an *occulting* of the Sun by the Moon; that is, the Moon comes between Earth and the Sun, temporarily blocking the Sun's light and darkening regions of Earth.

ocean remote sensing Covering about 70 percent of Earth's surface, the oceans are central to the continued existence of life on our planet. The oceans are where life first appeared on Earth. Today the largest creatures on the planet (whales) and the smallest creatures (bacteria and viruses) live in the global oceans. Important processes between the atmosphere and the oceans are linked. Oceans store energy. The wind roughens the ocean surface, and the ocean surface, in turn, extracts more energy from the wind and puts it into wave motion, currents, and mixing. When ocean currents change, they cause changes in global weather patterns and can cause droughts, floods, and storms.

However, our knowledge of the global oceans is still limited. Ships, coastlines, and islands provide places from which we can observe, sample, and study small portions of the oceans. But from Earth's surface, we can look at only a very small part of the global ocean. Satellites orbiting Earth can survey an entire ocean in less than an hour. Sensors on these satellites can "look" at clouds to study the

Important Data Provided by Ocean Remote Sensing from Space			
Sensor	*Data*	*Science Question*	*Application*
Ocean-color sensor	Ocean color	Phytoplankton concentration, ocean currents; ocean surface temperature; pollution and sedimentation	Fishing productivity, ship routing; monitoring coastal pollution
Scatterometer	Wind speed; wind direction	Wave structure; currents; wind patterns	Ocean waves; ship routing; currents; ship, platform safety
Altimeter	Altitude of ocean surface; wave height; wind speed	El Niño onset and structure	Wave and current forecasting
Microwave imager	Surface wind speed; ice edge; precipitation	Thickness, extent of ice cover, internal stress of ice; ice growth and ablation rates	Navigation information; shiprouting; wave and surf forecasting
Microwave	Sea-surface temperature	Ocean-air interactions	Weather forecasting

Source: U.S. Congress, Office of Technology Assessment, 1994.

weather or at the sea surface to measure its temperature, wave heights, and the direction of the waves. Some satellite sensors use microwave (radar) wavelengths to look through the "clouds" at the sea surface. One other important characteristic that we can see from space is the color of the ocean. Changes in color of ocean water over time or across a distance on the surface provide additional valuable information.

When we look at the ocean from space, we see many shades of blue. Using instruments that are more sensitive than the human eye, we can measure the subtle array of colors displayed by the global ocean. To skilled remote-sensing analysts and oceanographers, different ocean colors can reveal the presence and concentration of phytoplankton, sediments, and dissolved organic chemicals. Phytoplankton are small, single-celled ocean plants, smaller than the size of a pinhead. These tiny plants contain the chemical chlorophyll. Plants use chlorophyll to convert sunlight into food using a process called *photosynthesis*. Because different types of phytoplankton have different concentrations of chlorophyll, they appear as different colors to sensitive satellite instruments. Therefore, looking at the color of an area of the ocean allows scientists to estimate the amount and general type of phytoplankton in that area, such information tells them about the health and chemistry of that portion of the ocean. Comparing images taken at different periods reveals any changes and trends in the health of the ocean that are occurring over time.

Why are phytoplankton so important? Besides acting as the first link in a typical oceanic food chain, phytoplankton are a critical part of ocean chemistry. Carbon dioxide in the atmosphere is in balance with carbon dioxide (CO_2) in the ocean. During photosynthesis, phytoplankton remove carbon dioxide from sea water and release oxygen as a by-product. This allows the oceans to absorb additional carbon dioxide from the atmosphere. If fewer phytoplankton existed, atmospheric carbon dioxide would increase.

Phytoplankton also affect carbon dioxide levels when they die. Phytoplankton, like plants on land, are composed of substances that contain carbon. Dead phytoplankton can sink to the ocean floor. The carbon in the phytoplankton is soon covered by other material sinking to the ocean bottom. In this way, the oceans serve as a reservoir (i.e., a place to dispose of global carbon), which otherwise would accumulate in the atmosphere as carbon dioxide.

Observation of the ocean from space with a variety of special instruments provides oceanographers and environmental scientists with very important data. The table summarizes some of the data that these satellite sensors can provide. *See also* AQUA; REMOTE SENSING.

oersted (symbol: Oe) The unit of magnetic field strength in the centimeter-gram-second (c.g.s.) system of units.

1 oersted = 79.58 amperes/meter

This unit is named in honor of Hans Christian Oersted (1777–1851), a Danish physicist who was first to demonstrate the relationship between electricity and magnetism.

ogive The tapered or curved front of a missile.

ohm (symbol: Ω) The SI UNIT of electrical resistance. It is defined as the resistance (R) between two points on a conductor produced by a current flow (I) of one ampere when there is a constant voltage difference (potential) (V) of one volt between these points. From Ohm's law, the resistance in a conductor is related to the voltage and the current by the equation: $R = V/I$, so that, 1 ohm (W) of resistance = 1 volt per ampere. The unit and physical law are named in honor of the German physicist George Simon Ohm (1787–1854).

Ohm's law Statement that the current (I) in a linear constant current electric circuit is inversely proportional to the resistance (R) of the circuit and directly proportional to the voltage difference (V) (i.e., the electromotive force) in the circuit.

Expressed as an equation, $V = I \cdot R$. First postulated by the German physicist George S. Ohm (1787–1854). *See also* OHM.

Olympus Mons A huge mountain on MARS about 650 KILOMETERs (404 mi) wide and rising 26 kilometers (16 mi) above the surrounding plains—the largest known single VOLCANO in the SOLAR SYSTEM.

omni- A prefix meaning "all," as in *omnidirectional.*

omnidirectional antenna An ANTENNA that radiates or receives RADIO FREQUENCY (RF) SIGNALS with about the same efficiency in or from all directions. *Compare with* DIRECTIONAL ANTENNA.

one-g The downward acceleration of gravity at Earth's surface; at sea level, "one-g" corresponds to an acceleration of 9.8 meters per second per second (m/s²), or about 32.1 per second per second (ft/s²). *See also* G.

one-way communications Communications mode consisting only of downlink received from a spacecraft.

one-way light time (OWLT) Elapsed time (in units of light-seconds or light-minutes) for a radio signal to travel (one way) between Earth and a spacecraft or solar system body.

Oort Cloud A large number, or "cloud," of comets theorized to orbit the Sun at a distance of between 50,000 and 80,000 astronomical units (AU) (i.e., out to the limits of the Sun's gravitational attraction). First proposed in 1950 by the Dutch astronomer Jan Hendrik Oort (1900–92).

open loop A control system operating without feedback or perhaps with only partial feedback. An electrical or mechanical system in which the response of the output to an input is preset; there is no feedback of the output for comparison and corrective adjustment.

open system In thermodynamics, a system as defined by a control volume in space across whose boundaries energy and mass can flow. *Compare with* CLOSED SYSTEM.

open universe The open or unbounded universe model in cosmology assumes that there is not enough matter in the universe to halt completely (by gravitational attraction) the currently observed expansion of the galaxies. Therefore, the galaxies will continue to move away from each other and the expansion of the universe (which started with the "Big Bang") will continue forever. *See also* "BIG BANG" THEORY.

operating life The maximum operating time (or number of cycles) that an item can accrue before replacement or refurbishment without risk of degradation of performance beyond acceptable limits.

operating pressure Nominal pressure to which the fluid-system components are subjected under steady-state conditions in service operations.

operational missile A missile that has been accepted by the armed services for tactical and/or strategic application.

operations planning Performing those tasks that must be done to ensure that the launch vehicle systems and ground-based flight control operations support flight objectives. Consumables analyses and the preparation of flight rules are a part of operations planning.

opposition In astronomy, the alignment of a superior planet (e.g., Mars, Jupiter, etc.) with the Sun so that they appear to an observer on Earth to be in opposite parts of the sky; that is, the planetary object has a celestial longitude differing from that of the Sun by 180 degrees. More simply put, these bodies lie in a straight line with Earth in the middle. *Compare with* CONJUNCTION.

optoelectronic device A device that combines optical (light) and electronic technologies, such as a fiber optics communications system.

orbit 1. In physics, the region occupied by an electron as it moves around the nucleus of an atom. 2. In astronomy and space science, the path followed by a satellite around an astronomical body, such as Earth or the Moon. When a body is moving around a primary body, such as Earth, under the influence of gravitational force alone, the path it takes is called its orbit. If a spacecraft is traveling along a closed path with respect to the primary body, its orbit will be a circle or an ellipse. Perfectly circular orbits are not achieved in practice. However, the ellipse approaches a circle when the eccentricity becomes small (i.e., approaches zero). The planets, for example, have the Sun as their primary body and follow nearly circular orbits. When a satellite makes a full trip around its primary body, it is said to complete a *revolution,* and the time required is called its *period* or the *period of revolution. See also* ORBITS OF OBJECTS IN SPACE.

orbital Taking place in orbit, as *orbital refueling;* or pertaining to an orbit, as *orbital plane.*

orbital elements A set of six parameters used to specify a Keplerian orbit and the position of a satellite in such an orbit at a particular time. These six parameters are: semimajor axis (a), which gives the size of the orbit; eccentricity (e), which gives the shape of the orbit; inclination angle (i), which gives the angle of the orbit plane with respect to the central (primary) body's equator; right ascension of the ascending node (Ω), which gives the rotation of the orbit plane from the reference axis; argument of perigee (ω) which is the angle from the line of the ascending node to perigee point, measured along the orbit in the direction of the satellite's motion; and true anomaly (θ), which gives the location of the satellite on the orbit. *See also* KEPLERIAN ELEMENTS.

orbital injection The process of providing a space vehicle with sufficient velocity to establish an orbit.

orbital maneuvering system (OMS) Two orbital maneuvering engines, located in external pods on each side of the aft fuselage of the space shuttle orbiter structure, which provide thrust for orbit insertion, orbit change, orbit transfer, rendezvous and docking, and deorbit. Each pod contains a high-pressure helium storage bottle, the tank, pressurization regulators and controls, a fuel tank, an oxidizer tank, and a pressure-fed regeneratively cooled rocket engine. Each OMS engine develops a vacuum thrust of 27 kilonewtons (6,000 pounds-force (lbf) and uses a hypergolic propellant combination of nitrogen tetroxide and monomethyl hydrazine. These propellants are burned at a nominal oxidizer-to-fuel mixture ratio of 1.65 and a chamber pressure of 860 kilonewtons per square meter (kN/m^2) [125 pounds-force per square inch (psi)]. The system can provide a velocity change of 305 meters per second (1,000 feet per second) when the orbiter carries a payload of 29,500 kilograms (65,000 pounds-mass [lbm]). *See also* SPACE TRANSPORTATION SYSTEM (STS).

orbital period The interval between successive passages of a satellite or spacecraft through the same point in its orbit. Often called *period. See also* ORBITS OF OBJECTS IN SPACE.

orbital plane The imaginary plane that contains the orbit of a satellite and passes through the center of the primary (i.e., celestial body being orbited). The angle of inclination (θ) is defined as the angle between Earth's equatorial plane and the orbital plane of the satellite. *See also* ORBITS OF OBJECTS IN SPACE.

orbital transfer vehicle (OTV) A propulsion system used to transfer a payload from one orbital location to another—as, for example, from LOW EARTH ORBIT (LEO) to geostationary Earth orbit (GEO). Orbital-transfer vehicles can be expendable or reusable; many involve chemical, nuclear, or electric propulsion systems. An expendable orbital transfer vehicle frequently is referred to as an upper-stage unit, while a reusable OTV is sometimes called a space tug. OTVs can

be designed to move people and cargo between different destinations in cislunar space. *See also* UPPER STAGE.

orbital velocity The average velocity at which a satellite, spacecraft, or other orbiting body travels around its primary.

orbit determination The process of describing the past, present, or predicted (i.e., future) position of a satellite or other orbiting body in terms of its orbital parameters.

orbiter (spacecraft) A SPACECRAFT especially designed to travel through INTERPLANETARY space, achieve a stable ORBIT around the TARGET PLANET (or other CELESTIAL BODY), and conduct a program of detailed scientific investigation.

Orbiter (space shuttle) The winged AEROSPACE VEHICLE portion of NASA's SPACE SHUTTLE. It carries ASTRONAUTs and PAYLOAD into ORBIT and returns from OUTER SPACE by gliding and landing like an airplane. The operational Orbiter Vehicle (OV) fleet includes *Discovery* (OV-103), *Atlantis* (OV-104), and *Endeavour* (OV-105). The *Challenger* (OV-99) was lost in a LAUNCH accident on January 28, 1986, that claimed the lives of all seven crewmembers, and the *Columbia* (OV-102) was destroyed during a reentry accident on February 1, 2003, that claimed the lives of all seven crewmembers.

orbit inclination The angle between an orbital plane and Earth's equatorial plane. *See also* ORBITS OF OBJECTS IN SPACE.

Orbiting Astronomical Observatory (OAO) A series of large astronomical observatories developed by NASA in the late 1960s to significantly broaden our understanding of the universe. The first successful large observatory placed in Earth orbit was the *Orbiting Astronomical Observatory 2 (OAO-2),* nicknamed *Stargazer,* which was launched on December 7, 1968. In its first 30 days of operation, *OAO-2* collected more than 20 times the celestial ultraviolet (UV) data than had been acquired in the previous 15 years of sounding rocket launches. *Stargazer* also observed Nova Serpentis for 60 days after its outburst in 1970. These observations confirmed that mass loss by the nova was consistent with theory. NASA's Orbiting Astronomical Observatory 3 (OAO-3), named *Copernicus* in honor of the famous Polish astronomer, was launched successfully on August 21, 1972. This satellite provided much new data on stellar temperatures, chemical compositions, and other properties. It also gathered data on the BLACK HOLE candidate Cygnus X-1, so named because it was the first X-ray source discovered in the constellation Cygnus.

Orbiting Geophysical Observatory (OGO)

A series of six NASA scientific spacecraft placed in Earth-orbit between 1961 and 1965. At the beginning of the U.S. civilian space program, data from the Orbiting Geophysical Observatory (OGO) spacecraft made significant contributions to an initial understanding of the near-Earth space environment and Sun-Earth interrelationships. For example, they provided the first evidence of a region of low-energy electrons enveloping the high-energy VAN ALLEN RADIATION BELT region.

orbiting laser power station

A large orbiting system that collects incoming solar radiation and concentrates it to operate a direct solar-pumped laser. The orbiting facility then directs the beam from this solar-pumped laser to a suitable collection facility (e.g., on the lunar surface), where photovoltaic converters tuned to the laser wavelength produce electric power with an efficiency of perhaps 50%. In one scheme for the proposed orbiting laser power station, a nearly parabolic solar collector, with a radius of about 300 meters, captures sunlight and directs it, in a line focus, onto a 10-meter-long laser, with an average concentration of several thousand solar constants. (A solar constant is the total amount of the Sun's radiant energy that crosses normal to a unit area at the top of Earth's atmosphere. At one astronomical unit from the Sun a solar constant is about 1371 ± 5 watts per

meter squared [(W/m^2].) An organic iodide gas lasant flows through the laser, circulated by a turbine-compressor combination. (A lasant is a chemical substance that supports stimulated light emission in a laser.) The hot lasant then is cooled and purified at the radiator. New lasant is added from the supply tanks to make up for the small amount of lasant lost in each pass through the laser. Power from the laser then is spread and focused by a combination of transmission mirrors to provide a one-meter-diameter spot at distances up to 10,000 kilometers (6,200 miles) away or more.

Orbiting Quarantine Facility (OQF)

A proposed Earth-orbiting laboratory in which soil and rock samples from other worlds—for example, Martian soil and rock specimens—could first be tested and examined for potentially harmful alien microorganisms, before the specimens are allowed to enter Earth's biosphere. A space-based quarantine facility provides several distinct advantages: (1) it eliminates the possibility of a sample-return spacecraft's crashing and accidentally releasing its potentially deadly cargo of alien microorganisms; (2) it guarantees that any alien organisms which might escape from the orbiting laboratory's confinement facilities cannot immediately enter Earth's biosphere; and (3) it ensures that all quarantine workers remain in total isolation during protocol testing of the alien soil and rock samples. Three hypothetical results of such protocol testing are: (1) no replicating alien organisms are discovered; (2) replicating alien organisms are discovered, but they are also found not to be a threat to terrestrial life-forms; or (3) hazardous replicating alien life-forms are discovered. If potentially harmful replicating alien organisms were discovered during these protocol tests, then orbiting quarantine facility workers would: render the sample harmless (e.g., through heat- and chemical-sterilization procedures); retain it under very carefully controlled conditions in the orbiting complex and perform more detailed studies on the alien life-forms; or properly dispose of the sample before the alien life-forms could enter Earth's biosphere and infect terrestrial life-forms. *See also* EXTRATERRESTRIAL CONTAMINATION.

Orbiting Solar Observatory (OSO) A series of eight NASA Earth-orbiting scientific satellites that were used to study the Sun from space, with emphasis on its electromagnetic radiation emissions in the range from ultraviolet to gamma rays. These scientific observatories were launched between 1962 and 1975. They acquired an enormous quantity of important solar data during an 11-year solar cycle, when solar activity went from low to high and then back to low again. *See also* SUN.

orbits of objects in space We must know about the science and mechanics of orbits to launch, control, and track spacecraft and to predict the motion of objects in space. An *orbit* is the path in space along which an object moves around a primary body. Common examples of orbits include Earth's path around its celestial primary (the Sun) and the Moon's path around Earth (its primary body). A single orbit is a complete path around a primary as viewed from space. It differs from a revolution. A single *revolution* is accomplished whenever an orbiting object passes over the primary's longitude or latitude from which it started. For example, the space shuttle *Discovery* completed a revolution whenever it passed over approximately 80 degrees west longitude on Earth. However, while *Discovery* was orbiting from west to east around the globe, Earth itself was also rotating from west to east. Consequently, *Discovery* period of time for one revolution actually was longer than its orbital period. (See Figure 52.) If, on the other hand, *Discovery* were orbiting from east to west (not a practical flight path from a propulsion-economy stand-point), then because of Earth's west-to-east spin, its period of revolution would be shorter than its orbital period. An east-to-west orbit is called a *retrograde orbit* around Earth; a west-to-east orbit is called a *posigrade orbit.* If *Discovery* were traveling in a north-south orbit, or *polar orbit,* it would complete a period of revolution whenever it passed over the latitude from which it started. Its orbital period would be about the same as the revolution period, but not identical, because Earth actually wobbles slightly north and south.

Other terms are used to describe orbital motion. The *apoapsis* is the farthest distance in an orbit from the primary; the *periapsis,* the shortest. For

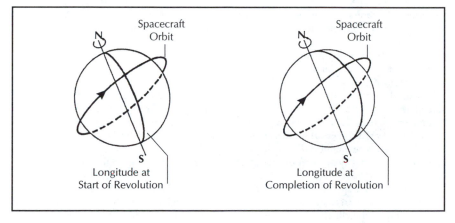

Figure 52. An illustration of a spacecraft's west-to-east orbit around Earth and how Earth's west-to-east rotation moves longitude ahead. As shown here, the period of one revolution can be longer than the orbital period.

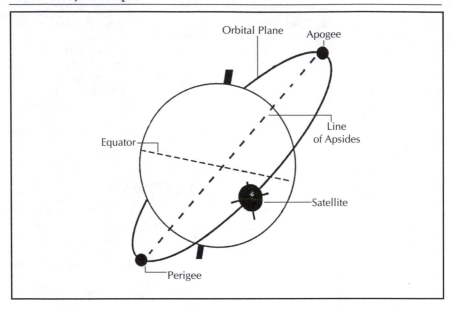

Figure 53. The terms *apogee* and *perigee* described in terms of a satellites's orbit around Earth.

orbits around planet Earth, the comparable terms are *apogee* and *perigee.* The *line of apsides* is the line connecting the two points of an orbit that are nearest and farthest from the center of attraction, as the perigee and apogee of a satellite in orbit around the Earth. (See Figure 53.)

For objects orbiting the Sun, *aphelion* describes the point on an orbit farthest from the Sun; *perihelion,* the point nearest to the Sun.

Another term we frequently encounter is the *orbital plane.* An Earth satellite's orbital plane can be visualized by thinking of its orbit as the outer edge of a giant, flat plate that cuts Earth in half. This imaginary plate is called the orbital plane.

Inclination is another orbital parameter. This term refers to the number of degrees the orbit is inclined away from the equator. The inclination also indicates how far north and south a spacecraft will travel in its orbit around Earth. If, for example, a spacecraft has an inclination of 56 degrees, it will travel around Earth as far north as 56 degrees north latitude and as far south as 56 degrees south latitude. Because of

Earth's rotation, it will not, however, pass over the same areas of Earth on each orbit. A spacecraft in a polar orbit has an inclination of about 90 degrees. As such, this spacecraft orbits Earth traveling alternately in north and south directions. A polar-orbiting satellite eventually passes over the entire Earth because Earth is rotating from west to east beneath it. The *Terra* spacecraft is an example of a spacecraft whose cameras and multispectral sensors observe the entire Earth from a nearly polar orbit, providing valuable information about the terrestrial environment and resource base.

A satellite in an equatorial orbit around Earth has zero inclination. The Intelsat communications satellites are examples of satellites in equatorial orbits. By their placement into near-circular equatorial orbits at just the right distance above Earth, these spacecraft can be made essentially to "stand still" over a point on Earth's equator. Such satellites are called *geostationary.* They are in *synchronous orbits,* meaning they take as long to complete an orbit around Earth as it takes for Earth to complete one rotation about its axis (i.e., approximately 24 hours). A satellite at the same "synchronous" altitude but in an inclined orbit also may be called synchronous. While this particular spacecraft would not move much east and west, it would move north and south over Earth to the latitudes indicated by its inclination. The terrestrial ground track of such a spacecraft resembles an elongated figure eight, with the crossover point on the equator.

All orbits are elliptical, in accordance with Kepler's first law of planetary motion (described shortly). However, a spacecraft generally is considered to be in a circular orbit if it is in an orbit that is nearly circular. A spacecraft is taken to be in an elliptical orbit when its apogee and perigee differ substantially.

Two sets of scientific laws govern the motions of both celestial objects and human-made spacecraft. One is Newton's law of gravitation; the other, Kepler's laws of planetary motion.

The brilliant English scientist and mathematician Sir Isaac Newton (1642–1727) observed the following physical principles:

1. All bodies attract each other with what we call gravitational attraction. This applies to the largest celestial objects and to the smallest particles of matter.
2. The strength of one object's gravitational pull upon another is a function of its mass—that is, the amount of matter present.
3. The closer two bodies are to each other, the greater their mutual attraction.

These observations can be stated mathematically as:

$$F = \frac{G\, m_1\, m_2}{r^2}$$

where F is the gravitational force acting along the line joining the two bodies (N); m1, m2 are the masses (in kilograms) of body one and body two, respectively; r is the distance between the two bodies (m); and G is the universal gravitational constant (6.6732 ¥ 10–11 newton-meter2/kilogram2).

Specifically, Newton's law of gravitation states that two bodies attract each other in proportion to the product of their masses

and inversely as the square of the distance between them. This physical principle is very important in launching spacecraft and guiding them to their operational locations in space and frequently is used by astronomers to estimate the masses of celestial objects. For example, Newton's law of gravitation tells us that for a spacecraft to stay in orbit, its velocity (and therefore its kinetic energy) must balance the gravitational attraction of the primary object being orbited. Consequently, a satellite needs more velocity in low than in high orbit. For example, a spacecraft with an orbital altitude of 250 kilometers (km) (150 miles [mi]) will have an orbital speed of about 28,000 (km) (17,500 mi) per hour.

Our Moon, on the other hand, which is about 442,170 km (238,857 mi) from Earth, has an orbital velocity of approximately 3,660 km (2,287 mi) per hour. Of course, to boost a payload from the surface of Earth to a high-altitude (versus low-attitude) orbit requires the expenditure of more energy, since we are in effect lifting the object farther out of Earth's "gravity well."

Any spacecraft launched into orbit moves in accordance with the same laws of motion that govern the motions of the planets around our Sun and the motion of the Moon around Earth. The three laws that describe these planetary motions, first formulated by the German astronomer Johannes Kepler (1571–1630), may be stated as follows:

1. Each planet revolves around the Sun in an orbit that is an ellipse, with the Sun as it focus, or primary body.
2. The radius vector—such as the line from the center of the Sun to the center of a planet, from the center of Earth to the center of the Moon, or from the center of Earth to the center (of gravity) of an orbiting spacecraft—sweeps out equal areas in equal periods of time.
3. The square of a planet's orbital period is equal to the cube of its mean distance from the Sun. We can generalize this last statement and extend it to spacecraft in orbit about Earth by saying that a spacecraft's orbital period increases with its mean distance from the planet.

In formulating his first law of planetary motion, Kepler recognized that purely circular orbits did not really exist—rather, only elliptical ones were found in nature, being determined by gravitational perturbations (disturbances) and other factors. Gravitational attractions, according to Newton's law of gravitation, extend to infinity, although these forces weaken with distance and eventually become impossible to detect. However, spacecraft orbiting Earth, while influenced primarily by the gravitational attraction of Earth (and anomalies in Earth's gravitational field), also are influenced also by the Moon and the Sun and possibly other celestial objects, such as the planet Jupiter.

Kepler's third law of planetary motion states that the greater a body's mean orbital altitude, the longer it will take for it to go around its primary. Let's take this principle and apply it to a rendezvous maneuver between a space shuttle orbiter and a satellite in LOW EARTH ORBIT (LEO). To catch up with and retrieve an uncrewed spacecraft in the same orbit, the space shuttle first must be decelerated. This action causes the orbiter vehicle to descend to a lower orbit. In this lower orbit, the shuttle's velocity would increase. When properly positioned in the vicinity of the target satellite, the orbiter then would be accelerated, raising its orbit and matching orbital velocities for the rendezvous maneuver with the target spacecraft.

Another very interesting and useful orbital phenomenon is the Earth satellite that appears to "stand still" in space with respect to a point on Earth's equator. Such satellites were first envisioned by the English scientist and writer Arthur C. Clarke (b. 1917), in a 1945 essay in *Wireless World*. Clarke described a system in which satellites carrying telephone and television would circle Earth at an orbital altitude of approximately 35,580 km (22,240 mi) above the equator. Such spacecraft move around Earth at the same rate that the Earth rotates on its own axis. Therefore, they neither rise nor set in the sky like the planets and the Moon but rather always appear to be at the same longitude, synchronized with Earth's motion. At the equator, Earth rotates about 1,600 km (1,000 mi) per hour. Satellites placed in this type of orbit are called *geostationary* or *geosynchronous* spacecraft.

It is interesting to note here that the spectacular Voyager missions to Jupiter, Saturn, and beyond used a "gravity-assist" technique to help speed them up and shorten their travel time. How can a spacecraft can be speeded up while traveling past a planet? A spacecraft increases in speed as it approaches a planet (due to gravitational attraction), but the gravity of the planet also should slow it down as it begins to move away again. So where does this increase in speed really come from?

There are three basic possibilities for a spacecraft trajectory when it encounters a planet. (See Figure 54.) The first possible trajectory involves a direct hit or hard landing. This is an *impact trajectory*. (See trajectory a.) The second type of trajectory is an *orbital-capture trajectory*. The spacecraft is simply "captured" by the gravitational field of the planet and enters orbit around it. (See trajectories b and c.) Depending on its precise speed and altitude (and other parameters), the spacecraft can enter this captured orbit from either the leading or trailing edge of the planet. In the third type of trajectory, a *flyby trajectory*, the spacecraft remains far enough away from the planet to avoid capture but passes close enough to be strongly affected by its gravity. In this case, the speed of the spacecraft will be increased if it approaches from the trailing side of the planet (see trajectory d) and diminished if it approaches from the leading side (see trajectory e). In addition to changes in speed, the direction of the spacecraft's motion also changes.

Thus the increase in speed of the spacecraft actually comes from a decrease in speed of the planet itself. In effect, the spacecraft is being "pulled" by the planet. Of course, this has been a greatly simplified discussion of complex encounter phenomena. A full account of spacecraft trajectories must consider the speed and actual trajectory of the spacecraft and planet, how close the spacecraft will come to the planet, and the size

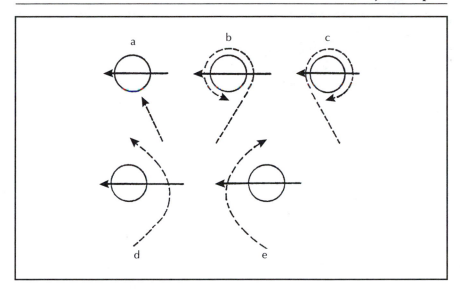

Figure 54. Possible trajectories of a spacecraft encountering a planet.

(mass) and speed of the planet in order to make even a simple calculation.

Perhaps an even better understanding of gravity assist can be obtained if we use vectors in a more mathematical explanation. The way in which speed is added to the flyby spacecraft during close encounters with the planet Jupiter is shown in Figure 55. During the time that spacecraft, such as *Voyager 1* and *2*, were near Jupiter, the heliocentric (Sun-centered) path they followed in their motion with respect to Jupiter closely approximated a hyperbola.

The heliocentric velocity of the spacecraft is the vector sum of the orbital velocity of Jupiter (V_j) and the velocity of the spacecraft with respect to Jupiter (i.e., tangent to its trajectory—the hyperbola). The spacecraft moves toward Jupiter along an asymptote, approaching from the approximate direction of the Sun and with asymptotic velocity (V_a). The heliocentric arrival velocity (V_1) is then computed by vector addition:

$$V1 = Vj + Va$$

The spacecraft departs Jupiter in a new direction, determined by the amount of bending that is caused by the effects of the gravitational attraction of Jupiter's mass on the mass of the spacecraft. The asymptotic departure speed (V_d) on the hyperbola is equal to the arrival speed. Thus, the length of V_a equals the length of V_d. For the heliocentric departure velocity, $V_2 = V_j + V_d$. This vector sum is also depicted in Figure 55.

During the relatively short period of time that the spacecraft is near Jupiter, the orbital velocity of Jupiter (V_j) changes very little, and we assume that V_j is equal to a constant.

The vector sums in Figure 55 illustrate that the deflection, or bending, of the spacecraft's trajectory caused by Jupiter's gravity results in an increase in the speed of the spacecraft along its hyperbolic path, as measured relative to the Sun. This increase in velocity reduces the total flight time necessary to reach Saturn and points beyond. This "indirect" type of deep-space mission to the outer planets saves two or three years of flight time when compared to "direct-trajectory" missions, which do not take advantage of gravity assist.

Of course, while the spacecraft gains speed during its Jovian encounter, Jupiter loses some of its speed. However, because

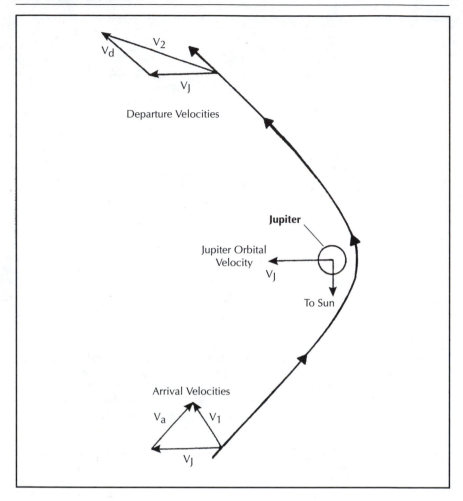

Figure 55. Velocity change during a Jupiter flyby.

of the extreme difference in their masses, the change in Jupiter's velocity is negligible.

orbit trim maneuvers Small changes in a spacecraft's orbit around a planet for the purpose of adjusting an instrument's field-of-view footprint, improving sensitivity of a gravity field survey, or preventing too much orbital decay. To make a change increasing the altitude of periapsis ("perigee" for an Earth-orbiting satellite), an orbit trim manever (OTM) would be designed to increase the spacecraft's velocity when it is at apoapsis (or "apogee" for an Earth-orbiting satellite). To decrease the apoapsis altitude, an OTM would be performed at periapsis, reducing the spacecraft's velocity. Slight changes in the orbital plane's orientation also can be made with orbit trim maneuvers. However, the magnitude of such changes is necessarily small due to the limited amount of "maneuver" propellant typically carried by a spacecraft.

order of magnitude A factor of 10; a value expressed to the nearest power of 10—for example, a cluster containing

9,450 STARS has approximately 10,000 stars in an order of magnitude estimate.

ordnance train A network of small explosive charges.

O-ring Sealing ring with a circular cross section, which may be either hollow or solid; generally made of elastomer or plastic but also can be metal.

Orion Nebula A bright NEBULA about 1,500 LIGHT-YEARS away in the CONSTELLATION Orion.

orthogonal At right angles; pertaining to or composed of right angles.

oscillating universe A CLOSED UNIVERSE in which gravitational collapse is followed by a new wave of expansion.

oscillation 1. Fluctuation or vibration on each side of a mean value or position 2. The variation, usually with time, of the magnitude of a quantity with respect to a specific reference when the magnitude is alternately greater and smaller than the reference.

oscillatory combustion Unstable combustion in a rocket engine or motor, characterized by pressure oscillations.

outer planets The planets in the solar system with orbits greater than the orbit of Mars. These are Jupiter, Saturn, Uranus, Neptune, and Pluto. Of these outer planets, all except Pluto also are called the *giant planets.*

outer space A general term referring to any region beyond Earth's atmospheric envelope. By informal international agreement, outer space usually is considered to begin at between 100 and 200 kilometers (km) (60 to 120 miles [mi]) altitude. Within the U.S. aerospace program, persons who have traveled beyond about 80 km altitude (50 mi) often are recognized as space travelers or astronauts.

outgassing Release of gas from a material when it is exposed to an ambient pressure lower than the vapor pressure of the gas. Generally refers to the gradual release of gas from enclosed surfaces when an enclosure is vacuum pumped or to the gradual release of gas from a spacecraft's surfaces and components when they are first exposed to the vacuum conditions of outer space following launch. Outgassing presents a problem to spacecraft designers because the released vapor might recondense on optical surfaces (instruments) or other special spacecraft surfaces where these unwanted material depositions can then degrade performance of the component or sensor. Aerospace engineers try to avoid such outgassing problems by carefully selecting the materials used in the spacecraft. In addition, certain (outgassing-prone) components also can be placed through a lengthy thermal vacuum ("bake-out") treatment to remove unwanted substances prior to final installation and launch.

overexpanding nozzle A nozzle in which the working fluid is expanded to a lower pressure than the external pressure; that is, an overexpanding nozzle has an exit area larger than the optimum exit area for the particular external pressure environment. For example, a nozzle designed to operate in space (where the ambient or external pressure is essentially zero) would overexpand the rocket's exhaust gases, if that particular rocket engine-nozzle configuration were used in Earth's atmosphere, where the ambient pressure is greater than the vacuum conditions found in space. *See also* NOZZLE.

overpressure The pressure resulting from the blast wave of an explosion. It is referred to an "positive" when it exceeds atmospheric pressure and "negative" during the passage of the wave when resulting pressures are less than atmospheric pressure.

oxidizer Material whose main function is to supply oxygen or other oxidizing materials for deflagration (burning) or a solid propellant or combustion of a liquid fuel.

P

pad The platform from which a rocket vehicle is launched. *See also* LAUNCHPAD.

pad deluge Water sprayed upon certain launch pads during the launch of a rocket so as to reduce the temperatures of critical parts of the pad or the rocket.

pair production The gamma ray interaction process in which a high-energy [>1.02 million electron volts (MeV)] photon is annihilated in the vicinity of the nucleus of an absorbing atom with the simultaneous formation of a negative and positive electron (positron) pair.

Pallas The large (about 540 KILOMETERS [335 mi] in diameter) MAIN-BELT ASTEROID discovered in 1802 by the German astronomer Heinrich Wilhelm Olbers (1758–1840). Pallas was the second asteroid found.

parabolic Pertaining to, or shaped like, a *parabola*. A parabola is a conic section with an eccentricity *(e)* equal to unity (i.e., $e = 1$). If a parabola is symmetrical about the x-axis and has its vertex at the origin, then it is defined by the equation: $y^2 = 4ax$, where a is the distance from the vertex to the focus and 4a is the length of the latus rectum. *See also* CONIC.

parabolic orbit An orbit shaped like a parabola; the orbit representing the least eccentricity for escape from an attracting body. *See also* ORBITS OF OBJECTS IN SPACE.

parabolic reflector A reflecting surface having a cross section along the axis in the shape of a parabola. Incoming parallel rays striking the reflector are brought to a focus point, or if the source of the rays is located at the focus, the rays are reflected as a parallel beam. Also called *parabolic effect. See also* CONIC; PARABOLIC.

parallax In general, the difference in the apparent position of an object when viewed from two different points. Specifically, the angular displacement in the apparent position of a celestial body when observed from two widely separated points (e.g., when Earth is the extremities of its orbit around the Sun). This difference is very small and expressed as an angle, usually in arc seconds (where 1 arc second = 1/3,600 of a degree). *See also* ASTROMETRY; PARSEC.

parking orbit The temporary (but stable) ORBIT of a SPACECRAFT around a CELESTIAL BODY. It is used for the assembly and/or transfer of equipment or to wait for conditions favorable for departure from that orbit.

parsec (symbol: pc) A unit of distance frequently encountered in astronomical studies. The parsec is defined as a parallax shift of one second of arc. The term itself is a shortened from of "parallax second." The parsec is the extraterrestrial distance at which the main radius of Earth's orbit (one astronomical unit [AU] by definition) subtends an angle of one arc-second. It is therefore also the distance at which a star would exhibit an annual parallax of one arc-second.

1 parsec = 3.2616 light-years
(or 206,265 AU)

The kiloparsec (kpc) represents a distance of 1,000 parsecs (or 3,261.6 light-years); and the megaparsec (Mpc), a distance of 1 million parsecs (or about 3.262 million light-years). *See also* PARALLAX.

partial pressure In general, the pressure exerted by a designated component or components of a gaseous mixture; as, for example, the partial pressure of carbon dioxide (CO_2) in the atmosphere. In the atmosphere, the partial pressure of CO_2 is defined as the pressure the CO_2 would exert if all other gases were removed. The sum of the partial pressure of all atmospheric gases then equals the atmospheric pressure. The partial pressure of CO_2 in the atmosphere is determined by the atmospheric CO_2 concentration and the atmospheric temperature.

particle A minute constituent of matter, generally one with a measurable mass. An elementary atomic particle such a proton, neutron, electron (beta particle), or alpha particle.

particle accelerator A device that accelerates charged nuclear and subnuclear particles by means of changing electromagnetic fields. Accelerators impart energies of millions of electron volts (MeV) and higher to particles, such as protons or electrons. Accelerators used in nuclear physics research have achieved energy levels of over 10^{12} electron volts. Accelerators useful in a BALLISTIC MISSILE DEFENSE (BMD) role, such as neutral particle beam (NPB) accelerators and free electron laser (FEL) devices, need only achieve particle energies of between 10^8 and 10^9 electron volts but generally require much higher beam currents.

particle beam A focused, collimated stream of nuclear or subnuclear particles (e.g., protons, electrons, or alpha particles) that have been accelerated to high velocities (often a considerable portion of the speed of light).

pascal (symbol: Pa) The SI UNIT of pressure. It is defined as the pressure that results from a force of one newton (N) acting uniformly over an area of 1 square meter.

$$1 \text{ pascal [Pa]} = 1 \text{ N/m}^2$$

This unit is named after the French scientist Blaise Pascal (1623–1662).

pass 1. A single circuit of Earth (or a planet) by a spacecraft or satellite. Passes start at the time the satellite crosses the equator from the Southern Hemisphere into the Northern Hemisphere (the ascending node). 2. The period of time a satellite or spacecraft is within telemetry range of a ground station.

passive Containing no power sources to augment output power or signal, such as a *passive electrical network* or a *passive reflector.* Applied to a device that draws all its power from the input signal. A dormant device or system, that is, one that is not active.

passive defense Measures taken to reduce the probability and to minimize the effects of damage caused by hostile action without the intention of taking the initiative (e.g., short of launching a preemptive first strike). *Compare with* ACTIVE DEFENSE.

passive homing guidance A system of homing guidance wherein the receiver in the missile uses radiation from the target. *Compare with* ACTIVE HOMING GUIDANCE. *See also* GUIDANCE SYSTEM.

passive satellite A satellite that does not transmit a signal. This type of "silent" human-made space object could be a dormant (but still functional) replacement satellite, a retired or decommissioned satellite, a satellite that has failed prematurely (and is now a piece of "space junk"), or possibly even a silent, stalking space mine.

passive sensor A sensor that detect radiation naturally emitted by (e.g., infrared) or reflected (e.g., sunlight) from a target.

passive target A "target" orbital payload or satellite that is stabilized along its three axes and is detected, acquired, and tracked by means of electromagnetic energy reflected from the target's surface (or skin).

path 1. Of a satellite or spacecraft, the projection of the orbital plane on Earth's surface. Since Earth is turning under the satellite, the path of a single orbital pass will not be a closed curve. The terms "path" and "track" are used interchangeably. On a cylindrical map projection, a satellite's path is a sine-shaped curve. 2. Of an aircraft or aerospace vehicle, the flight path. 3. Of a meteor, the projection of the trajectory on the celestial sphere as seen by an observer. *See also* GROUND-TRACK; ORBITS OF OBJECTS IN SPACE.

Patriot The U.S. Army's Patriot missile system provides high- and medium-ALTITUDE defense against AIRCRAFT and tactical BALLISTIC MISSILES. The combat element of the Patriot missile system is the fire unit that consists of a RADAR set, an engagement control station (ECS), an equipment power plant (EPP), an ANTENNA mast group (AMG), and eight remotely located LAUNCHERs. The single phased-array radar provides the following tactical functions: airspace surveillance, TARGET DETECTION and TRACKING, and missile guidance. The ECS provides the human interface for command and control operations. Each firing battery launcher contains four ready-to-fire missiles sealed in canisters that serve a dual purpose as shipping containers and launch tubes. The Patriot's fast reaction capability, high firepower, ability to track up to 50 targets simultaneously with a maximum RANGE of 68.5 KILOMETERS (42.6 mi), and ability to operate in a severe electronic countermeasures (ECM) environment are features not available in previous AIR DEFENSE systems. The U.S. Army received the Patriot in 1985, and the system gained notoriety during the Persian Gulf War of 1991 as the "SCUD killer."

The Patriot Advanced Capability-3 (PAC-3) upgrade program is currently being managed by the MISSILE DEFENSE AGENCY (MDA) with the Department of Defense. PAC-3 is a terminal defense system being built upon the previous Patriot and missile defense infrastructure. The PAC-3 missile is a high-VELOCITY HIT-TO-KILL VEHICLE that represents the latest generation of Patriot missile. It is being devel-oped to provide an increased capability against tactical ballistic missiles, CRUISE MISSILES, and hostile aircraft. The PAC-3 missile provides the range, accuracy, and lethality to effectively defend against THE-ATER BALLISTIC MISSILES (TBMs) that may be carrying either conventional HIGH EXPLOSIVE (HE), biological, chemical, or nuclear WARHEADS. Unlike previous Patriot missiles, the PAC-3 uses an ACTIVE radar seeker and closed loop guidance to directly hit the target.

With the PAC-3 missile and ground system, threat missiles can be engaged and destroyed at higher altitudes and greater ranges with better lethality. The system can operate despite electronic countermeasures. Due to the active seeker and closed loop guidance, a greater number of interceptors can be controlled at one time than with earlier Patriot missiles. *See also* BALLISTIC MISSILE DEFENSE; THEATER HIGH ALTITUDE AREA DEFENSE.

payload 1. Originally, the revenue-producing portion of an aircraft's load, such as passengers, cargo, and mail. By extension, the term "payload" has become applied to that which a rocket, aerospace vehicle, or spacecraft carries over and above what is necessary for the operation of the vehicle during flight. 2. With respect to the Space Transportation System (or space shuttle), the total complement of specific instruments, space equipment, support hardware and consumables carried in the orbiter vehicle (especially in the cargo bay) to accomplish a discrete activity in space but not included as part of the basic orbiter payload support (e.g., the remote manipulator system, or RMS). 3. With respect to the space station, an aggregate of instruments and software for the performance of specific scientific or applications investigations, or for commercial activities. Payloads may be located inside the pressurized modules of the space station, attached to the space station structure, or attached to a platform, or they may be free-flyers. 4. With respect to a military missile, the payload generally consists of a warhead, its protective container, and activating devices.

payload assist module (PAM) A family of commercially developed UPPER STAGE ROCKET vehicles intended for use with NASA's SPACE SHUTTLE or with EXPENDABLE LAUNCH VEHICLES, such as the DELTA rocket.

payload bay The 4.57-meter- (150-foot) diameter by 18.3-meter- (60-foot-) long enclosed volume within the space shuttle Orbiter vehicle designed to carry payloads, upper-stage vehicles with attached payloads, payload support equipment, and associated mounting hardware. Also called the *cargo bay*. See also SPACE TRANSPORTATION SYSTEM.

payload buildup The process by which the instrumentation (e.g., sensors, detectors, etc.) and equipment, including the necessary mechanical and electronic subassemblies, are combined into a complete operational package capable of achieving the objectives of a space particular mission.

payload canister Environmentally controlled transporter for use at the space shuttle launch site (Complex 39 A/B, Kennedy Space Center). It is the same size and configuration as the orbiter vehicle's cargo bay. See also SPACE TRANSPORTATION SYSTEM.

payload carrier With respect to the Space Transportation System (STS), one of the major classes of standard payload carriers certified for use with the space shuttle to support payload operations. Payload carriers are identified as: habitable modules (e.g., spacelab) and attached but uninhabitable modules and equipment (e.g., pallets, free-flying systems, satellites, and propulsive upper stages). *See also* SPACE TRANSPORTATION SYSTEM.

payload changeout room An environmentally controlled room on a movable support structure (including a manipulator system) at the space shuttle launch pad (Complex 39 A/B at the Kennedy Space Center). It used for inserting payloads vertically into the Orbiter vehicle's cargo bay at the launch pad and supports payload

transfer between the transport canister and the cargo bay. See also SPACE TRANSPORTATION SYSTEM.

payload integration The compatible installation of a complete payload package into the spacecraft and space launch vehicle.

payload retrieval mission A space shuttle mission in which an orbiting payload is captured, secured in the cargo bay, and then returned to Earth. See also SPACE TRANSPORTATION SYSTEM.

payload servicing mission A space shuttle mission in which an orbiting payload is given inspection, maintenance, modification, and/or repair. This type of mission generally involves rendezvous and capture operations as well as extravehicular activity (EVA), during which shuttle astronauts perform repair and servicing activities while outside of orbiter vehicle's pressure cabin. Sometimes called *payload repair mission*. See also HUBBLE SPACE TELESCOPE; SPACE TRANSPORTATION SYSTEM.

payload specialist The noncareer astronaut who flies as a space shuttle passenger and is responsible for achieving the payload/experiment objectives. He/she is the onboard scientific expert in charge of payload/experiment operations. The payload specialist has a detailed knowledge of the payload instruments and their subsystems, operations, requirements, objectives and supporting equipment. As such, he/she is either the principal investigator (PI) conducting experiments while in orbit or the direct representative of the principal investigator. Of course, the payload specialist also must be knowledgeable about certain basic orbiter systems, such as food and hygiene accommodations, life support systems, hatches, tunnels, and caution and warning systems.

payload station Location on the orbiter vehicle aft flight deck from which payload-specific operations are performed, usually by the payload specialist or mission specialist. Also the locations in the orbiter's flight decks and cargo bay at which payloads are mounted. *See also* SPACE TRANSPORTATION SYSTEM.

Peacekeeper (LGM-118A) The Peacekeeper (also called LGM-118A) is America's newest intercontinental ballistic missile (ICBM). Its deployment filled a key goal of the strategic modernization program and increased the strength and credibility of the ground-based leg of the U.S. strategic (nuclear) triad. The Peacekeeper is capable of delivering ten independently targeted warheads with greater accuracy than any other ballistic missile. It is a three-stage rocket ICBM system consisting of three major sections: the boost system, the post-boost vehicle system, and the reentry system.

The boost system consists of four rocket stages that launch the missile into space. These rocket stages are mounted atop one another and fire successively. Each of the first three stages exhausts its solid-propellant materials through a single moveable nozzle that guides the missile along its flight path. The fourth stage post-boost vehicle is made up of a maneuvering rocket and a guidance and control system. The reentry vehicle system consists of the deployment module, containing up to 10 cone-shaped reentry vehicles and a protective shroud.

The air force achieved initial operational capability of 10 deployed Peacekeeper missiles at F.E. Warren Air Force Base, Wyoming in December 1986. Full operational capability was achieved in December 1988 with the establishment of a squadron of 50 missiles. However, with the end of the cold war, the United States has begun to revise its strategic nuclear policy and has agreed to eliminate the multiple reentry vehicle Peacekeeper ICBMs by the year 2007 as part of the Strategic Arms Reduction Treaty II. *See also* UNITED STATES STRATEGIC COMMAND.

Pegasus (space launch vehicle) The Pegasus air-launched space booster is produced by Orbital Sciences Corporation and Hercules Aerospace Company to provide small satellite users with a cost-effective, flexible, and reliable method for placing payloads into LOW EARTH ORBIT (LEO). Launching a Pegasus rocket from an airplane flying at an altitude of approximately 12 kilometers (40,000 feet)—the same altitude most commercial jet airlines fly—reduces the amount of thrust needed to overcome Earth's gravity by 10 to 15%. *See also* LAUNCH VEHICLE.

Peltier effect The production or absorption of thermal energy (heat) that occurs at the junction of two dissimilar metals or semiconductors when an electric current flows through the junction. If thermal energy is generated by the flow of current in one direction, it will be absorbed when the current flow is reversed. This phenomenon, discovered by the French physicist J. C. A. Peltier (1785–1845), can be used in direct conversion cooling applications (e.g., infrared sensor arrays), especially for small volumes or regions where regular refrigeration and cooling techniques are impractical. *See also* DIRECT CONVERSION.

penetration aid In general, techniques and/or devices employed by offensive aerospace weapon systems to increase the probability of penetration of enemy defenses. With respect to ballistic missiles, a device, or group of devices that accompanies a reentry vehicle (RV) during its flight to misdirect defenses and thereby allow the RV to reach its target and detonate its warhead. *See also* BALLISTIC MISSILE DEFENSE.

penetrator Space scientists have concluded that experiments performed from a network of penetrators can provide essential facts needed to begin understanding the evolution, history, and nature of a planetary body, such as Mars. The scientific measurements performed by penetrators might include seismic, meteorologic, and local site characterization studies involving heat flow, soil moisture content, and geochemistry. A typical penetrator system consists of four major subassemblies: (1) the launch tube, (2) the deployment motor, (3) the decelerator (usually a two-stage device), and (4) the penetrator itself. The launch tube attaches to the host spacecraft and houses the penetrator, deployment motor, and two-stage decelerator. The deployment motor is based on well-proven solid rocket motor technology and provides the required deorbit velocity. If the planetary body has an atmosphere, the two-stage decelerator includes a furlable umbrella heat shield for the first stage of hypersonic deceleration. The penetrator itself is a steel device, shaped like a rocket, with a blunt ogive (curved) nose and conical-flared body. The aftbody (or afterbody) of the penetrator remains at the planet's surface, with the forebody penetrating the subsurface material. Communications with scientists on Earth from the surface/subsurface penetrator sites is accomplished by means of an orbiting mothership, which interrogates each penetrator at regular intervals.

A network of penetrators can be used to study planets with solid surfaces (e.g., Mercury, Venus, Mars, and Pluto) as well as many of the interesting moons in the solar system (e.g., Titan, Io, Callisto, Ganymede, Triton, and Charon). The Jovian moon, Europa, is an especially interesting target because of the possibility of a liquid water ocean underneath its smooth, icy surface. A network of penetrators also can be deployed on Earth's own Moon to conduct a variety of scientific experiments, including in-situ evaluation and assessment of surface deposits of frozen volatiles now suspected to occur in certain permanently shadowed polar regions.

perfect cosmological principle The postulation that at all times the UNIVERSE appears the same to all observers. *See also* COSMOLOGY.

perfect fluid In simplifying assumptions made as part of preliminary engineering analyses, a fluid chiefly characterized by a lack of viscosity and, usually, by incompressibility. Also called an *ideal fluid* or an *inviscid fluid*.

perfect gas A gas that obeys the following equation of state: pv = RT, where p is pressure, v is specific volume, T is absolute temperature, and R is the gas constant. *See also* IDEAL GAS.

perforation With respect to a solid-propellant rocket, the central cavity or hole down the center of a propellant grain. This perforation can take different shapes, ranging from a right circular cylinder to a complex star pattern. The geometry and dimensions of the perforation affect the solid-propellant burn rate and, consequently, the thrust delivered by the rocket as a function of time. *See also* ROCKET.

peri- A prefix meaning near, as in *perigee*.

periapsis The point in an orbit closest to the body being orbited. *See also* ORBITS OF OBJECTS IN SPACE.

periastron That point of the orbit of one member of a binary star system at which the stars are nearest to each other. The opposite of *apastron*. *See also* BINARY STARS.

pericynthian That point in the trajectory of a vehicle that is closest to the Moon.

perigee In general, the point at which a satellite's orbit is the closest to the primary (central body); the minimum altitude attained by an Earth-orbiting object. *See also* ORBITS OF OBJECTS IN SPACE.

perigee propulsion A programmed-thrust technique for escape from a planet by an orbiting space vehicle. The technique involves the intermittent application of thrust at perigee (when the object's orbital velocity is high) and coasting periods.

perihelion The place in a solar orbit that is nearest the Sun, when the Sun is the center of attraction. *Compare with* APHELION.

perilune In an orbit around the Moon, the point in the orbit nearest the lunar surface.

period (common symbol: T) 1. In general, the time required for one full cycle of a regularly repeated series of events. 2. The time taken by a satellite to travel once around its orbit. Also called the *orbital period*. 3. The time between two successive swings of a pendulum or between two successive crests in a wave, such as a radio wave. 4. The time required for one oscillation; consequently, the reciprocal of frequency. 5. In nuclear engineering, the time required for the power level of a reactor to change by the factor *e* (2.718, the base of natural logarithms). Also called the *reactor period*.

periodic comet A COMET with a period of less than 200 years. Also called a SHORT PERIOD COMET.

permanently crewed capability (PCC) The capability to operate a space station (or a planetary surface base) with a human crew on board, 24 hours a day, 365 days a year. Sometimes called *permanently manned capability (PMC)*.

Pershing (MGM-31A) A U.S. Army mobile surface-to-surface inertially guided missile of a two-stage solid-propellant type. It possesses a nuclear warhead capability and is designed to support ground forces by attacking long-range ground targets.

perturbation 1. Any departure introduced into an assumed steady-state system, or a small departure from the nominal path, such as a specified trajectory. 2. In astronomy, a disturbance in the regular motion of a celestial body, resulting from a force that is additional to that which causes the regular motion, specifically a gravitational force. For example, the orbit of an asteroid or comet is strongly *perturbed* when the small celestial object passes near a major planetary body such as Jupiter.

phase 1. The fractional part of a periodic quantity through which the particular value of an independent variable has advanced as measured from an arbitrary reference. 2. The physical state of a substance, such as solid, liquid, or gas. 3. The extent that the disk of the Moon or a planet, as seen from Earth, is illuminated by the Sun.

phase change The physical process that occurs when a substance changes between the three basic phases (or states) of matter: solid, liquid, and vapor. *Sublimation* is a change of phase directly from the solid phase to the vapor phase. *Vaporization* is a change of phase from the liquid phase to the vapor (gaseous) phase. *Condensation* is the change of phase from the vapor phase to the liquid phase (or from the vapor phase directly to the solid phase). *Fusion* is the change of phase from the liquid phase to the solid phase, while *melting* is a change of phase from the solid phase to the liquid phase. Phase change is a thermodynamic process involving the addition or removal of heat (thermal energy) from the substance. The temperature at which these processes occur is dependent on the pressure.

phase modulation (PM) A type of modulation in which the relative phase of the carrier wave is modified or varied in accordance to the amplitude of the signal. Specifically, a form of angle modulation in which the angle of a sine-wave carrier is caused to depart from the carrier angle by an amount proportional to the instantaneous value of the modulating wave. *See also* AMPLITUDE MODULATION; FREQUENCY MODULATION.

phases of the Moon The changing illuminated appearance of the NEARSIDE of the MOON to an observer on EARTH. The major phases include the new Moon (not illuminated), first quarter, full Moon (totally illuminated), and last (third) quarter. (See Figure 56.)

Phobos The larger of the two small moons of Mars. Phobos has a mean diameter of 22 kilometers, a period of rotation of 0.3188 days, and a mean distance from the planet of 9,380 kilometers. *See also* MARS.

Phoenix (AIM-54A) A long-range air-to-air missile with electronic guidance/homing developed by the U.S. Navy. Usually carried in clusters of up to six missiles on an aircraft such as the Tomcat, this missile is an airborne weapons control system with multiple-target—handling capabilities. It is used

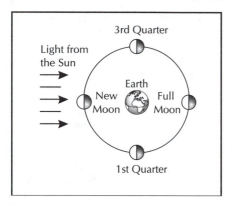

Figure 56. Phases of the Moon.

to kill multiple air targets with conventional warheads. Near-simultaneous launch is possible against up to six targets in all weather and heavy jamming environments. This missile is powered by a solid-propellant rocket motor. It is 3.9 meters (13 feet) long, 38.1 centimeters (15 inches) wide, and has a wing span of 0.9 meters (3 feet). It has a range in excess of 184 kilometers (115 miles). Its conventional warhead has a mass of 60.75 kilograms (135 pounds-mass), while the entire missile has a total mass of 460.8 kilograms (1,024 pounds-mass).

photoelectric effect   The emission of an electron from a surface as the surface absorbs a photon of electromagnetic radiation. Electrons so emitted are called *photoelectrons*. The effectiveness of the process depends on the surface metal involved and the wavelength of the radiant energy to which it is exposed. For example, cesium will emit photoelectrons when exposed to visible radiation. The energy of the photoelectron produced is equal to the energy of the incident photon (hv) minus the amount of work (Φ) needed to raise the electron to sufficient energy level to free it from the surface. This relationship is expressed in Einstein's photoelectric effect equation:
$$E_{photoelectron} = hv - \Phi$$
where h is the Planck constant, v is the frequency of the incident photon, and Φ is the work function of the solid (metal) surface. In 1905 Albert Einstein (1879–1955)

published an important physics paper that described the photoelectric effect.

photoionization   The IONIZATION of an ATOM or MOLECULE caused by collisions with an energetic PHOTON.

photometer   An INSTRUMENT that measures LIGHT intensity and the brightness of celestial objects, such as STARS.

photon   According to quantum theory, the elementary bundle or packet of electromagnetic radiation, such as a photon of light. Photons have no mass and travel at the speed of light. The energy *(E)* of the photon is equal to the product of the frequency (v) of the electromagnetic radiation and Planck's constant *(h)*:
$$E = hv$$
where *h* is equal to 6.626×10^{-34} joule-sec; and v is the frequency (hertz).

photon engine   A conceptual reaction engine in which thrust would be obtained by emitting photons, such as light rays. Although the thrust from such an engine would be very small, it would be applied continuously in the vacuum of outer space (if a suitable power source became available) and eventually might achieve significant speeds—perhaps even supporting interstellar flight. *See also* LASER PROPULSION SYSTEM.

photosphere   The intensely bright, visible portion (surface) of the Sun. *See also* SUN.

photovoltaic conversion   A form of DIRECT CONVERSION in which a PHOTOVOLTAIC MATERIAL converts incoming PHOTONS of visible LIGHT directly into ELECTRICITY. The SOLAR CELL is an example.

photovoltaic material   A substance (semiconductor material) that converts sunlight directly into electricity, using the photovoltaic effect. This effect involves the production of a voltage difference across a p–n junction, resulting from the absorption of a photon of energy. The p–n junction is the region at which two dissimilar types of semiconductor materials (i.e., materials of opposite polarity) meet. The voltage difference in the semiconductor material is caused by the internal drift of holes and electrons.

Crystalline silicon (Si) and gallium arsenide (GaAs) are the photovoltaic materials often used in spacecraft power system applications. The typical silicon *solar cell* delivers about 11.5% solar energy-to-electric energy conversion efficiency, while the newer but more expensive gallium arsenide cell promises to provide about 18% or more conversion efficiency. Solar panels and solar arrays are constructed in a variety of configurations from large numbers of individual solar cells that are made of these photovoltaic materials. The series-parallel pattern of electrical connections between the cells determines the total output voltage of the solar panel or array.

On orbit, prolonged exposure to sunlight causes the performance of these photovoltaic materials to degrade, typically by 1 or 2% per year. Exposure to radiation (both the permanently trapped radiation belts surrounding Earth and transitory solar flare radiations) can cause even more rapid degradation in the performance of photovoltaic materials. Solar cells constructed from gallium arsenide are more radiation resistant that silicon solar cells. *See also* SOLAR CELL.

piggyback experiment An experiment that rides along with the primary experiment on a space-available basis without interfering with the mission of the primary experiment.

pilot 1. In general, the person who handles the controls of an aircraft, aerospace vehicle, or spacecraft from within the craft and, in so doing, guides or controls it in three-dimensional flight. 2. An electromechanical system designed to exercise control functions in an aircraft, aerospace vehicle, or spacecraft; for example, the *automatic pilot*. 3. (verb) To operate, control, or guide an aircraft, aerospace vehicle, or spacecraft from within the vehicle so as to move in three-dimensional flight

through the air or space. 4. With respect to the U.S. Space Transportation System (STS), the second in command of a space shuttle flight. He or she assists the commander as required in conducting all phases of the orbiter vehicle flight. The pilot has such authority and responsibilities as are delegated him/her by the commander. The pilot normally operates the remote manipulator system (RMS) at the Orbiter's payload handling station, using the RMS to deploy, release, or capture payloads. The pilot also is generally the second crewperson (behind the mission specialist) for extravehicular activity (EVA). *See also* SPACE TRANSPORTATION SYSTEM.

Pioneer spacecraft NASA's solar-orbiting Pioneer spacecraft have contributed enormous amounts of data concerning the solar wind, solar magnetic field, cosmic radiation, micrometeoroids, and other phenomena of interplanetary space. *Pioneer 4,* launched March 3, 1959, was the first U.S. spacecraft to go into orbit around the Sun. It provided excellent radiation data. *Pioneer 5,* launched March 11, 1960, confirmed the existence of interplanetary magnetic fields and helped explain how solar flares trigger magnetic storms and the northern and southern lights (auroras) on Earth. *Pioneers 6* through *9* were launched between December 1965 and November 1968. These spacecraft provided large quantities of valuable data concerning the solar wind, magnetic and electrical fields, and cosmic rays in interplanetary space. Data from these spacecraft helped space scientists draw a new picture of the Sun as the dominant phenomenon of interplanetary space. *See also* PIONEER 10, 11; PIONEER VENUS MISSION.

Pioneer 10, 11 The *Pioneer 10* and *11* spacecraft, as their names imply, are true deep-space explorers—the first human-made objects to navigate the main asteroid belt, the first spacecraft to encounter Jupiter and its fierce radiation belts, the first to encounter Saturn, and the first spacecraft to leave the solar system. These spacecraft also investigated magnetic fields, cosmic rays, the solar wind, and the interplanetary dust concentrations as they flew through interplanetary space.

At Jupiter and Saturn, scientists used the spacecraft to investigate the giant planets and their interesting complement of moons in four main ways: (1) by measuring particles, fields, and radiation; (2) by spin-scan imaging the planets and some of their moons; (3) by accurately observing the paths of the spacecraft and measuring the gravitational forces of the planets and their major satellites acting on them; and (4) by observing changes in the frequency of the S-band radio signal before and after occultation (the temporary "disappearance" of the spacecraft caused by their passage behind these celestial bodies) to study the structures of their ionospheres and atmospheres. (See Figure 57.)

The *Pioneer 10* spacecraft was launched from Cape Canaveral Air Force Station, Florida by an Atlas-Centaur rocket on March 2, 1972. It became the first spacecraft to cross the main asteroid belt and the first to make close-range observations of the Jovian system. Sweeping past Jupiter on December 3, 1973 (its closest approach to the giant planet), it discovered no solid surface under the thick layer of clouds enveloping the giant planet—an indication that Jupiter is a liquid hydrogen planet. *Pioneer 10* also explored the giant Jovian magnetosphere, made close-up pictures of the intriguing Red Spot, and observed at

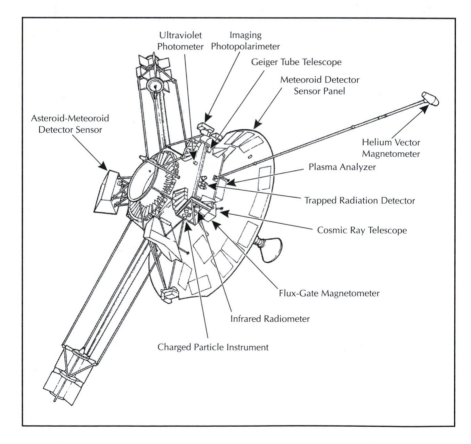

Figure 57. The *Pioneer 10* (and *11*) spacecraft with its complement of scientific instruments. Power was provided by a long-lived radioisotope-thermoelectric generator (RTG).

relatively close range the Galilean satellites Io, Europa, Ganymede, and Callisto. When *Pioneer 10* flew past Jupiter, it acquired sufficient kinetic energy to carry it completely out of the solar system.

Departing Jupiter, *Pioneer 10* continued to map the heliosphere (the Sun's giant magnetic bubble, or field, drawn out from it by the action of the solar wind). Then, on June 13, 1983, *Pioneer 10* crossed the orbit of Neptune, which at the time was (and until 1999 will be) the planet farthest out from the Sun, due to the eccentricity in Pluto's orbit, which now takes it inside that of Neptune. This historic date marked the first passage of a human-made object beyond the known planetary boundary of the solar system. Beyond this solar system boundary, *Pioneer 10* measured the extent of the heliosphere as it flies through interstellar space. Along with its sister ship, *Pioneer 11,* this spacecraft is still helping scientists investigate the deep space environment.

The *Pioneer 10* spacecraft is heading generally toward the red star Aldebaran. It is more than 68 light-years away from Aldebaran, and the journey will require about 2 million years to complete. Budgetary constraints forced the termination of routine tracking and project data processing operations on March 31, 1997. However, occasional tracking of *Pioneer 10* continued beyond that date. The last successful data acquisition from *Pioneer 10* by NASA's DEEP SPACE NETWORK (DSN) occurred on March 3, 2002—30 years after launch—and again on April 27, 2002. The spacecraft signal was last detected on January 23, 2003 after an uplink message was transmitted to turn off the last operational experiment, the Geiger Tube Telescope. However, no downlink data signal was achieved, and by early February no signal at all was detected. NASA personnel concluded that the spacecraft's RTG power supply had finally fallen below the level needed to operate the onboard transmitter. Consequently, no further attempts were made to communicate with *Pioneer 10.*

The *Pioneer 11* spacecraft was launched on April 5, 1973, and swept by Jupiter at an encounter distance of only 43,000 kilometers (26,720 miles) on December 2, 1974. It provided additional detailed data and pictures of Jupiter and its moons, including the first views of Jupiter's polar regions. Then, on September 1, 1979, *Pioneer 11* flew by Saturn, demonstrating a safe flight path through the rings for the more sophisticated Voyager spacecraft to follow. *Pioneer 11* (officially renamed *Pioneer Saturn)* provided the first close-up observations of Saturn, its rings, satellites, magnetic field, radiation belts, and atmosphere. It found no solid surface on Saturn but discovered at least one additional satellite and ring. After rushing past Saturn, *Pioneer 11* also headed out of the solar system toward the distant stars.

The *Pioneer 11* spacecraft has operated on a backup transmitter since launch. Instrument power sharing began in February 1985 due to declining RTG power output. Science operations and daily telemetry ceased on September 30, 1995, when the RTG power level became insufficient to operate any of the spacecraft's instruments. All contact with *Pioneer 11* ceased at of the end of 1995. At that time, the spacecraft was 44.7 astronomical units (AU) away from the Sun and traveling through interstellar space at a speed of about 2.5 AU per year.

Finally, both Pioneer spacecraft carry a special message (the "Pioneer plaque") for any intelligent alien civilization that might find them wandering through the interstellar void millions of years from now. This message is a drawing map, engraved on an anodized aluminum plaque. The plaque depicts the location of Earth and the solar system, a man and a woman, and other points of science and astrophysics that should be decipherable by a technically intelligent civilization. *See also* JUPITER; SATURN.

Pioneer Venus mission The *Pioneer Venus* mission consisted of two separate spacecraft launched by the United States to the planet Venus in 1978. The *Pioneer Venus* orbiter spacecraft (also called *Pioneer 12)* was a 553-kilogram (1220 pounds-mass

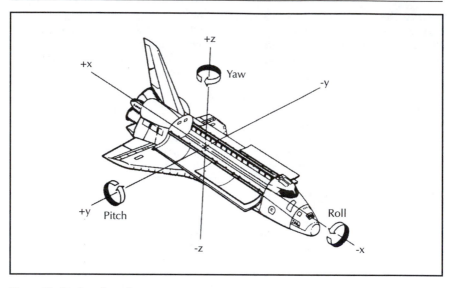

Figure 58. Pitch, roll, and yaw.

[lbm]) spacecraft that contained a 45-kilogram (100 lbm) payload of scientific instruments. It was launched on May 20, 1978, and placed into a highly eccentric orbit around Venus on December 4, 1978. For 14 years (1978–1992) the Orbiter spacecraft gathered a wealth of scientific data about the atmosphere and ionosphere of Venus and their interactions with the solar wind as well as details about the planet's surface. Then in October 1992, this spacecraft made an intended, final entry into the Venusian atmosphere, gathering data up to its final fiery plunge and dramatically ending the operations portion of the Pioneer Venus mission.

The *Pioneer Venus Multiprobe* spacecraft (also called *Pioneer 13*) consisted of a basic bus spacecraft, a large probe, and three identical small probes. The Multiprobe spacecraft was launched on August 8, 1978, and separated about three weeks before entry into the Venusian atmosphere. The four (now-separated) probes and their (spacecraft) bus successfully entered the Venusian atmosphere at widely dispersed locations on December 9, 1978, and returned important scientific data as they plunged toward the planet's surface. Although the probes were not

designed to survive landing, one hardy probe did and transmitted data for about an hour after impact.

The *Pioneer Venus* orbiter and multi-probe spacecraft provided a wealth of scientific data about Venus, its surface, atmosphere, and interaction with the solar wind. For example, the orbiter spacecraft made an extensive radar map, covering about 90% of the Venusian surface. Using its radar to look through the dense Venusian clouds, this spacecraft revealed that the planet's surface was mostly gentle, rolling plains with two prominent plateaus: Ishtar Terra and Aphrodite Terra. This highly successful, two-spacecraft mission also provided important groundwork for NASA's subsequent *Magellan* mission to Venus. *See also MAGELLAN MISSION; VENUS.*

pitch 1. In acoustics, the "highness" or "lowness" of a sound. Pitch depends mainly on the frequency of the sound stimulus but also on the pressure and waveform of the stimulus. 2. In aerospace technology, the rotation (angular motion) of an aircraft, aerospace vehicle, or spacecraft about its lateral axis. (See Figure 58.) 3. In engineering, the distance between corresponding points on adjacent teeth of a gear or an adjacent blades on a turbine wheel, as measured along a prescribed arc, called the *pitchline.*

pitch angle The angle between an aircraft's or missile's longitudinal axis and Earth's horizontal plane. Also called *inclination angle.*

pitchover The programmed turn from the vertical that a rocket or launch vehicle (under power) takes as it describes an arc and points in a direction other than vertical.

pitot tube An open-ended tube that, when immersed in a moving fluid with its mouth pointed upstream, can be used to measure the stagnation pressure of the fluid for subsonic flow or the stagnation pressure behind the tube's normal shock wave for supersonic flow.

pitting Creation of surface voids by mechanical erosion, chemical corrosion, or cavitation.

pixel Contraction for *picture element;* the smallest unit of information on a screen or in an image; the more pixels, the higher the potential resolution of the video screen or image.

plage A bright patch in the SUN's CHROMOSPHERE.

Planck, Max Karl (1858–1947) German physicist who introduced QUANTUM THEORY in 1900, a powerful new theory concerned with the transport of ELECTROMAGNETIC RADIATION in discrete ENERGY packets, or QUANTA. His work represents one of the two great pillars of modern physics (the other being Albert Einstein's [1879–1955] RELATIVITY). Planck received the 1918 Nobel Prize in physics for this important accomplishment. *See also* PLANCK'S RADIATION LAW.

Planck's constant (symbol: h) A fundamental constant in modern physics that is equal to the ratio of the energy (E) of a quantum of energy to the frequency (v) of this quantum (or photon); that is,

$$h = E/v = 6.626 \times 10^{-34} \text{ joule-second.}$$

This unit is named in honor of the German physicist Max Planck (1858–1947), who first developed quantum theory.

Planck's radiation law The fundamental physical law that describes the distribution of energy radiated by a BLACKBODY. Introducing the concept of the *quantum* (or photon) as a small unit of energy transfer, the German physicist Max Planck (1858–1947) first proposed this law in 1900. It is an expression for the variation of monochromatic radiant flux per unit area of source as a function of wavelength (λ) of blackbody radiation at a given thermodynamic temperature (T). Mathematically, Planck's law can be expressed as:

$$dw = [2\pi hc^2]/\{\lambda^5 [e^{hc/\lambda kT} - 1]\}d\lambda$$

where *dw* is the radiant flux from a blackbody in the wavelength interval *dλ*, centered around wavelength λ, per unit area of blackbody surface at thermodynamic temperature T. In this equation, c is the velocity of light, h is a constant (later named the *Planck constant*), and k is BOLTZMANN'S CONSTANT.

planet A nonluminous celestial body that orbits around the Sun or some other star. There are nine such large objects, or "major planets," in the solar system, and numerous "minor planets," or asteroids. The distinction between a planet and a satellite may not always be clear-cut, except for the fact that a satellite orbits around a planet. For example, our Moon is nearly the size of the planet Mercury and is very large in comparison to its parent planet, Earth. In some cases, Earth and the Moon can be treated almost as a "double-planet system"; the same is true for icy Pluto and its large satellite, or moon, Charon.

The largest planet is Jupiter, which has more mass than all the other planets combined. Mercury is the planet nearest the Sun, while (on the average) Pluto is the farthest away. At perihelion (the point in an orbit at which a celestial body is nearest the Sun), Pluto actually is closer to the Sun than Neptune. Saturn is the least dense planet in the solar system. If we could find some giant cosmic swimming pool, Saturn would float, since it is less dense than water. Seven of the nine planets have satellites, or moons, some of which are larger than the planet Mercury. *See also* ASTEROID; EARTH; JUPITER; MARS; MERCURY; NEPTUNE; PLUTO; SATURN; URANUS; VENUS.

planetary albedo The fraction of incident solar radiation that is reflected by a planet (and its atmosphere) and returned to space. For example, Earth has a planetary albedo of approximately 0.3—that that is, it reflects about 30% of the incident solar radiation back to space, mostly by backscatter from clouds in the atmosphere.

planetary orbital operations Planetary spacecraft can engage in two general categories of orbital operations: exploration of the planetary system and mapping of the planet. Exploring a planetary system includes making observations of the target planet and its system of satellites and rings. A mapping mission generally is concerned with acquiring large amounts of data about the planet's surface features. An orbit of low inclination at the target planet usually is well suited to a planetary system exploration mission, since it provides repeated exposure to satellites (moons) orbiting with the equatorial plane as well as adequate coverage of the planet and its magnetosphere. However, an orbit of high inclination is much better suited for a mapping mission, because the target planet will fully rotate below the spacecraft's orbit, thereby providing eventual exposure to every portion of the planet's surface. During either type of mission, the orbiting spacecraft is involved in an extended encounter period with the target planet and requires continuous (or nearly continuous) support from the flight team members at mission control on Earth. Galileo is an example of a planetary system exploration mission, while Magellan is an example of a planetary mapping mission. *See also* GALILEO PROJECT; MAGELLAN MISSION.

planetary orbit insertion The same type of highly precise navigation and course correction procedures used in flyby missions are applied during the cruise phase for a planetary orbiter mission. This process then places the spacecraft at precisely the correct location at the correct time to enter into an orbit about the target planet. However, orbit insertion requires not only the precise position and timing of a flyby mission but also a controlled deceleration. As the spacecraft's trajectory is bent by the planet's gravity, the command sequence aboard the spacecraft fires its retroengine(s) at the proper moment and for the proper duration. Once this retroburn (or retrofiring) has been completed successfully, the spacecraft is captured into orbit by its target planet. If the retroburn fails (or is improperly sequenced), the spacecraft will continue to fly past the planet. It is quite common for this retroburn to occur on the far side of a planet as viewed from Earth—requiring this portion of the orbit insertion sequence to occur essentially automatically (based on onboard commands) and without any interaction with the flight controllers on Earth.

planetary probe An INSTRUMENT-containing SPACECRAFT deployed in the ATMOSPHERE or on the surface of a planetary body in order to obtain in situ scientific data and environmental information about the CELESTIAL BODY.

planetesimals Small celestial objects in the solar system, such as asteroids, comets, and moons.

planet fall The landing of a spacecraft or space vehicle on a planet.

planetoid An asteroid, or minor planet. *See also* ASTEROID.

plasma An electrically neutral gaseous mixture of positive and negative ions. Sometimes called the "fourth state of matter," since a plasma behaves quite differently from solids, liquids, or gases.

plasma engine A reaction (rocket) engine using electromagnetically accelerated plasma as propellant. *See also* ELEC-TRIC PROPULSION.

plasma instruments Plasma detectors on a scientific spacecraft measure the density, composition, temperature, velocity, and three-dimensional distribution of plasmas that exist in interplanetary space and within planetary magnetospheres. Plasma detectors are typically sensitive to solar and planetary plasmas, and they can observe the solar wind and its interactions with a planetary system. *See also* MAGNETOSPHERE; SOLAR WIND.

plasma rocket A rocket engine in which the ejection of plasma generates thrust; a rocket using a plasma engine. *See also* ELECTRIC PROPULSION.

plasticity The tendency of a loaded body to assume a (deformed) state other than its original state, when the load is removed.

platform 1. An uncrewed (unmanned) orbital element of the SPACE STATION program that provides standard support services to payloads not attached to the SPACE STATION. 2. With respect to ballistic missile defense, a satellite in space that is used as a carrier for weapons, sensors, or both. *See also* BALLISTIC MISSILE DEFENSE.

Plesetsk The northern Russian LAUNCH SITE about 300 KILOMETERS (185 mi) south of Archangel that supports a wide variety of military space launches, BALLISTIC MISSILE testing, and space lift services for scientific or civilian SPACECRAFT requiring a POLAR ORBIT.

plug nozzle An annular nozzle that discharges exhaust gas with a radial inward component; a truncated aerospike. *See also* NOZZLE.

plume 1. In aerospace, the hot, bright exhaust gases from a rocket. 2. In environmental studies, a visible or measurable discharge of a contaminant from a given point of origin; can be visible or thermal in water, or visible in the air (e.g., a plume of smoke).

plutino (little Pluto) Any of numerous, small (~100-KILOMETER [62-mi] diameter), icy celestial bodies that occupy the inner portions of the KUIPER BELT and whose orbital motion resonance with NEPTUNE resembles that of PLUTO—namely, that each icy object completes two ORBITS around the SUN in the time Neptune takes to complete three orbits. *See also* TRANS-NEPTUNIAN OBJECT.

Pluto Pluto, the smallest planet in the solar system, has remained a mystery since its discovery by astronomer Clyde Tombaugh in 1930. Pluto is the only planet not yet viewed close-up by spacecraft, and given its great distance from the Sun and tiny size, study of the planet continues to challenge and extend the skills of planetary astronomers. In fact, most of what scientists know about Pluto has been learned since the late 1970s. Such basic characteristics as the planet's radius and mass were virtually unknown before the discovery of Pluto's moon Charon in 1978. Since then observations and speculations about the Pluto-Charon system—it is now considered a "double planet" system—have progressed steadily to a point where many of the key questions remaining about the system must await the close-up observation of a flyby space mission, such as NASA's planned PLUTO-KUIPER BELT MISSION.

For example, there is a strong variation in brightness, or albedo, as Pluto rotates, but planetary scientists do not know if what they are observing is a system of varying terrains, areas of different composition, or both. Scientists know there is a dynamic, largely nitrogen and methane atmosphere around Pluto that waxes (grows) and wanes (diminishes) with the planet's elliptical orbit around the Sun, but they still need to understand how the Plutonian atmosphere arises, persists, is again deposited on the surface, and how some of it escapes into space.

Telescopic studies (both Earth-based telescopes and NASA's Earth-orbiting HUBBLE SPACE TELESCOPE) indicate that Pluto and Charon are very different bodies. Pluto is more rock, while Charon appears more icy. How and when the two bodies in this interesting double-planet system could have evolved so differently is another key question that awaits data from a close-up observation. Data about Pluto and Charon, as gathered using ground-based and Earth-orbiting observatories, have helped improve our fundamental understanding of these planetary bodies. The most recent of these data are presented in Table 1 for Pluto and Table 2 for Charon.

Because of Pluto's highly eccentric orbit, its distance from the Sun varies from about 4.4 to 7.4 billion kilometers, or some 29.5 to 49.2 astronomical units (AU). For most of its orbit, Pluto is the outermost of the planets. However, from 1979 it actually orbited closer to the Sun than Neptune. This condition remained until 1999, when Pluto again became the outermost planet in the solar system. Upon its discovery in 1930, Clyde Tombaugh (keeping with astronomical tradition) named the planet Pluto—after the god of the underworld in Roman mythology. Inci-

dentally, this discovery was not a real astronomical surprise, since astronomers at the turn of the century had predicted Pluto's existence from perturbations (disturbances) observed in the orbits of both Uranus and Neptune.

However, the discovery of Charon by James W. Christy of the U.S. Naval Observatory in 1978, triggered a revolution in our understanding of Pluto. Charon was the boatman in Greek and Roman mythology who ferried the dead across the River Styx to the underworld, ruled by Pluto ("Hades" in Greek myth). This large moon has a diameter of about 1,270 kilometers, compared to tiny Pluto's diameter of 2,290 kilometers. These similar sizes encourage many planetary scientists to regard the pair as the only true double-planet system in our solar system. Charon circles Pluto at a distance of 19,405 kilometers. Its orbit is gravitationally synchronized with Pluto's rotation period (approximately 6.4 days) so that both the planet and its moon keep

Table 1. Dynamic Properties and Physical Data for Pluto	
Diameter	2,290 km
Mass	1.25×10^{22} kg
Mean Density	2.05 g/cm^3
Albedo (visual)	0.3
Surface Temperature (average)	~40K
Surface gravity	0.4–0.6 m/s^2
Escape velocity	1.1 km/s
Atmosphere (a transient phenomenon)	Nitrogen (N_2) and methane (CH_4)
Period of rotation	6.387 days
Inclination of axis (of rotation)	119.6°
Orbital period (around Sun)	248 years (90,591 days)
Orbit inclination	17.15°
Eccentricity of orbit	0.2482
Mean orbital velocity	4.75 km/s
Distance from Sun	
Aphelion	7.38×10^9 km (49.2 AU) (409.2 light-min)
Perihelion	4.43×10^9 km (29.5 AU) (245.3 light-min)
Mean Distance	5.91×10^9 km (39.5 AU) (328.5 light-min)
Solar flux (at 30 AU ~perihelion)	1.5 W/m^2
Number of known natural satellites	1

(Note: Some of these data are speculative.)

Source: NASA.

Table 2. Dynamic Properties and
Physical Data for Charon

Diameter	1,270 km
Mass	1.7×10^{21} kg
Mean density	1.8 g/cm^3
Surface gravity	0.2 m/s^2
Escape velocity	.58 m/s
Albedo (visual)	0.375
Mean distance from Pluto	19,405 km
Sidereal period (gravitational synchronized orbit)	6.387 days
Orbital inclination	98.8°
Orbital eccentricity	0

(Note: Some of these data are speculative.)
Source: NASA.

the same hemisphere facing each other at all times.

Planetary scientists currently believe that Pluto possesses a very thin atmosphere that contains nitrogen (N_2) and methane (CH_4). Pluto's atmosphere is unique in the solar system in that it undergoes a formation-and-decay cycle each orbit around the Sun. The atmosphere begins to form several decades before perihelion (the planet's closet approach to the Sun) and then slowly collapses and freezes out decades later, as the planet's orbit takes it farther and farther away from the Sun to the frigid outer extremes of the solar system. In September 1989 Pluto experienced perihelion. Several decades from now its thin atmosphere will freeze out and collapse, leaving a fresh layer of nitrogen and methane snow on the planet's surface.

Pluto-Kuiper Belt Mission Express

Originally conceived as the Pluto Fast Flyby (PFF), NASA's Pluto-Kuiper Belt Mission (also called the New Horizons Mission) is scheduled for launch in early 2006. This reconnaissance-type exploration MISSION will help scientists understand the interesting yet poorly understood worlds at the edge of the SOLAR SYSTEM. The first SPACECRAFT FLYBY of PLUTO and CHARON—the frigid double planet system—will take place as early as 2015. The mission will then continue beyond Pluto and visit one or more KUIPER BELT objects by 2026. The spacecraft's long journey will help resolve some basic questions about the surface features and properties of these icy bodies as well as their geology, interior makeup, and ATMOSPHERES.

With respect to the Pluto-Charon system, some of the major scientific objectives include the characterization of the global geology and geomorphology of Pluto and Charon, the mapping of the composition of Pluto's surface, and the determination of the composition and structure of Pluto's transitory atmosphere. It is intended that the spacecraft reach Pluto before the tenuous Plutonian atmosphere can refreeze onto the surface as the planet recedes from its 1989 PERIHELION. Studies of the double-planet system will actually begin some 12 to 18 months before the spacecraft's closest approach to Pluto in 2015. The modest-sized spacecraft will have no deployable structures and will receive all its electric power from long-lived RADIOISOTOPE THERMOELECTRIC GENERATORS (RTGs) that are similar in design to those used on the CASSINI MISSION to SATURN.

This important mission will complete the initial scientific reconnaissance of the solar system with ROBOT SPACECRAFT. At present, Pluto is the most poorly understood planet in the solar system. As some scientists speculate, the tiny planet may even be considered the largest member of the family of primitive icy objects that reside in the Kuiper belt. In addition to the first close-up view of Pluto's surface and atmosphere, the spacecraft will obtain gross physical and chemical surface properties of Pluto, Charon, and (possibly) several Kuiper belt objects. The CELESTIAL MECHANICS opportunity to launch this mission to Pluto by way of a GRAVITY-ASSIST from JUPITER occurs in January 2006.

pneumatic   Operated, moved, or effected by a pressurized gas (typically air) that is used to transmit energy.

pod   An enclosure, housing, or detachable container of some kind; for example, an *instrument pod* or an *engine pod.*

pogo   A term developed within the aerospace industry to describe the longitudinal (vertical) dynamic oscillations (vibrations) generated by the interaction of the launch vehicle's structural dynamics with the propellant and engine combustion process. The name appears to have originated from the similarity of this bumping phenomenon to the up-and-down bumping motion of a pogo stick.

pogo suppressor   A device within a liquid rocket engine that absorbs any vibration of the vehicle's structure during the engine combustion process. *See also* POGO.

point of impact   The point at which a projectile, bomb, or reentry vehicle inpacts or is expected to impact.

point of no return   1. The point along the launch track of an aerospace vehicle or space vehicle from which it can no longer return to its launch base. 2. For a human surface expedition on Mars, the point at which the team can no longer return to its surface base with the remaining supplies (e.g., air, food, water, or fuel). 3. For an aircraft, a point along its flight track

beyond which its endurance will not permit return to its own or some other associated base with its remaining fuel supply.

poise (symbol: P)   Unit of dynamic viscosity in the centimeter-gram-second (cgs) unit system. It is defined as the tangential force per unit area (e.g., dynes/cm^2) required to maintain a unit difference in velocity (e.g., 1 cm/sec) between two parallel plates in a liquid that are separated by a unit distance (e.g., 1 centimeter).

1 poise = (1 dyne-second)/(centimeter)^2 = 0.1 (newton-second)/(meter)^2

The unit is named after the French scientist Jean Louis Poiseuille (1799–1869). The centipoise, or 0.01 poise, is encountered often. For example, the dynamic viscosity of water at 20° Celsius is approximately 1 centipoise.

polar coordinates   A coordinate system in which a point (P) that is defined as P(x, y) in two-dimensional Cartesian coordinates is now represented as P(r, θ), where x = r cosθ and y = r sinθ. Physically, in the polar coordinate system: r is the radial distance in the x-y plane from the origin (O) to point P and θ is the angle formed between the x-axis and the radial vector (i.e., the line of length r from the origin O to point P).

In three dimensions, the Cartesian coordinate system point P(x,y,z) becomes P(r,θ,z) in *cylindrical polar coordinates*—where the point P (r,θ,z) is now regarded as lying on the surface of a cylinder. The terms r and θ are as previously defined, while z represents the height above (or below) the x-y plane. *See also* CARTESIAN COORDINATES; CYLINDRICAL COORDINATES.

Polaris (UGM-27)   An underwater/surface-launched, surface-to-surface, solid-propellant ballistic missile with inertial guidance and nuclear warhead. The Polaris (version UGM-27C) has a range of 4,600 kilometers (2,500 nautical miles). *See also* FLEET BALLISTIC MISSILE.

polarization   A distinct orientation of the wave motion and direction of travel of ELECTROMAGNETIC RADIATION, including LIGHT. Along with brightness and color, polarization is a special quality of light. It represents a condition in which the planes of vibration of the various rays in a BEAM of light are at least partially (if not completely) aligned.

polar orbit An ORBIT around a PLANET (or PRIMARY BODY) that passes over or near its POLES; an orbit with an INCLINATION of about 90°. (See Figure 59.)

polar orbiting platform (POP) An uncrewed (unmanned) spacecraft in polar or near-polar inclination that is operated from ground stations but dependent on the space station program to provide services for its complement of payloads. *See also* SPACE STATION.

Polar-orbiting Operational Environmental Satellite (POES) The National Oceanic and Atmospheric Administration's (NOAA's) Polar-orbiting Operational Environmental Satellites follow orbits that pass close to the North and South Pole as Earth rotates beneath them. Orbiting at an altitude of about 840 kilometers (520 miles), these satellites provide continuous global coverage of the state of Earth's atmos-

phere, including such essential information as atmospheric temperature, humidity, cloud cover, ozone concentration, and Earth's energy (radiation) budget. They also provide important surface data, such as sea-ice and sea-surface temperature and snow and ice coverings. *See also* NATIONAL POLAR-ORBITING OPERATIONAL ENVIRONMENTAL SATELLITE SYSTEM (NPOESS).

Polar Space Launch Vehicle (PSLV) A launch vehicle developed by the Indian Space Research Organization (ISRO). It consists of two solid stages, two hypergolic stages, and six strap-on solid rocket motors. This vehicle can place up to 3,000 kilograms into LOW EARTH ORBIT (LEO). It was first launched in 1993. *See also* LAUNCH VEHICLE.

poles The poles for a rotating CELESTIAL BODY are located at the ends (usually called north and south) of the body's AXIS of rotation.

popping Sudden, short-duration surges of pressure in a combustion chamber.

Population I stars Hot, luminous young STARS, including those like the SUN, that reside in the DISK of a SPIRAL GALAXY and are higher in heavy ELEMENT content (about 2% abundance) than POPULATION II STARS.

Population II stars Older STARS that are lower in heavy ELEMENT content than POPULATION I STARS and reside in GLOBULAR CLUSTERS as well as in the halo of a GALAXY—that is, the distant spherical region that surrounds a galaxy.

port 1. An opening; a place to access an aerospace system through which energy, data, or fluids may be supplied to or withdrawn from the system. For example, an *observation port,* a *refueling port,* or a *communications port.* 2. The left-hand side of an aerospace vehicle as a person looks from the rear of the craft to the nose.

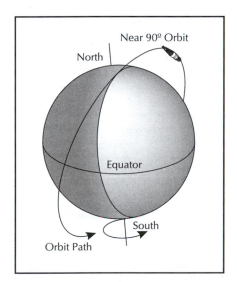

Figure 59. A spacecraft in polar orbit around Earth (or other planet).

posigrade rocket An auxiliary rocket that fires in the direction in which the vehicle is pointed, used, for example, in separating two stages of a vehicle. A posigrade course correction adds to the spacecraft's speed, in contrast to a *retrograde* correction, which slows it down. *See also* ORBITS OF OBJECTS IN SPACE.

positron (symbol: β⁺) An elementary particle with the mass of an electron, but charged positively. It is the "antielectron." A positron is emitted in some types of radioactive decay and also is formed in pair production by the interaction of high-energy gamma rays with matter. *See also* ANTIMATTER.

post-boost phase The phase of a missile trajectory, after the booster's stages have finished firing, in which the various reentry vehicles (RVs) are independently placed on ballistic trajectories toward their targets. In addition, penetration aids can be dispensed from the post-boost vehicle during this phase, which typically lasts between three and five minutes. *See also* BALLISTIC MISSILE DEFENSE.

post-boost vehicle (PBV) The portion of a ballistic missile payload that carries the multiple warheads and has maneuvering capability to place each warhead on its final trajectory to a target. Also referred to as a *bus*. *See also* BALLISTIC MISSILE DEFENSE.

potential energy (PE) Energy possessed by an object by virtue of its position in a gravity field. In general, the potential energy (PE) of an object can be expressed as:

$$PE = m \cdot g \cdot z$$

where m is the mass of the object, g is the acceleration due to gravity, and z is the height of the object above some reference position or datum.

pound (symbol: lb) 1. pound (mass) [lbm]: The unit of mass in the foot-pound-second (or U.S. customary) unit system. 1 lbm = 0.45359237 kilogram. 2. pound (force) [lbf]: The unit of force in the foot-pound-second (or U.S. custom-

ary) unit system equal to the force required to give a 1 pound mass at sea level on Earth an acceleration of 32.175 feet/second². 1 lbf = 4.448 newtons.

power (symbol: P) 1. In physics and engineering, the rate with respect to time at which work is done or energy is transformed or transferred to another location. In the SI UNIT system, power typically is measured in multiples of the watt, which is defined as one joule (of energy) per second. 2. In mathematics, the number of times a quantity is multiplied by itself; for example, b^3 means that b has been raised to the *third power*, or simply $b^3 = b \cdot b \cdot b$.

power plant In the context of aerospace engineering, the complete installation of an engine (or engines) with supporting subsystems (e.g., cooling subsystem, ignition subsystem) that generates the motive power for a self-propelled vehicle or craft, such as an aircraft, rocket, or aerospace vehicle.

preburners Some liquid-hydrogen–fueled rocket engines, such as the space shuttle main engines (SSMEs), have fuel and oxidizer preburners that provide hydrogen-rich hot gases at approximately 1030 kelvins (K) (760° Celsius) (1400° Fahrenheit). These gases then are used to drive the fuel and oxidizer high-pressure turbopumps.

precession The gradual periodic change in the direction of the axis of rotation of a spinning body due to the application of an external force (torque).

precession of equinoxes The slow westward motion of the EQUINOX points across the ECLIPTIC relative to the STARS of the ZODIAC caused by the slight wobbling of EARTH about its AXIS of rotation.

precombustion chamber In a liquid-propellant rocket engine, a chamber in which the propellants are ignited and from which the burning mixture expands torchlike to ignite the mixture in the main combustion chamber.

preflight In aerospace operations, something that occurs before launch vehicle liftoff.

preset guidance A technique of missile control wherein a predetermined flight path is set into the control mechanism and cannot be adjusted after launching. *See also* GUIDANCE SYSTEM.

pressurant Gas that provides ullage pressure in a propellant tank.

pressure (symbol: p) A thermodynamic property that two systems have in common when they are in mechanical equilibrium. Pressure can be defined as the normal component of force per unit area exerted by a fluid on a boundary. According to the *Pascal Principle*, the pressure at any point in a fluid at rest is the same in all directions. (Blaise Pascal [1623–62] was the French scientist who discovered this hydrostatic principle in the 1650s.) The fundamental principle of hydrostatics is:

$$dp/dz = -\rho g$$

where dp is the change in pressure corresponding to a change in height or depth (dz), ρ is the fluid density, and g is the acceleration due to gravity. Assuming that the density (ρ) of the fluid is constant, we can compute the hydrostatic pressure of a liquid from the following relationship:

$$p(z) = \rho\, g\, z$$

where p is the pressure and z is the height or depth with respect to some reference or datum, such as sea level.

In aerostatics, density varies (decreases) with height in the atmosphere, and, therefore, pressure cannot be evaluated as a function of height until the density is specified in some manner. The ideal gas law often can be used to describe the thermodynamic properties of a gas, such as air. Using this relationship

$$p = \rho\, R\, T$$

where p is the pressure of the gas, ρ is the density, T is the absolute temperature, and R is the specific gas constant.

The pascal (named in honor of Blaise Pascal) is the SI UNIT of pressure. It is defined as:

$$1 \text{ pascal} = 1 \text{ newton}/(\text{meter})^2$$

pressure-actuated seal A seal designed such that the pressure of the fluid being sealed activates or increases the sealing action.

pressure fed In rocketry, a propulsion system in which tank ullage pressure expels the propellants from the tanks into the combustion chamber of the engine.

pressure-laden sequence In rocketry, a method to accomplish fail-safe liquid-propellant engine starts by sequencing the operation of rocket engine control valves; the sequencing is achieved by vent mechanisms on the control system or propellant feed system, or both, that are triggered by pressure changes.

pressure regulator A pressure control valve that varies the volumetric flow rate through itself in response to a downstream pressure signal so as to maintain the downstream pressure nearly constant.

pressure suit In general, a garment designed to provide pressure upon the body so that the respiratory and circulatory functions may continue normally, or nearly so, under low-pressure conditions, such as occur at high altitudes or in space without benefit of a pressurized module or crew compartment. 1. (partial) A skintight suit that does not completely enclose the body but that is capable of exerting pressure on the major portion of the body in order to counteract an increased intrapulmonary oxygen pressure. 2. (full) A suit that completely encloses the body and in which a gas pressure, sufficiently above ambient pressure for maintenance of respiratory and circulatory functions, may be sustained.

pressurization For a launch vehicle prior to liftoff, the sequence of operations that increases the propellant tank ullage pressure to the desired level just before the main sequence of propellant flow and engine firing.

pressurization system The set of fluid system components that provides and maintains a controlled gas pressure in the

ULLAGE space of the launch vehicle propellant tanks.

pressurized habitable environment Any module or enclosure in space in which an astronaut may perform activities in a "shirt-sleeve" environment.

prestage A step in the action of igniting a large liquid-propellant rocket taken prior to the ignition of the full flow. It involves igniting a partial flow of propellants into the thrust chamber.

primary body The celestial body about which a satellite, moon, or other object orbits or from which it is escaping or toward which it is falling. For example, the primary body of Earth is the Sun; the primary body of the Moon is Earth.

primitive atmosphere The atmosphere of a planet or other celestial object as it existed in the early stages of its formation. For example, the primitive atmosphere of Earth some 3 billion years ago was thought to consist of water vapor, carbon dioxide, methane, and ammonia.

principal investigator (PI) The research scientist who is responsible for a space experiment and reporting the results.

prism A block (often triangular) of transparent material that disperses an incoming BEAM of white LIGHT into the visible SPECTRUM of rainbow colors—that is, red, orange, yellow, green, blue, indigo, and violet in order of decreasing WAVELENGTH.

probe 1. Any device inserted in an environment for the purpose of obtaining information about the environment. 2. An instrumented spacecraft or vehicle moving through the upper atmosphere or outer space or landing on another celestial body in order to obtain information about the specific environment, as, for example, a *deep-space probe,* a *lunar probe,* a *Jovian atmosphere probe.* 3. A slender device (e.g., a pitot tube) projected into a moving fluid for measurement purposes. 4. A ground-based set of

sensors that could be launched rapidly into space on warning of attack and then function as tracking and acquisition sensors to support weapon allocation and firing by BALLISTIC MISSILE DEFENSE (BMD) weapons against enemy intercontinental ballistic missiles (ICBMs) and reentry vehicles (RVs). 5. An extended structure on one spacecraft that is inserted into a compatible receptacle (called a *drogue*) on another spacecraft, creating a linkage (docking) of the two spacecraft.

Prognoz A series of scientific Russian spacecraft placed in highly elliptical Earth orbits for the purpose of investigating the Sun, the effects of the solar wind on Earth's magnetosphere, and solar flares. The first spacecraft in this series, *Prognoz 1,* was launched from Tyuratam on April 14, 1972. This 845-kilogram pound-mass spacecraft was placed in a highly elliptical Earth orbit that had a perigee of 950 kilometers and an apogee of 200,000 kilometers, an inclination of 65 degrees, and a period of 4.04 days.

prograde orbit An orbit having an inclination of between 0 and 90 degrees. *See also* ORBITS OF OBJECTS IN SPACE.

program In an aerospace context, a major activity that involves the human resources, material, funding, and scheduling necessary to achieve desired space exploration or space technology goals. For example, the *space shuttle program,* the *space station program,* and the *Hubble Space Telescope program.*

Progress An uncrewed supply spacecraft, configured to perform automated rendezvous and docking with orbiting Russian space stations, such as the *Salyut* and the *Mir.* After the Russian cosmonauts have unloaded the *Progress* cargo ship, it separates from the space station, deorbits, and then burn up in the upper regions of Earth's atmosphere.

progressive burning The condition in which the burning area of a solid-propellant rocket grain increases with time, thereby increasing pressure and thrust as the burning progresses.

project A planned undertaking of something to be accomplished, produced, or constructed, having a finite beginning and a finite ending; for example, the *Mercury Project* or the *Apollo-Soyuz Test Project*.

projectile An object propelled by an applied exterior force and continuing in motion by virtue of its own inertia, such as a bullet, shell, or missile.

prominence A cloud of cooler PLASMA extending high above the SUN's visible surface, rising above the PHOTOSPHERE into the CORONA.

propellant In general, a material, such as a fuel, an oxidizer, an additive, a catalyst, or any compound or mixture of these, carried in a rocket vehicle that releases energy during combustion and thus provides thrust to the vehicle. Propellants commonly are in either liquid or solid form. Modern launch vehicles use three types of liquid propellants: petroleum-based, cryogenic, and hypergolic.

The petroleum used as a rocket fuel is a type of kerosene similar to the compound burned in heaters and lanterns. However, the rocket petroleum is highly refined and is called RP-1 (Refined Petroleum). It usually is burned with liquid oxygen (LOX) to produce thrust. However, the RP-1/LOX propellant combination delivers considerably less *specific impulse* (I_{sp}) than that delivered by cryogenic propellant combinations.

The specific impulse is an important parameter in describing the performance of a rocket engine and the propellant combination being used. It is defined as:

$$I_{sp} = \text{(thrust force)}/$$
(mass flow rate of propellant)

In the SI UNIT system, specific impulse has the units of: newtons/(kilogram/second), which then reduces to meter/second. In the U.S. standard (or engineering) unit system, this important propellant performance parameter has the units of: pounds-force [lbf] per pounds-mass [lbm] per second (lbf/[lbm/s]), which often is simplified to just "second(s)," since 1 lbf = 1 lbm at sea level on Earth. This is a somewhat confusing situation, but as long as the basic definition is used consistently within a system of units, the specific impulse (expressed as lbf/(lbm/s), "seconds", or m/s) will describe the relative performance capabilities of different types of rocket propellant combinations.

The most important cryogenic propellants are liquid oxygen (LOX), which serves as the oxidizer, and liquid hydrogen (LH_2), which serves as the fuel. The word "cryogenic" is a derivative of the Greek word "kruos" (κρυοζ) meaning "very cold." Oxygen remains in a liquid state at temperatures of –183° Celsius (C) (–298° Fahrenheit [F]) and below. Hydrogen remains in a liquid state at temperatures of –253°C (–423°F) and below.

In gaseous form, oxygen and hydrogen have such low densities that extremely large tanks would be required to store them onboard a rocket or launch vehicle. But cooling and compressing these propellants into liquids greatly increases their densities, making it possible to store large quantities in much smaller volume tanks. However, the tendency of liquid cryogenic propellants to return to gaseous form unless kept very cold makes them difficult to store over long periods of time. Therefore, cryogenic propellant generally are not considered suitable, for military rockets, which must be kept launch-ready for months at a time.

But the high-performance capability of the LH_2/LOX combination makes this low-temperature storage problem acceptable when the launch vehicle reaction time and propellant storability issues are not too critical. The LOX/LH_2 combination of cryogenic propellants has been used for a variety of U.S. rockets including: the RL-10 engines on the Centaur upper stage, the J-2 engines used on the Saturn V rocket vehicle's second and third stages, and the space shuttle's main engines (SSMEs).

Using the simplified U.S. customary unit of specific impulse for comparison purposes, the Centaur RL-10 engine has a specific impulse of 444 seconds, the J-2 engines of the Saturn V upper stages had specific impulse ratings of 425 seconds, and the space shuttle main engines have a specific impulse rating of 455 seconds. In contrast, the giant cluster of F-1 engines in the Saturn V vehicle first stage burned LOX and kerosene and had a specific impulse rating of just 260 seconds. This same propellant combination was used by the booster stages of the Atlas/Centaur rocket combination and yielded a specific impulse rating of 258 seconds in the Atlas' booster engine and a 220 seconds rating in its sustainer engine.

Hypergolic propellants are fuels and oxidizers that ignite on contact with each other and, therefore, need no ignition source. This easy start/restart capability makes them attractive for both crewed and uncrewed spacecraft maneuvering systems. Another advantage of hypergolic propellants is the fact that they do not have the extremely low storage temperature requirements needed for cryogenic propellants.

Typically, monomethyl hydrazine (MMH) is used as the hypergolic fuel and nitrogen tetroxide (N_2O_4) as the oxidizer. Hydrazine is a clear, nitrogen/hydrogen compound with a "fishy" smell; it is similar to ammonia. Nitrogen tetroxide is a reddish fluid that has a pungent, sweetish smell. Both of these fluids are highly toxic and must be handled under the most stringent safety precautions.

Hypergolic propellants are used in the core liquid propellant stages of the Titan family of launch vehicles and on the second stage of the Delta launch vehicle. The space shuttle orbiter uses hypergolic propellants (or "hypergols") in its orbital maneuvering system (OMS), which is used for orbital insertion, major orbital maneuvers, and deorbit operations. The orbiter's reaction control system (RCS) also uses hypergols to accomplish attitude control.

The solid-propellant rocket motor is the oldest and simplest of all forms of rocketry, dating back to the ancient Chinese. In its simplest form, the solid-propellant rocket is just a casing (usually steel) that is filled with a mixture of chemicals (fuel and oxidizer) in solid form that burn at a rapid rate after ignition, expelling

hot gases from the nozzle to produce thrust. Solid-propellant rocket motors do not require turbopumps or complex propellant feed systems. A simple squib device at the top of the motor directs a high-temperature flame along the surface of the propellant grain, igniting it instantaneously.

Generally, solid propellants are stable and storable. Unlike a liquid-propellant rocket engine, however, a solid-propellant motor cannot be shut down. Once ignited, it will burn until all the propellant is exhausted.

Solid-propellant rockets have a variety of uses in space operations. Small solids often power the final stage of a launch vehicle or are attached to payload elements to boost satellites and spacecraft to higher orbits.

Medium-size solid-propellant rocket motors are used in special upper-stage systems that are designed specifically to boost a payload from LOW EARTH ORBIT (LEO) to geosynchronous Earth orbit (GEO) or to place a spacecraft on an interplanetary trajectory. The Payload Assist Module (PAM) and the Inertial Upper Stage (IUS) are examples of upper-stage systems.

The Scout launch vehicle is a four-stage rocket used to place small satellites into orbit. All four stages are solid-propellant rockets.

Finally, liquid-propellant launch vehicles such as the Titan, Delta, and space shuttle use solid-propellant rockets to provided added thrust at liftoff. The space shuttle, for example, uses the largest solid rocket motors ever built and flown. Each reusable solid rocket booster (SRB) contains 453,600 kilograms (1.1 million-pounds-mass) of propellant, in the form of a hard, rubbery substance with a consistency like that of a pencil eraser. A solid-propellant rocket motor always contains its own oxygen supply. The oxidizer in the shuttle's solid-rocket booster is ammonium perchlorate, which forms 69.93% of the mixture. The fuel is a form of powdered aluminum (16%), with an iron oxidizer powder (0.07%) as a catalyst. The binder that holds the mixture together is polybutadiene acrylic acid acrylonitrile (12.04%). In addition, the mixture contains an epoxy-curing agent (1.96%). The binder and epoxy also burn as fuel, adding thrust. The specific impulse of the shuttle's solid rocket booster is 242 seconds at sea level.

propellant mass fraction (symbol: ζ) In rocketry, the ratio of the propellant mass (m_p) to the total initial mass (m_o) of the vehicle before operation, including propellant load, payload, and structure.

$$\zeta = m_p/m_o$$

proper motion (symbol: μ) The apparent angular displacement of a STAR with respect to the CELESTIAL SPHERE. BARNARD'S STAR has the largest known proper motion.

propulsion system The launch vehicle or space vehicle system that includes the rocket engines, propellant tanks, fluid lines, and all associated equipment necessary to provide the propulsive force as specified for the vehicle. For example, typical launch vehicles, expendable or reusable, depend on chemical propulsion systems (i.e., solid-propellant and/or liquid-propellant rockets) to take a payload from the surface of Earth into orbit (generally LOW EARTH ORBIT). Upper-stage chemical rockets also serve as the propulsion system used to take certain payloads from low Earth orbit to geosynchronous Earth orbit or to place a spacecraft on an interplanetary trajectory. However, nuclear propulsion systems (thermal or electric) also might be used to take a payload/spacecraft from Earth orbit to Mars, to the outer reaches of the solar system and even beyond. Solar-electric propulsion systems can be used to take nonpriority payloads from Earth orbit to various locations in the inner solar system. Finally, very exotic propulsion systems, based on nuclear fusion reactions or even matter-antimatter annihilation reactions, have been suggested (stretching our current knowledge of physics and engineering) as the propulsive systems necessary for interstellar travel. *See also* ELECTRIC PROPULSION; LAUNCH VEHICLE; ROCKET; SPACE NUCLEAR PROPULSION; UPPER STAGE.

protium The most abundant ISOTOPE of HYDROGEN—ordinary hydrogen consist-

ing of one PROTON in the NUCLEUS surrounded by one ELECTRON. *See also* DEUTERIUM; TRITIUM.

protogalaxy A galaxy at the early stages of its evolution.

proton (symbol: p) A stable elementary nuclear particle with a single positive charge and a rest mass of about 1.672×10^{-27} kilograms, which is about 1,836 times the mass of an electron. A single proton makes up the nucleus of an ordinary (or) light hydrogen atom. Protons are also constituents of all other nuclei. The atomic number (Z) of an atom is equal to the number of protons in its nucleus.

Proton A Russian liquid-propellant expendable launch vehicle. The four-stage configuration of this vehicle often is used for interplanetary missions. It consists of three hypergolic stages and one cryogenic stage and is capable of placing 4,790 kilograms (kg) into geostationary transfer orbit (GTO), that is, an elliptical orbit with an apogee of 35,400 kilometers and a perigee of 725 kilometers. The Proton D-l vehicle consists of three hypergolic stages and can place up to 20,950 kg into LOW EARTH ORBIT. The three-stage Proton vehicle configuration was used to place *Salyut* and *Mir* space station modules into orbit. *See also* LAUNCH VEHICLE.

proton-proton chain reaction The series of THERMONUCLEAR FUSION reactions in stellar interiors by which four HYDROGEN nuclei are fused into a HELIUM NUCLEUS. This is the main ENERGY liberation mechanism in STARs like the SUN.

proton storm The burst or flux of protons sent into space by a solar flare. *See also* SUN.

protoplanet Any of a star's planets as such planets emerge during the process of accretion (accumulation), in which planetesimals collide and coalesce into large objects.

protostar A star in the making. Specifically, the stage in a young star's evolution after it has separated from a gas cloud but prior to its collapsing sufficiently to support thermonuclear reactions.

protosun The Sun as it emerged in the formation of the solar system.

prototype 1. Of any mechanical device, a production model suitable for complete evaluation of mechanical and electrical form, design, and performance. 2. The first of a series of similar devices. 3. A physical standard to which replicas are compared. 4. A spacecraft or component that has passed or is undergoing tests, qualifying the design for fabrication of complete flight units or components thereof. 5. A model suitable for evaluation of design, performance, and production potential.

Proxima Centauri The closest STAR to the SUN—the third member of the ALPHA CENTAURI triple star system. It is some 4.2 LIGHT-YEARS away.

Ptolemaic system The ancient Greek model of an EARTH-centered (GEOCENTRIC) UNIVERSE as described by PTOLEMY in the *ALMAGEST*. *Compare with* COPERNICAN SYSTEM.

Ptolemy (Claudius Ptolemaeus) (c. 100– c. 170 C.E.) Greek astronomer who lived in Alexandria, Egypt, and wrote (in about 150 C.E.) *Syntaxis* (The Great Mathematical Compilation)—a compendium of astronomical and mathematical knowledge from all the great Greek philosophers and astronomers. His book preserved the Greek geocentric cosmology in which EARTH was considered the unmoving center of the UNIVERSE while the wandering PLANETs and FIXED STARS revolved around it. Arab astronomers translated the book in about 820 C.E., calling it the *ALMAGEST* ("The Greatest"). The PTOLEMAIC SYSTEM remained essentially unchallenged in western thinking until NICHOLAS COPERNICUS and the start of the scientific revolution in the 16th century.

pulsar A stellar radio source that emits radio waves in a pulsating rhythm; believed to be a rotating neutron star.

pulse The variation of a quantity whose value is normally constant; this variation is characterized by a rise and a decay, and has a finite duration. This definition is so broad that it covers almost any transient phenomenon, such as the short burst of electromagnetic energy associated with a radar pulse. The only features common to all pulses are rise, finite duration, and decay (fall). The rise, duration, and decay must be of a quantity that is constant (not necessarily zero) for some time before the pulse and must have the same constant value for some time afterward.

pulse code modulation (PCM) The transmission of information by controlling the amplitude, position, or duration of a series of pulses. An analog signal (e.g., an image, music, a voice, etc.) is broken up into a digital signal (i.e., binary code) and then transmitted in this series of pulses via telephone line or radio waves.

pulsejet A jet engine containing neither compressor or turbine in which combustion takes place intermittently, producing thrust by a series of "miniexplosions," or combustion pulses. This type of engine is equipped with valves or shutters in the front that open and shut to take in air for the periodic combustion pulses. The World War II German V-1 "buzz-bomb" contained a pulsejet engine.

pump A machine for transferring mechanical energy from an external source to the fluid flowing through it. The increased energy is used to lift the fluid, to increase its pressure, or to increase its rate of flow.

pump-fed Term used to describe a propulsion system that incorporates a pump that delivers propellant to the combustion chamber at a pressure greater than the tank ullage pressure.

purge To rid a line or tank of residual fluid; especially of fuel or oxide in the tanks or lines of a rocket after a test firing or simulated test firing.

pyrogen A small rocket motor used to ignite a larger rocket motor.

pyrolysis Chemical decomposition of a substance due to the application of intense heat (thermal energy).

pyrometer An instrument for the remote (noncontact) measurement of temperatures. This term is generally applied to instruments that measure temperatures above 600° Celsius (1112° Fahrenheit).

pyrophoric fuel A fuel that ignites spontaneously in air. *Compare with* HYPERGOLIC FUEL (HYPERGOL).

pyrotechnic As an adjective, term used to describe a mixture of chemicals that, when ignited, is capable of reacting exothermically to produce light, heat, smoke, sound, or gas and also may be used to introduce a delay into an explosive train because of its known burning time. The term includes propellants and explosives; for example, *pyrotechnic device* and *pyrotechnic delay.*

pyrotechnics As a noun, igniters (other than pyrogens) in which solid explosives or energetic propellantlike chemical formulations are used as the heat-producing materials.

pyrotechnic subsystems Space system designers use electrically initiated pyrotechnic devices to open certain valves, ignite solid rocket motors, explode bolts to separate or jettison hardware, and to deploy spacecraft appendages. Pyrotechnic subsystems receive their electrical power from a bank of capacitors that are charged from the spacecraft's main power bus a few minutes prior to the planned detonation (activation) of a particular pyrotechnic device.

q In AEROSPACE engineering, the symbol commonly used for DYNAMIC PRESSURE. For example, after LIFTOFF the LAUNCH VEHICLE encounters "max q"—that is, maximum dynamic pressure—at 45 seconds into the flight.

quad A unit in the U.S. Standard System used to express very large amounts of energy. One quad is defined as a quadrillion (i.e., 10^{15}) British thermal units (BTUs). As a point of reference, the total annual energy consumption of the United States is less than 100 quads.

1 quad = 1×10^{15} BTU = 1.055×10^{18} joules = 2.93×10^{11} kilowatt-hours

quantity/distance (Q/D) Aerospace safety term for criteria that specify safe distances for the location of buildings that process or store rocket propellants or propellant ingredients (liquid or solid).

quantum (plural: quanta) In general, a discrete small unit; in any given physical process, it is considered the minimum permissible unit of energy. 2. In modern physics, the discrete bundle of energy possessed by a photon (e.g., a light wave or a gamma ray). The energy of the photon *(E)* is related to its frequency (v) by the equation:

$$E = h \cdot v$$

where *E* is the photon energy (joules), v is the frequency (hertz), and *h* is Planck's constant (6.626×10^{-34} joule-seconds).

quantum mechanics The physical theory that emerged from MAX PLANCK's original QUANTUM THEORY and developed into WAVE theory, matrix mechanics, and relativistic quantum mechanics in the 1920s and 1930s.

quantum theory The theory, first stated in 1900 by German physicist Max Planck (1858–1947), that all electromagnetic radiation is emitted and absorbed in "quanta," or discrete energy packets, rather than continuously. Each quantum of energy has a magnitude, hv, where h is the Planck constant (a quantity having the dimensions of energy x time, namely h = 6.626×10^{-34} joule-seconds) and v is the frequency. Planck's classic equation is, therefore, E = hv.

In his theory, Planck further postulated that a system (e.g., a BLACKBODY radiator) could change its energy only in discrete packets or an integral number of quanta (i.e., by hv, 2hv, 3hv, . . . nhv). Albert Einstein (1879–1955) used this quantum of energy concept in 1905 to explain the photoelectric effect, by assuming that light was radiated in quanta or photons. In 1913 the Danish physicist Niels Bohr (1885–1962), developed his theory of atomic spectra, by assuming that an atom (in this case, the hydrogen atom) can exist only in certain energy states and that photons (light) are emitted or absorbed when the atom changes from one energy state to another. Bohr proposed that the angular momentum of an orbiting electron can assume only an integral value (i.e., nh/2π, where n = 0,1, 2, 3 . . . and h is the Planck constant).

In quantum theory, therefore, physical quantities are "quantized"—that is, certain restrictions are imposed on these physical quantities limiting them to a set of discrete values. Quantum changes in energy are small and, subsequently, observable only on the atomic scale. Classical mechanics (as developed from Newton's laws) can adequately describe the behavior of large-scale (nonrelativistic) systems. The original quantum theory applied the "quantum of energy" concept to problems in classical mechanics. More precise developments and refinements, involving the formulation of a new system of mechanics, resulted in the creation of *quantum mechanics,* in the 1920s. Finally, the extension of quantum mechanics to include Einstein's special theory of relativity resulted in the formulation of relativistic quantum mechanics.

Quaoar The largest object found in the SOLAR SYSTEM since the discovery of PLUTO in 1930. First observed in June 2002, Quaoar is an icy world with a diameter of about 1,250 KILOMETERS (780 mi), making it about half the size of Pluto. It is an object located in the KUIPER BELT about 1.6 billion kilometers (1.0 × 10^9 mi) beyond Pluto and about 6.4 billion kilometers (4 × 10^9 mi) away from EARTH. The name Quaoar (pronounced kwah-o-wahr) comes from the creation mythology of the Tongva, a Native American people who inhabited the Los Angeles, California, area before the arrival of European explorers and settlers.

quasar A mysterious, very distant object with a high REDSHIFT—that is, traveling away from EARTH at great speed. These objects appear almost like STARS but are far more distant than any individual star now observed. They might be the very luminous centers of distant active galaxies. When first identified in 1963, they were called *quasi-stellar* radio sources, or quasars.These objects emit tremendous quantities of ENERGY from very small volumes. Some of the most distant quasars observed are so far away that they are receding at more than 90% of the SPEED OF LIGHT. Also called quasi-stellar object (QSO).

quiet Sun The collection of SOLAR phenomena and features, including the PHOTOSPHERE, the solar SPECTRUM, and the CHROMOSPHERE, that are always present. *Compare with* ACTIVE SUN.

rad The traditional unit in radiation protection and nuclear technology for absorbed dose of ionizing radiation. A dose of 1 rad means the absorption of 100 ergs of ionizing radiation energy per gram of absorbing material (or 0.01 joule per kilogram in SI UNITS). The term is an acronym derived from *radiation absorbed dose*. *See also* ABSORBED DOSE.

radar An active form of remote sensing generally used to detect objects in the atmosphere and space by transmitting electromagnetic waves (e.g., radio or microwaves) and sensing the waves reflected by the object. The reflected waves (called "returns" or "echos") provide information on the distance to the object (or target). The and the velocity of the object (or target). The reflected waves also can provide information about the shape of the object (or target). The term is an acronym for *radio detection and ranging*.

Satellite-borne radar systems (called *synthetic aperture radar* or *SAR systems*) can be used to study the features of a planet's surface when conditions (e.g., clouds or night time) obscure the surface from observation in the visible portion of the electromagnetic spectrum. *See also* RADAR ASTRONOMY; RADAR IMAGING; RADARSAT, REMOTE SENSING.

radar altimeter An active instrument, carried onboard an aircraft, aerospace vehicle, or spacecraft, used for measuring the distance (or altitude) of the vehicle or craft above the surface of a planet. An accurate determination of altitude is obtained by carefully timing the travel of a radar pulse down to the surface and back.

radar astronomy The use of radar by astronomers to study objects in our solar system, such as the Moon, the planets, asteroids, and even planetary ring systems. For example, a powerful radar telescope, such as the Arecibo Observatory, can hurl a radar signal through the "opaque" Venusian clouds (some 80 kilometers thick) and then analyze the faint return signal to obtain detailed information for the preparation of high-resolution surface maps. Radar astronomers can precisely measure distances to celestial objects, estimate rotation rates, and also develop unique maps of surface features, even when the actual physical surface is obscured from view by thick layers of clouds. *See also* ARECIBO OBSERVATORY; *MAGELLAN* MISSION.

radar imaging An active remote sensing technique in which a radar antenna first emits a pulse of microwaves that illuminates an area on the ground (called a *footprint*). Any of the microwave pulse that is reflected by the surface back in the direction of the imaging system is then received and recorded by the radar antenna. An image of the ground thus is made as the radar antenna alternately transmits and receives pulses at particular microwave wavelengths and polarizations. Generally, a radar imaging system operates in the wavelength range of one centimeter to one meter, which corresponds to a frequency range of about 300 megahertz (MHz) to 30 gigahertz (GHz). The emitted pulses can be polarized in a single vertical or horizontal plane. About 1,500 high-power pulses per second are transmitted toward the target or surface area to be imaged, with each pulse having a pulse width (i.e., pulse duration) of between 10 to 50 microseconds (μs). The pulse typically involves a small band of frequencies, centered on the operating frequency selected for the radar. The bandwidths used in imaging radar systems generally range from 10 to 200 MHz.

At the surface of Earth (or other planetary body, such as cloud-enshrouded Venus), the energy of this radar pulse is scattered in all directions, with some reflected back toward the antenna. The roughness of the surface affects radar backscatter. Surfaces whose roughness is much less than the radar wavelength scatter in the specular direction. Rougher surfaces scatter more energy in all directions, including the direction back to the receiving antenna.

As the radar imaging system moves along its flight path, the surface area illuminated by the radar also moves along the surface in a swath, building the image as a result of the radar platform's motion. The length of the radar antenna determines the resolution in the azimuth (i.e., along-track) direction of the image. The longer the antenna, the finer the spatial resolution in this dimension. The term "synthetic aperture radar" (SAR) refers to a technique used to synthesize a very long antenna by electronically combining the reflected signals received by the radar as it moves along its flight track.

As the radar imaging system moves, a pulse is transmitted at each position and the return signals (or echoes) are recorded. Because the radar system is moving relative to the ground, the returned signals (or echoes) are Doppler-shifted. This Doppler shift is negative as the radar system approaches a target and positive as it moves away from a target. When these Doppler-shifted frequencies are compared to a reference frequency, many of the returned signals can be focused on a single point, effectively increasing the length of the antenna that is imaging the particular point. This focusing operation, often referred to as SAR processing, is done rapidly through the use of high-speed digital computers and requires a very precise knowledge of the relative motion between the radar platform and the objects (surface) being imaged.

The synthetic aperture radar is now a well-developed technology that can be used to generate high-resolution radar images. The SAR imaging system is a unique remote sensing tool. Since it provides its own source of illumination, it can image at any time of the day or night, independent of the level of sunlight available. Because its radio-frequency wavelengths are much longer than those of visible light or infrared radiation, the SAR imaging system can penetrate through clouds and dusty conditions, imaging surfaces that are obscured to observation by optical instruments.

Radar images are composed of many picture elements, or pixels. Each dot or picture element in a radar image represents the radar backscatter from that area of the surface. Typically, dark areas in a radar image represent low amounts of backscatter (i.e., very little radar energy being returned from the surface), while bright areas indicate a large amount of backscatter (i.e., a great deal of radar energy being returned from the surface).

A variety of conditions determine the amount of radar signal backscatter from a particular target area. These conditions include: the geometric dimensions and surface roughness of the scattering objects in the target area, the moisture content of the target area, the angle of observation, and the wavelength of the radar system. As a general rule, the greater the amount of backscatter from an area (i.e., the brighter it appears in an image), the rougher the surface being imaged. Therefore, flat surfaces that reflect little or no radar (microwave) energy back toward the SAR imaging system usually appear dark or black in the radar image. In general, vegetation is moderately rough (with respect to most radar imaging system wavelengths) and consequently appears as gray or light gray in a radar image. Natural and human-made surfaces that are inclined toward the imaging radar system will experience a higher level of backscatter (i.e., appear brighter in the radar image) than similar surfaces that slope away from the radar system. Some areas in a target scene (e.g., the back slope of mountains) are shadowed and do not receive any radar illumination. These shadowed areas also will appear dark or black in the image. Urban areas provide interesting radar image results. When city streets are lined up in such a manner that the incoming radar pulses can bounce off the streets and then bounce off nearby buildings (a "double-bounce") directly back toward the radar system, the streets appear very bright (i.e., white) in a radar image. However, open roads and highways are generally physically flat surfaces that reflect very little radar signal, so often they appear dark in a radar image. Buildings that do not line up with an incoming radar pulse so as to reflect the pulse straight back to the imaging system actually appear light gray in a radar image, because buildings behave like very rough (diffuse) surfaces.

The amount of signal backscattered also depends on the electrical properties of the target and its water content. For example, a wetter object will appear bright, while a drier version of the same object will appear dark. A smooth body of water is the exception to this general rule. Since a smooth body of water behaves like a flat surface, it backscatters very little signal and will appear dark in a radar image.

Spacecraft-carried radar imaging systems have been used to explore and map the cloud-enshrouded surface of Venus. Earth-orbiting radar systems, such as NASA's Spaceborne Imaging Radar-C/X-band Synthetic Aperture Radar (SIR-C/X-SAR), which is carried into space in the space shuttle's cargo bay, support a variety of contemporary Earth resource observation and monitoring programs. *See also* MAGELLAN MISSION; PIONEER VENUS MISSION; *RADARSAT*; REMOTE SENSING.

Radarsat Canada's first remote sensing satellite, *Radarsat* was placed in an 800-kilometer polar orbit by a U.S. Delta II expendable launch vehicle on November 4, 1995 from Vandenberg Air Force Base in California. The principal instrument onboard this spacecraft is an advanced synthetic aperture radar (SAR), which produces high-resolution surface images

of Earth despite clouds and darkness. The *Radarsat*'s SAR is an active sensor that transmits and receives a microwave signal (variations of C-band) that is sensitive to the moisture content of vegetation and soil and supports crop assessments in Canada and around the world. SAR imagery can provide important geological information for mineral and petroleum exploration. The instrument also delineates ice cover and its extent (e.g., first-year ice versus heavier multiyear ice), thereby permitting the identification of navigable Arctic sea routes. *Radarsat* was developed under the management of the Canadian Space Agency (CSA) in cooperation with the U.S. National Aeronautics and Space Administration (NASA), the National Oceanic and Atmospheric Administration (NOAA), the provincial governments of Canada, and the private sector. The U.S. supplied the launch vehicle for the 2,700-kilogram satellite in exchange for 15% of its viewing time. Canada has 51% of the observation time. The rest of the observation time is being marketed by Radarsat International. The spacecraft has exceeded its five-year life expectancy. *See also* RADAR IMAGING; REMOTE SENSING.

radial-burning In rocketry, a solid-propellant grain that burns in the radial direction, either outwardly (e.g., an internal-burning grain) or inwardly (e.g., an internal-external burning tube or rod and tube grain). *See also* ROCKET.

radian A unit of angle. One radian is the angle subtended at the center of a circle by an arc equal in length to a radius of the circle.

$$1 \text{ radian} = 360°/(2\pi) = 57.2958°$$

radiance (symbol: L_e or L) For a point source of radiant energy, radiance is defined as the radiant intensity (I_e) in a specified direction per unit projected area; namely, $L_e = [\, d\, I_e/d\, A] \cdot \cos\theta$, where A is the area and θ is the angle between the surface and the specified direction. Radiance is expressed in watts per steradian per square meter (W sr^{-1} m^2).

radiant flux (symbol: Φ_e or Φ) The total power (i.e., energy per unit time) emitted or received by a body in the form of electromagnetic radiation. Generally expressed in watts (W).

radiant heat transfer The transfer of thermal energy (heat) by electromagnetic radiation that arises due to the temperature of a body. Most energy transfer of this type is in the infrared portion of the electromagnetic spectrum. However, if the emitting object has a high enough temperature, it also will radiate in the visible spectrum and beyond (e.g., the Sun). The term "thermal radiation" often is used to distinguish this form of electromagnetic radiation from other forms, such as radio waves, light, X rays, and gamma rays. Unlike convection and conduction, radiant heat transfer takes place in and through a vacuum.

radiant intensity (symbol: I_e or I) The radiant flux (Φ_e) emitted by a point source per unit solid angle; measured in watts per steradian (W sr^{-1}).

radiation The propagation of energy by electromagnetic waves (photons) or streams of energetic nuclear particles. The energy so propagated. *Nuclear radiation* generally is emitted from atomic nuclei (as a result of various nuclear reactions) in the form of alpha particles, beta particles, neutrons, protons, and/or gamma rays.

radiation belt The region(s) in a PLANET's MAGNETOSPHERE where there is a high DENSITY of trapped atomic PARTICLES from the SOLAR WIND. *See also* EARTH'S TRAPPED RADIATION BELTS.

radiation cooling In rocketry, the process of cooling a combustion chamber or nozzle in which thermal energy losses by radiation heat transfer from the outer combustion chamber or nozzle wall (to the surroundings) balances the heat gained (on the inner wall) from the hot combustion gases. As a result of radiation cooling, the combustion chamber or nozzle operates in a state of thermal equilib-

rium. To accommodate such heat transfer processes, radiatively cooled rocket engine components often are made of refractory metals or graphite.

radiation hardness The ability of electronic components and systems to function in high-intensity nuclear (ionizing) radiation environments. Spacecraft passing through or operating in a planet's trapped radiation belts would experience high-intensity nuclear radiation environments. Similarly, spacecraft exposed to solar flare conditions also can experience high-intensity radiation environments, depending on their distance from the Sun and the relative strength of the solar flare event. Techniques for increasing the radiation hardness of electronic components and systems include use of semiconductors that are less susceptible to ionizing radiation upset, shielding, reduction in size of the equipment/components, and redundancy.

radiation laws The collection of empirical and theoretical laws that describe radiative transport phenomena. The four fundamental laws that together provide a fundamental description of the behavior of blackbody radiation are: (1) *Kirchoff's law of radiation*—which relationship between emission and absorption at any given wavelength and a given temperature; (2) *Planck's radiation law*—which describes the variation of intensity of a blackbody radiator at a given temperatures as a function of wavelength; (3) the *Stefan-Boltzmann law*—which relates the time rate of radiant energy emission from a blackbody to the fourth power of its absolute temperature; and (4) *Wien's displacement law*—which relates the wavelength of maximum intensity emitted by a blackbody to its absolute temperature. *See also* BLACKBODY; KIRCHOFF'S LAW (OF RADIATION); PLANCK'S RADIATION LAW; STEFAN-BOLTZMANN LAW; WIEN'S DISPLACEMENT LAW.

radiation sickness A potentially fatal illness resulting from excessive exposure to IONIZING RADIATION. *See also* ACUTE RADIATION SYNDROME.

radiator 1. Any source of radiant energy. 2. In aerospace engineering, a device that rejects waste heat from a spacecraft or satellite to outer space by radiant heat transfer processes. Radiator design depends on both operating temperature and the amount of thermal energy to be rejected. The amount of waste heat that can be radiated to space by a given surface area is determined by the Stefan-Boltzmann law and is proportional to the fourth power of the radiating surface temperature. The surface area and mass of the thermal radiator are very sensitive to the heat rejection temperature. Higher heat rejection temperatures correspond to smaller radiator areas and, therefore, lower radiator masses. For radioisotope thermoelectric generator (RTG) power systems, for example, radiator temperatures are typically about 575 kelvins (K).

Radiators can be fixed or deployable. Flat plate and cruciform configurations often are used. The radiator can be solid, all-metal construction, or else contain embedded coolant tubes and passageways to assist in the transport of thermal energy to all portions of the radiator surface. However, radiators with embedded coolant tubes and channels must be sufficiently thick ("armored") to protect against meteoroid impact damage and the subsequent loss of coolant. Heat pipe radiator configurations also have been considered. *See also* STEFAN-BOLTZMANN LAW; SPACE NUCLEAR POWER.

radioactivity The spontaneous decay or disintegration of an unstable atomic nucleus, usually accompanied by the emission of ionizing radiation, such as alpha particles, beta particles, and gamma rays. The radioactivity, often shortened to just "activity," of natural and human-made (artificial) radioisotopes decreases exponentially with time, governed by the fundamental relationship:

$$N = N_0 \, e^{-\lambda t}$$

where N is the number of radionuclides (of a particular radioisotope) at time (t), N_0 is the number of radionuclides of that particular radioisotope at the start of the count (i.e., at t = 0), λ is the decay constant of the radioisotope, and t is time.

The decay constant (λ) is related to the half-life ($T_{1/2}$) of the radioisotope by the equation:

$$\lambda = (\ln 2)/T_{1/2} = 0.69315/T_{1/2}$$

The half-lives for different radioisotopes vary widely in value, from as short as about 10^{-8} seconds to as long as 10^{10} years and more. The longer the half-life of the radioisotope, the more slowly it undergoes radioactive decay. For example, the natural radioisotope uranium-238 has a half-life of 4.5×10^9 years and therefore is only slightly radioactive (emitting an alpha particle when it undergoes decay).

Radioisotopes that do not normally occur in nature but are made in nuclear reactors or accelerators are called *artificial radioactivity* or *human-made radioactivity*. Plutonium-239, with a half-life of 24,400 years, is an example of artificial radioactivity.

radio astronomy Branch of astronomy that collects and evaluates radio signals from extraterrestrial sources. Radio astronomy is a relatively young branch of astronomy. It was started in the 1930s, when Karl Jansky (1905–1950), an American radio engineer, detected the first extraterrestrial radio signals. Until Jansky's discovery, astronomers had used only the visible portion of the electromagnetic spectrum to view the universe. The detailed observation of cosmic radio sources is difficult, however, because these sources shed so little energy on Earth. But starting in the mid-1940s with the pioneering work of the British astronomer Sir Alfred Charles Bernard Lovell (b. 1913), at the United Kingdom's Nuffield Radio Astronomy Laboratories at Jodrell Bank, the radio telescope has been used to discover some extraterrestrial radio sources so unusual that their very existence had not even been imagined or predicted by scientists.

One of the strangest of these cosmic radio sources is the pulsar—a collapsed giant star that emits pulsating radio signals as it spins. When the first pulsar was detected in 1967, it created quite a stir in the scientific community. Because of the regularity of its signal, scientists thought they had just detected the first interstellar signals from an intelligent alien civilization.

Another interesting celestial object is the quasar, or quasi-stellar radio source. Discovered in 1964, quasars now are considered to be entire galaxies in which a very small part (perhaps only a few light-days across) releases enormous amounts of energy—equivalent to the total annihilation of millions of stars. Quasars are the most distant known objects in the universe; some of them are receding from us at over 90% of the speed of light. *See also* ARECIBO OBSERVATORY; PULSAR; QUASAR.

radio frequency (RF) In general, a frequency at which electromagnetic radiation is useful for communication purposes; specifically, a frequency above 10,000 hertz and below 3×10^{11} hertz. One hertz is defined as 1 cycle per second. *See also* ELECTROMAGNETIC SPECTRUM.

radio galaxy A galaxy (often exhibiting a dumbbell-shaped structure) that produces very strong signals at radio wavelengths. Cygnus A, one of the closest bright radio galaxies, is about 650 million light-years away. *See also* RADIO ASTRONOMY.

radioisotope A radioactive isotope. An unstable isotope of an element that spontaneously decays or disintegrates, emitting nuclear radiation (e.g., an alpha particle, beta particle, or gamma ray). More than 1,300 natural and artificial radioisotopes have been identified.

radioisotope thermoelectric generator (RTG) A space power system in which thermal energy (heat) deposited by the absorption of alpha particles from a radioisotope source (generally plutonium-238) is converted directly into electricity. Radioisotope thermoelectric generators, or RTGs, have been used in space missions where long life, high reliability, operation independent of the distance or orientation to the Sun, and operation in severe environments (e.g., lunar night, Martian dust storms) are critical. The first RTG used by the United States in space was the SNAP 3B design for the U.S. Navy's *Transit 4A*

and *4B* satellites, which were launched successfully into orbit around Earth in June and November 1961, respectively. SNAP is an acronym that stands for *Systems for Auxiliary Nuclear Power*.

SNAP-27 units provided electric power to instruments left on the lunar surface by the Apollo astronauts. SNAP-19 units provided electric power to the two highly successful Viking lander spacecraft that touched down on Mars in 1976. The *Pioneer 10/11* spacecraft that visited Jupiter and Saturn (*Pioneer 11* only) also were powered by SNAP-19 units. The *Voyager 1/2* spacecraft that explored the outer regions of our solar system, including flybys of Jupiter, Saturn, Uranus (*Voyager 2* only) and Neptune (*Voyager 2* only) were powered by an RTG unit called the Multihundred Watt (MHW) Generator. Finally, an RTG unit called the General Purpose Heat Source (GPHS) is providing electric power to more recent spacecraft, such as *Galileo Ulysses,* and *Cassini.*

The radioisotope thermoelectric generators currently designed for space missions (i.e., the GPHS) contain several kilograms of an isotopic mixture of plutonium (primarily the isotope plutonium-238) in the form of an oxide, pressed into a ceramic pellet. The pellets are arranged in a converter housing and function as a heat source to generate the electricity provided by the RTG. The radioactive decay of plutonium-238, which has a half-life of 87.7 years, produces heat, some of which is converted into electricity by an array of thermocouples made of silicon-germanium junctions. Waste heat then is radiated into space from an array of metal fins.

Plutonium, like all radioactive substances and many nonradioactive materials, can be a health hazard under certain circumstances, especially if inadvertently released into the terrestrial environment. Consequently, RTGs fueled by plutonium are designed under rigorous safety standards with the specific goal that these devices be capable of surviving all credible launch accident environments without releasing their plutonium content.

Prior to launch, a nuclear-powered spacecraft undergoes a thorough safety analysis and review by various agencies of the government, including the Department of Energy. The results of that safety analysis and review also are evaluated by an independent panel of experts. Both these safety reviews are then sent to the White House; where the staff makes a final evaluation concerning the overall risk presented by the mission. Finally, presidential approval is required before a spacecraft carrying an RTG can be launched by the United States. *See also* SPACE NUCLEAR POWER; THERMOELECTRIC CONVERSION.

radiometer   An instrument for detecting and measuring radiant energy, especially infrared radiation. *See also* REMOTE SENSING.

radionuclide   A radioactive isotope characterized according to its atomic mass (A) and atomic number (Z). Radionuclides experience spontaneous decay in accordance with their characteristic half-life and can be either naturally occurring or human-made.

radio telescope   A large, metallic device, generally parabolic (dish-shaped), that collects radio wave signals from extraterrestrial objects (e.g., active galaxies and pulsars) or from distant spacecraft and focuses these radio signals onto a sensitive radio-frequency (RF) receiver. (See Figure 60.) *See also* ARECIBO OBSERVATORY; RADIO ASTRONOMY.

radio waves   Electromagnetic waves of wavelength between about 1 millimeter (0.001 meter) and several thousand kilometers; and corresponding frequencies between 300 gigahertz and a few kilohertz. The higher frequencies are used for spacecraft communications. *See also* ELECTROMAGNETIC SPECTRUM.

rail gun   A device that uses the electromagnetic force experienced by a moving current in a transverse magnetic field to accelerate very small objects to very high velocities. *See also* GUN-LAUNCH TO SPACE.

ramjet   A reaction (jet-propulsion) engine containing neither compressor nor turbine

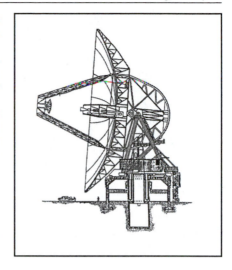

Figure 60. A large parabolic radio telescope, or "dish," as it is often called.

that depends for its operation on the air compression accomplished by the forward motion of the vehicle. The ramjet has a specially shaped tube or duct open at both ends into which fuel is fed at a controlled rate. The air needed for combustion is shoved, or "rammed," into the duct and compressed by the forward motion of the vehicle/engine assembly. This rammed air passes through a diffuser and is mixed with fuel and burned. The combustion products then are expanded in a nozzle. A ramjet cannot operate under static conditions. The duct geometry depends on whether the ramjet operates at subsonic or supersonic speeds.

range   1. The distance between any given point and an object or target. 2. Extent or distance limiting the operation or action of something, such as the *range* of a surface-to-air missile. 3. An area in and over which rockets, missiles, or aerospace vehicles are fired for testing.

Ranger Project   The Ranger spacecraft were the first U.S. robot spacecraft sent toward the Moon in the early 1960s to pave the way for the Apollo Project's human landings at the end of that decade.

Ranger Mission Summary (1961–1965)			
Mission	*Launch Date*	*Objective*	*Result(s)*
Ranger 1	08/23/61	Lunar mission prototype	Launch failure
Ranger 2	11/18/61	Lunar mission prototype	Launch failure
Ranger 3	01/26/62	Lunar (impact) probe	Spacecraft failed; missed Moon
Ranger 4	04/23/62	Lunar (impact) probe	Spacecraft failed; impact farside
Ranger 5	10/18/62	Lunar (impact) probe	Spacecraft failed; missed Moon
Ranger 6	01/30/64	Lunar (impact) probe	Successful impact; Cameras failed
Ranger 7	07/28/64	Lunar (impact) probe	Successful impact; numerous images
Ranger 8	02/17/65	Lunar (impact) probe	Successful impact; numerous images
Ranger 9	03/21/65	Lunar (impact) probe	Successful impact; numerous images

Source: NASA data.

The Rangers were a series of fully attitude-controlled spacecraft designed to photograph the lunar surface at close range before impacting. *Ranger 1* was launched on August 23, 1961 and set the stage for the rest of the Ranger missions, by testing spacecraft navigational performance. *Ranger 2* through *9* were launched from November 1961 through March 1965. (See Table) All of the early Ranger missions (*Ranger 1* through 6)suffered setbacks of one type or another. Finally, *Ranger 7, 8*, and *9* succeeded with flights that returned many thousands of images (before impact) and greatly advanced scientific knowledge about the lunar surface. *See also* APOLLO PROJECT; MOON; SURVEYOR.

Rankine cycle A fundamental thermodynamic heat engine cycle in which a working fluid is converted from a liquid to a vapor by the addition of heat (in a boiler), and the hot vapor then is used to spin a turbine, thereby providing a work output. This work output often is used to rotate a generator and produce electric power. After passing through the turbine, the vapor is cooled in a condenser (rejecting heat to the surroundings). While being cooled, the working fluid changes phase (from vapor to liquid) and then passes through a pump to start the cycle again. *See also* BRAYTON CYCLE; CARNOT CYCLE; HEAT ENGINE; THERMODYNAMICS.

Rankine temperature scale (symbol: R) An absolute temperature scale with the degree-interval of the Fahrenheit (F) temperature scale and the zero point at absolute zero. Consequently, 0°R is equivalent to −459.67°F; the normal boiling point of water is 671.67°R or 212°F.

raw data Data that have not been reduced or processed.

Rayleigh scattering Selective scattering process caused by spherical particles (e.g., gas molecules) whose characteristic dimensions are about one-tenth (or less) of the wavelength (λ) of the radiation being scattered. Rayleigh scattering is inversely proportional to the fourth power of the wavelength—that is, this scattering process goes as λ^{-4}. For the scattering of visible light, this means that blue light ($\lambda = 0.4$ μm) is scattered much more severely by Earth's atmosphere than longer-wavelength red light ($\lambda = 0.7$ μm). For a clear daytime atmosphere, this selective scattering causes the sky to appear blue to an observer who is not along the direction of illumination. However, at sunrise and sunset, sunlight must pass through a much thicker column of the atmosphere and the preferential scattering blue light is so complete that essentially only red, orange, and yellow light reaches an observer on Earth. Named after the British physicist Lord Rayleigh (1842–1919).

Another selective scattering process is called *Mie scattering*. It involves scattering of radiation by particles whose characteristic size is between one-tenth (0.1) and 10 times the wavelength of the incident radiation. Smoke, fumes, and haze in Earth's atmosphere result in this type of scattering, which is inversely proportional to the wavelength of the radiation (i.e., the process goes as λ^{-1}).

Nonselective scattering is caused by particles (such as found in clouds and fog) with characteristic dimensions of more than about 10 times the wavelength of the radiation being scattered. These larger particles scatter all wavelengths of light equally well, so that clouds and fog appear white, even though they contain particles of water, which is actually colorless. *See also* MIE SCATTERING.

rays (lunar) Bright streaks extending across the surface from young IMPACT CRATERS on the MOON; also observed on MERCURY and on several of the large MOONS of the OUTER PLANETS.

reaction 1. In chemistry, a chemical change that occurs when two or more substances are mixed, usually in solution; chemical activity between substances. 2. In physics, the equal in magnitude but opposite in direction force that occurs in response to the action of some other force in accordance with Newton's Third Law of Motion (i.e., the action-reaction principle). *See also* NEWTON'S LAWS OF MOTION.

reaction-control jets Small propulsion units on a spacecraft or aerospace vehicle that are used to rotate the craft or to accelerate it in a specific direction.

reaction control system (RCS) The collection of thrusters on the space shuttle orbiter vehicle that provides attitude control and three-axis translation during orbit insertion, on-orbit operations, and reentry. *See also* SPACE TRANSPORTATION SYSTEM.

reaction engine An engine that develops thrust by its physical reaction to the ejection of a substance (including possibly photons and nuclear radiations) from it; commonly, the reaction engine ejects a stream of hot gases created by combusting a propellant within the engine.

A reaction engine operates in accordance with Newton's Third Law of Motion (i.e., the action-reaction principle). Both rocket engines and jet engines are reaction engines. Sometimes called a *reaction motor*. *See also* NEWTON'S LAWS OF MOTION; ROCKET.

reaction time In human factors engineering, the interval between an input signal (physiological) or a stimulus (psychophysiological) and a person's response to it. Normally it takes a human being at least three-tenths of a second (0.3 sec) to begin to respond to a crisis or emergency situation (e.g., applying the brakes of an automobile). Training and special alarms and signals can reduce this reaction time a bit, while inexperience, panic, or fear can significantly lengthen it.

readout 1. (verb) The action of a spacecraft's transmitter sending data that either are being instantaneously acquired or else extracted from storage (often by playing back a magnetic tape upon which the data have been recorded previously). 2. (noun) The data transmitted by the action described in sense 1. 3. (verb) In computer operations, to extract information from storage.

readout station A recording (or receiving) station at which the data-carrying radio-frequency signals transmitted by a spacecraft or space probe are acquired and initially processed.

real-time Time in which reporting on or recording events is simultaneous with the events; essentially, "as it happens."

real-time data Data presented in usable form at essentially the same time the event occurs.

reconnaissance satellite An Earth-orbiting military satellite intended to perform reconnaissance missions against enemy nations and potential adversaries. In the 1960s the United States developed and flew its first generation of photoreconnaissance satellites, called the CORONA, ARGON, and LANYARD systems.

In 1958 during the cold war, President Dwight D, Eisenhower approved a program that would answer questions about Soviet missile capability and replace risky U-2 reconnaissance flights over Soviet territory. The Central Intelligence Agency (CIA) and the U.S. Air Force would jointly develop satellites to photograph denied areas from space. The program had both a secret mission and a secret name—CORONA.

The CIA and the air force developed this first-generation space program with great speed and tight secrecy. In August 1960, after numerous unsuccessful attempts, a CORONA satellite was launched successfully and its film capsule was recovered from space. During the next 12 years, CORONA satellites ushered in a new era of technical intelligence and a new era of space "firsts" that also contributed to advancements in other areas of the national space program. For example, the CORONA program provided valuable experience on how to recover objects from orbit—experience that was later used to recover astronauts from orbit during NASA's Mercury, Gemini, and Apollo projects. CORONA also provided a fast and relatively inexpensive way to map Earth from space.

But the most important contribution of the CORONA system to national security came from the intelligence it provided. During the cold war, CORONA looked through the Iron Curtain and helped lay the groundwork for important disarmament agreements. For example, with reconnaissance satellites, the United States could verify reductions in missiles without on-site inspections. Satellite imagery gave U.S. leaders the confidence to enter into negotiations and to sign important arms control agreements with the Soviet Union. Successor programs continued to monitor intercontinental ballistic missile (ICBM) sites and verify strategic arms agreements and the Nuclear Nonproliferation Treaty.

The CORONA program operated from August 1960 to May 1972 and collected both intelligence and mapping imagery. ARGON was a mapping system that used the organizational framework of CORONA and achieved 7 of 12 successful missions from May 1962 to August 1964. Finally, LANYARD was an attempt to gain higher-resolution imagery. It flew one successful mission in 1963.

Early imagery collections were driven, for the most part, by the pressures of the cold war and the needs of the United States to confirm suspected developments in Soviet strategic arms capabilities. Worldwide photographic coverage also was used to produce maps and charts for the Department of Defense and other U.S. government mapping programs.

A presidential executive order (dated February 24, 1995) has declassified more than 800,000 images collected by these early photoreconnaissance systems from 1960 to 1972. The historic, declassified imagery (some with a resolution of 1.82 meters [6 feet]) is being made available to assist environmental scientists improve their understanding of global environmental processes and to develop a baseline in the 1960s for assessing important environmental changes. *See also* NATIONAL RECONNAISSANCE OFFICE (NRO).

recovery The procedure or action that occurs when the whole of a satellite, spacecraft, scientific instrument package, or component of a rocket vehicle is recovered after a launch or mission; the result of this procedure.

rectifier A device for converting alternating current (a.c.) into direct current (d.c.); usually accomplished by permitting current flow in one direction only.

recycle In aerospace launch operations, to stop the count and to return to an earlier point in the countdown; as, for example, in "we have *recycled* and are now at T minus two hours and counting."

red dwarf (star) Reddish MAIN-SEQUENCE STARS (spectral type K and M) that are relatively cool (~ 4000 K surface TEMPERATURE) and have low MASS (about 0.5 SOLAR MASSES or less). These faint, low-LUMINOSITY stars are inconspicuous, yet they represent the most common type of star in the UNIVERSE and the longest lived. BARNARD'S STAR is an example.

Redeye (FIM-43) A U.S. Army, lightweight, person-portable, shoulder-fired air defense artillery weapon for low-altitude air defense of forward combat area troops.

red fuming nitric acid (RFNA) Concentrated nitric acid (HNO_3) in which nitrogen dioxide (NO_2) has been dissolved. This toxic and corrosive mixture typically contains between 5 and 20% nitrogen dioxide. It is used as an oxidizer in liquid-propellant rockets.

red giant (star) A large, cool STAR with a surface TEMPERATURE of about 2,500 K and a diameter 10 to 100 times that of the SUN. This type of highly luminous star is at the end of its evolutionary life, departing from the main sequence after exhausting the HYDROGEN in its CORE. It is often a VARIABLE STAR. Some 5 billion years from now, the Sun will evolve into a massive red giant. See also MAIN-SEQUENCE STAR.

redline 1. (noun) Term denoting a critical value for a parameter or a condition that, if exceeded, threatens the integrity of a system, the performance of a vehicle, or the success of a mission. 2. (verb) To establish a critical value as described in sense 1.

Red Planet The PLANET MARS—so named because of its distinctive reddish soil.

redshift The apparent increase in the wavelength of a light source caused by the receding motion of the source. The Doppler-effect shift of the visible spectra of distant galaxies toward red light (i.e., longer wavelength) indicates that these galaxies are receding. The greater redshift observed in more distant galaxies has been interpreted that the universe expanding. Compare with BLUESHIFT.

Redstone An early, liquid-propellant, medium-range surface-to-surface missile used by the U.S. Army. The Redstone served as the first stage of the Jupiter-C vehicle that launched the first American satellite, *Explorer-1*, in January 1958. It also served as the launch vehicle for Astronaut Alan Shepard's suborbital flight on May 5, 1961, in the Mercury *Freedom 7* capsule—the first American human space mission. See also EXPLORER I; MERCURY PROJECT.

reentry The return of objects, originally launched from Earth, back into the sensible (measurable) atmosphere; the action involved in this event. The major types of reentry are: ballistic, gliding, and skip. To perform a safe controlled reentry, a spacecraft or aerospace vehicle must be capable of achieving a controlled dissipation of its kinetic and potential energies. The kinetic and potential energy values are determined by the returning object's velocity and altitude at the atmospheric reentry interface. Successful reentry, culminating in a safe landing on the surface of Earth (land or water), requires a very carefully designed and maintained flight trajectory.

Random or uncontrolled reentry, as might occur with a derelict satellite or piece of space debris, normally results in excessive aerodynamic heating and the burnup of the object in the upper atmosphere. On certain occasions, however, natural or human-made objects have undergone uncontrolled reentry and actually survived the fiery plunge through the atmosphere, impacting on Earth. Also called *entry* in the aerospace literature.

reentry body That part of a rocket or space vehicle that reenters Earth's atmosphere after flight above the sensible (measurable) atmosphere.

reentry corridor The narrow region or pathway, established primarily by velocity and altitude, along which a spacecraft or aerospace vehicle must travel in order to return safely to Earth's surface. The design of the vehicle and the operational circumstances (e.g., normal end-of-mission return to Earth, emergency reentry, etc.) also contribute to the dimensions of the reentry corridor.

reentry nose cone A nose cone designed especially for reentry; usually consisting of one or more chambers protected by a specially designed outer surface that serves as both an aerodynamic surface and a heat shield.

reentry phase That portion of the trajectory of a ballistic missile or aerospace vehicle where there is a significant interaction of the vehicle and Earth's sensible (measurable) atmosphere, following flight in outer space. *See also* BALLISTIC MISSILE DEFENSE.

reentry trajectory The trajectory or pathway followed by an aerospace vehicle, spacecraft, or other object during reentry. If the object is unguided after passing the reentry interface at the top of the sensible (measurable) atmosphere, its trajectory will be essentially ballistic. With controlled aerodynamic maneuvering and guidance during reentry, however, the object would follow a glide or skip trajectory as it descends to the surface of Earth.

reentry vehicle (RV) 1. In general, that part of a space vehicle designed to reenter Earth's atmosphere in the terminal portion of its trajectory. 2. In the context of ballistic missile defense (BMD), that part of a ballistic missile (or post-boost vehicle) that carries the nuclear warhead to its target. The reentry vehicle is designed to enter Earth's atmosphere in the terminal portion of its trajectory and proceed to its target. The reentry vehicle is designed to survive rapid heating during high-velocity flight through the atmosphere and to protect its warhead until the nuclear weapon can detonate at the target. *See also* BALLISTIC MISSILE DEFENSE.

reentry window The area at the limits of Earth's sensible (measurable) atmosphere through which an aerospace vehicle or spacecraft in a given trajectory should pass to accomplish a successful reentry.

reflectance (symbol: ρ) In radiation heat transfer, the reflectance (ρ) of a body is defined as the ratio of the incident radiant energy reflected by the body to the total radiant energy falling on the body. The reflecting surface of the body may be *specular* (mirrorlike) or *diffuse*. For the special case of an ideal BLACKBODY, all the radiant energy incident upon this blackbody is absorbed and none is reflected, regardless of wavelength or direction. Therefore, the reflectance for a blackbody has a value of zero, that is, $\rho_{blackbody} = 0$. All other "real-world" solid objects have a reflectance of greater than (or perhaps approximately equal to) zero. Also called *reflectivity*. *Compare with* ABSORPTANCE and TRANSMITTANCE.

reflecting telescope A telescope that collects and focuses light from distant objects by means of a mirror (called the *primary mirror*). The first reflecting telescope was introduced by Sir Isaac Newton (1642–1727) in 1670, and his design is sometimes referred to as a *Newtonian reflector*. *See also* TELESCOPE.

reflection The return of all or part of a BEAM of LIGHT when it encounters the interface (boundary) between two different media, such as AIR and water. A MIRROR-like surface reflects most of the light falling upon it.

refracting telescope A telescope that collects and focuses light from distant objects by means of a lens (called the *objective lens*) or system of lenses. Also called a *refractor*. *See also* TELESCOPE.

refraction The deflection or bending of electromagnetic waves when these waves pass from one type of transparent medium into another. For example, visible light, passing through a prism at a suitable angle, is dispersed into a continuum of colors. This happens because of refraction. When visible light waves cross an interface between two transparent media of different densities (e.g., from air into glass) at an angle other than 90 degrees, the light waves are bent (or refracted). Different wavelengths of visible light are bent different amounts; this causes them to be dispersed into a continuum of colors.

The *index of refraction* is defined as the ratio of the speed of light in a vacuum to the speed of light in the transparent substance of the observed medium. Each type of transparent substance lowers the speed of light slightly and by a different amount; low-density air has a different index of refraction from high-density air and from water, glass, carbon dioxide (gas), and so on.

regenerative cooling A common approach to cooling large liquid-propellant rocket engines and nozzles, especially those that must operate for an appreciable period of time. In this cooling technique, one of the liquid propellants (e.g., liquid oxygen) is first sent through special cooling passages in the thrust chamber and nozzle walls before being injected into the combustion chamber.

regenerative life support system A controlled ecological life support system in which biological and physiochemical subsystems produce plants for food and process solid, liquid, and gaseous wastes for reuse in the system.

regenerator A device used in a thermodynamic process for capturing and returning to the process thermal energy (heat) that otherwise would be lost. The use of a regenerator helps increase the thermodynamic efficiency of a heat engine cycle like the RANKINE CYCLE.

regolith The lunar regolith is the unconsolidated mass of surface debris that overlies the Moon's bedrock. This blanket of pulverized lunar "dust and soil" was created by eons of meteoric and cometary impacts. The "fines" are the fraction of the lunar regolith containing particles that are less than one millimeter in diameter. *See also* MOON.

regressive burning With respect to a solid-propellant rocket, the condition in which the thrust, pressure, or burning surface decreases with time or with the amount of web burned. *See also* WEB.

regulator Flow-control device that adjusts the pressure and controls the flow of fluid (propellant) to meet the demands of a liquid-propellant rocket engine.

reheating In thermodynamics, the addition of thermal energy (heat) to a working fluid in a heat engine after a partial expansion in a multistage turbine. Reheating is used to increase the thermodynamic efficiency of a cycle like the RANKING CYCLE.

relative atomic mass (symbol: A) The total number of NUCLEONs (that is, both PROTONs and NEUTRONs) in the NUCLEUS of an ATOM. Also called the ATOMIC MASS or sometimes *atomic mass number*. For example, the relative atomic mass of the ISOTOPE carbon-12 is twelve. *See also* ATOMIC MASS UNIT.

relative state In aerospace operations, the position and motion of one spacecraft relative to another.

relativistic In general, pertaining to an object (including nuclear particles) moving at speeds that are an appreciable fraction of the speed of light (c).

relativity The theory of space and time developed by Albert Einstein (1879–1955), which has become one of the foundations of modern physics. Einstein's theory of relativity often is discussed in two general categories: the *special theory of relativity,* which he first proposed in 1905, and the *general theory of relativity,* which he presented in 1915.

The special theory of relativity is concerned with the laws of physics as seen by observers moving relative to one another at constant velocity—that is, by observers in nonaccelerating or inertial reference frames. Special relativity has been well demonstrated and verified by many types of experiments and observations.

Einstein proposed two fundamental postulates in formulating special relativity: (1) *First Postulate of Special Relativity:*

The speed of light (c) has the same value for all (inertial-reference-frame) observers, regardless and independent of the motion of the light source or the observers. (2) *Second Postulate of Special Relativity:* All physical laws are the same for all observers moving at constant velocity with respect to each other.

From the theory of special relativity, scientists now conclude that only a "zero-rest-mass" particle, such as a photon, can travel at the speed of light. Another major consequence of special relativity is the equivalence of mass and energy, which is expressed in Einstein's famous formula:

$$E = \Delta m \, c^2$$

where E is the energy equivalent of an amount of matter (Δm) that is annihilated or converted completely into pure energy and c is the speed of light.

In 1915 Einstein introduced his general theory of relativity. He used this development to describe the space-time relationships developed in special relativity for cases where there was a strong gravitational influence—such as white dwarf stars, neutron stars, and black holes. One of Einstein's conclusions was that gravitation is not really a force between two masses (as Newtonian mechanics indicates) but rather arises as a consequence of the curvature of space and time. In a four-dimensional universe (x, y, z, and time), space-time becomes curved in the presence of matter.

The fundamental postulate of general relativity is also called *Einstein's principle of equivalence:* The physical behavior inside a system in free-fall is indistinguishable from the physical behavior inside a system far removed from any gravitating matter (i.e., the complete absence of a gravitational field).

Several experiments have been performed to confirm the general theory of relativity. These experiments included observation of the bending of electromagnetic radiation (starlight and radiowave transmissions from the Viking Project on Mars) by the Sun's immense gravitational field and recognizing the subtle perturbations (disturbances) in the orbit (at perihelion) of the planet Mercury as caused by the curvature of space-time in the vicinity of the Sun. While some scientists do not think that these experiments have demonstrated the validity of general relativity conclusively, data from more powerful space-based observatories in the upcoming years should provide additional experimental evidence.

reliability The probability of specified performance of a piece of equipment or system under stated conditions for a given period of time.

relief valve Pressure-relieving device that opens automatically when a predetermined pressure is reached.

rem In radiation protection and nuclear technology, the traditional unit for dose equivalent (symbol: H). The dose equivalent in rem is the product of the absorbed dose (symbol: D) in rad and the quality factor (QF) as well as any other modifying factors considered necessary to characterize and evaluate the biological effects of ionizing radiation doses received by human beings or other living creatures. The term is an acronym derived from the expression: *r*oentgen *e*quivalent *m*an. The rem is related to the *sievert,* the SI UNIT of dose equivalent, as follows: 100 rem = 1 sievert. *See also* DOSE EQUIVALENT.

remaining body That part of an expendable rocket vehicle that remains after the separation of a fallaway section or companion body. In a multistage rocket, the remaining body diminishes in size as each section or part is cast away and successively becomes a different body.

remote control Control of an operation from a distance, especially by means of telemetry and electronics; a controlling switch, level, or other device used in this type of control, as in *remote-control arming switch. See also* TELEOPERATION.

remotely piloted vehicle (RPV) An aircraft or aerospace vehicle whose pilot does not fly on board but rather controls it at a distance (i.e., remotely) using a telecommunications link from a crewed aircraft, aerospace vehicle, or ground station. RPVs often are used on extremely hazardous missions or on long-duration missions involving extended loitering and surveillance activities. Also called an *unmanned* (or *uncrewed*) *aerial vehicle (UAV)*.

remote manipulator system (RMS)
The Canadian-built, 15.2-meter- (50-foot-) long articulating arm that is remotely controlled from the aft flight deck of the space shuttle orbiter. The elbow and waist movements of the RMS permit payloads to be grappled for deployment out of the cargo bay or to be retrieved and secured in the cargo bay for on-orbit servicing or return to Earth. Because the RMS can be operated from the shirtsleeve environment of the orbiter cabin, an extravehicular activity (EVA) often is not required to perform certain orbital operations. *See also* SPACE TRANSPORTATION SYSTEM.

remote sensing The sensing of an object, event, or phenomenon without having the sensor in direct contact with the object being studied. Information transfer from the object to the sensor is accomplished through the use of the electromagnetic spectrum. Remote sensing can be used to study Earth in detail from space or to study other objects in the solar system, generally using flyby and orbiter spacecraft. Modern remote sensing technology uses many different portions of the electromagnetic spectrum, not just the visible portion we see with our eyes. As a result, often very different and very interesting "images" are created by these new remote sensing systems.

For example, Figure 61 is a radar image showing the volcanic island of Reunion, about 700 kilometers (434 miles) east of Madagascar in the southwest Indian Ocean. The southern half of the island is dominated by the active volcano Piton de la Fournaise. This is one of the world's most active volcanoes, with more than 100 eruptions in the last 300 years. The most recent activity occurred in the vicinity of Dolomieu Crater, shown in the lower center of the image within a horseshoe-shaped collapse zone. The radar illumination is from the left side of the image and dramatically emphasizes the precipitous cliffs at the edges of the central canyons of the island. These canyons are remnants from the collapse of formerly active parts of the volcanoes that built the island. This image was acquired by the Spaceborne Imaging Radar-C/X-Band Synthetic Aperture Radar (SIR-C/X-SAR) flown aboard the space shuttle *Endeavour* on October 5, 1994. The SIR-C/X-SAR is part of the Mission to Planet Earth (MTPE) program of the National Aeronautics and Space Administration (NASA). The radars illuminate Earth with microwaves, allowing detailed observations at any time regardless of weather or sunlight conditions. SIR-C/X-SAR, a joint mission of the German (DARA), Italian (ASI), and American (NASA) space agencies, uses three microwave wavelengths: L-band (24 centimeters [cm]), C-band (6 cm), and X-band (3 cm). The international scientific community is using these multifrequency radar imagery data to better understand the global environment and how it is changing.

Earth receives and is heated by energy in the form of electromagnetic radiation from the Sun. Some of this incoming solar

Figure 61. A radar image (C and X band) of the volcanic island of Réunion, which lies about 700 kilometers east of Madagascar in the southwest Indian Ocean. This image was acquired by NASA's Spaceborne Imaging Radar-C/X Band Synthetic Aperture Radar (SIR-C/X-SAR) system flown onboard the space shuttle *Endeavour* on October 5, 1994. *(Courtesy of NASA/JPL)*

radiation is reflected by the atmosphere, while most penetrates the atmosphere and subsequently is reradiated by atmospheric gas molecules, clouds, and the surface of Earth itself (including, e.g., oceans, mountains, plains, forests, ice sheets, and urbanized areas). All remote sensing systems (including those used to observe Earth from space) can be divided into two general classes: *passive sensors* and *active sensors*. Passive sensors observe reflected solar radiation (or emissions characteristic of and produced by the target itself), while active sensors (like a radar system) provide their own illumination on the target. Both passive and active remote sensing systems can be used to obtain images of the target or scene or else simply to collect and measure the total amount of energy (within a certain portion of the spectrum) in the field of view.

Passive sensors collect reflected or emitted radiation. Types of passive sensors include: *imaging radiometers* and *atmospheric sounders*. Imaging radiometers sense the visible, near-infrared, thermal-infrared, or ultraviolet wavelength regions and provide an image of the object or scene being viewed. Atmospheric sounders collect the radiant energy emitted by atmospheric constituents, such as water vapor or carbon dioxide, at infrared and microwave wavelengths. These remotely sensed data are then used to infer temperature and humidity throughout the atmosphere.

Active sensors provide their own illumination (radiation) on the target and then collect the radiation reflected back by the object. Active remote sensing systems including: *imaging radar, scatterometers, radar altimeters,* and *lidar altimeters.* An imaging radar emits pulses of microwave radiation from a radar transmitter and collects the scattered radiation to generate an image. (Look again at Figure 61 for an example of a detailed radar image collected from space.) Scatterometers emit microwave radiation and sense the amount of energy scattered back from the surface over a wide field of view. These type of instruments are used to measure surface wind speeds and direction and to determine cloud content. Radar altimeters

emit a narrow pulse of microwave energy toward the surface and accurately time the return pulse reflected from the surface, thereby providing a precise measurement of the distance (altitude) above the surface. Similarly, lidar altimeters emit a narrow pulse of laser light (visible or infrared) toward the surface and time the return pulse reflected from the surface.

Today, remote sensing of Earth from space provides scientific, military, governmental, industrial, and individual users with the capacity to gather data to perform a variety of important tasks. These tasks include: (1) simultaneously observing key elements of an interactive Earth system; (2) monitoring clouds, atmospheric temperature, rainfall, wind speed, and direction; (3) monitoring ocean surface temperature and ocean currents; (4) tracking anthropogenic and natural changes to the environment and climate; (5) viewing remote or difficult-to-access terrain; (6) providing synoptic views of large portions of Earth's surface without being hindered by political boundaries or natural barriers; (7) allowing repetitive coverage of the same area over comparable viewing conditions to support change detection and long-term environmental monitoring; (8) identifying unique surface features (especially with the assistance of multispectral imagery); and (9) performing terrain analysis and measuring moisture levels in soil and plants.

Space-based remote sensing is a key technology in our intelligent stewardship of planet Earth, its resources, and interwoven biosphere in the 21st century. Monitoring of the weather and climate supports accurate weather forecasting and identifies trends in the global climate. Monitoring of the land surface assists in global change research, the management of known natural resources, the exploration for new resources (e.g., oil, gas, and minerals), detailed mapping, urban planning, agriculture, forest management, water resource assessment, and national security. Monitoring of the oceans helps determine such properties as ocean productivity, the extent of ice cover, sea-surface winds and waves, ocean currents and circulation, and ocean-surface temperatures. These type of ocean data have particular value to scientists as well as to the fishing and shipping industries.

In 1999 two successful satellite launches dramatically changed the civilian (that is, publicly accessible) use of remote sensing of Earth from space. Since September 1999 and the launch of Space Imaging's *Ikonos* satellite, excellent quality, high-RESOLUTION (one-meter or better) imagery from space has become commercially available from a private firm. Prior to that launch, there was a clear distinction between tightly controlled, high-resolution military imaging (photo-reconnaissance) satellites and civilian (government-owned and sponsored) EARTH OBSERVING SATELLITES (such as the LANDSAT family) that collected openly available but lower-resolution multispectral images of Earth. A presidential directive in March 1994 (called PDD-23) blurred that long-standing distinction within the U.S. government.

The launch of another nonmilitary Earth observing satellite, called TERRA, also helped to accelerate the growth of the orbiting information revolution, an ongoing process sometimes referred to as the *transparent globe*. In December 1999 NASA successfully placed the new *Terra* spacecraft into orbit around Earth. It carries a payload of five STATE OF THE ART (SOA) sensors that are simultaneously collecting data about Earth's atmosphere, lands, oceans, and solar energy balance.

NASA further accelerated the use of remote sensing in support of EARTH SYSTEM SCIENCE with the successful launch of the AQUA spacecraft on May 4, 2002. As the technical sibling to the *Terra* spacecraft, this new Earth observing satellite carries six state of the art remote sensing instruments that simultaneously collect data about the role and movement of water in the Earth system.

rendezvous The close approach of two or more spacecraft in the same orbit, so that docking can take place. These objects meet at a preplanned location and time with essentially zero relative velocity. A

rendezvous would be involved, for example, in the construction, servicing, or resupply of a space station, or when the space shuttle orbiter performs on-orbit repair/servicing of a satellite. The term is also applied to a space mission, such as the *Near Earth Asteroid Rendezvous (NEAR) mission,* in which a scientific spacecraft is maneuvered so as to fly alongside a target body, such as a comet or asteroid, at zero relative velocity. *See also* NEAR EARTH ASTEROID RENDEZVOUS (NEAR) MISSION.

repressurization Sequence of operations during a rocket vehicle or spacecraft flight that uses an onboard pressurant supply to restore the ullage pressure to the desired level after an engine burn period.

resilience The property of a material that enables it to return to its original shape and size after deformation. For example, the resilience of a sealing material is the property that makes it possible for a seal to maintain sealing pressure despite wear, misalignment, or out-of-round conditions.

resistance (symbol: R) 1. Electrical resistance (R) is defined as the ratio of the voltage (or potential difference) (V) across a conductor to the current (I) flowing through it. In accordance with Ohm's law, $R = V/I$. The SI UNIT of resistance is the ohm (Ω), where 1 ohm = 1 volt per ampere. 2. Mechanical resistance is the opposition by frictional effects to forces tending to produce motion. 3. Biological resistance is the ability of plants and animals to withstand poor environmental conditions and/or attacks by chemicals or disease. This ability may be inborn or developed.

resolution 1. In general, a measurement of the smallest detail that can be distinguished by a sensor system under specific conditions. 2. The degree to which fine details in an image or photograph can be seen as separated or resolved. Spatial resolution often is expressed in terms of the most closely spaced line-pairs per unit distance that can be distinguished. For example, when the resolution is said to be 10 line-pairs per millimeter, this means that a standard pattern of black-and-white lines whose line plus space width is 0.1 millimeter is barely resolved by an optical system, finer patterns are not resolved, and coarser patterns are more clearly resolved.

restart In aerospace operations, the act of firing a spacecraft's or space vehicle's "restartable" rocket engine after a previous thrust period (powered flight) and a coast phase in a parking or transfer orbit.

rest mass (symbol: m_0) The mass that an object has when it is at (absolute) rest. From Einstein's special relativity theory, the mass of an object increases when it is in motion according to the formula:
$$m = m_0/\sqrt{(1 - [V^2 / c^2])}$$
where *m* is its mass in motion, m_0 is its mass at rest, v is the speed of the object, and c is the speed of light. This is especially significant at speeds approaching a considerable fraction of the speed of light. Newtonian physics, in contrast, makes no distinction between rest mass *(m_0)* and motion mass *(m)*. This is acceptable in classical physics, as long as the object under consideration is traveling at speeds well below the speed of light—that is, when v<<c.

restricted propellant A solid propellant that has only a portion of its surface exposed for burning; the other propellant surfaces are covered by an inhibitor.

restricted surface The surface of a solid-propellant grain that is prevented from burning through the use of inhibitors.

retrieval With respect to the space shuttle, the process of using the remote manipulator system (RMS) and/or other handling aids to return a captured payload to a stowed or berthed position. No payload is considered retrieved until it is fully stowed for safe return or berthed for repair and maintenance. *See also* REMOTE MANIPULATOR SYSTEM; SPACE TRANSPORTATION SYSTEM.

retrofire To ignite a retrorocket.

retrofit The modification of or addition to a spacecraft, aerospace vehicle, or expendable launch vehicle after it has become operational.

retrograde In a reverse or backward direction.

retrograde motion Motion in an orbit opposite to the usual orbital direction of celestial bodies within a given system. Specifically, of a satellite, motion in a direction opposite to the direction of rotation of the primary.

retrograde orbit An orbit having an inclination of more than 90 degrees. In a retrograde orbit, the spacecraft or satellite has a motion that is opposite in direction to the rotation of the primary body (such as Earth). *Compare with* PROGRADE ORBIT. *See also* ORBITS OF OBJECTS IN SPACE.

retroreflection Reflection in which the reflected rays return along paths parallel to those of their corresponding incident rays.

retroreflector A mirrorlike instrument, usually a corner reflector design, that returns light or other electromagnetic radiation (e.g., an infrared laser beam) in the direction from which it comes. *See also* LASER GEODYNAMICS SATELLITE (LAGEOS).

retrorocket A small rocket engine on a satellite, spacecraft, or aerospace vehicle used to produce a retarding thrust or force that opposes the object's forward motion. This action reduces the system's velocity.

retrothrust Thrust used for a braking maneuver; a reverse thrust.

reusable launch vehicle (RLV) A launch vehicle that incorporates simple, fully reusable designs for airline-type operations using advanced technology and innovative operational techniques.

reverse thrust Thrust applied to a moving object in a direction to oppose the object's motion; used to decelerate a spacecraft or aerospace vehicle.

reversible process   In thermodynamics, a process that goes in either the forward or the reverse direction without violating the SECOND LAW OF THERMODYNAMICS; a constant entropy process. For example, the entropy of a control mass experiencing a reversible, adiabatic process will not change. Such a constant entropy process is also called an *isentropic process.* However, the "reversible process" is just an idealization used in engineering analyses to help approximate the behavior of certain realworld systems. Some of the more commonly encountered reversible process idealizations are: frictionless motion, current flow without resistant, and "frictionless" pulleys. In contrast, irreversible processes, such as molecular diffusion, a spontaneous chemical reaction, or motion with friction, all produce entropy. *See also* ENTROPY; THERMODYNAMICS.

revetment   In aerospace operations and safety, a wall of concrete, earth, or sandbags installed for the protection of launch crew personnel and equipment against the blast and flying shrapnel from an exploding rocket during a launch abort near the pad.

revolution   1. Orbital motion of a celestial body or spacecraft about a center of gravitational attraction, such as the Sun or a planet, as distinct from rotation around an internal axis. For example, Earth *revolves* around the Sun annually and *rotates* daily about its axis. 2. One complete cycle of the movement of a spacecraft or a celestial object around its primary. *See also* ORBITS OF OBJECTS IN SPACE.

rift valley   A depression in a PLANET's surface due to crustal MASS separation.

rill   A deep, narrow depression on the LUNAR surface that cuts across all other types of topographical features on the MOON. From the German word *rille,* meaning groove.

ring (planetary)   A DISK of matter that encircles a PLANET. Such rings usually contain ice and dust PARTICLES ranging in size from microscopic fragments up to chunks that are tens of METERS in diameter.

ringed world   A PLANET with a RING or set of rings encircling it. In the SOLAR SYSTEM, JUPITER, SATURN, URANUS, and NEPTUNE all have ring systems of varying degrees of composition and complexity. Ring systems may be a common feature of EXTRASOLAR JOVIAN PLANETS.

ring seal   Piston-ring type of seal that assumes its sealing position under the pressure of the fluid to be sealed.

robotics   Robotics is basically the science and technology of designing, building, and programming robots. Robotic devices, or robots as they are usually called, are primarily "smart machines" with manipulators that can be programmed to do a variety of manual or human labor tasks automatically. A robot, therefore, is simply a machine that does mechanical, routine tasks on human command. The expression "robot" is attributed to Czech writer Karel Capek, who wrote the play *R.U.R. (Rossum's Universal Robots.)* This play first appeared in English in 1923 and is a satire on the mechanization of civilization. The word "robot" is derived from *robata,* a Czech word meaning "compulsory labor" or "servitude."

A typical robot consists of one or more manipulators (arms), end effectors (hands), a controller, a power supply, and possibly an array of sensors to provide information about the environment in which the robot must operate. Because most modern robots are used in industrial applications, their classification is based on these industrial functions. Terrestrial robots frequently are divided into the following classes: nonservo (or pick-and-place), servo, programmable, computerized, sensory, and assembly robots.

The *nonservo robot* is the simplest type. It picks up an object and places it at another location. The robot's freedom of movement usually is limited to two or three directions.

The *servo robot* represents several categories of industrial robots. This type of robot has servo-mechanisms for the manipulator and end effector to enable it to change direction in midair (or mid stroke) without having to strip or trigger a mechanical limit switch. Five to seven directions of motion are common, depending on the number of joints in the manipulator.

The *programmable robot* is essentially a servo robot that is driven by a programmable controller. This controller memorizes (stores) a sequence of movements and then repeats these movements and actions continuously. Often this type of robot is programmed by "walking" the manipulator and end effector through the desired movement.

The *computerized robot* is simply a servo robot run by computer. This kind of robot is programmed by instructions fed into the controller electronically. These "smart robots" may even have the ability to improve upon their basic work instructions.

The *sensory robot* is a computerized robot with one or more artificial senses to observe and record its environment and to feed information back to the controller. The artificial senses most frequently employed are sight (robot or computer vision) and touch.

Finally, the *assembly robot* is a computerized robot, generally with sensors, that is designed for assembly-line and manufacturing tasks, both on Earth and eventually in space.

In industry, robots are designed mainly for manipulation purposes. The actions that can be produced by the end effector or hand include: (1) motion (from point to point, along a desired trajectory or along a contoured surface); (2) a change in orientation; and (3) rotation.

Nonservo robots are capable of point-to-point motions. For each desired motion, the manipulator moves at full speed until the limits of its travel are reached. As a result, nonservo robots often are called "limit sequence," "bang-bang," or "pick-and-place" robots. When nonservo robots reach the end of a particular motion, a mechanical stop or limit switch is tripped, stopping the particular movement.

Servo robots are also capable of point-to-point motions; but their manipulators move with controlled variable velocities and trajectories. Servo robot

motions are controlled without the use of stop or limit switches.

Four different types of manipulator arms have been developed to accomplish robot motions. These are the rectangular, cylindrical, spherical, and anthropomorphic (articulated or jointed arm). Each of these manipulator arm designs features two or more degrees of freedom (DOF), a term that refers to the direction a robot's manipulator arm is able to move. For example, simple straight line or linear movement represents one DOF. If the manipulator arm is to follow a two-dimensional curved path, it needs two degrees of freedom: up and down and right and left. Of course, more complicated motions will require many degrees of freedom. To locate an end effector at any point and to orient this effector in a particular work volume requires six DOF. If the manipulator arm needs to avoid obstacles or other equipment, even more degrees of freedom will be required. For each DOF, one linear or rotary joint is needed. Robot designers sometimes combine two or more of these four basic manipulator arm configurations to increase the versatility of a particular robot's manipulator.

Actuators are used to move a robot's manipulator joints. Three basic types of actuators currently are used in contemporary robots: pneumatic, hydraulic, and electrical. Pneumatic actuators employ a pressurized gas to move the manipulator joint. When the gas is propelled by a pump through a tube to a particular joint, it triggers or actuates movement. Pneumatic actuators are inexpensive and simple, but their movement is not precise. Therefore, this kind of actuator usually is found in nonservo or pick-and-place robots. Hydraulic actuators are quite common and capable of producing a large amount of power. The main disadvantages of hydraulic actuators are their accompanying apparatus (pumps and storage tanks) and problems with fluid leaks. Electrical actuators provide smoother movements, can be controlled very accurately, and are very reliable. However, these actuators cannot deliver as much power as hydraulic actuators of comparable mass. Nevertheless, for modest power actuator functions, electrical actuators often are preferred.

Many industrial robots are fixed in place or move along rails and guideways. Some terrestrial robots are built into wheeled carts, while others use their end effectors to grasp handholds and pull themselves along. Advanced robots use articulated manipulators as legs to achieve a walking motion.

A robot's end effector (hand or gripping device) generally is attached to the end of the manipulator arm. Typical functions of this end effector include grasping, pushing and pulling, twisting, using tools, performing insertions, and various types of assembly activities. End effectors can be mechanical, vacuum or magnetically operated, can use a snare device or have some other unusual design feature. The final design of the end effector is determined by the shapes of the objects that the robot must grasp. Usually most end effectors are some type of gripping or clamping device.

Robots can be controlled in a wide variety of ways, from simple limit switches tripped by the manipulator arm to sophisticated computerized remote sensing systems that provide machine vision, touch, and hearing. In the case of computer-controlled robots, the motions of the manipulator and end effector are programmed; that is, the robot "memorizes" what it is supposed to do. Sensor devices on the manipulator help to establish the proximity of the end effector to the object to be manipulated and feed information back to the computer controller concerning any modifications needed in the manipulator's trajectory.

Another interesting type of robot system, the *field robot,* has become practical recently. A field robot is a robot that operates in unpredictable, unstructured environments, typically outdoors (on Earth) and often operates autonomously or by tele-operation over a large workspace (typically a square kilometer or more). For example, in surveying a potentially dangerous site, a human operator will stay at a safe distance away in a protected work environment and control (by cable or radio frequency link) the field robot, which then actually operates in the hazardous environment. These terrestrial field robots can be considered as "technological first cousins" to the teleoperated robot planetary rovers that will roam the Moon, Mars, and other planetary bodies in the next century.

Robotic systems have played and will continue to play a major role in our exploration and development of the solar system. Early American space robots included: NASA's Surveyor spacecraft, which soft-landed on the lunar surface starting in 1966 operating soil scoops and preparing the way for the Apollo astronauts (see Figure 62); and NASA's Viking Lander spacecraft, which explored and examined the Martian surface for signs of microbial lifeforms starting in 1976. The Soviet Lunakhod remotely controlled robot moon rovers roamed across the lunar surface in the early 1970s, conducting numerous experiments and soil property investigations.

The space shuttle orbiter's remote manipulator system (RMS) was first flown in 1981. This versatile "robot arm" was designed to handle spacecraft deployment and retrieval operations, as well as to permit the assembly of large structures (e.g., a permanent space station). It is installed in the space shuttle orbiter's port (left) side cargo bay door hinges and is operated by a "shirt-sleeved" astronaut from inside the crew cabin. The RMS was designed and built by the National Research Council of Canada. It is a highly sophisticated robotic device that is similar to a human arm. The 15-meter-long (49-feet-long) RMS features a shoulder, wrist, and hand—although its "hand" does not look like a human hand. The skeleton of this mechanical arm is made of lightweight graphite composite materials. Covering the skeleton are skin layers consisting of thermal blankets. The muscles driving the joints are electric actuators (motors). Built-in sensors act like nerves and sense joint positions and rotation rates.

The RMS includes two closed-circuit television cameras, one at the wrist and one at the elbow. These cameras allow an

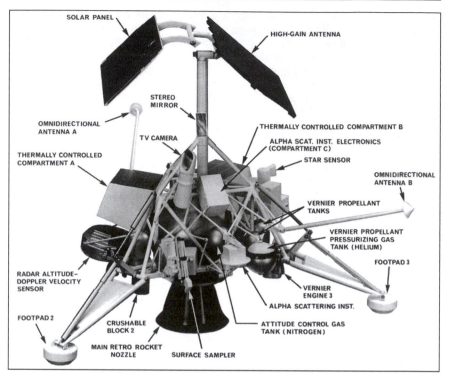

Figure 62. NASA's *Surveyor* spacecraft performed robotic exploration of the lunar surface from 1966 to 1968 in preparation for the landings by the Apollo astronauts (1969–1972). *(Courtesy of NASA)*

astronaut, who is operating the RMS from the orbiter's aft flight deck, to see critical points along the arm and the target toward which the arm is moving.

The INTERNATIONAL SPACE STATION *(ISS)*, currently under development in low Earth orbit, will have several robots to help astronauts and cosmonauts complete their tasks in space. Japan is developing a remote manipulator system for the Japanese Experiment Module (JEM). The EUROPEAN SPACE AGENCY and the Russian Space Agency are developing the European robotic arm. Finally, building upon space shuttle RMS technology and experience, Canada and the United States are developing the space station's mobile servicing system (MSS). The MSS is the most complex robotic system on the *ISS*. It consists of the space station remote manipulator system (SSRMS), the mobile remote servicer base system (MBS),

the special purpose dexterous manipulator (SPDM), and the mobile transporter (MT). The MSS is controlled by an astronaut or cosmonaut at one of two robotics work stations inside the *ISS*. The primary functions of the MSS robotic system on the *ISS* are to assist in the assembly of the main elements of the station, to handle large payloads, to exchange orbital replacement units, to support astronaut extravehicular activities, to assist in station maintenance, and to provide transportation around the station outside the PRESSURIZED HABITABLE ENVIRONMENT.

Expanding on its previous space robot experience, NASA is now pursuing "intelligent teleoperation" technology as an interim step toward developing the technology needed for truly autonomous space robots. The goals of NASA's current space telerobotics program are to develop, integrate, and demonstrate the science and technology of remote telerobotics, leading to increases in operational capability, safety, cost effectiveness, and probability of success of NASA missions. Space telerobotics technology requirements can be characterized by the need for manual and automated control, nonrepetitive tasks, time delay between operator and manipulator, flexible manipulators with complex dynamics, novel locomotion, operations in the space environment, and the ability to recover from unplanned events. There are three specific areas of focus: on-orbit servicing, science payload maintenance, and exploration robotics.

The on-orbit servicing telerobotics program is concerned with the development of space robotics for the eventual application to on-orbit satellite servicing by both free-flying and platform attached servicing robots. Relevant technologies include virtual reality telepresence, advanced display technologies, proximity sensing for perception technologies, and robotic flaw detection. Potential mission applications include: repair of free-flying small satellites, ground-based control of robotic servicers, and servicing of external space platform payloads (including payloads externally mounted on the international space station).

The science payload maintenance telerobotics program is intended to mature technologies for robotics that will be used inside pressurized living space to maintain and service payloads. Once developed, this telerobotic capability will off-load the requirements of intensive astronaut maintenance of these payloads and permit the operation of the payloads during periods when the astronauts may not be present. Relevant technologies include lightweight manipulators, redundant robotic safety systems, and self-deploying systems. One particular mission application involves intravehicular activity (IVA) robots for the space station.

The exploration robotics program involves the development of robot systems for the surface exploration of the Moon and Mars, including robotic reconnaissance and surveying systems as precursors to eventual human missions. During such surface exploration missions, these robots will explore potential landing sites and areas of scientific interest, deploy scientific instruments, gather samples for in-situ analysis or possible return to Earth, acquire and transmit video imagery, and provide the images needed to generate "virtual environments" of the lunar and Martian surface. The robotic systems for these operations will need high levels of local autonomy, including the ability to perform local navigation, identify areas of potential scientific interest, regulate on-board resources, and schedule activities—all with limited ground command intervention.

The *science rover/remote geologist robot* represents a 20-kilogram (44-pound-mass) class future microrover that can autonomously traverse many kilometers on the surface of Mars and perform scientist-directed experiments and then return relevant data back to Earth.

An *aerobot* is an autonomous robotic aerovehicle (e.g., a free-flying balloon or a specially designed "aeroplane") that is capable of flying in the atmospheres of Venus, Mars, Titan, or the outer planets. For Martian or Venusian aerobots, the robotic system would be capable of one or more of the following activities: autonomous state determination; periodic altitude variations; altitude control and the ability to follow a designated flight path within a planetary atmosphere using prevailing planetary winds; and landing at a designated surface location.

Future exploration of interplanetary small bodies, such as comets and asteroids, requires telerobotic technology developments in a variety of areas. Landing and surface operations in the very low-gravity environment of small interplanetary bodies (where the acceleration due to gravity is typically 10^{-4} to 10^{-2} meter per second-squared) is an extremely challenging problem. The robotic comet or asteroid explorer must have mechanisms and autonomous control algorithms to perform landing, anchoring, surface/subsurface sampling and sample manipulation for a complement of scientific instruments. A robotic lander might use, for example, crushable material on the underside of a base plate design to absorb almost all of the landing kinetic energy. An anchoring, or attachment system, then would be used to secure the lander and compensate for the reaction forces and moments generated by the sample acquisition mechanisms.

Recent advances in microtechnology and mobile robotics have made it possible to consider the creation and use of extremely small automated or remote-controlled vehicles, called *nanorovers*, in planetary surface exploration missions. A *nanorover* is a robotic vehicle with a mass of between 10 and 50 grams. One or several of these tiny robots could be used to survey areas around a lander and to look for a particular substance, such as water ice or microfossils. The nanorover would then communicate its scientific findings back to Earth via the lander spacecraft (possibly in conjunction with an orbiting "mothership").

Eventually, space robots will achieve higher levels of artificial intelligence, autonomy, and dexterity, so that servicing and exploration operations would become less and less dependent on a human operator being present in the control loop. These robots would be capable of interpreting very high level command structures and executing commands without human intervention. Erroneous command structures, incomplete task operations, and the resolution of differences between the robot's built-in "world model" and the real-world environment it is encountering would be handled autonomously. This is especially important as more sophisticated robots are sent deeper into the outer solar system and telecommunications time delays of minutes become hours. This higher level of autonomy is also very important in the development and operation of permanent lunar or Martian surface bases, where smart machines become our permanent partners in the development of the space frontier.

These increasingly more sophisticated space robots will have working lifetimes of decades with little or no maintenance. Some space planners envision robots capable of repairing themselves or other robots—again with little or no direct human supervision. The brilliant Hungarian-American mathematician John von Neumann (1903–1957) was the first person to seriously explore the problem of *self-replicating systems (SRSS)*— that is, robots smart and dexterous enough to make copies of themselves. From von Neumann's work and the more recent work of other investigators, five broad classes of SRS behavior have been suggested:

1. *Production.* The generation of useful output from useful input. In the production process, the unit machine remains unchanged. Production is a simple behavior demonstrated by all working machines, including SRS devices.
2. *Replication.* The complete manufacture of a physical copy of the original machine unit by the machine unit itself.
3. *Growth.* An increase in the mass of the original machine unit by its own actions while still retaining the integrity of its original design. For example, the machine might add an additional set of storage compartments in which to keep a larger supply of parts or constituent materials.
4. *Evolution.* An increase in the complexity of the unit machine's function or structure. This is accomplished by additions or deletions to existing subsystems, or by changing the characteristics of these subsystems.
5. *Repair.* Any operation performed by a unit machine on itself that helps reconstruct, reconfigure, or replace existing subsystems but does not change the SRS unit population, the original unit mass, or its functional complexity.

In theory, such replicating systems can be designed to exhibit any or all of these machine behaviors. When such machines are actually built (perhaps in the mid- to late 21st century), a particular SRS unit most likely will emphasize just one or several kinds of machine behavior, even if it were capable of exhibiting all of them. For example, a particular SRS unit might be the fully autonomous, general-purpose, self-replicating lunar factory that first makes a sufficient number of copies of itself and then sets about harvesting lunar resources and converting these resources into products needed to support a permanently inhabited (by humans) lunar settlement.

Once we have developed the sophisticated robotic devices needed for the detailed investigation of the outer regions of this solar system, the next step becomes quite obvious. Sometime in the 21st century, humankind will build and launch its first fully automated robot explorer to a nearby star system. This interstellar explorer will be capable of searching for extrasolar planets around other suns, targeting any suitable planets for detailed investigation, and then initiating the search for extraterrestrial life. Light-years away, terrestrial scientists will wait for faint distant radio signals by which the robot starship describes any new worlds it has encountered, perhaps shedding light on the greatest cosmic mystery of all: Does life exist elsewhere in the universe? *See also* ARTIFICIAL INTELLIGENCE; *CASSINI* MISSION LUNOKHOD; *MARS PATHFINDER*; MARS SAMPLE RETURN MISSION; MARS SURFACE ROVER(S); 2003 MARS EXPLORATION ROVER MISSION; TELEOPERATION; TELEPRESENCE; THOUSAND ASTRONOMICAL UNIT (TAU) MISSION; VIKING PROJECT.

robot spacecraft A semiautomated or fully automated SPACECRAFT capable of executing its primary exploration MISSION with minimal or no human supervision. *See also* ROBOTICS.

Roche limit As postulated by the French mathematician Edouard Albert Roche (1820–1883) in the 19th century, the smallest distance from a PLANET at which gravitational FORCES can hold together a natural SATELLITE or MOON that has the same average DENSITY as the PRIMARY BODY. If the moon's ORBIT falls within the Roche limit, it will be torn apart by tidal forces.

rocket In general, a completely self-contained projectile, pyrotechnic device, or flying vehicle propelled by a reaction (rocket) engine. Since it carries all of its propellant, a rocket vehicle can function in the vacuum of outer space and represents the key to space travel. Rockets obey Newton's Third Law of Motion, which states. "For every action there is an equal and opposite reaction." Rockets can be classified by the energy source used by the reaction engine to accelerate the ejected matter that creates the vehicle's thrust, as, for example: chemical rocket, nuclear rocket, and electric rocket. Chemical rockets, in turn, often are divided into two general subclasses: solid-propellant rockets and liquid-propellant rockets.

A *solid-propellant chemical rocket* is the simplest type of rocket. It can trace its technical heritage all the way back to the gunpowder-fueled "fire-arrow" rockets of ancient China (circa 1045 C.E.). Those in use today generally consist of a solid propellant (i.e., fuel and oxidizer compound) with the following associated hardware: case, nozzle, insulation, igniter, and stabilizers. (See Figure 63.)

Solid propellants, commonly referred to as the "grain," are basically a chemical mixture or compound containing a fuel and oxidizer that burn (combust) to produce very hot gases at high pressure. The important feature here is that these propellants are self-contained and can burn *without* the introduction of outside oxygen sources (such as air from Earth's atmosphere). Consequently, the solid-propellant rocket, and its technical sibling, the liquid-propellant rocket (which is discussed shortly), can operate in outer space. In fact, a rocket vehicle performs the best in space, as was originally postulated in 1919 by the brilliant American rocket scientist Dr. Robert H. Goddard (1882–1945). Goddard is often called the father of American rocketry.

Solid propellants often are divided into three basic classes: monopropellants, double-base, and composites. Monopropellants are energetic compounds such as nitroglycerin or nitrocellulose. Both of these compounds contain fuel (carbon and hydrogen) and oxidizer (oxygen). Monopropellants are rarely used in modern rockets. Doublebase propellants are mixtures of monopropellants, such as nitroglycerin and nitrocellulose. The nitrocellulose adds physical strength to the grain, while the nitroglycerin is a high-performance, fast-burning propellant. Usually double-base propellants are mixed together with additives that

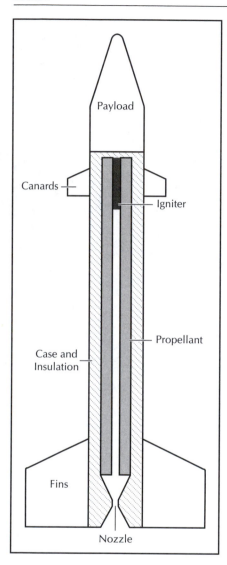

Figure 63. The basic components of a solid-propellant rocket.

other as the oxidizer. For example, the propellants used in the solid rocket boosters (SRBs) of the space shuttle fall into this category. The propellant type used in this case is known as PBAN, which means polybutadiene acrylic acid acrylonitrile terpolymer. This somewhat exotic-sounding chemical compound is used as a binder for ammonium perchlorate (oxidizer), powdered aluminum (fuel), and iron oxide (an additive). The cured propellant looks and feels like a hard rubber eraser.

The thrust produced by the propellants is determined by the combustive nature of the chemicals used and by the shape of their exposed burning surfaces. A solid propellant will burn at any point that is exposed to heat or hot gases of the right temperature. Grains usually are designed to be either end-burning or internal-burning. (See Figure 64.)

End-burning grains burn the slowest of any grain design. The propellant is ignited close to the nozzle, and the burning proceeds the length of the propellant load. The area of the burning surface is always at a minimum. While the thrust produced by end-burning is lower than for other grain designs, the thrust is sustained over longer periods.

Much more massive thrusts are produced by internal-burning. In this design, the grain is perforated by one or more hollow cores that extend the length of the case. With an internal-burning grain, the burning surface of the exposed cores is much larger than the surface exposed in an end-burning grain. The entire surfaces of the cores are ignited at the same time, and burning proceeds from the inside out. To increase the surface available for burning, a core may be shaped into a cruciform or star design.

By varying the geometry of the core design, the thrust produced by a large internal-burning grain also can be customized as a function of time to accommodate specific mission needs. For example, the massive solid rocket boosters (SRBs) used by the space shuttle feature a single core that has an 11-point star design in the forward section. At 65 seconds into

improve the burning characteristics of the grain. The mixture becomes a puttylike material that is loaded into the rocket case.

Composite solid propellants are formed from mixtures of two or more unlike compounds that by themselves do not make good propellants. Usually one compound serves as the fuel and the

the launch, the star points are burned away and thrust temporarily diminishes. This coincides with the passage of the space shuttle vehicle through the sound barrier. Buffeting occurs during this passage and the reduced SRB thrust helps alleviate strain on the vehicle.

The rocket case is the pressure and load-carrying structure that encloses the solid propellant. Cases are usually cylindrical, but some are spherical in shape. The case is an inert part of the rocket, and its mass is an important factor in determining how much payload the rocket can carry and how far it can travel. Efficient, high-performance rockets require that the casing be constructed of the lightest materials possible. Alloys of steel and titanium often are used for solid rocket casings. Upper-stage vehicles may use thin metal shells that are wound with fiberglass for extra strength.

Unless protected by insulation, the solid rocket motor case will lose strength rapidly and burst or burn through. Therefore, to protect the casing, insulation is bonded to the inside wall of the case before the propellant is loaded. The thickness of this insulation is determined by its thermal properties and how long the casing will be exposed to the high-pressure, very hot combustion gases. A frequently used insulation for solid rockets is an asbestos-filled rubber compound that is thermally bonded to the casing wall.

During the combustion process, the resulting high temperature and high-pressure gases exit the rocket through a narrow opening, called the nozzle, that is located at the lower end of the motor. The most efficient nozzles are convergent-divergent designs. During operation, the exhaust gas velocity in the convergent portion of the nozzle is subsonic. The gas velocity increases to sonic speed at the throat and then to supersonic speeds, as it flows through and exits the divergent portion of the nozzle. The narrowest part of the nozzle is the throat. Escaping gases flow through this constricted region with relatively high velocity and temperature. Excessive heat transfer to the nozzle wall at the throat is a great problem. Often thermal protection for the throat consists of a liner that either withstands these high temperatures (for a brief period of operation) or

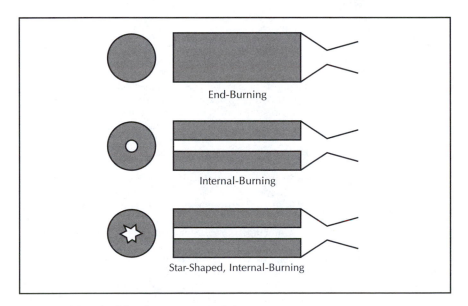

End-Burning

Internal-Burning

Star-Shaped, Internal-Burning

Figure 64. Typical solid-rocket motor grain designs.

else ablates (i.e., intentionally erodes carrying heat away). Large solid rocket motors, such as the space shuttle's solid rocket booster, generally rely on ablative materials to protect the nozzle's throat. Smaller solid rocket motors, such as might be used in an air-to-air combat missile or in a short-range surface-to-surface military missile, often use high-temperature-resistant materials to protect the nozzle. These temperature-resistant materials (possibly augmented by a thin layer of heat-resistant liner) can protect the nozzle's throat sufficiently, since the rocket's burn period is quite short (typically a few seconds).

To ignite the propellants of a solid rocket, the grain surface must be saturated with hot gases. The igniter is usually a rocket motor itself, but much smaller in size. The igniter can be placed inside the upper end of the hollow core, at the lower end of the core, or even completely outside the solid rocket motor. In the latter case, the exhaust of the igniter is directed into the nozzle of the larger solid rocket motor. An electrical circuit with a hot-wire resistor or an exploding bridgewire starts the igniter. The initial part of the ignition sequence begins with fast-burning pellets that, in turn, fire up the main igniter propellants.

Active directional control of solid-propellant rockets in flight generally is accomplished by one of two basic approaches. First, fins and possibly canards (small fins on the front end of the casing) can be mounted on the rocket's exterior. During flight in the atmosphere, these structures tilt to steer the rocket in much the same way that a rudder operates on a boat. Canards have an opposite effect on the directional changes of the rocket from the effect produced by the (tail) fins. Second, directional control also may be accomplished by using a gimbaled nozzle. Slight changes in the direction of the exhaust gases are accomplished by moving the nozzle from side to side. Large solid rocket motors, such as the space shuttle's solid rocket boosters, use the gimbaled-nozzle approach to help steer the vehicle.

When compared to liquid-propellant rocket systems, solid-propellant rockets offer the advantage of simplicity and reliability. With the exception of stability controls, solid-propellant rockets have no moving parts. When loaded with propellant and erected on a launch pad (or placed in an underground strategic missile silo or inside a launch tube in a ballistic missile submarine), solid rockets stand ready for firing at a moment's notice. In contrast, liquid-propellant rocket systems require extensive prelaunch preparations.

Solid rockets generally have an additional advantage in that a greater portion of their total mass can consist of propellants. Liquid-propellant rocket systems require fluid feed lines, pumps, and tanks, all adding additional (inert) mass to the vehicle. In fact, aerospace engineers often describe the performance of a rocket vehicle in terms of the *propellant mass fraction*. The propellant mass fraction is defined as the rocket vehicle's propellant mass divided by its total mass (i.e., propellant, payload, structure, and supporting systems mass). A well-designed, "ideal" rocket vehicle should have a propellant mass fraction of between 91 and 93% (i.e., 0.91 to 0.93). This means that 91 to 93% of the rocket's total mass is propellant. Large liquid-propellant rocket vehicles generally achieve a mass fraction about 90% or less.

The principal disadvantage of solid-propellant rockets involves the burning characteristics of the propellants themselves. Solid propellants are generally less energetic (i.e., deliver less thrust per unit mass consumed) than the best liquid propellants. Also, once ignited, solid-propellant motors burn rapidly and are extremely difficult to throttle or extinguish. In contrast, liquid-propellant rocket engines can be started and stopped at will.

Today, solid-propellant rockets are used for strategic nuclear missiles (e.g., Minuteman), for tactical military missiles (e.g., Sidewinder), for small expendable launch vehicles (e.g., Scout), and as strap-on solid boosters for a variety of liquid-propellant launch vehicles, including the reusable space shuttle and the expendable Titan-IV. Solid-rocket motors also are used in small sounding rockets and in many types of upper-stage vehicles, such as the *Intertial Upper Stage.*

The liquid-propellant rocket engine was invented in 1926 by Dr. Goddard. So important were his contributions to the field of rocketry that it is even appropriate to state: "Every liquid propellant rocket is a Goddard rocket."

Figure 65 describes the major components of a typical liquid-propellant chemical rocket. Here, the propellants (liquid hydrogen for the fuel and liquid oxygen for the oxidizer) are pumped to the combustion chamber, where they begin to react. The liquid fuel often is passed through the tubular walls of the combustion chamber and nozzle to help cool them and prevent high-temperature degradation of their surfaces.

Liquid-propellant rockets have three principal components in their propulsion system: propellant tanks, the rocket engine's combustion chamber and nozzle assembly, and turbopumps. The propellant tanks are load-bearing structures that contain the liquid propellants. There is a separate tank for the fuel and for the oxidizer. The combustion chamber is the region into which the liquid propellants are pumped, vaporized, and reacted (combusted), creating the hot exhaust gases, which then expand through the nozzle generating thrust. The turbopumps are fluid-flow machinery that deliver the propellants from the tanks to the combustion chamber at high pressure and sufficient flow rate. (In some liquid-propellant rockets, the turbopumps are eliminated by using an "over-pressure" in the propellant tanks to force the propellants into the combustion chamber.) A simplified liquid-propellant rocket engine is illustrated in Figure 66.

Propellant tanks serve to store one or two propellants until needed in the combustion chamber. Depending on the type of liquid propellants used, the tank may be nothing more than a low-pressure envelope, or it may be a pressure vessel capable of containing propellants under high pressure. In the case of cryogenic (extremely low temperature) propellants, the tank has to be an extremely well-insulated structure to prevent the very cold liquids from boiling away.

As with all rocket vehicle components, the mass of the propellant tanks is an important design factor. Aerospace engineers fully recognize that the lighter they can make the propellant tanks, the more payload that can be carried by the rocket or the greater its range. Many liquid-propellant tanks are made out of very thin metal or are thin metal sheaths wrapped with high-strength fibers and cements. These tanks are stabilized by the internal pressure of their contents, much the same way the wall of a balloon gains strength from the gas inside (at least up to a certain level of internal pressure). However, very large propellant tanks and tanks that contain cryogenic propellants require additional strengthening or layers. Structural rings and ribs are used to strengthen tank walls, giving the tanks the appearance of an aircraft frame. With cryogenic propellants,

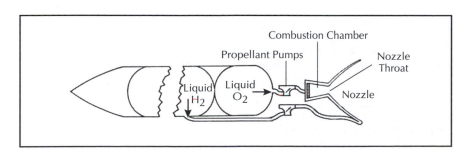

Figure 65. Basic hardware associated with a bipropellant (here LH_2 and LO_2) liquid rocket.

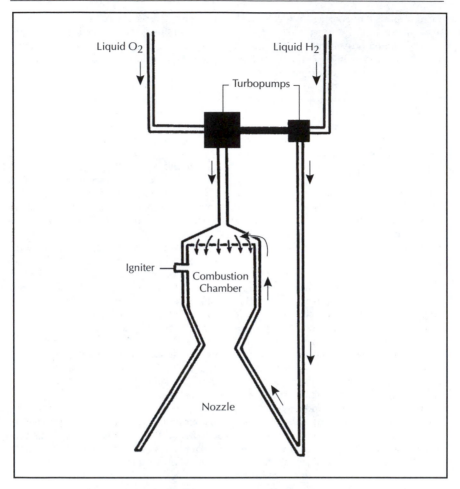

Figure 66. A simplified liquid-propellant rocket engine that employs regenerative cooling of the combustion chamber and nozzle walls.

extensive insulation is needed to keep the propellants in their liquefied form. Unfortunately, even with the best available insulation, cryogenic propellants are difficult to store for long period of time and eventually will boil off (i.e., vaporize). For this reason, cryogenic propellants are not used in liquid-propellant military rockets, which must be stored for months at a time in a launch-ready condition.

Turbopumps provide the required flow of propellants from the low-pressure propellant tanks to the high-pressure combus-tion chamber. Power to operate the tur-bopumps is produced by combusting a fraction of the propellants in a preburner. Expanding gases from the burning propel-lants drive one or more turbines, which, in turn, drive the turbopumps. After pass-ing through the turbines, these exhaust gases are either directed out of the rocket through a nozzle or are injected, along with liquid oxygen, into the combustion chamber for more complete burning.

The combustion chamber of a liquid-propellant rocket is a bottle-shaped container with openings at opposite ends. The openings at the top inject propellants into the chamber. Each opening consists of a small nozzle that injects either fuel or oxidizer. The main purpose of the injectors is to mix the propellants to ensure smooth and complete combustion and to avoid detonations. Combustion chamber injectors come in many designs, and one liquid-propellant engine may have hundreds of injectors. (See Figure 67.)

After the propellants have entered the combustion chamber, they must be ignited. Hypergolic propellant combinations ignite on contact, but other propellants need a starter device, such as a spark plug. Once combustion has started, the thermal energy released continues the process.

The opening at the opposite (lower) end of the combustion chamber is the throat or narrowest part of the nozzle. Combustion of the propellants builds up gas pressure inside the chamber, which then exhausts through this nozzle. By the time the gas leaves the exit cone (widest part of the nozzle), it achieves supersonic velocity and imparts forward thrust to the rocket vehicle.

Because of the high temperatures produced by propellant combustion, the chamber and nozzle must be cooled. For example, the combustion chambers of the space shuttle's main engines (i.e., the SSMEs) reach 3,590 kelvins (K) (3,317° Celsius) (6,000 ° Fahrenheit) during firing. All surfaces of the combustion chamber and nozzle need to be protected from the eroding effects of the high-temperature, high-pressure gases.

Two general approaches can be taken to cool the combustion chamber and nozzle. One approach is identical to the cooling approach taken with many solid-propellant rocket nozzles. The surface of the nozzle is covered with an ablative material that sacrificially erodes when exposed to the high-temperature gas stream. This intentional material erosion process keeps the surface underneath cool, since the ablated material carries away a large amount of thermal energy. However, this cooling approach adds extra mass to a liq-

uid-propellant engine, which in turn reduces payload and range capability of the rocket vehicle. Therefore, ablative cooling is used only when the liquid-propellant engine is small or when a simplified engine design is more important than high performance. In considering such technical choices, an aerospace engineer is making "design tradeoffs."

The second method of cooling is called *regenerative cooling*. A complex plumbing arrangement inside the combustion chamber and nozzle walls circulates the fuel (in the case of the space shuttle's main engines, very cold (cryogenic) liquid hydrogen fuel) before it is sent through the preburner and into the combustion chamber. This circulating fuel then absorbs some of the thermal energy entering the combustion chamber and nozzle walls, providing a level of cooling. Although more complicated than ablative cooling, regenerative cooling reduces the mass of large rocket engines and improves flight performance.

Propellants for liquid rockets generally fall into two categories: monopropellants and bipropellants. Monopropellants consist of a fuel and an oxidizing agent stored in one container. They can be two premixed chemicals, such as alcohol and hydrogen peroxide, or a homogeneous chemical, such as nitromethane. Another chemical, hydrazine, becomes a monopropellant when first brought into contact with a catalyst. The catalyst initiates a reaction that produces heat and gases from the chemical decomposition of the hydrazine.

Bipropellants have the fuel and oxidizer separate from each other until they are mixed in the combustion chamber. Commonly used bipropellant combinations include: liquid oxygen (LO_2) and kerosene, liquid oxygen (LO_2) and liquid hydrogen (LH_2), and monomethylhydrazine (MMH) and nitrogen tetroxide (N_2O_4). The last bipropellant combination, MMH and N_2O_4, is hypergolic—meaning these two propellants ignite spontaneously when brought into contact with each other. Hypergolic propellants are especially useful for attitude control rockets where frequent firings and high reliability are required.

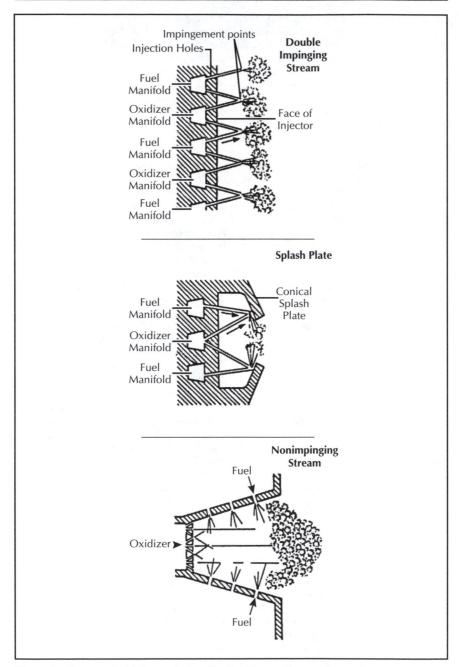

Figure 67. Some typical injector designs for liquid-propellant rocket engines.

Aerospace engineers must consider many factors in selecting bipropellant combinations for a particular rocket system. For example, LH_2 and N_2O_4 would make a good combination based on propellant performance, but their widely divergent storage temperatures (cryogenic and room temperature, respectively) would require the use of large quantities of thermal insulation between the two tanks, adding considerable mass to the rocket vehicle. Another important factor is the toxicity of the chemicals used. MMH and N_2O_4 are both highly toxic. Rocket vehicles that use this propellant combination require special propellant handing and prelaunch preparation. *See also* ELECTRIC PROPULSION; LAUNCH VEHICLE; PROPELLANT; ROCKETRY; SPACE NUCLEAR PROPULSION.

rocket engine 1. In general, a reaction engine that contains within itself (or carries along with itself) all the substances (i.e., matter and energy sources) necessary for its operation. A rocket engine does not require the intake of any outside substance and is therefore capable of operating in outer space. The rocket engine converts energy from some suitable source and uses this energy to eject stored matter at high velocity. The ejected matter is called the rocket engine's *propellant*. Rocket engines often are classified according to the type of energy sources used to manipulate the ejected matter, as, for example: chemical rocket engines, nuclear rocket engines, and electric rocket engines. Very commonly, however, the term "rocket engine" is used to refer to chemical propulsion system engines as a class. Chemical propulsion system engines are further classified by the type of propellant liquid-propellant rocket engine or solid-propellant rocket engine. 2. Specifically, that portion of a rocket's chemical propulsion system in which combustible materials (propellants) are supplied to a chamber and burned under specified conditions. The thermal energy released in the chemical combustion reactions is converted into the kinetic energy of hot combustion gases that are ejected through a nozzle, creating the thrust (reaction force) that then propels the vehicle to which the rocket engine is attached. In common aerospace practice, the term "rocket engine" usually is applied to a device (machine) that burns liquid propellants and therefore requires a rather complex system of propellant tanks, ducts, pumps, flow control devices, and so on. The term "rocket motor" is customarily applied to a device (machine) that burns solid propellants and therefore is relatively simple, normally requiring just the solid-propellant grain within a case, an igniter, and a nozzle. *See also* ELECTRIC PROPULSION; ROCKET; SPACE NUCLEAR PROPULSION.

rocket propellant 1. In general, the matter ejected by a rocket (reaction) engine. 2. Any agent used for consumption or combustion in a rocket engine and from which the rocket derives its thrust, such as a chemical fuel, oxidizer, catalyst, or any compound or mixture of these. Chemical rocket propellants can be either liquid or solid propellants. For example, liquid hydrogen (fuel) and liquid oxygen (oxidizer) often are used as the cryogenic liquid propellants in modern, high-performance launch vehicles. Nuclear thermal rockets use a monopropellant, such as liquid hydrogen, which is heated to extremely high temperatures by a nuclear reactor; electric rockets use readily ionizable elements such as cesium, mercury, argon, or xenon as the propellant, which is accelerated and ejected at extremely high velocities by electromagnetic processes. *See also* ELECTRIC PROPULSION; PROPELLANT; ROCKET; SPACE NUCLEAR PROPULSION.

rocket propulsion Reaction propulsion by means of a rocket engine. Note: Rocket propulsion differs from "jet propulsion" in that a rocket propulsion system is completely self-contained (i.e., it carries all the substances necessary for its operation, such as chemical fuel and oxidizer), while a jet propulsion system needs external materials to achieve reactive propulsion (e.g., air from the atmosphere to support the chemical combustion processes). Rocket propulsion works both within and outside Earth's atmosphere; jet propulsion works only up to certain altitudes within Earth's atmosphere. *See also* JET PROPULSION; ROCKET; X-15.

rocket ramjet A ramjet engine that has a rocket mounted within the ramjet duct. This rocket is used to bring the ramjet up to the necessary operating speed. *See also* RAMJET.

rocketry The art or science of making rockets; the branch of science that deals with rockets, including theory, research, development, experimentation, and application.

According to certain historical records, the Chinese were the first to use gunpowder rockets, which they called "fire arrows," in military applications. In the battle of K'ai-Fung-Foo (1232 C.E.) these rocket fire-arrows helped the Chinese repel Mongol invaders. About 1300, the rocket migrated to Europe, where over the next few centuries (especially during the Renaissance) it ended up in most arsenals as a primitive bombardment weapon. However, artillery improvements eventually made the cannon more effective in battle, and military rockets remained in the background.

Rocketry legend suggests that around 1500, a Chinese official named Wan-Hu (also spelled Wan-Hoo) conceived of the idea of flying through the air in a rocket-propelled chair-kite assembly. Serving as his own test pilot, he vanished in a bright flash during the initial flight of this rocket-propelled device.

The rocket also found military application in the Middle East and in India. For example, in the late 1700s, Rajah Hyder Ali, Prince of Mysore, India, used iron-case stick rockets to defeat a British military unit. Profiting from their adverse rocket experience in India, the British, led by Sir William Congrieve, improved the design of these Indian rockets and developed a series of more efficient bombardment rockets, which ranged in mass from about 8 to 136 kilograms (18 to 300 pounds-mass [lbm]). Perhaps the most famous application of Congrieve's rockets was the British bombardment of the American Fort McHenry in the War of 1812. This rocket attack is now immortalized in the "rockets' red glare" phrase of "The Star-Spangled Banner."

An American named William Hale attempted to improve the inherently low accuracy of 19th century military rockets through a technique called spin stabilization. U.S. Army units used bombardment rockets during the Mexican-American War (e.g., the siege of Veracruz in March 1847) and during the U.S. Civil War. However, improvements in artillery technology again outpaced developments in military rocketry; by the beginning of the 20th century, rockets remained more a matter of polite speculation, rather than a widely accepted technology.

The Russian schoolteacher Konstantin Tsiolkovsky (1857–1935) wrote a series of articles about the theory of rocketry and space flight at the turn of the century. Among other things, his visionary works suggested the necessity for liquid-propellant rockets—the very devices that American physicist Dr. Robert H. Goddard would soon develop. (Because of the geopolitical circumstances in tsarist Russia, Goddard and many other scientists outside of Russia were unaware of Tsiolkovsky's work.) Today Tsiolkovsky is regarded as the Father of Russian Rocketry and one of the founders of modern rocketry. His tombstone bears the prophetic inscription: "Mankind will not remain tied to the Earth forever!"

Similarly, the brilliant physicist Dr. Robert H. Goddard (1882–1945) is regarded as the Father of American Rocketry and the developer of the practical modern rocket. He started working with rockets by testing solid-fuel models. In 1917 when the United States entered World War I, Goddard worked to perfect rockets as weapons. One of his designs became the technical forerunner of the *bazooka*—a tube-launched, antitank rocket. Goddard's device (about 45.7 centimeters [cm] long and 2.54 cm in diameter) (about 18 inches [in] and 1 inch in diameter) was tested in 1918, but the war ended before it could be used against enemy tanks.

In 1919 Goddard published the important technical paper, "Method of Reaching Extreme Altitudes," in which he concluded that the rocket actually would work better in the vacuum of outer space than in Earth's atmosphere. At the time, Goddard's "radical" (but correct) suggestion cut sharply against the popular (but wrong) belief that a rocket needed air to "push against." He also suggested that a multistage rocket could reach very high altitude and even attain sufficient velocity to "escape from Earth." Unfortunately, the press scoffed at his ideas and the general public failed to appreciate the great scientific merit of this paper. Despite this adverse publicity, Goddard continued to experiment with rockets, but now he intentionally avoided publicity. On March 16, 1926, he launched the world's first liquid-fueled rocket from a snow-covered field at his Aunt Effie Goddard's farm in Auburn, Massachusetts. This simple gasoline-and-liquid-oxygen-fueled device rose to a height of just 12 meters (39 ft) and landed about 56 meters (190 ft) away in a frozen field. Regardless of its initial "range," Goddard's liquid-propellant rocket successfully flew, and the world would never be quite the same. The technical progeny of this simple rocket have taken human beings into Earth orbit and to the surface of the Moon. They also have sent sophisticated space probes throughout the solar system and beyond.

After this initial success, Goddard flew other rockets in rural Massachusetts, at least until they started crashing into neighbors' pastures. After the local fire marshal declared that his rockets were a fire hazard, Goddard terminated his New England test program. The famous aviator Charles Lindbergh came to Goddard's rescue and helped him receive a grant from the Guggenheim Foundation. With this grant, Goddard moved to sparsely populated Roswell, New Mexico, where he could experiment without disturbing anyone. At his Roswell test complex, Goddard developed the first gyro-controlled rocket guidance system. He also flew rockets faster than the speed of sound and at altitudes up to 2,300 meters (7,500 ft). Yet, despite his numerous technical accomplishments in rocketry, the U.S. government never really

developed an interest in his work. In fact, only during World War II did he receive any government funding; and that was for him to design small rockets to help aircraft take off from navy carriers. By the time he died in 1945, Goddard held more than 200 patents in rocketry. It is essentially impossible to design, construct, or launch a modern liquid-propellant rocket without using some idea or device that originated from Goddard's pioneering work in rocketry.

While Goddard worked essentially unnoticed in the United States, a parallel group of "rocketeers" thrived in Germany, centered originally around the German Rocket Society. In 1923 Dr. Hermann Oberth published a highly prophetic book entitled *The Rocket Into Interplanetary Space*. In this important work, he proved that flight beyond the atmosphere was possible. One of the many readers inspired by this book was a brilliant young teenager named Werhner von Braun. Oberth published another book in 1929 entitled *The Road to Space Travel*. In this work, he proposed liquid-propellant rockets, multistage rockets, space navigation, and guided reentry systems. From 1939 to 1945 Oberth, along with other German rocket scientists (including von Braun), worked in the military rocket program. Under the leadership of von Braun this program produced a number of experimental designs, the most famous of which was the large, liquid-fueled A-4 rocket. The German military gave this rocket its more sinister and well-recognized name, V-2, for "vengeance weapon number two." It was the largest rocket vehicle at the time, being about 14 meters (46 feet) long and 1.7 meters (5.5 ft) in diameter and developing some 249,000 newtons (56,000 pounds-force [lbf]) of thrust.

At the end of World War II, the majority of the German rocket development team from the Peenemunde rocket test site, led by von Braun, surrendered to the Americans. This team of German rocket scientists and captured V-2 rocket components were sent to White Sands Missile Range (WSMR), New Mexico, to initiate an American military rocket program. The first reassembled V-2 rocket was launched by a combined American-German team on April 16, 1946. A V-2, assembled and launched on this range, was America's first rocket to carry a heavy payload to high altitude. Another V-2 rocket set the first high-altitude and velocity record for a single-stage missile, and still another V-2 rocket was the first large missile to be controlled in flight.

Stimulated by the cold war, the American and German scientists at White Sands worked to develop a variety of missile and rocket systems including the Corporal, Redstone, Nike, Aerobee, and, Atlas.

The need for more room to fire rockets of longer range became evident in the late 1940s. In 1949 the Joint Long Range Proving Ground was established at a remote, deserted location on Florida's eastern coast known as Cape Canaveral. On July 24, 1950, a two-stage Bumper rocket became the first rocket vehicle to be launched from this now world-recognized location. The Bumper consisted of a V-2 first stage and a WAC-Corporal rocket second stage.

The Missile Race and the Space (Moon) Race of the cold war era triggered tremendous developments in rocketry during the second half of the 20th century. On October 4, 1957, the Russians launched *Sputnik 1,* the first artificial satellite to orbit Earth. The United States quickly responded on January 31, 1958 with the launching of *Explorer 1,* the first American satellite. The U.S. Army Ballistic Missile Agency (including von Braun's team of German rocket scientists then located at the Redstone Arsenal in Huntsville, Alabama) modified a Redstone-derived booster into a four-stage launch vehicle configuration, called the Juno I. (This Juno I launch vehicle was the satellite-launching version of the Army's Jupiter C rocket.) The Jet Propulsion Laboratory (JPL) was responsible for the fourth stage, which included America's first satellite.

Another descendant of the V-2 rocket, the mighty Saturn V launch vehicle, carried U.S. astronauts to the surface of the Moon between 1969 and 1972. The first flight of the space shuttle on April 12, 1981 opened up the era of aerospace vehicles and interest in reusable space transportation systems.

With the end of the cold war in 1989, missile and space confrontation races have been replaced by cooperative programs that are truly international in their perspective. For example, during the construction and operation of the international space station, both American and Russian rockets carry people and equipment.

As we enter the next millennium, the dream machines advocated by such visionaries as Tsiolkovsky, Oberth, Goddard and von Braun at the beginning of the 20th century have now become the well-recognized vehicles that will help fulfill our destiny among the stars. *See also* CAPE CANAVERAL AIR FORCE STATION; GODDARD, ROBERT HUTCHINGS; LAUNCH VEHICLE; ROCKET; SPACE TRANSPORTATION SYSTEM; WHITE SANDS MISSILE RANGE.

rocketsonde A rocket-borne instrument package for the measurement and transmission of upper-air meteorological data (up to about 76 kilometers [47 miles] altitude), especially that portion of the atmosphere inaccessible to radiosonde techniques. A meteorological rocket.

rocket thrust The thrust of a rocket engine, usually expressed in newtons (N) or pounds-force. One pound-force is equal to 4.448 newtons.

rocket vehicle A vehicle propelled by a rocket engine. *See also* ROCKET.

rockoon A high-altitude sounding system consisting of a small solid-propellant rocket carried aloft by a large plastic balloon. The rocket is fired near the maximum altitude of the balloon flight. It is relatively mobile sounding system and has been used extensively from ships.

roentgen A unit of exposure to ionizing radiation, especially X rays or gamma rays. The roentgen is defined as the quantity of X rays or gamma rays required to produce a charge of 2.58×10^{-4} coulomb in a kilogram of dry air under standard conditions. The unit is named after Wilhelm Roentgen (1845–1923), the German scientist who discovered X rays in 1895.

Roentgen Satellite (ROSAT) An Earth-orbiting X-ray observatory successfully launched by an expendable Delta II rocket on June 1, 1990 and placed into a 580-kilometer altitude, 53-degree inclination orbit. This mission represents a cooperative effort by the United States (NASA), the United Kingdom, and Germany. Germany constructed the spacecraft and main X-ray telescope; the United Kingdom contributed a wide field camera (WFC); and NASA provided the high-resolution imager (HRI), the launch vehicle, and launch services. Germany also performed the mission operations, including data collection and initial processing. During its first six months in orbit, ROSAT was dedicated to an all-sky survey and detected some 60,000 X-ray and extreme ultraviolet (EUV) sources. The spacecraft's primary mission (all-sky X-ray source survey) was completed by February 1992. *See also* X-RAY ASTRONOMY.

rogue star A wandering STAR that passes close to a SOLAR SYSTEM, disrupting the celestial bodies in the system and triggering cosmic catastrophes on life-bearing PLANETS.

roll The rotational or oscillatory movement of an aircraft, aerospace vehicle, or missile about its longitudinal (lengthwise) axis. *See also* PITCH; YAW.

rollout That portion of the landing of an aircraft or aerospace vehicle following touchdown.

Rosetta Mission A planned COMET exploration mission sponsored by the EUROPEAN SPACE AGENCY to RENDEZVOUS with comet 46 P/Wirtanen and perform REMOTE SENSING investigations. *Rosetta* will also carry a PROBE that lands on the surface of the comet's NUCLEUS and performs in situ scientific measurements.

Rossi X-ray Timing Explorer (RXTE) A NASA ASTROPHYSICS mission designed to study the temporal and broadband spectral phenomena associated with stellar and galactic systems containing compact X-RAY emitting objects. The X-ray ENERGY RANGE observed by this SPACECRAFT extended from 2 to 200 kiloELECTRON VOLTS (keV), and the time scales monitored varied from MICROSECONDS to YEARS. The 3,200-KILOGRAM (7,040-lbm) spacecraft carried three special INSTRUMENTS to measure X-ray emissions from DEEP SPACE—the proportional counter array (PCA), the high-energy X-ray timing experiment (HEXTE), and the all sky monitor (ASM). The PCA and the HEXTE worked together to form a large X-ray OBSERVATORY that was sensitive to X-rays from 2 to 200 keV. The ASM instrument observed the long-term behavior of X-ray sources and also served as a sentinel, monitoring the sky and enabling the spacecraft to swing rapidly to observe TARGETs of opportunity with its other two instruments. Working together, these instruments gathered data about interesting X-ray emissions from the vicinity of BLACK HOLES and from NEUTRON STARS and WHITE DWARFs, along with telltale energetic ELECTROMAGNETIC RADIATION from exploding STARs and active galactic nuclei.

NASA successfully launched this spacecraft into a 580-KILOMETER (360-mi) ALTITUDE circular ORBIT around EARTH on December 30, 1995, with an expendable DELTA II ROCKET from CAPE CANAVERAL AIR FORCE STATION, Florida. Following launch, NASA renamed the spacecraft the *Rossi X-ray Timing Explorer (RXTE)* in honor of Professor Bruno B. Rossi (1905–1993), the distinguished Italian-American physicist who helped pioneer the field of X-RAY ASTRONOMY. This spacecraft is also referred to as *Explorer 69* and the *NASA X-Ray Timing Explorer.*

rotary seal Mechanical seal that rotates with the shaft and is used with a stationary mating ring.

rotate To turn about an internal axis. Said especially of celestial bodies; for example, Earth rotates or spins on its axis once every 24 hours (approximately).

rotating service structure (RSS) An environmentally controlled facility at the launch pad that is used for inserting payloads vertically into the space shuttle orbiter's cargo bay. *See also* SPACE TRANSPORTATION SYSTEM.

rotation 1. The turning of an object (especially a celestial body) about an axis within the body, as the daily rotation of Earth. 2. One turn (i.e., a rotation) of a body about an internal axis. For example, Earth has a sidereal rotation period of 23.9345 hours. *See also* ORBITS OF OBJECTS IN SPACE.

round off To adjust or delete less significant digits from a number and possibly apply some rule of correction to the part retained.

rover A crewed or robotic vehicle used to explore a planetary surface. *See also* LUNAR ROVER(S); MARS SURFACE ROVER(S).

Rover Program The name given to the overall U.S. nuclear rocket development program conducted from 1959 to 1973. At the time, it was envisioned that nuclear (thermal) rockets, placed in Earth orbit by the giant Saturn V launch vehicle, would propel a human expedition to Mars (sometime in the 1980s) and safely return the astronauts back to Earth. *See also* NUCLEAR ROCKET; SPACE NUCLEAR PROPULSION.

Royal Greenwich Observatory (RGO) The famous British OBSERVATORY founded by King Charles II in 1675 at Greenwich in London. The original structure is now a museum, while the RGO is presently located at Cambridge and serves as the principle astronomical institute of the United Kingdom. Until 1971, the ASTRONOMER ROYAL also served as the RGO director, but these positions are now separate appointments.

RP-1 Rocket propellant number one (RP-1)—a common hydrocarbon-based, liquid-propellant rocket fuel that is essentially a refined kerosene mixture. It is relatively inexpensive and generally burned with liquid oxygen (LOX). *See also* PROPELLANT; ROCKET.

rubber-base propellant A solid-propellant mixture in which the oxygen supply is obtained from a perchlorate and the fuel is provided by a synthetic rubber compound. *See also* PROPELLANT; ROCKET.

rumble In rocketry, a form of combustion instability, especially in a liquid-propellant rocket engine, characterized by a low-pitched, low-frequency rumbling noise.

runaway greenhouse An environmental catastrophe during which the GREENHOUSE EFFECT produces excessively high global TEMPERATURES that cause all the liquid (surface) water on a life-bearing PLANET to evaporate permanently. *Compare with* ICE CATASTROPHE.

S

safe/arm (S/A) system Mechanism in a solid-propellant rocket motor igniter that, when in the "SAFE" condition, physically prevents the initiating charge from propagating to the energy release system, which would then ignite the main propellant charge. It is an aerospace safety device intended to prevent an accidental or premature ignition of a solid rocket motor.

Safeguard An early, midcourse, and terminal-phase ballistic missile defense system deployed by the United States in North Dakota in 1975. It was dismantled in 1976 because of its limited cost effectiveness. *See also* BALLISTIC MISSILE DEFENSE.

safety analysis The determination of potential sources of danger and recommended resolutions in a timely manner. A safety analysis addresses those conditions found in either the hardware/software systems, the human-machine interface, or the human/environment relationship (or combinations thereof) that could cause the injury or death of aerospace personnel, damage or loss of the aerospace system, injury or loss of life to the public, or harm to the environment.

safety critical 1. (noun) Aerospace facility, support, test, and flight systems containing: (1) pressurized vessels, lines, and components; (2) propellants, including cryogenic propellants; (3) hydraulics and pneumatics; (4) high voltages; (5) ionizing and nonionizing radiation sources; (6) ordnance and explosive devices or devices used for ordnance and explosive checkout; (7) flammable, toxic, cryogenic, or reactive elements or compounds; (8) high temperatures; (9) electrical equipment that operates in the area where flammable fluids or

solids are located; (10) equipment used for handling aerospace program hardware; and (11) equipment used by personnel for walking or as work platforms. 2. (adjective) Describes something (e.g., a function, a piece of equipment, a set of data) that may affect the safety of aerospace ground and flight personnel, the aerospace system or vehicle, payloads, the general public, or the environment.

safety device A device that prevents unintentional functioning.

safing 1. An action taken to retreat from an armed condition. 2. Actions taken to eliminate or control hazards.

SAINT An early U.S. satellite inspector system designed to demonstrate the feasibility of intercepting, inspecting, and reporting on the characteristics of satellites in orbit.

Salyut A series of Earth-orbiting space stations launched by the former Soviet Union (now Russian Federation) in the 1970s and early 1980s to support a variety of civilian and military missions. The name Salyut means "salute." *Salyut 1 to 5* often are regarded as first-generation crewed spacecraft in this series, while *Salyut 6* and *7* are considered to be an evolved, second-generation design. One major design difference is the fact that the first generation had just one docking port, while the second generation had two docking ports.

Salyut 1 was launched into Earth orbit on April 19, 1971. It functioned in a civilian research capacity as an observation platform, gathering data in the fields of astronomy, Earth resources monitoring, and meteorology. On June 29, the three cosmonauts, who had just spent 24 days aboard the space station, died during reentry operations. Cosmonauts Georgi Dobrovolsky, Vladislav Volkov, and Victor Patseyev suffocated when a vent valve on their *Soyuz 11* spacecraft opened and air rushed out of the capsule as they separated from the space station. Recovery crews found the cosmonauts dead after the capsule touched down. The *Salyut 1* station decayed from orbit on October 11, 1971.

Salyut 2, considered the first "military-mission" Salyut, was launched on April 3, 1973. However, before a cosmonaut crew could be sent to this station, it suffered a catastrophic explosion on April 14. This explosion tore away the space station's solar panels, telecommunications equipment, and docking apparatus. The derelict space station then tumbled out of orbit and on about May 28 burned up in Earth's atmosphere. *Salyut 3,* the first operational military Salyut, was launched into a low (219-kilometer by 270 kilometer) orbit in late June 1974. Apparently the cosmonaut crews performed military photoreconnaissance operations, although little information has been released about this spacecraft or the other military space station, *Salyut 5.*

Salyut 4, launched on December 26, 1974, was similar to *Salyut 1.* Its crews performed civilian missions including astronomical, Earth resources and biomedical observations, and materials processing experiments.

Salyut 5, the last station in this series dedicated primarily to military activities, was launched on June 22, 1976. In addition to military mission equipment, this station also had materials processing equipment.

Salyut 6 was launched on September 29, 1977. Its design represented an evolution of the earlier generation of civilian *Salyut* spacecraft. This station functioned successfully from 1977 to 1981, more than twice its design lifetime of 18 months. The addition of a second docking port gave this second-generation space station much greater flexibility. The use of the Progress supply ship also improved orbital operations. Civilian research activities included

microgravity materials processing, Earth resource photography, and astronomical observations. The last cosmonaut crew left the *Salyut 6* on April 25, 1977 and the station reentered the atmosphere and was destroyed in July 1982.

Salyut 7 was launched on April 19, 1982 and, like the *Salyut 6* crew visit program, enjoyed a pattern of progressively longer flights by the resident cosmonaut crew. However, in early 1985, when it was not occupied by a crew, this station suffered several major technical problems. In June 1985 a cosmonaut crew (*Soyuz T-13* mission) was sent to repair and reoccupy the derelict space station. Repairs enabled it to perform one additional long-duration flight. However, the final *Salyut 7* crew suddenly departed in their spacecraft on November 11, 1985, terminating the mission early because of a serious illness involving one cosmonaut.

Building on the experience of the Salyut space stations, the Russian launched their third-generation space station, called Mir, on February 19, 1986. In March a crew from the Mir space station flew to the *Salyut 7,* visited for about six weeks, and then returned to the Mir. After this "space station" to "space station" visit, the *Salyut 7* was placed in a parking orbit for possible reuse. However, the uncrewed space station experienced a more rapid orbital decay than anticipated, was permanently abandoned, and reentered Earth's atmosphere in February 1991. *See also* MIR; SPACE STATION.

satellite A smaller (secondary) body in orbit around a larger (primary) body. Earth is a *natural satellite* of the Sun, and the Moon is a natural satellite of Earth. Human-made spacecraft placed in orbit around Earth are called *artificial satellites,* or more commonly just *satellites.*

satellite laser ranging (SLR) In satellite laser ranging (SLR), ground-based stations transmit short, intense laser pulses to a retroreflector-equipped satellite, such as the Laser Geodynamics Satellite (LAGEOS). The roundtrip time of flight of the laser pulse is measured precisely and corrected for atmospheric delay to obtain a geometric range. Transmitting pulses of light to these retroflector-equipped spacecraft with a global network of laser ranging stations (both fixed and mobile) allows scientists to determine both the precise orbit of a satellite and the position of the individual ground stations. By monitoring the carefully measured position of these ground stations over time (e.g., several months to years), researchers can deduce the motion of the ground site locations due to plate tectonics or other geodynamic processes, such as subsidence. In fact, a global network of more than 30 SLR stations already has provided a basic framework for determining plate motion, confirming the anticipated motion for most plates.

The theory of plate tectonics tries to explain how the continents arrived at their current positions and to predict where the continents will be in the future. Scientists theorize that up to about 200 million years ago, one giant continent existed where the Atlantic Ocean is today. It is called *Pangaea,* meaning "all lands." Then, about 180 million years ago, Pangaea started to break up into several continents. The plates carrying the continents drifted away from each other, and a low layer of rock formed between the plates. At present, the plates comprise the solid outer 100 kilometers (60 miles) of Earth. These plates move slowly, typically not faster than about 15 centimeters (cm) (6 inches [in.]) a year. For example, North America and Europe are moving apart at a rate of about 3 cm (1.2 in) per year. Although the accumulated motion of the plates is slow when averaged over time, the effects of their short-term drastic movements (such as during earthquakes) can be catastrophic. Plates may bump into one another, spread apart, or move horizontally past one another, in the process causing earthquakes, building mountains, or triggering volcanoes.

Scientists have used satellite laser ranging techniques for the past two decades to study the motions of Earth. These geodynamic studies include measurements of: global tectonic plate motions, regional crustal deformation near plate boundaries,

Earth's gravity field, and the orientation of its polar axis and its spin rate. In the future, scientists will also "invert" the traditional SLR system by placing the laser ranging hardware onboard a satellite and then pulsing selected retroreflecting objects on Earth's surface from space. This "inverted" SLR technology, sometimes called a *laser altimeter system (LAS),* can be used to measure ice sheet topography and temporal changes, cloud heights, planetary boundary heights, aerosol vertical structure, and land and water topography. A laser altimeter system also can be used to perform similar laser ranging measurements on other planets, such as Mars. *See also* LASER GEODYNAMICS SATELLITE (LAGEOS).

satellite power system (SPS) A conceptual CONSTELLATION of very large (kilometers on a side) GEOSTATIONARY ORBIT space structures (constructed in space) that continuously harvest SOLAR ENERGY and transmit it as MICROWAVE RADIATION to special receiver/converter stations on EARTH's surface.

satellite reconnaissance The use of specially designed Earth-orbiting spacecraft to collect intelligence data. Photoreconnaissance satellites are an example. *See also* NATIONAL RECONNAISSANCE OFFICE; RECONNAISSANCE SATELLITE.

saturated liquid In thermodynamics, a substance that exists as a liquid at the saturation temperature and pressure of a mixture of vapor and liquid.

saturated vapor In thermodynamics, a substance that exists as a vapor at the saturation temperature and pressure of a mixture of vapor and liquid.

saturation A term in thermodynamics that defines a condition in which a mixture of vapor and liquid can exist together at a given temperature and pressure. The vapor dome on a traditional pressure (p)–specific volume (v) plot for a simple compressible substance is bounded by the saturated liquid line on the left and the saturated vapor line on the right. The point where two lines meet at the peak of this dome-shaped region is called the *critical point*. It represents the highest pressure and temperature at which distinct liquid and vapor (gas) phases of a substance can coexist.

Saturn Family of expendable launch vehicles developed by NASA (Dr. Wernher von Braun's team at the Marshall Space Flight Center in Huntsville) to support the Apollo Project. The Saturn 1B was used initially to launch Apollo lunar spacecraft into Earth orbit to help the astronauts train for the manned flights to the Moon. The first launch of a Saturn 1B vehicle with an unmanned Apollo spacecraft took place in February 1966. A Saturn 1B vehicle launched the first crewed Apollo flight, *Apollo 7,* on October 11, 1968. After completion of the Apollo Project, the Saturn 1B was used to launch the three crews (three astronauts each crew) for the *Skylab* space station in 1973. Then, in 1975, a Saturn 1B launched the American astronaut crew for the Apollo-Soyuz Test Project, a joint American-Russian docking mission. With an Apollo spacecraft on top, the Saturn 1B vehicle was approximately 69 meters (233 feet) tall. This expendable launch vehicle developed 7.1 million newtons (1.6 million pounds-force) of thrust at liftoff.

The Saturn V rocket, America's most powerful staged rocket, successfully carried out the ambitious task of sending astronauts to the Moon during the Apollo Project. The first launch of the Saturn V vehicle, the unmanned *Apollo 4* mission, occurred on November 9, 1967. The first crewed flight of the Saturn V vehicle, the *Apollo 8* mission, was launched in December 1968. This historic mission was the first human flight to the Moon. The three astronauts aboard *Apollo 8* circled the Moon (but did not land) and then returned safely to Earth. On July 16, 1969, a Saturn V vehicle sent the *Apollo 11* spacecraft and its crew on the first lunar landing mission. The last crewed mission of the Saturn V vehicle occurred on December 7, 1972, when the *Apollo 17* mission lifted off on the final human expedition to the Moon this century. The Saturn

V vehicle flew it last mission on May 14, 1973, when it lifted the unmanned *Skylab* space station into Earth orbit. *Skylab* later was occupied by three different astronaut crews for a total period of 171 days.

All three stages of the Saturn V vehicle used liquid oxygen (LO_2) as the oxidizer. The fuel for the first stage was kerosene, while the fuel for the upper two stages was liquid hydrogen (LH_2). The Saturn V vehicle, with the Apollo spacecraft and its small emergency escape rocket on top, stood 111 meters (363 feet) tall and developed 34.5 million newtons (7.75 million pounds-force) of thrust at liftoff. (See Figure 68.) *See also* APOLLO PROJECT; LAUNCH VEHICLE.

Saturn Saturn is the sixth planet from the Sun and to many the most beautiful celestial object in the solar system. To the naked eye the planet is yellowish in color. This planet is named after the elder god and powerful Titan of Roman mythology (Cronus in Greek mythology), who ruled supreme until he was dethroned by his son Jupiter (Zeus to the ancient Greeks).

Composed mainly of hydrogen and helium, Saturn (with an average density of just 0.7 grams per cubic centimeter) is so light that it would float on water, if there were some cosmic ocean large enough to hold it. The planet takes about 29.5 Earth years to complete a single orbit around the Sun. But a Saturnian day is only approximately 10 hours, 30 minutes long.

The first telescope observations of the planet were made in 1610 by Galileo Galilei (1564–1642). The existence of its magnificent ring system was not known until the Dutch astronomer Christian Huygens (1629–1695), using a better-resolution telescope, properly identified the ring system in 1655. Galileo had seen the rings but mistook them for large moons on either side of the planet. Huygens is also credited with the discovery of Saturn's largest moon, Titan, in 1655.

Astronomers had very little information about Saturn, its rings, and its constellation of moons, until the *Pioneer 11* spacecraft (September 1, 1979), *Voyager 1* spacecraft (November 2, 1980), and *Voyager 2*

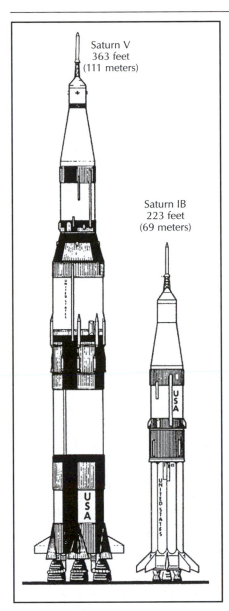

Saturn V
363 feet
(111 meters)

Saturn IB
223 feet
(69 meters)

USA

USA

Figure 68. A side-by-side comparison of the Saturn V and Saturn IB expendable launch vehicles.

spacecraft (August 26, 1981) encountered the planet. These spacecraft encounters revolutionized our understanding of Saturn and have provided the bulk of the current information about this interesting giant planet, its beautiful ring system, and its large complement of moons (30 [6 major, 24 minor] discovered and identified at present). The planet now is being observed, on occasion, by the *Hubble Space Telescope* and will be visited again for a detailed scientific mission by the *Cassini* spacecraft (and its Huygens probe for Titan) in July 2004.

Saturn is a giant planet, second in size only to Jupiter. Like Jupiter, it has a stellar-type composition, rapid rotation, strong magnetic field, and intrinsic internal heat source. Saturn has a diameter of approximately 120,540 kilometers at it equator, but 10% less at the poles because of its rapid rotation. Saturn has a mass of approximately 5.68×10^{26} kilograms, the lowest of any planet in the solar system; this fact indicates that much of Saturn is in a gaseous state. Table 1 lists contemporary data for the planet.

Although the other giant planets—Jupiter, Uranus and Neptune—all have rings, Saturn has an unrivaled, complex yet delicate ring system of billions of icy particles whirling around the planet in an orderly fashion. The main ring areas stretch from about 7,000 kilometers (km) above Saturn's atmosphere out to the F Ring—a total span of about 74,000 km. Within this vast region, the icy particles generally are organized into ringlets, each typically less than 100 km wide. Beyond the F Ring lies the G and E rings, with the latter extending some 180,000 to 480,000 km from the planet's center. The complex ring system contains a variety of interesting physical features, including: kinky rings, clump rings, resonances, spokes, shepherding moons (moons that keep the icy particles in an organized structure), and probably additional as-yet undiscovered moonlets.

Scientists think that the Saturnian rings resulted from one of three basic processes: a small moon venturing too close to the planet and ultimately getting torn apart by large gravitationally induced tidal forces; some of the planet's primordial material failing to coalesce into a moon; or collisions among several larger objects that orbited the planet.

```
┌─────────────────────────────────────────────────────────────────────────┐
│                    Table 1.  Physical and Dynamic Properties              │
│                              of the Planet Saturn                         │
│                                                                           │
│  Diameter (equatorial)          120,540 km                                │
│  Mass                           5.68 × 10²⁶ kg                            │
│  Density                        0.69 g/cm³                                │
│  Surface gravity (equatorial)   9.0 m/s² (approx)                         │
│  Escape velocity                35.5 km/s (approx)                        │
│  Albedo (visual geometric)      0.5                                       │
│  Atmosphere                     Hydrogen (89%), helium (11%), small        │
│                                   amounts of methane, ammonia, and ethane │
└─────────────────────────────────────────────────────────────────────────┘
```

Table 1. Physical and Dynamic Properties of the Planet Saturn

Diameter (equatorial)	120,540 km
Mass	5.68×10^{26} kg
Density	0.69 g/cm^3
Surface gravity (equatorial)	9.0 m/s^2 (approx)
Escape velocity	35.5 km/s (approx)
Albedo (visual geometric)	0.5
Atmosphere	Hydrogen (89%), helium (11%), small amounts of methane (CH_4), ammonia (NH_3), and ethane (C_2H_6); water-ice aerosols.
Natural satellites	30
Rings (thousands)	Complex system
Period of rotation (a Saturnian day)	0.44 days (approx)
Average distance from the Sun	14.27×10^8 km (9.539 AU) [79.33 light-min]
Eccentricity	0.056
Period of revolution around the Sun (a Saturnian year)	29.46 years
Mean orbital velocity	9.6 km/s
Solar flux at planet (at top of clouds)	15.1 watts/m^2 (at 9.54 AU)
Magnetosphere	Yes (strong)
Temperature (blackbody)	77 K

Source: NASA.

Saturn has 18 significant satellites (identified in Tables 2 and 3) plus many small satellites that bring the total number of observed moons to at least 30. These moons form a diverse and remarkable constellation of celestial objects. The largest satellite, Titan, is in a class by itself because of its size and dense atmosphere. The six other major satellites (Iapetus, Rhea, Dione, Tethys, Enceladus, and Mimas) are somewhat similar; all are of intermediate size (some 400 to 1,500 km in diameter) and consist mainly of water ice. Saturn also has many smaller moons, ranging in size from the irregularly shaped Hyperion (about 350 by 200 km across) to tiny Pan (about 20 km in diameter). Hyperion orbits Saturn in a random, chaotic tumbling motion, an orbital condition perhaps indicative of an ancient, shattering collision. Phoebe, the outermost moon of Saturn, orbits in a retrograde direction (opposite the direction of the other moons) in a plane much closer to the ecliptic than Saturn's equatorial plane.

Titan is the largest and perhaps most interesting of Saturn's satellites. It is the second largest moon in the solar system and the only one known to have a dense atmosphere. The atmospheric chemistry now taking place on Titan may be similar to those processes that occurred in Earth's atmosphere several billion years ago.

Larger in size than the planet Mercury, Titan's density appears to be about twice that of water ice. Scientists believe, therefore, that it may be composed of nearly equal amounts of rock and ice. Titan's surface is hidden from spacecraft cameras by a dense, optically thick photochemical haze whose main layer is about 300 km above its surface. The Titanian atmosphere is mostly nitrogen. Carbon-nitrogen compounds may exist on Titan because of

Table 2. Physical and Dynamic Properties
of the More Significant Moons of Saturn

Moon	Diameter (km)	Semimajor Axis of Orbit (km)	Period of Rotation (days)
Phoebe	220 (approx.)	12,952,000	550.5 (retrograde)
Iapetus	1,440	3,561,300	79.33
Hyperion	350 × 200	1,481,000	21.28
Titan	5,150	1,221,850	15.95
Rhea	1,530	527,040	4.52
Helene	40 (approx.)	377,400	2.74
Dione	1,120	377,400	2.74
Calypso	30 (approx.)	294,660	1.89
Tethys	1,050	294,660	1.89
Telesto	25 (approx.)	294,660	1.89
Enceladus	500	238,020	1.37
Mimas	390	185,520	0.942
Janus	220 × 160	151,470	0.695
Epimetheus	140 × 100	151,420	0.694
Pandora	110 × 70	141,700	0.629
Prometheus	140 × 80	139,350	0.613
Atlas	40 (approx.)	137,640	0.602
Pan	20 (approx.)	133,580	0.575

Source: NASA.

the great abundance of both nitrogen and hydrocarbons.

What does the surface of Titan look like? Scientists currently speculate that there must be large quantities of methane on the surface, enough perhaps to form methane rivers or even a methane sea. The temperature on the surface is about 91 kelvins (K) (−182° Celsius), which is close enough to the temperature at which methane can exist as a liquid or solid under the atmosphere pressure near the surface. Some researchers have further speculated that Titan is like Earth (but

Table 3. Physical Data for the Larger Moons of Saturn

Moon	Diameter (km)	Mass (kg)	Density (g/cm3)	Albedo (visual geometric)
Iapetus	1,440	1.6×10^{21}	1.0	0.05–0.5
Titan	5,150	1.35×10^{23}	1.9	0.2
Rhea	1,530	2.3×10^{21}	1.2	0.7
Dione	1,120	1.05×10^{21}	1.4	0.7
Tethys	1,050	6.2×10^{20}	1.0	0.9
Enceladus	500	7.3×10^{19}	1.1	~1.0
Mimas	390	3.8×10^{19}	1.2	0.5

Source: NASA.

colder), with methane playing the role that water plays on Earth. This analogy, if correct, leads to visions of methane-filled seas near Titan's equator and frozen methane ice caps in its polar regions. Titan's surface also might experience a constant rain of organic compounds from the upper atmosphere, perhaps creating up to a 100-meter-thick layer of tarlike materials. It is hoped that the Cassini spacecraft and the Huygens probe that will be released into the Titanian atmosphere in November 2004 will help answer this question. *See also* CASSINI MISSION; *PIONEER 10, 11;* VOYAGER.

Saturnian Of or relating to the planet Saturn.

scalar Any physical quantity whose field can be described by a single numerical value at each point in space. A scalar quantity is distinguished from a vector quantity by the fact that a scalar quantity possesses only magnitude, while a vector quantity possesses both magnitude and direction.

scar Aerospace jargon for design features to accommodate the addition or upgrade of hardware at some future time.

scarp A cliff produced by erosion or faulting.

scattering 1. (particle) A process that changes a particle's trajectory. Scattering is caused by particle collisions with atoms, nuclei, and other particles or by interactions with fields of magnetic force. If the scattered particle's internal energy (as contrasted with its kinetic energy) is unchanged by the collision, *elastic scattering* prevails; if there is a change in the internal energy, the process is called *inelastic scattering.* 2. (photon) Scattering also may be viewed as the process by which small particles suspended in a medium of a different refractive index diffuse a portion of the incident radiation in all directions. In scattering, no energy transformation results, only a change in the spatial distribution of the radiation. Along with absorption, scat-

tering is a major cause of the attenuation of radiation by the atmosphere. Scattering varies as a function of the ratio of the particle diameter to the wavelength of the radiation. When this ratio is less than about one-tenth, *Rayleigh scattering* occurs in which the scattering process varies inversely as the fourth power of the wavelength. At larger values of the ratio of particle diameter to wavelength, the scattering varies in a complex fashion described by *Mie scattering* theory; at a ratio of the order of 10, the laws of geometric optics begin to apply. *See also* MIE SCATTERING; RAYLEIGH SCATTERING.

schlieren (German: streaks, striae) 1. Regions of different density in a fluid, especially as shown by special apparatus. 2. Pertaining to a method or apparatus for visualizing or photographing regions of varying density in a fluid field. *See also* SCHLIEREN PHOTOGRAPHY.

schlieren photography A method of photography for flow patterns that takes advantage of the fact that light passing through a density gradient in a gas is refracted as if it were passing through a prism. Schlieren photography allows the visualization of density changes, and therefore shock waves, in fluid flow. Schlieren techniques have been used for decades in laboratory wind tunnels to visualize supersonic flow around special scale-model aircraft, missiles, and aerospace vehicles. *See also* WIND TUNNEL.

Schwarzschild black hole An uncharged BLACK HOLE that does not rotate; the German astronomer Karl Schwarzschild (1873–1916) hypothesized this basic model of a black hole in 1916.

Schwarzschild radius The radius of the EVENT HORIZON of a BLACK HOLE. Named after the German astronomer Karl Schwarzschild (1873–1916), who applied RELATIVITY theory to very high-DENSITY objects and point masses (singularities).

science payload The complement of scientific instruments on a spacecraft, including both remote-sensing and direct-sensing devices that together cover large portions of the electromagnetic spectrum, large ranges in particle energies, or a detailed set of environmental measurements.

scientific airlock An opening in a crewed spacecraft or space station from which experiment and research equipment can be extended outside (into space) while the interior of the vehicle retains its atmospheric integrity (i.e., remains pressurized).

scientific notation A method of expressing powers of 10 that greatly simplifies writing large numbers. In scientific notation, a number expressed in a positive power of 10 means the decimal point moves to the right (e.g., $3 \times 10^6 = 3,000,000$); a number expressed in a negative power of 10 means that the decimal moves to the left (e.g., $3 \times 10^{-6} = 0.000003$).

Scout A four-stage, all solid-propellant expendable launch vehicle developed and used by NASA to place small payloads into Earth orbit and to place probes on suborbital trajectories. The first Scout (Solid Controlled Orbital Utility Test) vehicle was launched on July 1, 1960, from the Mark 1 Launcher at the NASA Goddard Space Flight Center's Wallops Flight Facility, Wallops Island, Virginia. The Scout's first-stage motor was based on an earlier version of the navy's Polaris missile motor; the second-stage solid propellant motor was derived from the army's Sergeant surface-to-surface missile; and the third- and fourth-stage motors were adapted by NASA from the navy's Vanguard missile.

The standard Scout vehicle is a slender, all solid-propellant, four-stage booster system, approximately 23 meters (75 feet) in length with a launch mass of 21,500 kilograms (47,400 pounds-mass). This expendable vehicle is capable of placing a 186-kilogram (400 pounds-mass) payload into a 560-kilometer (350-mile) orbit around Earth. *See also* LAUNCH VEHICLE; ROCKET.

Scud A family of mobile, short-range tactical BALLISTIC MISSILES developed by

the former Soviet Union (Russia) during the COLD WAR. Scud GUIDED MISSILES are single STAGE, LIQUID-PROPELLANT ROCKETS about 11.25-meters (37 ft) long, except for the Scud-A, which has a length of 10.25 meters [33.6 ft]). The Scud A through D models were designed to carry various WARHEADS ranging from conventional HIGH EXPLOSIVE (HE) weapons to nuclear, chemical, or biological weapons. The Scud-B (first deployed in 1965) had a maximum RANGE of about 300 KILOMETERS (185 mi), while the Scud-D (first deployed in the 1980s) had a maximum range of about 700 kilometers (435 mi). During the cold war, the Russian government exported numerous nonnuclear warhead models of the Scud to Warsaw Pact and non–Warsaw Pact nations, such as Iraq.

Iraqi Scud missiles earned some degree of notoriety during the Persian Gulf War of 1991. All the Iraqi variants of the Russian Scud B guided missile used KEROSENE as the FUEL and some form of red fuming nitric acid as the OXIDIZER. With a maximum range of about 300-kilometers (185 mi), the 1991-era Iraqi Scuds exhibited notoriously poor accuracy—the farther they flew the more inaccurate they became. Relatively unsophisticated GYROS guided the missile and only during the powered phase of the flight, a period generally lasting about 80 seconds. Once the Iraqi Scud's engine shut down, the entire spent missile body with warhead attached followed a simple BALLISTIC TRAJECTORY to the target area. Upon REENTRY into the SENSIBLE ATMOSPHERE, the dynamically awkward missile body–warhead combination became unstable and often disintegrated before impacting within the intended target area.

screaming A form of combustion instability, especially in a liquid-propellant rocket engine, of relatively high frequency and characterized by a high-pitched noise.

screeching A form of combustion instability, especially in an afterburner, of relatively high frequency and characterized by a harsh, shrill noise.

scrub To cancel or postpone a rocket firing, either before or during the countdown.

sealant Liquid/solid mixture installed at joints and junctions of components to prevent leakage of fluid (especially gas) from the joint or junction.

sealed cabin The crew-occupied space of an aircraft, aerospace vehicle, or spacecraft characterized by walls that do not allow any gaseous exchange between the cabin (inner) atmosphere and its surroundings and containing its own mechanisms for maintenance of the cabin atmosphere.

sea-level engine A rocket engine designed to operate at sea level; that is, the exhaust gases achieve complete expansion at sea-level ambient pressure. *See also* ROCKET; ROCKET ENGINE.

seal extrusion Permanent displacement (under the action of fluid pressure) of part of a seal into a gap provided for such displacement.

sealing rings Mechanical devices designed to fit together tightly when two spacecraft or a spacecraft and a space station are docked so that their cabin atmospheres will not leak out during crew and equipment transfers.

search for extraterrestrial intelligence (SETI) The major aim of SETI programs is to listen for evidence of radio-frequency (microwave) signals generated by intelligent extraterrestrial civilizations. This search is an attempt to answer an important philosophical question: Are we alone in the universe? The classic paper by Giuseppe Cocconi and Philip Morrison entitled "Searching for Interstellar Communications" (*Nature,* 1959) often is regarded as the start of modern SETI. With the arrival of the space age, the entire subject of extraterrestrial intelligence (ETI) has left the realm of science fiction and now is regarded as a scientifically respectable (although currently speculative) field of endeavor.

The current understanding of stellar formation leads scientists to think that planets are normal and frequent companions of most stars. As interstellar clouds of dust and gas condense to form stars, they appear to leave behind clumps of material that form into planets. The Milky Way Galaxy contains at least 100 billion to 200 billion stars.

Current theories on the origin and chemical evolution of life indicate that it probably is not unique to Earth but may be common and widespread throughout the galaxy. Further, some scientists believe that life on alien worlds could have developed intelligence, curiosity, and the technology necessary to build the devices needed to transmit and receive electromagnetic signals across the interstellar void. For example, an intelligent alien civilization might, like humans, radiate electromagnetic energy into space. This can happen unintentionally (as a result of planetary-level radio-frequency communications) or intentionally (through the beaming of structured radio signals out into the galaxy in the hope some other intelligent species can intercept and interpret these signals against the natural electromagnetic radiation background of space).

SETI observations may be performed using radio telescopes on Earth, in space, or even (some day) on the far side of the Moon. Each location has distinct advantages and disadvantages.

Until now only very narrow portions of the electromagnetic spectrum have been examined for "artifact signals" (i.e., those generated by intelligent alien civilizations). Human-made radio and television signals—the kind radio astronomers reject as clutter and interference—actually are similar to the signals SETI researchers are hunting for.

The sky is full of radio waves. In addition to the electromagnetic signals we generate as part of our technical civilization (e.g., radio, TV, radar, etc.), the sky also contains natural radiowave emissions from such celestial objects as the Sun, the planet Jupiter, radio galaxies, pulsars, and quasars. Even interstellar space is characterized by a constant, detectable radio-noise spectrum.

And just what would a radio-frequency signal from an intelligent extraterrestrial civilization look like? Figure 69 presents a spectrogram that shows a simulated "artifact signal" from outside the solar system. This particular signal was sent by *Pioneer 10* from beyond the orbit of Neptune and was received by a Deep Space Network (DSN) radio telescope at Goldstone, California, using a 65,000-channel spectrum analyzer. The three signal components are quite visible above the always-present background radio noise. The center spike appearing in the figure has a transmitted signal power of approximately 1 watt—about half the power of a miniature Christmas tree light. SETI scientists are looking for a radio-frequency signal that might appear this clearly, or for one that may actually be quite difficult to distinguish from the background radio noise. To search through this myriad of radio-frequency signals, SETI scientists have developed state-of-the-art spectrum analyzers that can sample millions of frequency channels simultaneously and identify candidate "artifact signals" automatically for further observation and analysis.

In October 1992, NASA started a planned decade-long SETI program called the High Resolution Microwave Survey (HRMS). The objective of HRMS was to search other SOLAR SYSTEMS for microwave signals, using radio telescopes at the National Science Foundation's ARECIBO OBSERVATORY in Puerto Rico, NASA's Goldstone DEEP SPACE COMMUNICATIONS COMPLEX in California, and other locations. Coupled with these telescopes were HRMS—dedicated high-speed digital data processing systems that contained off-the-shelf HARDWARE and specially designed SOFTWARE.

The search proceeded in two different modes: a Targeted Search and an All-Sky Survey. The Targeted Search focused on about 1,000 nearby stars that resembled our Sun. In a somewhat less sensitive search mode, the All-Sky Survey was planned to search the entire CELESTIAL SPHERE for unusual radio signals. However, severe budget constraints and refocused national objectives resulted in the premature termination of the NASA's HRMS program in 1993—after just one year of observation.

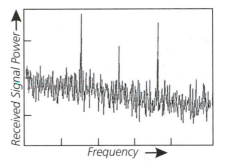

Figure 69. A simulated signal from an extraterrestrial civilization (using the *Pioneer 10* spacecraft transmitting an "artifact" signal from beyond the orbit of Neptune). *(Courtesy of NASA)*

Since then, while NASA has remained deeply interested in searching for life within our solar system, specific SETI projects no longer receive government funding.

Today, privately funded organizations and foundations (such as the SETI Institute in Mountain View, California—a nonprofit corporation that focuses on research and educational projects relating to the search for extraterrestrial life) are conducting surveys of the heavens in search of radio signals from intelligent alien civilizations.

If an alien signal is ever detected and decoded, then the people of Earth would face another challenging question: Do we answer? For the present time, SETI scientists are content to listen passively for "artifact signals" that might arrive across the interstellar void. *See also* EXTRATERRESTRIAL LIFE; RADIO ASTRONOMY.

Seasat A NASA Earth observation spacecraft designed to demonstrate techniques for global monitoring of oceanographic phenomena and features, to provide oceanographic data, and to determine key features of an operational ocean-dynamics system. *Seasat-A* was launched on June 26, 1978 by an expendable Atlas-Agena launch vehicle from Vandenberg Air Force Base, California. It was placed into a near-circular (~800 kilometer (~500 miles) altitude) polar orbit. The major difference between *Seasat-A* and previous Earth observation satellites was the use of active and passive microwave sensors to achieve an all-weather capability. After 106 days of returning data, contact with *Seasat-A* was lost when a short circuit drained all power from its batteries. *See also* AQUA SPACECRAFT; OCEAN REMOTE SENSING; REMOTE SENSING.

Sea Sparrow (RIM-7M) The U.S. Navy's Sea Sparrow missile, designated RIM-7, is a radar-guided, air-to-air missile with a high-explosive (HE) warhead. It has a cylindrical body with four wings (canards) at midbody and four tail fins. A solid-propellant rocket motor powers this versatile all-weather, all-altitude operational weapon system that can attack high-performance aircraft and missiles from any direction. The navy also uses the Sea Sparrow aboard ships as a surface-to-air antimissile defense system. *See also* SPARROW.

second 1. The SI UNIT of time (symbol: s), now defined as the duration of 9,192,631,770 periods of radiation corresponding to the transition between two hyperfine levels of the ground state of the cesium-133 atom. Previously, this unit of time had been based on astronomical observations. 2. A unit of angle (symbol: ″) equal to 1/3600 of a degree or 1/60 of a minute of angle.

secondary crater The CRATER formed when a large chunk of material from a primary-IMPACT CRATER strikes the surrounding planetary surface.

secondary radiation Electromagnetic (photons) or particulate ionizing radiation resulting from the absorption of other (primary) ionizing radiation in matter. For example, the ionizing occurring within a spacecraft, when high-energy cosmic rays impact the wall of a spacecraft and remove electrons or nuclei from the matter in the wall. These removed electrons or nuclei are called secondary radiations.

Second Law of Thermodynamics An inequality asserting that it is impossible to transfer thermal energy (heat) from a colder to a warmer system without the occurrence of other simultaneous changes in the two systems or in the environment.

It follows from this important physical principle that during an adiabatic process, ENTROPY cannot decrease. For reversible adiabatic processes, entropy remains constant; while for irreversible adiabatic processes, it increases.

An equivalent formulation of the law is that it is impossible to convert the heat of a system into work without the occurrence of other simultaneous changes in the system or its environment. This version of the Second Law, which requires a heat engine to have a cold sink as well as a hot source, is particularly useful in engineering applications.

Another important statement of the Second Law is that the change in entropy (ΔS) for an isolated system is greater than or equal to zero. Mathematically,

$$(\Delta S)_{isolated} \geq 0$$

See also FIRST LAW OF THERMODYNAMICS; THERMODYNAMICS.

section One of the cross-section parts that a rocket vehicle is divided into, each adjoining another at one or both of its ends. Usually described by a designating word, as in *nose section, aft section, center section, tail section, thrust section, propellant tank section*.

Seebeck effect The establishment of an electric potential difference that tends to produce a flow of current in a circuit of two dissimilar metals, the junctions of which are at different temperatures. It is the physical phenomenon involved in the operation of a thermocouple and is named for the German scientist Thomas Seebeck (1770–1831), who first observed it in 1822. *See also* THERMOCOUPLE; THERMOELECTRIC CONVERSION.

Selenian Of or relating to Earth's Moon. Once a permanent lunar base is established, a resident of the Moon. *See also* MOON.

selenocentric 1. Relating to the center of Earth's Moon; referring to the Moon as a center. 2. Orbiting about the Moon as a central body.

selenodesy That branch of science (applied mathematics) that determines, by observation and measurement, the exact positions of points and the figures and areas of large portions of the Moon's surface, or the shape and size of the Moon.

selenoid A satellite of Earth's Moon; a spacecraft in orbit around the Moon, as, for example, NASA's lunar orbiter spacecraft. *See also* LUNAR ORBITER.

selenography The branch of astronomy that deals with Earth's Moon, its surface, composition, motion, and the like. *Selene* is Greek for "Moon."

self-cooled Term applied to a combustion chamber or nozzle in which temperature is controlled or limited by methods that do not involve flow within the wall of coolant supplied from an external source.

self-pressurization Increase of ullage pressure by vaporization or boil-off of contained fluid without the aid of additional pressurant.

self-replicating system (SRS) An advanced robot system, first postulated by the Hungarian-born German-American mathematician John Von Neumann (1903–1957) that would be capable of gathering materials, repairing and maintaining itself, manufacturing desired products, and even making copies of itself (self-replication). *See also* ROBOTICS.

semiactive homing guidance A system of homing guidance in which the receiver in the missile uses radiations reflected from a target that has been illuminated by a source outside the missile. For example, a pilot in an aircraft can use a laser designator to mark, or "paint," a particular target. Some of this laser light then is reflected back to the missile, which follows it to the target. *See also* GUIDANCE SYSTEM.

semimajor axis One-half the major AXIS of an ELLIPSE. For a PLANET, this corresponds to its average orbital distance from the SUN.

sensible atmosphere That portion of a planet's atmosphere that offers resistance to a body passing through it.

sensible heat The thermal energy (heat) that, when added to or removed from a thermodynamic system, causes a change in temperature. *Compare with* LATENT HEAT.

sensor In general, a device that detects and/or measures certain types of physically observable phenomena. More specifically, that part of an electronic instrument that detects electromagnetic radiations (or other characteristic emissions, such as nuclear particles) from a target or object at some distance away and then converts these incident radiations (or particles) into a quantity (i.e., an internal electronic signal) that is amplified, measured (quantified), displayed, and/or recorded by another part of the instrument. A passive sensor uses characteristic emissions from the object or target as its input signal. In contrast, an active sensor (like a radar) places a burst of electromagnetic energy on the object or target being observed and then uses the reflected signal as its input. *See also* REMOTE SENSING.

Sentinel A former antiballistic missile (ABM) system that was designed for light-area defense against a low-level ballistic missile attack on the United States. It was developed into the Safeguard system in the late 1960s. *See also* BALLISTIC MISSILE DEFENSE; SAFEGUARD.

separation 1. The action of a fallaway section or companion body as it casts off from the remaining body of a vehicle, or the action of the remaining body as it leaves a fallaway section behind it. 2. The moment at which this action occurs.

separation velocity The velocity at which a rocket, aerospace vehicle, or spacecraft is moving when some part or section is separated from it; specifically, the velocity of a space probe or satellite at the time of separation from the launch vehicle.

Sergeant (MGM-29A) A U.S. Army mobile, inertially guided, solid-propellant, surface-to-surface missile with nuclear capability, designed to attack targets up to a range of about 140 kilometers (85 miles).

service module (SM) The large part of the Apollo spacecraft that contained support equipment; it was attached to the command module (CM) until just before the CM (carrying three astronauts) reentered Earth's atmosphere at the end of a mission. *See also* APOLLO PROJECT.

servo A device that helps control (usually by hydraulic means) a large moment of inertia by the application of a relatively small moment of inertia. Also called a *servomechanism. See also* ROBOTICS.

Seyfert galaxy A type of SPIRAL GALAXY with a very bright GALACTIC NUCLEUS—first observed in 1943 by the American astronomer Carl K. Seyfert (1911–1960).

shaft A bar (almost always cylindrical) used to support rotating pieces or to transmit power or motion by rotation.

shake-and-bake test A series of prelaunch tests performed on a spacecraft to simulate the launch vibrations and thermal environment (i.e., temperature extremes) it will experience during the mission. A typical test plan might involve the use of enormous speakers to blast the spacecraft with acoustic vibrations similar to those it will encounter during launch. As part of a typical shake-and-bake test plan, aerospace engineers also will place the spacecraft in a special environmental chamber and subject it to the extreme hot and cold temperatures it will experience during space flight.

shake table Device for subjecting components or assemblies to vibration in order to reveal vibrational mode patterns.

shear strength In materials science, the stress required to produce fracture in the plane of cross section, the conditions of loading being such that the directions of force and of resistance are parallel and opposite, although their paths are offset a specified minimum amount.

shelf life Storage time during which an item, such as a battery or a propellant compound, remains serviceable, that is, will operate satisfactorily when put to use.

shepherd moon A small inner MOON (or pair of moons) that shapes and forms a particular RING around a RINGED PLANET. For example, the shepherds Ophelia and Cordelia tend the Epsilon Ring of URANUS.

shield A quantity of material used to reduce the passage of ionizing radiation. The term "shielding" refers to the arrangement of shields used for a particular nuclear radiation protection circumstance. *See also* SPACE RADIATION ENVIRONMENT.

shield volcano A wide, gently sloping VOLCANO formed by the gradual outflow of molten rock; many occur on VENUS.

Shillelagh (MGM-51) A U.S. Army missile system that is mounted on a main battle tank or assault reconnaissance vehicle for use against enemy armor, troops, and field fortifications.

shirt-sleeve environment A space station module or spacecraft cabin in which the atmosphere is similar to that found on the surface of Earth, that is, it does not require a pressure suit.

shock 1. A SHOCK WAVE. 2. A blow, impact, collision, or violent jar. 3. The sudden agitation of a person's mental or emotional state; or the event causing it. 4. The sudden (sometimes fatal) stimulation caused by an electrical discharge on a human being or animal. Also called an *electric shock.*

shock diamonds The diamond-shaped (or wedge-shaped) shock waves that sometimes become distinctly visible in the exhaust of a rocket vehicle during launch ascent.

shock front 1. A SHOCK WAVE regarded as the forward surface of a fluid region having characteristics different from those of the region ahead of it. 2. The boundary between the pressure disturbance created by an explosion (in air, water, or earth) and the ambient atmosphere, water, or earth.

shock isolator   A resilient support that tends to isolate a system from applied shock. Also called *shock mount*.

shock layer   The layer of fluid between the SHOCK WAVE and the vehicle in which the velocity of the fluid changes from supersonic (or hypersonic) to subsonic.

shock tube   A relatively long tube in which brief, high-velocity gas flows are produced by the sudden release of very high-pressure gas into the low-pressure portion of the tube. This action produces a traveling SHOCK WAVE along the tube with high-speed flow behind the shock.

shock wave   1. A surface of discontinuity (i.e., of abrupt changes in conditions) set up in a supersonic field of flow, through which the fluid undergoes a finite decrease in velocity accompanied by a marked increase in pressure, density, temperature, and ENTROPY. Sometimes called a *shock*. 2. A pressure pulse in air, water, or earth, propagated from an explosion. It has two distinct phases: in the first, or positive, phase, the pressure rises sharply to a peak, then subsides to the normal pressure of the surrounding medium; in the second, or negative, phase, the pressure falls below that of the medium, then returns to normal. A shock wave in air is also called a *blast wave*.

short-period comet   A COMET with an ORBITAL PERIOD of less than 200 YEARS.

short-range attack missile (SRAM) (AGM-69)   A supersonic U.S. Air Force air-to-surface missile that can be armed with a nuclear warhead and can be launched from the B-52 and the B-1B aircraft. The missile's range, speed, and accuracy allow the carrier aircraft to "stand off" from its intended targets and launch this missile while still outside enemy defenses. The SRAM is used by bombers to neutralize enemy air defenses and to strike heavily defended targets.

short-range ballistic missile   A ballistic missile with a range capability up to about 1,100 kilometers (690 miles).

shot   A colloquial aerospace term used to describe the act or instance of firing a rocket, especially from Earth's surface; for example, the launch crew prepared for the next *shot* of the space shuttle.

Shrike (AGM-45)   An air-launched, antiradiation missile used by the U.S. Navy to home in on and destroy hostile antiaircraft radars. The missile uses a solid-propellant rocket motor, is 3.05 meters (10 feet) long, has a mass of 177 kilograms (390 pounds-mass). It employs a passive radar homing guidance system and contains a conventional warhead.

shroud   1. In general, the protective outer covering for a payload that is being transported into space by a launch vehicle. Once the launch vehicle has left Earth's sensible atmosphere, the shroud is discarded. 2. In the context of BALLISTIC MISSILE DEFENSE (BMD), a shroud would be used with a reentry vehicle (RV) or a postboost vehicle (PBV) to confuse the missile defense system. For example, the RV shroud is a thin envelope that would enclose a reentry vehicle, interfering with the infrared radiation it normally would emit; a PBV shroud is a loose conical device that would be positioned behind a PBV to conceal the deployment of reentry vehicles and decoys.

shutdown   1. In aerospace engineering, the process of decreasing the thrust of a rocket engine to zero. 2. In nuclear engineering, the process of reducing the power level of a nuclear reactor to zero (i.e., the process of stopping the neutron chain reaction).

shutoff valve   Valve that terminates the flow of fluid; usually a two-way valve that is either fully open or fully closed.

Shuttle Commonly encountered, shortened name for NASA's space shuttle vehicle or the U.S. SPACE TRANSPORTATION SYSTEM.

Shuttle glow A phenomenon first observed on early space shuttle missions in which a visible glow, caused by the optical excitation of the tenuous residual atmosphere of atomic oxygen and molecular nitrogen found between 250 to 300 kilometers (km) (155 to 186 miles [mi]) altitude, is seen around the leading edges of the vehicle. As the space shuttle or other spacecraft flies through these thin gases at orbital velocities (about 8 km [5 mi] per second), the impact of these gases actually gives rise to two phenomena: Some portions of the space vehicle glow while in orbit, while other portions of the space vehicle can erode away slowly due to a chemical interaction with the highly reactive residual atmosphere (i.e., atomic oxygen chemically attacks the surfaces of an orbiting spacecraft).

sideband 1. Either of the two frequency bands on both sides of the carrier frequency within which fall the frequencies of the wave produced by the process of modulation. 2. The wave components lying within such a band.

sidereal Of or pertaining to the stars.

sidereal day The duration of one rotation of Earth on its axis, with respect to the vernal equinox. It is measured by successive transits of the vernal equinox over the upper branch of a meridian. Because of the procession of the equinoxes, the sidereal day so defined is slightly less than the period of rotation with respect to the stars, but the difference is less than 0.01 second. The length of the mean sidereal day is 24 hours of sidereal time or 23 hours 56 minutes 4.091 seconds of mean solar time.

sidereal month The average amount of time the MOON takes to complete one orbital REVOLUTION around EARTH when using the FIXED STARS as a reference; approximately 27.32 days.

sidereal period The period of time required by a CELESTIAL BODY to complete one REVOLUTION around another celestial body with respect to the FIXED STARS.

sidereal time Time measured by the rotation of Earth with respect to the stars, which are considered "fixed" in position. Sidereal time may be designated as local or Greenwich, depending on whether the local meridian or the Greenwich meridian is used as the reference.

sidereal year The period of one apparent revolution of Earth around the Sun, with respect to a referenced ("fixed") star. A sidereal year averages 365.25636 days in duration. Because of the precession of the equinoxes, this is about 20 minutes longer than a tropical year. The tropical year is defined as the average period of time between successive passages of the Sun across the vernal equinox. *See also* YEAR.

Sidewinder (AIM-9) A supersonic, heat-seeking, air-to-air missile carried by U.S. Navy, Marine, and Air Force fighter aircraft. The Sidewinder has the following main components: an infrared homing guidance section, an active optical target detector, a high-explosive warhead, and a solid-propellant rocket motor. The heat-seeking infrared guidance system enables the missile to home on a target air-craft's engine exhaust. This infrared seeker also permits the pilot to launch the missile, then leave the area or take evasive action, while the missile guides itself to the target. The AIM-9A, prototype of Sidewinder, was fired successfully first in September 1953. Since then the design of this missile has evolved continuously to where it has become the most widely used air-to-air missile by the United States and its Western allies.

siemens (symbol: S) The SI UNIT of electrical conductance. It is defined as the conductance of an electrical circuit or element that has a resistance of one ohm. In the past this unit sometimes was called the *mho* or *reciprocal ohm*. The unit is named in honor of the German scientist Ernst Werner von Siemens (1816–1892).

sievert (symbol: Sv) The special SI UNIT for dose equivalent (H). In radiation protection, the dose equivalent in sieverts (Sv) is the product of the absorbed dose (D) in grays (Gy) and the quality factor (QF) as well as any other modifying factors considered necessary to characterize and evaluate the biological effects of the ionizing radiation received by human beings or other living creatures. The sievert is related to the traditional dose equivalent unit (rem) as follows: 1 sievert = 100 rem. *See also* DOSE EQUIVALENT; SPACE RADIATION ENVIRONMENT.

signal 1. A visible, audible, or other indication used to convey information. 2. Information to be transmitted over a communications system. 3. Any carrier of information; as opposed to noise. 4. In electronics, any transmitted electrical impulse. The variation of amplitude, frequency, and waveform are used to convey information.

signal-to-noise ratio (SNR) The ratio of the amplitude of the desired signal to the amplitude of noise signals at a given point in time. The higher the signal-to-noise ratio, the less interference with reception of the desired signal.

signature In general, the set of characteristics by which an object or target can be identified; for example, the distinctive type of radiation emitted or reflected by the target that can be used to recognize it. A target's signature might include a characteristic infrared emission, an unusual pattern of motion, or a distinctive radar return.

simulation 1. In general, the art of replicating relevant portions of the "real-world" environment to test equipment, train mission personnel, and prepare for emergencies. Simulations can involve the use of physical mass and energy replicants, high-fidelity (i.e., reasonably close to the original) hardware, and supporting software. With the incredible growth in computer and display technologies, computer-based simulations are assuming an ever-increasing role as aerospace design tools and astronaut training aids. In fact, today's computer-based simulations often are referred to as "reality in a box." Virtual reality systems are extremely versatile simulation tools that can be operated to test equipment and operating procedures and to train personnel safely using a variety of highly interactive mission scenarios. *See also* VIRTUAL REALITY. 2. In a military context, the art of making a decoy look like a more valuable strategic target.

simulator A heavily computer-dependent training facility that imitates flight-hardware responses and can be used to train flight personnel.

single-event upset (SEU) A bit flip (i.e., a "0" is changed to a "1," or vice versa) in a digital micro-electronic circuit. The single-event upset (SEU) is caused by the passage of high-energy ionizing radiation through the silicon material of which the semiconductors are made. These space radiation-induced bit flips (or SEUs) can damage data stored in memory, corrupt operating software, cause the central processing unit (CPU) to write over critical data tables or to halt, and even trigger an unplanned event involving some piece of computer-controlled hardware (e.g., fire a thruster) that then severely impacts the mission. Aerospace engineers deal with the SEU problem in a variety of ways, including: the use of additional shielding around sensitive spacecraft electronics components, the careful selection of electronic components (i.e., using more radiant resistant parts), the use of multiple-redundancy memory units and "polling" electronics, and the regular resetting of the spacecraft's onboard computers. *See also* SPACE RADIATION ENVIRONMENT.

single failure point (SFP) A single element of hardware, the failure of which would lead directly to loss of life, vehicle, or mission. Where safety conditions dictate that abort be initiated when a redundant element fails, that element also is considered a single failure point.

single-stage rocket A rocket vehicle provided with a single rocket propulsion system.

single-stage-to-orbit (SSTO) vehicle

Since the early 1960s, aerospace engineers have envisioned building reusable launch vehicles because they offer the potential of relative operational simplicity and reduced costs, when compared with expendable launch vehicles. Until recently, however, the necessary technologies were not available. Now, thanks to aerospace technology improvements, many engineers believe that it is technically feasible to design and produce a single-stage-to-orbit (SSTO) launch vehicle with sufficient payload capacity to meet most government and commercial space transportation requirements.

But despite these recent technology advances, the development of an SSTO space transportation system involves considerable risk. For example, because an SSTO launch vehicle will have no expendable components, it will need to carry more fuel than would be necessary otherwise, if it were shedding mass by dropping (expended) stages during launch. Achieving the propellant mass fraction necessary to reach orbit with a useful payload will require numerous technological advances that improve fuel efficiency and lower structural mass without compromising structural integrity. In addition, completely reusable launch vehicles are technologically much more difficult to achieve because components must be capable of resisting deterioration and surviving multiple launches and reentries.

The difficulties of a horizontal takeoff-horizontal landing vehicle are known from the National Aerospace Plane (NASP) Program. Other candidate SSTO vehicle configurations include vertical takeoff/vertical landing; vertical takeoff/horizontal landing (winged body); and vertical takeoff/horizontal landing (lifting body). NASA's former (now cancelled) X-33 program focused SSTO vehicle development using a vertical-takeoff/horizontal-landing (lifting body) approach.

singularity

The hypothetical central point in a BLACK HOLE at which the curvature of space and time becomes infinite; a theoretical point that has infinite DENSITY and zero VOLUME. The *principle of cosmic censorship*—a theorem of black hole physics—states that an event horizon always conceals a singularity, preventing that singularity from communicating its existence to observers elsewhere in the UNIVERSE.

SI unit(s)

The Systéme International d'Unités (International System of Units, or SI units) is the internationally agreed-upon system of coherent units that is now in use throughout the world for scientific and engineering purposes. The metric (or SI) system was developed originally in France in the late 18th century. Its basic units for length, mass, and time—the meter (m), the kilogram (kg), and the second (s)—were based on natural standards. The modern SI units are still based on natural standards and international agreement, but these standards now are ones that can be measured with greater precision than the previous natural standards. For example, the *meter (m)* [British spelling: *metre*] is the basic SI unit of length. Currently it is defined as being equal to 1,650,763.73 wavelengths in a vacuum of the orange-red line of the spectrum of krypton-86. (Previously the meter had been defined as one ten-millionth the distance from the equator to the North Pole along the meridian nearest Paris.)

Metric measurements play an important part in the global aerospace industry. Unfortunately, a good deal of American aerospace activity still involves the use of another set of units, the English system of

Fundamental SI Units	
Name (and Symbol)	*Physical Quantity Represented*
meter (m)	length
kilogram (kg)	mass
second (s)	time
ampere (A)	electric current
candela (cd)	luminous intensity
kelvin (K)	thermodynamic temperature
mole (mol)	amount of substance

Derived and Supplementary SI Units	
Name (and symbol)	*Physical Quantity Represented*
becquerel (Bq)	radioactivity
coulomb (C)	electric charge
farad (F)	electric capacitance
gray (Gy)	absorbed dose of ionizing radiation
henry (H)	inductance
hertz(Hz)	frequency
joule(J)	energy
lumen (lm)	luminous flux
lux (lx)	illuminance
newton (N)	force
ohm (Ω)	electric resistance
pascal (Pa)	pressure
radian (rad)	plane angle
siemens (S)	electric conductance
sievert (Sv)	ionizing radiation dose equivalent
steradian (sr)	solid angle
tesla (T)	magnetic flux density
volt (V)	electric potential difference
watt (W)	power
weber (Wb)	magnetic flux

measurement (sometimes referred to as the American Engineering system of units). One major advantage of the metric system over the English system is that the metric system uses the number 10 as a base. Therefore, multiples or submultiples of SI units are reached by multiplying or dividing by 10.

The metric system uses the kilogram as the basic unit of mass. In contrast, the English system has a unit called the pound-mass (lbm), which is a *unit of mass,* and another unit, the pound-force (lbf) which is a *unit of force.* Furthermore, these two "pound" units are related by an arbitrary (pre–space age) definition within the English system of measurement that states that a 1 pound-mass object "weighs" 1 pound-force at sea level on Earth. Novices and professionals alike all too often forget that this "pound" equivalency is valid *only* at sea level on the surface of Earth. Therefore, this particular

English unit arrangement can cause much confusion in working aerospace engineering problems. For example, in the metric system, a kilogram is a kilogram, whether on Earth or on Mars. However, in the English system, if an object "weighs" one pound* on Earth, it only "weighs" 0.38 pounds* on Mars. Of course, from the basic physical concept of mass, one pound-mass (lbm) on Earth is still one-pound mass (lbm) on Mars. What changes the object's "weight" (as calculated from Newton's second law of motion: force = mass × acceleration) is the difference in the local acceleration due to gravity. On Earth, the sea-level acceleration due to gravity (g) is equal to 9.8 meters per second per second (m/s^2) (or about 32.2 feet per second per second [ft/s^2]); while on the surface of Mars, the local acceleration due to gravity (g_{Mars}) is only 3.73 meters per second per second (m/s^2) (or about 12.2 feet per second per second [ft/s^2]). Confusing? That is why the metric system is used extensively throughout this book and why the companion English units (wherever appropriate) are included in parentheses. A special effort has been made to identify the pound-mass and the pound-force, as separate and distinct units within the English system of measurements. The metric companion unit to the pound-mass is the kilogram (kg), namely 1 kg = 2.205 pounds-mass; while the metric companion unit to the pound-force is the newton (N), namely 1 N = 0.2248 pound-force.

Appendix I provides a summary of contemporary SI units and their conversion factors, the recommended SI unit prefixes, special units used in astronomical investigations, and common metric/English conversion factors.

skin The outer covering of a rocket, missile, or aerospace vehicle.

skirt The lower outer part of a missile or rocket vehicle.

skirt fog The cloud of steam and water that often surrounds the engines of a rocket being launched from a wet emplacement.

* To be perfectly correct, we must specify the object's "weight" in pound-force (lbf) units.

Skylab (SL) The first U.S. space station, which was placed in orbit in 1973 by a two-stage configuration of the Saturn V expendable launch vehicle and then visited by three astronaut crews who worked on scientific experiments in space for a total of approximately 172 days, with the last crew spending 84 days in Earth orbit.

Skylab (SL) was composed of five major parts: the Apollo telescope mount (ATM); the multiple docking adapter (MDA); the airlock module (AM); the instrument unit (IU); and the orbital workshop (OWS), which included the living and working quarters. The ATM was a solar observatory, and it provided attitude control and experiment pointing for the rest of the cluster. The retrieval and installation of film used in the ATM was accomplished by the astronauts during extravehicular activity (EVA). The MDA served as a dock for the modified Apollo spacecraft that taxied the crews to and from the space station. The airlock module (AM) was located between the docking port (MDA) and the living and working quarters, and contained controls and instrumentation. The instrument unit (IU), which was used only during launch and the initial phases of operation, provided guidance and sequencing functions for the initial deployment of the ATM, its solar arrays, and the like. The orbital workshop (OWS) was a modified Saturn IV-B stage that had been converted into a "two-story" space laboratory with living quarters for a crew of three. This orbital laboratory was capable of unmanned, in-orbit storage, reactivation, and reuse.

There were four launches in the Skylab Program from Complex 39 at the Kennedy Space Center. The first launch was on May 14, 1973. A two-stage Saturn V vehicle placed the unmanned 90-metric ton (100-ton) Skylab space station in an initial 435-kilometer (270-mile) orbit around Earth. As the rocket accelerated past 7,620 meters (25,000 feet), atmospheric drag began clawing at Skylab's meteoroid/Sun shield. This cylindrical metal shield was designed to protect the orbital workshop from tiny particles and the Sun's scorching heat. Sixty-three seconds after launch, the shield ripped away from the spacecraft, trailing an aluminum strap that caught on one of the unopened solar wings. The shield became tethered to the laboratory while at the same time prying the opposite solar wing partly open. Minutes later, as the booster rocket staged, the partially deployed solar wing and meteoroid/Sun shield were flung into

Skylab Mission Summary (1973–1974)				
Mission	*Dates*	*Crew*	*Mission Duration*	*Remarks*
Skylab 1	Launched 05/14/73	Unmanned	Re-entered atmosphere 07/11/79	90-metric ton space Station visited by three astronaut crews
Skylab 2	05/25/73 to 06/22/73	Charles Conrad, Jr Paul J. Weitz Joseph P. Kerwin (M.D.)	28 days 49 min	Repaired Skylab; 392 hrs experiments; 3 EVAs
Skylab 3	07/28/73 to 09/25/73	Alan L. Bean Jack R. Lousma Owen K. Garriott (Ph.D.)	59 days 11 hrs	Performed maintenance; 1,081 hrs experiments; 3 EVAs
Skylab 4	11/16/73 to 02/08/74	Gerald P. Carr William R. Pogue Edward G. Gibson (Ph.D.)	84 days 1 hr	Observed Comet Kohoutek; 4 EVAs; 1,563 hrs experiments

space. With the loss of the meteoroid/Sun shield, temperatures inside Skylab soared, rendering the space station uninhabitable and threatening the food, medicine, and film stored onboard. The Apollo Telescope Mount (ATM), the major piece of scientific equipment, did deploy properly, however, an action that included the successful unfolding of its four solar panels.

The countdown for the launch of the first Skylab crew was halted. NASA engineers worked quickly to devise a solar parasol to cover the workshop and to find a way to free the remaining stuck solar wing. On May 25, 1973, astronauts Charles "Pete" Conrad. Jr., Dr. Joseph P. Kerwin, and Paul J. Weitz were launched by a Saturn 1B rocket toward Skylab.

After repairing Skylab's broken docking mechanism, which had refused to latch, the astronauts entered the space station and erected a mylar solar parasol through a space access hatch. It shaded part of the area where the protective meteoroid/Sun shield had been ripped away. Temperatures within the spacecraft immediately began dropping, and Skylab soon became habitable without space suits. But the many experiments on board demanded far more electric power than the four ATM solar arrays could generate. Skylab could fulfill its scientific mission only if the first crew freed the remaining crippled solar wing. Using equipment that resembled long-handled pruning shears and a prybar, the astronauts pulled the stuck solar wing free. The space station was now ready to meet its scientific mission objectives.

The duration of the first crewed mission was 28 days. The second astronaut crew— Alan Bean, Jack Lousma, and Dr. Owen Garriott—was launched on July 28, 1973; mission duration on the space station was approximately 59 days and 11 hours. The third Skylab crew—Gerald Carr, William Pogue, and Dr. Edward Gibson—was launched November 16, 1973; mission duration was a little over 84 days. Saturn IB rockets launched all three crews in modified Apollo spacecraft, which also served as their return-to-Earth vehicle. The third and final manned Skylab mission ended with

splashdown in the Pacific Ocean on February 8, 1974. (See the table.)

After the last astronaut crew departed the space station in February 1974, it orbited Earth as an abandoned derelict. Unable to maintain its original altitude, the station finally reentered the atmosphere on July 11, 1979, during orbit 34,981. While most of the station was burned up during reentry, some pieces survived and impacted in remote areas of the Indian Ocean and sparsely inhabited portions of Australia. *See also* SPACE STATION.

slenderness ratio A dimensionless number that expresses the ratio of a rocket vehicle's length to its diameter.

slew To change the position of a sensor or an antenna assembly by injecting a signal into the positioning servo-mechanism.

slip flow Flow in the transition regime of gas dynamics, wherein the mean free path of the gas molecules is of the same order of magnitude as the thickness of the boundary layer. The gas in contact with a body surface immersed in the flow is no longer at rest with respect to the surface.

slipstream The flow of fluid around a structure that is moving through the fluid.

sliver The portion of solid-propellant grain remaining at the time of web burnout. *See also* ROCKET.

sloshing The back-and-forth movement of a liquid rocket propellant in its tank(s), creating problems of stability and control in the vehicle. Aerospace engineers often use antislosh baffles in a rocket vehicle's propellant tanks to avoid this problem.

slug A unit of mass in the standard system of measurement (i.e., the foot-pound-second system of units) approximately equal to 32.17 pounds-mass (lbm). It is defined as the amount of mass that, if acted upon by a one-pound force (lbf), would experience an acceleration of one foot per second per second. This unit physically represents an amount of mass (in pound-mass units) that is *numerically* equal to the value of the standard acceleration of gravity at sea level on the surface of Earth (namely, 32.17 feet per second per second).

slurry fuel   A rocket fuel consisting of a suspension of fine solid particles in a liquid.

Small Magellanic Cloud (SMC)   An IRREGULAR GALAXY about 9,000 LIGHT-YEARS in diameter and 180,000 light-years from EARTH. *See also* MAGELLANIC CLOUDS.

SNAP   An acronym for *Systems for Nuclear Auxiliary Power.* The SNAP program was created by the U.S. Atomic Energy Commission (forerunner of the U.S. Department of Energy) to develop small auxiliary nuclear power sources for specialized outer space, remote land, and sea applications. Two general approaches were pursued. The first approach used the thermal energy from the decay of a radioisotope source to produce electricity directly by thermoelectric or thermionic methods. The second approach used the thermal energy liberated in a small nuclear reactor to produce electricity by either direct conversion methods or dynamic conversion methods. *See also* SPACE NUCLEAR POWER.

Snark   A surface-to-surface guided missile developed by the U.S. Air Force in the early 1950s. This now obsolete winged missile had a range of about 8,000 kilometers (5,000 miles) and was capable of carrying a nuclear warhead.

snubber   A device used to increase the stiffness of an elastic system (usually by a large factor), whenever the displacement of the system becomes larger than a specified amount.

soft landing   The act of landing on the surface of a planet without damage to any portion of the space vehicle or payload, except possibly the landing gear. NASA's Surveyor spacecraft and Viking lander spacecraft were designed for soft landings

on the Moon and Mars, respectively. *See also* SURVEYOR; VIKING PROJECT.

soft target   In strategic nuclear warfare, any target that cannot be hardened (protected) in order to survive a nearby nuclear detonation. Cities, airfields, and large industrial centers are examples of soft targets.

software   The programs (that is, sets of instructions and ALGORITHMs) and data used to operate a DIGITAL COMPUTER. *Compare with* HARDWARE.

soft X rays   Photons with energy from 100 electron volts (eV) to 10 kiloelectron volts (keV).

Sol   The Sun.

sol   A Martian day (about 24 hours 37 minutes 23 seconds in duration); seven sols equal approximately 7.2 Earth days.

solar   1. Of or pertaining to the Sun or caused by the Sun, as *solar radiation, solar atmospheric tide* 2. Relative to the Sun as a datum or reference, such as *solar time.*

solar activity   Any type of variation in the appearance or energy output of the Sun.

Solar and Heliospheric Observatory (SOHO)   The primary scientific aims of the European Space Agency's (ESA's) *Solar and Heliospheric Observatory* (SOHO) are to investigate the physical processes that form and heat the Sun's corona, maintain it, and give rise to the expanding solar wind; and to study the interior structure of the Sun. SOHO is part of the International Solar-Terrestrial Physics Program (ISTP) and involves NASA participation. The 1,350-kilogram (on-orbit dry mass) spacecraft was launched on December 2, 1995, and placed in a halo orbit at the Earth-Sun Lagrangian libration point one (L1) to obtain uninterrupted sunlight. It had a two-year design life, but onboard consumables are sufficient for an extra four years of operation. The spacecraft carries a complement of 12 scientific instruments.

In April 1998, SOHO successfully completed its nominal two-year mission to study the Sun's ATMOSPHERE, surface, and interior. The major science highlights of this mission include the detection of rivers of PLASMA beneath the surface of the Sun and the initial detection of SOLAR FLARE–induced solar quakes. In addition, the SOHO spacecraft discovered more than 50 sun-grazing COMETs. Then, on June 24, 1998, during routine HOUSEKEEPING and maintenance operations, contact was lost with the SOHO spacecraft. After several anxious weeks, ground controllers were able to reestablish contact on August 3. They then proceeded to recommission various defunct subsystems and to perform an ORBIT correction maneuver that brought SOHO back to a normal operational mode with all its scientific instruments properly functioning by November 4. *See also* EUROPEAN SPACE AGENCY; LAGRANGIAN LIBRATION POINTS; SUN.

solar cell   Solar cells are proven direct energy conversion devices that have been used for over three decades to provide electric power for spacecraft. In a direct energy conversion (DEC) device, electricity is produced directly from the primary energy source without the need for thermodynamic power conversion cycles involving the heat engine principle and the circulation of a working fluid. A solar cell or "photovoltaic system" turns sunlight directly into electricity. The solar cell has no moving parts to wear out and produces no noise, fumes, or other polluting waste products. However, the space environment, especially trapped radiation belts and the energetic particles released in solar flares, can damage solar cells used on spacecraft and reduce their useful lifetime. *See also* PHOTOVOLTAIC MATERIAL; SATELLITE POWER SYSTEM.

solar constant   The total amount of the Sun's radiant energy that normally crosses perpendicular to a unit area at the top of Earth's atmosphere (i.e., at one astronomical unit from the Sun). The currently used value of the solar constant is 1,371 ± 5 watts per square meter. The spectral distribution

of the Sun's radiant energy resembles that of a BLACKBODY radiator with an effective temperature of 5,800 kelvins (K). This means that the most of the Sun's radiant energy lies in the visible portion of the electromagnetic spectrum, with a peak value near 0.45 micrometer (μm). *See also* SUN.

solar cycle   The periodic (semiregular) change in the number of sunspots. It is the interval between successive minima and is about 11.1 years (unless the magnetic polarities of the Northern and Southern Hemisphere sunspots are considered; then the 11-year interval is actually part of a 22-year magnetic cycle). *See also* SUN; SUNSPOT.

solar electric propulsion (SEP)   A low-thrust propulsion system in which the electricity required to power the ion engines (or other type of electric rocket engines) is generated either by a solar-thermal conversion system or by a solar-photovoltaic conversion system. *See also* ELECTRIC PROPULSION.

solar energy   ENERGY from the SUN; radiant energy in the form of sunlight. *See also* SOLAR CONSTANT; SOLAR RADIATION.

solar flare   A highly concentrated explosive release of energy within the solar atmosphere. It appears as a sudden, short-lived brightening of a localized area within the Sun's chromosphere. The electromagnetic radiation output from a solar flare ranges from radio to X-ray frequencies. Energetic nuclear particles, primarily electrons, protons, and a few alpha particles, also are ejected during a solar flare. *See also* SUN.

solar mass   The mass of the Sun, namely 1.99×10^{30} kilograms; it is commonly used as a unit in comparing stellar masses. *See also* SUN.

Solar Maximum Mission (SMM)   A NASA spacecraft designed to provide coordinated observations of solar activity, especially solar flares, during a period of maximum solar activity. The 2,315-kilogram (5,100 pounds-mass) spacecraft was placed in orbit on February 14, 1980. Its scientific payload consisted of seven instruments specifically selected to study the short-wavelength and coronal manifestations of solar flares. Data were obtained on the storage and release of flare energy, particle acceleration, formation of hot plasma, and mass ejection.

In April 1984 space shuttle mission STS 41-C involved on-orbit repair of the SMM spacecraft. During this shuttle mission, the astronauts rendezvoused with the spacecraft while performing an extravehicular activity (EVA) from the orbiter and successfully repaired it. The SMM spacecraft then continued to collect data until November 24, 1989. It reentered Earth's atmosphere on December 2, 1989.

solar nebula The cloud of dust and gas from which the Sun, the planets, and other minor bodies of the solar system are postulated to have formed (condensed). *See also* SUN.

solar panel A winglike set of solar cells used by a spacecraft to convert sunlight directly into electric power; also called a *solar array. See also* SOLAR CELL; SOLAR PHOTOVOLTAIC CONVERSION.

solar photovoltaic conversion The direct conversion of sunlight (solar energy) into electrical energy by means of the photovoltaic effect. A single photovoltaic (PV) converter cell is called a *solar cell,* while a combination of cells, designed to increase the electric power output, is called a *solar array* or a *solar panel.*

Since 1958, solar cells have been used to provide electric power for a wide variety of spacecraft. The typical spacecraft solar cell is made of a combination of n-type (*nega*tive) and p-type (*positive*) semiconductor materials (generally silicon). When this combination of materials is exposed to sunlight, some of the incident electromagnetic radiation removes bound electrons from the semiconductor material atoms, thereby producing free electrons. A hole (positive charge) is left at each location from which a bound electron has been removed. Consequently, an equal number of free electrons and holes are formed. An electrical barrier at the p-n junction causes the newly created free electrons near the barrier to migrate deeper into the n-type material and the matching holes to migrate further into the p-type material.

If electrical contacts are made with the n- and p-type materials and these contacts connected through an external load (conductor), the free electrons will flow from the n-type material to the p-type material. Upon reaching the p-type material, the free electrons will enter existing holes and once again become bound electrons. The flow of free electrons through the external conductor represents an electric current that will continue as long as more free electrons and holes are being created by exposure of the solar cell to sunlight. This is the general principle of solar photovoltaic conversion. *See also* PHOTOVOLTAIC MATERIAL; SOLAR CELL.

solar power satellite (SPS) *See* SATELLITE POWER SYSTEM.

solar radiation The total electromagnetic radiation emitted by the Sun. The Sun is considered to radiate as a BLACKBODY at a temperature of about 5,770 kelvins (K). Approximately 99.9% of this radiated energy lies within the wavelength interval from 0.15 to 4.0 micrometers (μm), with some 50% lying within the visible portion of the electromagnetic spectrum (namely 0.4 to 0.7 μm) and most of the remaining radiated energy in the near-infrared portion of the spectrum. *See also* ELECTROMAGNETIC SPECTRUM; SUN.

solar storm A disturbance in the space environment triggered by an intense solar flare (or flares) that produces bursts of electromagnetic radiation and charged particles. A solar storm can adversely impact human space operations, cause damage to orbiting spacecraft (especially geostationary satellites that are outside of the Earth's magnetosphere), and affect Earth's upper atmosphere and magnetic field. Solar storms are associated with unexpected surges in solar activity.

Periods of intense solar activity present several major problems for orbiting spacecraft. First, a solar storm heats Earth's atmosphere, causing it to expand. This atmospheric expansion increases the drag experienced by low-Earth-orbiting (LEO) satellites, forcing them to use additional altitude control propellant and shortening their operating lifetimes. Second, during a solar storm more energetic charged particles bombard spacecraft, causing electronic upsets and damage to sensors and electronic equipment. Finally, a solar storm involves bursts of electromagnetic radiation that upset Earth's ionosphere and interfere with radiowave communications. Also called a *geomagnetic storm*. *See also* SOLAR FLARE; SUN.

solar system The Sun and the collection of celestial objects that are bound to it gravitationally. These celestial objects include the nine major planets, over 60 known moons, more than 2,000 minor planets or asteroids, and a very large number of comets. Except for the comets, all of these celestial objects orbit around the Sun in the same direction and their orbits lie close to the plane defined by Earth's own orbit and the Sun's equator.

The nine major planets can be divided into two general categories: (1) the terrestrial or Earth-like planets, consisting of Mercury, Venus, Earth and Mars; and (2) the outer or Jovian planets, consisting of the gaseous giants Jupiter, Saturn, Uranus, and Neptune. Tiny Pluto is currently regarded as a

"frozen snowball" in a class by itself. Because of the size of its moon, Charon, some astronomers like to consider Pluto and Charon as forming a double-planet system. (See Figure 70.)

As a group, the terrestrial planets are dense, solid bodies with relatively shallow or no atmosphere. In contrast, the Jovian planets are believed to contain modest-size rock cores, surrounded by concentric layers of frozen hydrogen, liquid hydrogen, and gaseous hydrogen, respectively. Their atmospheres also contain such gases as helium, methane and ammonia.

When used in the lower case, *solar system* can refer to any star and its gravitationally bound collection of planets, asteroids, and comets. *See also* ASTEROID; COMET; EARTH; JUPITER; MARS; MERCURY; NEPTUNE; PLUTO; SATURN; SUN; URANUS; VENUS.

solar thermal conversion The conversion of sunlight (solar energy) into electricity by means of a thermodynamic cycle involving a heat engine. In general, incoming solar energy is concentrated and focused to heat a working fluid. The solar-heated working fluid undergoes a series of changes in its thermodynamic state during which mechanical energy (work) is extracted from the hot fluid. This mechanical work then is used to spin a generator to produce electricity. Thermodynamic cycles that could be used in solar thermal conversion include the Brayton, Rankine, and Stirling cycles. *See also* BRAYTON CYCLE; HEAT ENGINE; RANKINE CYCLE; STIRLING CYCLE.

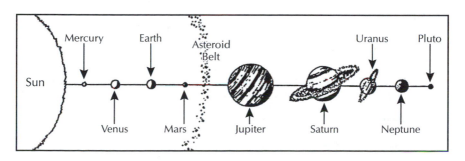

Figure 70. The major components of our solar system (not to scale).

solar wind The variable stream of electrons, protons, and various atomic nuclei (e.g., alpha particles) that flows continuously outward from the Sun into interplanetary space. The solar wind has typical speeds of a few hundred kilometers per second. *See also* MAGNETOSPHERE; SUN.

solenoid Helical coil of insulated wire that, when conducting electricity, generates a magnetic field that actuates a movable core.

solid A state of matter characterized by a three-dimensional regularity of structure. When a solid substance is heated beyond a certain TEMPERATURE, called the melting point, the FORCES between its ATOMS or MOLECULES can no longer support the characteristic lattice structure, causing it to break down as the solid material transforms into a LIQUID or (more rarely) transforms directly into a VAPOR (SUBLIMATION).

solid angle (symbol: Ω) Three-dimensional angle formed by the vertex of a cone; that portion of the whole of space about a given point, bounded by a conical surface with its vertex at that point and measured by the area cut by the bounding surface from the surface of a sphere of unit radius centered at that point. The steradian (sr) is the SI UNIT of solid angle.

solid Earth The lithosphere portion of the EARTH SYSTEM, including this PLANET's CORE, mantle, crust, and all surface rocks and unconsolidated rock fragments.

solid lubricant A dry film lubricant. *See also* SPACE TRIBOLOGY.

solid propellant A rocket propellant in solid form, usually containing both fuel and oxidizer combined or mixed, and formed into a monolithic (not powdered or granulated) grain. *See also* PROPELLANT; ROCKET.

solid-propellant rocket A rocket propelled by a chemical mixture or compound of fuel and oxidizer in solid form. Also called a *solid rocket*. *See also* PROPELLANT; ROCKET.

solid rocket booster (SRB) Two solid rocket boosters (SRBs) operate in parallel to augment the thrust of the space shuttle main engines (SSMEs) from the launch pad through the first two minutes of powered flight. These boosters also assist in guiding the entire vehicle during the initial ascent. Following separation, they are recovered (after a parachute-assisted splashdown in the ocean) for refurbishment and reuse. In addition to its basic component, the solid rocket motor (SRM), each booster contains several subsystems: structural, thrust vector control (TVC), separation, recovery, and electrical and instrumentation.

The heart of the solid rocket booster is the solid rocket motor, the first ever developed for use on a crewed launch vehicle. The huge solid rocket motor is composed of a segmented motor case loaded with solid propellants, an ignition system, a movable nozzle, and the necessary instrumentation and integration hardware.

Each SRB develops about 11,790 kilonewtons (2,650,000 pounds-force) at liftoff and contains more than 500,000 kilograms (1.1 million pounds-mass) of propellant. The type of propellant used in the solid rocket motor is known as PBAN, which stands for polybutadiene acrylic acid acrylonitrile terpolymer. In addition to the PBAN (which serves as the binder), the propellant consists of approximately 70% ammonium perchlorate (the oxidizer), 16% powdered aluminum (the fuel), and a trace of iron oxide (a catalyst) to control the burning rate. The cured solid propellant looks and feels like a hard rubber eraser. *See also* PROPELLANT; ROCKET; SPACE TRANSPORTATION SYSTEM.

solid-state device A device that uses the electric, magnetic, and photonic properties of solid materials, mainly semiconductors. It contains no moving parts and depends on the internal movement of charge carriers (i.e., electrons and "positive" holes) for its operation.

solidus temperature The temperature at which melting starts.

solstice The two times of the YEAR when the SUN's position in the sky is the most distant from the celestial EQUATOR. For the Northern Hemisphere, the summer solstice (longest day) occurs about June 21 and the winter solstice (shortest day) about December 21.

sonic 1. In aerodynamics, of or pertaining to the speed of sound; that which moves at acoustic velocity, as in *sonic flow;* designed to operate or perform at the speed of sound, as in a *sonic leading edge.* 2. Of or pertaining to sound, as in *sonic amplifier.* In this sense, "acoustic" is preferred to "sonic."

sonic boom A thunderlike noise caused by a shock wave that emanates from an aircraft, aerospace vehicle, or flying object that is traveling at or above sonic velocity in Earth's atmosphere. As supersonic objects travel through the air, the air molecules are pushed aside with great force, and this action forms a SHOCK WAVE. The bigger and heavier the supersonic object, the more the air is displaced and the stronger the resultant shock waves. Several factors can influence sonic booms—the mass, size, and shape of the supersonic vehicle or object; its altitude, attitude, and flight path; and local weather or atmospheric conditions.

The shock wave forms a cone of pressurized air molecules that moves outward and rearward in all directions and extends to the ground. As the cone spreads across the landscape along the flight path, it creates a continuous sonic boom along the full width of its base.

sorbent A substance or material that takes up gas by absorption, adsorption, chemisorption, or any combination of these processes. *See also* ABSORPTION; ADSORPTION.

sorption The taking up of gas by absorption, adsorption, chemisorption, or any combination of these processes. *See also* ABSORPTION; ADSORPTION.

sound In physics, a vibration in an elastic medium that is at a frequency and intensity that can be heard by a human being. The normal human ear can respond to sounds in the frequency range from approximately 20 to 20,000 hertz (Hz). Vibrations lower than this frequency range are called *infrasounds,* while those above this frequency range are called *ultrasounds.*

sounding In geophysics, any penetration of the natural environment for scientific observation. 2. In meteorology, an observation of the upper air; also, a single complete radiosonde observation. A radiosonde is a balloon-borne device for the simultaneous measurement and transmission of meteorological data up to a height of about 30,000 meters (100,000 ft).

sounding rocket A rocket, usually with a solid-propellant motor, used to carry scientific instruments on parabolic trajectories into the upper regions of Earth's sensible atmosphere (i.e., beyond the reach of aircraft and scientific balloons) and into near-Earth space. A sounding rocket basically is divided into two major components: the solid rocket motor and the payload. Payloads typically include: the experiment package, the nose cone, a telemetry system, an attitude control system, a radar tracking beacon, the firing despin module, and the recovery section. Many sounding rocket payloads are recovered for refurbishment and reuse.

Sounding rockets fly vertical trajectories from 48 kilometers (km) (30 mi [mi]) to over 1,285 km (800 mi) in altitude. The flight normally lasts less than 30 minutes. Sounding rockets come in a wide variety of sizes and types. For example, NASA's sounding rocket fleet ranges from the single-stage Super Arcas (which stands 3 meters [7 ft] high and 11 centimeters [4.5 in] in diameter) to the single-stage Aries (which stands 11 meters [35 ft] high and 111 centimeters [44 in] in diameter). The Super Arcas has been used by NASA since 1962 for carrying meteorological measuring devices. It can carry a 4.5-kilogram (10 pound-mass) payload to an altitude of 91 km (57 mi). The four-stage Black Brant XII, the tallest in NASA's fleet, is 20 meters (67 ft) high.

All of NASA's sounding rockets use solid-propellant rocket motors. Furthermore, all of these rocket vehicles are unguided, except the Aries and those that use the S-19 Boost Guidance System. During flight, all sounding rockets except the Aries are imparted with a spinning motion to reduce potential dispersion of the flight trajectory due to vehicle misalignments. NASA conducts between 30 to 35 suborbital sounding rocket flights per year, providing low-cost, quick-response flight opportunities to scientists in the areas of physics and astronomy, microgravity materials processing research, and instrument development. Most of these sounding rocket launches take place at NASA's Wallops Flight Facility, Wallops Island, Virginia. *See also* ROCKET.

South Atlantic Anomaly (SAA) A region of Earth's trapped radiation particle zone that dips close to the planet in the South Atlantic Ocean southeast of the Brazilian coast. This region represents the most significant source of ionizing radiation for space travelers in low-inclination, low-altitude orbits around Earth. *See also* EARTH'S TRAPPED RADIATION BELTS; SPACE RADIATION ENVIRONMENT.

Soyuz launch vehicle The "workhorse" Soviet (and later Russian) launch vehicle that was first used in 1963. With its two cryogenic stages and four cryogenic strap-ons, this vehicle is capable of placing up to 6,900 kilograms into LOW EARTH ORBIT (LEO). At present, it is the most frequently flown launch vehicle in the world. Since 1964 it has been used to launch every Russian human crew space mission. *See also* LAUNCH VEHICLE.

Soyuz spacecraft An evolutionary family of crewed spacecraft that have been used by the Soviet Union and later the Russian Federation on a wide variety of space missions. The word *soyuz* is Russian for "union". The first Soyuz spacecraft, called *Soyuz 1*, was launched in April 1967. Unfortunately, upon reentry a parachute failed to open properly and the spacecraft was destroyed on impact and its occupant,

Cosmonaut Vladimir Komarov, killed. The second Russian space tragedy occurred at the end of the *Soyuz 11* mission (June 1971), when a valve malfunctioned as the spacecraft was separating from the *Salyut 1* space station, allowing all the air to escape from the crew compartment. This particular early version of the Soyuz spacecraft did not have sufficient room for the crew to wear their pressure suits during reentry; consequently, the three cosmonauts, Georgi Dobrovolsky, Victor Patseyev, and Vladislav Volkov, suffocated during the reentry operation. They were found dead by the Russian recovery team after touchdown. In July 1975 the *Soyuz 19* spacecraft was used successfully by Cosmonauts Alexei Leonov and Valeri Kubasov in the Apollo-Soyuz Test Project—an international rendezvous and docking mission. The next major variant of this versatile spacecraft, called the Soyuz-T (with the "T" standing for transport), was first flown in December 1979. The Soyuz-TM is a modernized version of the Soyuz-T. It was flown in May 1986 and can be used to ferry crew and supplies to an orbiting station such as the Mir or the International Space Station.

The Soyuz TMA-1 is a Russian automatic passenger spacecraft designed for LAUNCH by a SOYUZ LAUNCH VEHICLE from the BAIKONUR COSMODROME. Following launch, the spacecraft proceeds in an automated fashion to RENDEZVOUS and dock with the INTERNATIONAL SPACE STATION (ISS). The Soyuz TMA-1 is a larger craft with a more comfortable interior than the previous Soyuz TM models. After docking, the spacecraft remains parked at the *ISS*, serving as an emergency escape spacecraft until it is relieved by the arrival of another Soyuz spacecraft. For example, in late October 2002, a Soyuz TMA-1 was launched from the Baikonur Cosmodrome and successfully carried three COSMONAUTS (two Russian and one Belgian) to the *ISS*. The Soyuz TMA-1 automatically docked with the *ISS*. After 10 days of MICROGRAVITY research, the three visiting cosmonauts departed from the *ISS* using a previously parked Soyuz TM-34 spacecraft. The Soyuz TMA-1 spacecraft that carried them into space remained behind as a lifeboat for the permanent crew of the *ISS*. *See also* APOLLO-SOYUZ TEST PROJECT; *MIR*; SALYUT.

space 1. Specifically, the part of the universe lying outside the limits of Earth's atmosphere. 2. More generally, the volume in which all celestial bodies, including Earth, move.

space base A large, permanently inhabited space facility located in orbit around a celestial body or on its surface. The space base would serve as a center of human operations in some particular region of the solar system—supporting exploration, scientific missions, and extraterrestrial resource applications. An orbiting facility also could serve as a space transportation hub, robot repair and maintenance facility, a space construction site, and even a recreation and medical services center for space travelers. With from 10 to perhaps 200 occupants, the space base would have a much larger human population than a space station. Modular construction, the use of extraterrestrial materials for radiation shielding, development of a closed regenerative life support system, and large solar or nuclear power supplies (e.g., 50 to 100 kilowatts-electric or more) are some characteristic design features. For an orbiting facility, artificial gravity would be provided by rotating the crew habitats and certain work areas. For surface bases on the Moon and Mars, the human inhabitants would experience the natural gravity of the particular celestial object. *See also* LUNAR BASE; MARS BASE; SPACE SETTLEMENT; SPACE STATION.

space-based astronomy Astronomical observations conducted by human-crewed or robotic spacecraft above Earth's atmosphere; for example, the excellent cosmic observations performed by the Hubble Space Telescope (a human-tended but remotely operated Earth-orbiting instrument platform) or the solar observations made by the crew of the U.S. Skylab space station (1973–1974). *See also* COSMIC BACKGROUND EXPLORER (COBE); HUBBLE SPACE TELESCOPE (HST); SKYLAB.

space-based high-energy laser system A proposed space-based, multimegawatt, high-energy laser (HEL) system that could be used in several modes of operation. In its "weapons mode" with the laser system on high power, it could attack ground, air, and space targets. In its "surveillance mode," it could operate either using the laser at low power levels for active-illumination imaging or with the laser inoperative for passive imaging. *See also* BALLISTIC MISSILE DEFENSE; HIGH-ENERGY LASER.

space-based interceptor (SBI) A kinetic energy kill rocket vehicle that is based in space. *See also* BALLISTIC MISSILE DEFENSE.

spaceborne imaging radar In general, a synthetic aperture radar (SAR) system that is placed on a spacecraft. The radar is an active remote sensing instrument that transmits pulses of microwave energy toward the planet's surface and measures the amount of energy that is reflected back to the radar antenna. A SAR system therefore provides its own illumination and can produce images of the planet's surface during the day or night, through cloud cover. These types of radar systems have been used to image the cloud-enshrouded surface of Venus and to help scientists study the planet Earth more effectively.

For example, the Spaceborne Imaging Radar-C/X Band Synthetic Aperture Radar (SIR-C/X-SAR) was flown in the space shuttle orbiter's cargo bay during the STS-59 mission (April 1994). This radar system, a joint project of NASA, the German Space Agency (DARA), and the Italian Space Agency (ASI), collected an enormous amount of radar imagery data in support of NASA's MISSION TO PLANET EARTH and other planetary study efforts. *See also* MAGELLAN MISSION; RADAR IMAGING; REMOTE SENSING.

space capsule A container used for conducting an experiment or operation in space, usually involving a human being or other living organisms. The Mercury Project spacecraft were small space capsules designed to carry a single American astronaut into space and return him safely to Earth. As American human-rated spacecraft became a little larger during the Gemini and Apollo projects, the term "space capsule" generally was abandoned in favor of "spacecraft." *See also* MERCURY PROJECT.

space colony An earlier term used to describe a large, permanent space habitat and industrial complex occupied by up to 10,000 persons. Currently the term "space settlement" is preferred. *See also* SPACE SETTLEMENT.

space commerce The commercial sector of space operations and activities. At least six major areas are now associated with the field commonly referred to as space commerce. These areas are: (1) space transportation; (2) satellite communications; (3) satellite-based positioning and navigational services; (4) satellite remote sensing (including geographic information system support); (5) space-based industrial facilities; and (6) materials research and processing in space.

space construction Large structures in space, such as modular space stations, global communication and information services platforms, and satellite power systems (SPSs), will all require on-orbit assembly operations by space construction workers. Space construction requires protection of the workforce and some materials from the hard vacuum, intense sunlight, and natural radiation environment encountered above Earth's protective atmosphere.

Outer space, however, is also an environment that in many ways is ideal for the construction process. First, because of the absence of significant gravitational force (i.e., the microgravity experienced by the free-fall condition of orbiting objects), the structural loads are quite small, even minute. Structural members may therefore much lighter than terrestrial structures of the same span and stiffness. Second, the absence of gravitational forces greatly facilitates the movement of material and equipment. On Earth, the movement of materials during a construction operation absorbs a large portion of the total work effort expended by construction personnel and their machines. Third, the absence of an atmosphere with its accompanying wind loads, inclement weather, and unpredictable changes permits space work to be planned accurately and executed readily without environmental interruptions (except perhaps due to solar flares, which would increase the radiation hazard).

Automated fabrication is considered to be a key requirement for viable space construction activities. Work in space will require the close interaction of astronaut (i.e., space worker) and very smart machine. For example, the advanced maneuverable spacesuit will be a versatile, self-contained, life-supporting backpack with gaseous-nitrogen-propelled jet thrusters that enable a space worker to travel back and forth to various space construction locations. The automated beam builder is a machine designed for fabricating "building-block" structural beams in space. Combined with a space structure fabrication system, the beam builder will allow space workers to manufacture and assemble structures in LOW EARTH ORBIT (LEO) using the international space station as an early "construction camp." Eventually, as the demand for more sophisticated space construction and assembly efforts grows, permanent space construction bases can be established in LEO and elsewhere in cislunar space. Remote astronaut workstations can be mounted on large manipulator arms attached to the space station or space base. These "open cherry pickers" would have a convenient tool and parts bin, a swing-away control and display panel, and lights for general and point illumination. The closed version of this cherry picker would involve a pressurized human-occupied remote workstation *(space construction module)* that contains life support equipment and controls and displays for operating dexterous manipulators. *See also* LARGE SPACE STRUCTURES; SATELLITE POWER SYSTEM.

space control operations Aerospace operations that provide freedom of action in space for friendly forces while, when directed, denying it to an enemy. These operations include the broad aspects of protection of U.S. and allied space systems and negation of enemy space systems. Space control operations also encompass all elements of the space defense mission. *See also* SPACE DEFENSE.

spacecraft In general, a human-occupied or uncrewed platform that is designed to be placed into an orbit about Earth or into a trajectory to another celestial body. The spacecraft is essentially a combination of hardware that forms a space platform. It provides structure, thermal control, wiring, and subsystem functions, such as attitude control, command, data handling, and power. Spacecraft come in all shapes and sizes, each often tailored to meet the needs of a specific mission in space. For example, spacecraft designed and constructed to acquire scientific data are specialized systems intended to function in a specific hostile environment. The complexity of these scientific spacecraft varies greatly. Often they are categorized according to the missions they are intended to fly.

Scientific spacecraft include: flyby spacecraft, orbiter spacecraft, atmospheric probe spacecraft, atmospheric balloon packages, lander spacecraft, surface penetrator spacecraft, and surface rover spacecraft.

Flyby spacecraft follow a continuous trajectory and are not captured into a planetary orbit. These spacecraft have the capability of using their onboard instruments to observe passing celestial targets (e.g., a planet, a moon, an asteroid), even compensating for the target's apparent motion in an optical instrument's field of view. They must be able to transmit data at high rates back to Earth and also must be capable of storing data onboard for those periods when their antennas are not pointing toward Earth. They must be capable of surviving in a powered-down, cruise mode for many years of travel through interplanetary space and then of bringing all their sensing systems to focus rapidly on the target object, during an encounter period that may last only for a

few crucial hours or minutes. NASA's *Pioneer 10* and *11* and the *Voyager 1* and *2* are examples of highly successful flyby scientific spacecraft. NASA uses the flyby spacecraft during the initial, or reconnaissance phase, of solar system exploration.

An *orbiter spacecraft* is designed to travel to a distant planet and then orbit around that planet. This type of scientific spacecraft must possess a substantial propulsive capability to decelerate at just the right moment in order to achieve a proper orbit insertion. Aerospace engineers design an orbiter spacecraft recognizing the fact that solar occultations will occur frequently as it orbits the target planet. During these periods of occultation, the spacecraft is shadowed by the planet, cutting off solar array production of electric power and introducing extreme variations of the spacecraft's thermal environment. Generally, a rechargeable battery system augments solar electric power. Active thermal control techniques (e.g., the use of tiny electric-powered heaters) are used to complement traditional passive thermal control design features. The periodic solar occultations also interrupt uplink and downlink communications with Earth, making onboard data storage a necessity. NASA uses orbiter spacecraft as part of the second, in-depth study phase of solar system exploration. The lunar orbiter, Magellan and Galileo spacecraft are examples of successful scientific orbiters.

Some scientific exploration missions involve the use of one or more smaller, instrumented spacecraft, called *atmospheric probe spacecraft*. These probes separate from the main spacecraft prior to closest approach to a planet in order to study the planet's gaseous atmosphere as they descend through it. Usually an atmospheric probe spacecraft is deployed from its "mothership" (i.e., the main spacecraft) by the release of springs or other devices that simply separate it from the mothership without making a significant modification of the probe's trajectory. Following probe release, the mothership usually executes a trajectory correction maneuver to prevent its own atmospheric entry and to help the main spacecraft continue on with its flyby or orbiter mission activities. NASA's *Pioneer Venus* (four probes) and Galileo (one probe) missions involved the deployment probes into the target planet's atmosphere (i.e., Venus and Jupiter, respectively). Similarly, the Cassini mission to Saturn will involve the deployment of a probe spacecraft into the murky atmosphere of the moon Titan. An aeroshell protects the atmospheric probe spacecraft from the intense heat caused by atmospheric friction during entry. At some point in the descent trajectory, the aeroshell is jettisoned and a parachute then is used to slow the probe's descent sufficiently so it can perform its scientific observations. Data usually are telemetered from the atmospheric probe to the mothership, which then either relays the data back to Earth in real time or else records the data for later transmission to Earth.

An *atmospheric balloon package* is designed for suspension from a buoyant gas–filled bag that can float and travel under the influence of the winds in a planetary atmosphere. Tracking of the balloon package's progress across the face of the target planet will yield data about the general circulation patterns of the planet's atmosphere. A balloon package needs a power supply and a telecommunications system (to relay data and support tracking). It also can be equipped with a variety of scientific instruments to measure the planetary atmosphere's composition, temperature, pressure, and density.

Lander spacecraft are designed to reach the surface of a planet and survive at least long enough to transmit back to Earth useful scientific data, such as: imagery of the landing site, measurement of the local environmental conditions, and an initial examination of soil composition. For example, the Russian Venera lander spacecraft have made brief scientific investigations of the infernolike Venusian surface. In contrast, NASA's Surveyor lander craft extensively explored the lunar surface at several landing sites in preparation for the human Apollo Project landing missions, while NASA's *Viking 1* and *2* lander craft investigated the surface conditions of Mars at two separate sites for many months.

A *surface penetrator spacecraft* is designed to enter the solid body of a planet, an asteroid, or a comet. It must survive a high-velocity impact and then transmit subsurface information back to an orbiting mothership.

Finally, a *surface rover spacecraft* is carried to the surface of a planet, soft-landed, and then deployed. The rover can either be semiautonomous or fully controlled (through teleoperation) by scientists on Earth. Once deployed on the surface, the electrically powered rover can wander a certain distance away from the landing site and take images and perform soil analyses. Data then are telemetered back to Earth by one of several techniques; via the lander spacecraft, via an orbiting mothership, or (depending on size of rover) directly from the rover vehicle. The Soviet Union deployed two highly successful robot surface rovers (called Lunokhod 1 and 2) on the Moon in the 1970s. NASA plans to explore the surface of Mars with a variety of small mobile robots over the next decade.

spacecraft charging In orbit or in deep space, spacecraft and space vehicles can develop an electric potential up to tens of thousands of volts relative to the ambient outer space plasma (the solar wind). Large potential differences (called *differential charging*) also can occur on the space vehicle. One of the consequences is electrical discharge or arcing, a phenomenon that can damage space vehicle surface structures and electronic systems. Many factors contribute to this complex problem, including the spacecraft configuration, the materials from which the spacecraft is made, whether the spacecraft is operating in sunlight or shadow, the altitude at which the spacecraft is performing its mission, and environmental conditions, such as the flux of high-energy solar particles and the level of magnetic storm activity.

Wherever possible, spacecraft designers use conducting surfaces and provide adequate grounding techniques. These design procedures can significantly reduce differential charging, which is generally a more serious problem than the development of a high spacecraft-to-space (plasma) electrical potential.

spacecraft clock Generally, the timing component within the spacecraft's command and data-handling subsystem; it meters the passing time during the life of the spacecraft and regulates nearly all activity within the spacecraft. *See also* CLOCK.

spacecraft drag A space vehicle or spacecraft operating at an altitude below a few thousand kilometers will encounter a significant number of atmospheric particles (i.e., the residual atmosphere) during each orbit of Earth. These encounters result in drag or friction on the spacecraft, causing it gradually to "slow down" and lose altitude, unless some onboard propellant is expended to overcome this drag and maintain the original orbital altitude. If the density of the residual atmosphere at the space vehicle's altitude increases, so will the drag on the vehicle. Any mechanism that can heat Earth's atmosphere (e.g., a geomagnetic storm) will create density changes in the upper atmosphere that can alter a spacecraft's orbit rapidly and significantly. When the residual upper atmosphere is heated by these solar disturbances, it expands outward and makes its presence felt at even higher (than normal) altitudes.

The significance and severity of spacecraft drag was demonstrated clearly by the rapid and premature demise of the abandoned U.S. Skylab space station. Atmospheric heating during a period of maximum solar activity caused the space station's drag to increase considerably, a situation that then resulted in a much more rapid rate of loss in orbital altitude than had been projected. As a result, the 90 metric ton station (last occupied by an astronaut crew in 1974) experienced a fiery reentry in July 1979, years before its originally projected demise. In fact, NASA had been considering using an early space shuttle mission to "rescue Skylab" by providing a reboost to higher altitude. However, the first shuttle mission (STS-1) was not flown until April 1981, almost two years after Skylab became a "front-page" victim of spacecraft drag. *See also* SKYLAB.

spacecraft navigation In general, navigating a spacecraft involves measuring its radial distance and velocity, the angular distance to the spacecraft, and its velocity in the plane of the sky. From these data, a mathematical model can be developed and maintained that describes the history of the spacecraft's location in three-dimensional space as a function of time. Any necessary corrections to a spacecraft's trajectory or orbit then can be identified based on this mathematical model. A spacecraft's navigational history often is reconstructed and incorporated in its scientific observations during a planetary encounter.

Within NASA, the art of spacecraft navigation draws on tracking data, which include measurements of the DOPPLER SHIFT of the downlink carrier and the pointing angles of antennas in the DEEP SPACE NETWORK (DSN). Such navigational data differ from the telemetry data associated with scientific instruments and spacecraft state-of-health sensors.

The distance to a spacecraft can be measured in the following manner. A uniquely coded ranging pulse can be added to the uplink communications sent to a spacecraft. The transmission time of this specially coded pulse is recorded. When the spacecraft receives the ranging pulse, it then returns the special pulse on its downlink communications transmission. The time it takes the spacecraft to turn the pulse around within its onboard electronics package is known from prelaunch calibration testing. Thus when the specially coded downlink pulse is received at one or several DSN sites, the true travel time elapsed can be determined and the spacecraft's distance from Earth then calculated. When several DSN sites have received this coded downlink transmission, the angular position of the spacecraft also can be determined, using triangulation techniques (since the angles at which the DSN antennas point are recorded with an accuracy of thousandths of a degree). Similarly, a spacecraft's velocity component can be determined by a careful measurement of an induced Doppler shift in the downlink frequency, when a two-way coherent communications mode is being used.

Spacecraft that are equipped with imaging instruments can employ these instruments to perform optical navigation. The imaging instrument observes the spacecraft's target (destination) planet against a known background star field. These images are called *optical navigation (OPNAV) images,* and their interpretation provides as very precise data set that is useful for refining knowledge of a spacecrafts's trajectory.

Once a spacecraft's solar or planetary orbital parameters are known, they are compared to the parameters desired for the flight. A *trajectory correction maneuver (TCM)* then can be planned and executed to correct any discrepancy between a spacecraft's actual and desired/planned trajectory. A TCM typically will involve a spacecraft velocity change (ΔV) on the order of just meters to tens of meters per second because of the limited propellant supply carried by a spacecraft.

Similarly, small changes in a spacecraft's orbit around a planet might be needed to adjust an instrument's field-of-view footprint, to improve sensitivity of a gravity field survey, or to prevent too much orbital decay (i.e., loss of altitude). These orbit changes are called orbit trim maneuvers (OTMs), and generally they are carried out in the same manner as TCMs. For example, to make a change that increases the altitude of perigee, an OTM would be performed that increased the spacecraft's velocity when the craft was at apogee. Finally, slight changes in a spacecraft's orbital plane orientation also can be accomplished using OTMs. However, the magnitude of the orbital plane change often is necessarily small, due to the limited amount of propellant a spacecraft carries for these navigational maneuvers.

space debris Space junk or derelict human-made space objects in orbit around Earth. Space debris represents a hazard to astronauts, spacecraft, and large space facilities such as space stations.

Since the start of the space age in 1957, the natural meteoroid environment has been a design consideration for spacecraft. Meteoroids are part of the inter-planetary environment and sweep through Earth orbital space at an average speed of about 20 kilometers (km) per second. Space science data indicate that, at any one moment, a total of approximately 200 kilograms (kg) of meteoroid mass is within some 2,000 km of Earth's surface—the region of space (called LOW EARTH ORBIT, or LEO) most frequently used. The majority of this mass is found in meteoroids about 0.01 centimeter (cm) diameter; however, lesser amounts of this total mass occur in meteoroid sizes both smaller and larger than 0.01 cm. The natural meteoroid flux varies in time as Earth travels around the Sun.

Human-made space debris is also called *orbital debris* and differs from natural meteoroids, because it remains in Earth orbit during its lifetime and is not a transient phenomenon like the meteoroid showers that occur as the Earth travels through interplanetary space around the Sun. The estimated mass of human-made objects orbiting the Earth within about, 2,000 km of its surface is about 3 million kilograms (or about 15,000 times more mass than represented by the natural meteoroid environment). These human-made objects are for the most part in high-inclination orbits and pass one another at an average relative velocity of 10 km per second. Most of this mass is contained in over 3,000 spent rocket stages, inactive satellites, and a comparatively few active satellites. A lesser amount of space debris mass (some 40,000 kg) is distributed in the over 4,000 smaller-size, orbiting objects currently being tracked by space surveillance systems. The majority of these smaller space debris objects are the by-products of over 130 on-orbit fragmentations (satellite breakup events). Recent studies indicate a total mass of at least 1,000 kg for orbital debris sizes of 1 cm or smaller, and about 300 kg for orbital debris smaller than 0.1 cm. The explosion or fragmentation of a large space object also has the potential of producing a large number of smaller objects, objects too small to be detected by contemporary ground-based space surveillance systems. Consequently, this orbital debris environment is now considered more hazardous than the natural meteoroid environment to spacecraft operating in Earth orbit below an altitude of 2,000 km.

Two general types of orbital debris are of concern: (1) large objects (greater than 10 cm in diameter) whose population, while small in absolute terms, is large relative to the population of similar masses in the natural meteoroid environment; and (2) a much greater number of smaller objects (less than 10 cm diameter), whose size distribution approximates the natural meteoroid population and whose numbers add to the "natural debris" environment in those size ranges. The interaction of these two general classes of space debris objects, combined with their long residence time in orbit, creates further concern that new collisions producing additional fragments and causing the total space debris population to grow are inevitable.

An orbiting object loses energy through frictional encounters with the upper limits of Earth's atmosphere and as a result of other orbit-perturbing forces (e.g., gravitational influences). Over time, the object falls into progressively lower orbits and eventually makes a final plunge toward Earth. Once an object enters the sensible atmosphere, atmospheric drag slows it down rapidly and causes it either to burn up completely or to fall through the atmosphere and impact on Earth's surface or in its oceans. A *decayed satellite* (or piece of orbital debris) is one that reenters Earth's atmosphere under the influence of natural forces and phenomena (e.g., atmospheric drag). Space vehicles and satellites that are intentionally removed from orbit are said to have been "deorbited."

One of the most celebrated reentries of a large human-made object occurred on July 11, 1979, when the then-decommissioned and abandoned first American space station, called SKYLAB, came plunging back to Earth over Australia and the Indian Ocean—a somewhat spectacular reentry event that occurred without harm to life or property. In fact, although human-made objects reenter from orbit on the average of more than one per day, only a very small percentage of these reentry events result in

debris surviving to reach Earth's surface. The aerodynamic forces and heating associated with reentry processes usually break up and vaporize most incoming space debris.

Solar activity greatly affects the natural decay of Earth-orbiting objects. High levels of solar activity heat Earth's upper atmosphere, causing it to expand further into space and to reduce the orbital lifetimes of space objects found at somewhat higher altitudes in the LEO regime. However, about 600 km altitude, the atmospheric density is sufficiently low and solar activity–induced atmospheric density increases do not noticeably affect the debris population lifetimes. This solar cycle–based natural cleansing process for space debris in LEO is extremely slow and by itself cannot offset the current rate of human-made space debris generation.

The effects of orbital debris impacts on spacecraft and space facilities depend on the velocity and mass of the debris. For debris sizes less than approximately 0.01 cm in diameter, surface pitting and erosion are the primary effects. Over a long period of time, the cumulative effect of individual particles colliding with a satellite could become significant, because the number of such small debris particles is very large in LEO.

For debris larger than about 0.1 cm in diameter, the possibility of structural damage to a satellite or space facility becomes an important consideration. For example, a 0.3-cm-diameter sphere of aluminum traveling at 10 kilometers per second has about the same kinetic energy as a bowling ball traveling at 100 kilometers per hour. Aerospace engineers anticipate significant structural damage to the satellite or space facility if such an impact occurs.

Space system engineers, therefore, find it helpful to distinguish three space debris size ranges in designing spacecraft. These are debris sizes 0.01 cm diameter and below, which produce surface erosion; debris sizes ranging from 0.01 to 1.0 cm diameter, which produce significant impact damage that can be quite serious; and (3) space debris objects greater than 1.0 cm diameter, which can readily produce catastrophic damage in a satellite or space facility.

Today, only about 5% of the cataloged objects in Earth orbit are active, operational spacecraft. The remainder of these human-made space objects represent various types of orbital debris. Space debris often is divided into four general categories: (1) operational debris (about 12%), objects intentionally discarded during satellite delivery or satellite operations (this category includes lens caps, separation and packing devices, spin-up mechanisms, payload shrouds, empty propellant tanks, and a few objects discarded or "lost" during extravehicular activities [EVAs] by astronauts or cosmonauts); (2) spent and intact rocket bodies (14%); (3) inactive (decommissioned or dead) payloads (20%); and (4) fragmentation (on-orbit space object breakup) (49%).

Aerospace engineers consider the growing space debris problem when they design new spacecraft. In an attempt to make these new spacecraft as "litter-free" as possible, new spacecraft are designed with provisions for retrieval or removal at the end of their useful operations. Telerobotic space debris collection systems also have been proposed. See also METEOROIDS.

space defense All defensive measures designed to destroy attacking enemy vehicles (including missiles) while they are in space, or to nullify or reduce the effectiveness of such an attack from (or through) space.

Space Detection and Tracking System (SPADATS) A network of space surveillance sensors operated by the U.S. Air Force. This ground-based network is capable of detecting and tracking space vehicles and objects in orbit around Earth. The orbital characteristics of such objects are reported to a central control facility. See also CHEYENNE MOUNTAIN AIR STATION.

space food Eating is a basic survival need that an astronaut or cosmonaut must accomplish in order to make space a suitable place to work and accomplish mission objectives. The space food (meals) he or she consumes must be nutritious, safe, lightweight, easily prepared, convenient to use, require little storage space, need no refrigeration and be psychologically acceptable (especially for crews on long-duration missions).

Since the beginning of human space flight in the early 1960s, eating in space has become more natural and "Earth-like," while better meeting the other criteria. Space flight feeding has progressed from squeezing pastelike foods from "toothpaste" tubes to eating a "sit-down" dinner complete with normal utensils, except for the addition of scissors, which today's astronauts use to cut open packages.

In general, there are five basic approaches to preparing food for use in space: rehydratable food, intermediate-moisture food, thermostabilized food, irradiated food, and natural-form food.

Rehydratable food has been dehydrated by a technique such as freeze drying. In the space shuttle program, for example, foods are dehydrated to meet launch vehicle mass and volume restrictions. They are rehydrated later in orbit when they are ready to be eaten. Water used for rehydration comes from the orbiter vehicle's fuel cells, which produce electricity by combining hydrogen and oxygen; water is the resultant by-product. More than 100 different food items, such as cereals, spaghetti, scrambled eggs, and strawberries, go through this dehydration/rehydration process. For example, when a strawberry is freeze-dried, it remains full size in outline, with its color, texture, and quality intact. The astronaut can then rehydrate the strawberry with either saliva (mouth moisture) as it is chewed or by adding water to the package.

Twenty varieties of drinks, including tea and coffee, also are dehydrated for use in space travel. But pure orange juice and whole milk cannot be included. If water is added to dehydrated orange juice, orange "rocks," form in water; they do not rehydrate. Dehydrated whole milk does not dissolve properly upon rehydration. It floats around in lumps and has a disagreeable taste. So skim milk must be used. Back in the 1960s, General Foods Corporation developed a synthetic orange juice product (called Tang) that could be used in place of orange juice.

Intermediate-moisture food is partially dehydrated food, such as dried apricots, dried pears, or dried peaches. *Thermostabilized food* is cooked at moderate temperatures to destroy bacteria and then sealed in cans or aluminum pouches. This type of space food includes tuna, canned fruit in heavy syrup, and ground beef. *Irradiated food* is preserved by exposure to ionizing radiation. Various types of meat and bread are processed in this manner. Finally, *natural-form food* is low in moisture and taken into space in much the same form as is found on Earth. Peanut butter, nuts, graham crackers, gum, and hard candies are examples. Salt and pepper are packaged in liquid form because crystals would float around the crew cabin and could cause eye irritation or contaminate equipment.

All food in space must be packaged in individual serving portions that allow easy manipulation in the microgravity environment of an orbiting spacecraft. These packages can be off-the-shelf thermostabilized cans, flexible pouches, or semi-rigid containers.

In the space shuttle program, the variety of food carried into orbit is sufficiently broad that crew members can enjoy a six-day menu cycle. A typical dinner might consist of a shrimp cocktail, steak, broccoli, rice, fruit cocktail, chocolate pudding, and grape drink.

When the *International Space Station (ISS)* becomes operational (sometime beyond 2004 because of the COLUMBIA ACCIDENT), the station's permanent crew of three might eventually grow in number to a maximum of seven persons. Their food and other supplies must be replenished at regular intervals. ISS residents are now using an extension of the joint U.S.-Russian food system that was developed during the initial SHUTTLE-MIR phase of the ISS program. ISS crewmembers have a menu cycle of eight days, meaning the menu repeats every eight days. Half of the food system is American and half is Russian. However, there are also plans to include foods of other ISS partner countries, including Japan and Canada. The packaging system for the daily menu food is based on single-service, disposable containers. Single-service containers eliminate the need for a dishwasher, and the disposal approach is quite literally out-of-this-world.

Since electrical power for the ISS is generated by SOLAR PANELS rather than FUEL CELLS (as on the SPACE SHUTTLE), there is no extra water generated onboard the station. Water is recycled from cabin AIR, but not enough for significant use in the food system. Consequently, the percentage of shuttle-era rehydratable foods is being decreased and the percentage of thermostabilized foods increased over time.

Generally, the American portion of the current ISS food system is similar to the shuttle food system. It uses the same basic types of foods—thermostabilized, rehydratable, natural form, and irradiated—and the same packaging methods and materials. As on the shuttle, beverages on the ISS are in powdered form. The water temperature is different on the station; unlike the shuttle, there is no chilled water. Crewmembers have only ambient, warm, and hot water available to them.

Space station crewmembers usually eat breakfast and dinner together. They use the food preparation area in the Russian Zvezda service module to prepare their meals. The module has a fold-down table designed to accommodate three astronauts or cosmonauts eating together under MICROGRAVITY conditions. Used food packaging materials are bagged and placed along with other trash in a PROGRESS supply vehicle. Then the ROBOT SPACECRAFT assumes a secondary mission as an extraterrestrial garbage truck. It is JETTISONed from the ISS and burns up upon REENTRY into EARTH'S ATMOSPHERE.

However, long-duration human missions beyond Earth orbit (where resupply is difficult or impossible) will require food supplies capable of extended storage. The menu cycle will have to be greatly expanded to support crew morale and nutritional well-being. A permanent lunar base could have a highly automated greenhouse to provide fresh vegetables and fruits. A rotating "space greenhouse" (to provide an appropriate level of artificial gravity for plant development) might accompany human expeditions to Mars and beyond. A permanent Martian surface base most likely would include an "agricultural facility" as part of its closed environment life support system (CELSS).

Space Infrared Telescope Facility (SIRTF) The final MISSION in NASA's Great Observatories Program—a family of four orbiting observatories each studying the UNIVERSE in a different portion of the ELECTROMAGNETIC SPECTRUM. Consisting of a 0.85-meter diameter TELESCOPE and three cryogenically cooled science INSTRUMENTs, SIRTF represents the most powerful and sensitive infrared telescope ever launched. SIRTF will obtain IMAGES and spectra of celestial objects at INFRARED RADIATION WAVELENGTHs between 3 and 180 MICRONS—an important spectral region of observation mostly unavailable to ground-based telescopes because of the blocking influence of EARTH's ATMOSPHERE. Following LAUNCH (August 25, 2003) from CAPE CANAVERAL AIR FORCE STATION by an expendable DELTA ROCKET, SIRTF traveled to an Earth-trailing HELIOCENTRIC ORBIT that allows the telescope to cool rapidly with a minimum expenditure of onboard cryogen (cryogenic coolant). With a projected mission lifetime of at least 2.5 YEARS, SIRTF will take its place alongside NASA's other great orbitableting observatories by collecting high-RESOLUTION infrared data that help scientists better understand how galaxies, STARS, and PLANETs form and develop. Other missions in this program include the *HUBBLE SPACE TELESCOPE (HST)*, the *COMPTON GAMMA RAY OBSERVATORY (CGRO)*, and the *CHANDRA X-RAY OBSERVATORY (CXO)*. See also INFRARED ASTRONOMY.

space junk A popular aerospace industry expression for space debris or orbital debris. *See also* SPACE DEBRIS.

Spacelab (SL) An orbiting laboratory facility delivered into space and sustained while in ORBIT within the huge CARGO BAY of the SPACE SHUTTLE ORBITER. Developed by the EUROPEAN SPACE AGENCY (ESA) in cooperation with NASA, Spacelab featured several interchangeable elements that were arranged in various configurations to meet the particular needs of a given flight. The major elements were a habitable module (short or long configuration) and pallets. Inside the pressurized habitable research module, ASTRONAUT scientists (PAYLOAD SPECIALISTs) worked in a SHIRT-SLEEVE ENVIRONMENT and performed a variety of experiments while orbiting EARTH in a MICROGRAVITY environment. Several platforms (called pallets) could also be placed in the Orbiter's cargo bay behind the habitable module. Any INSTRUMENTs and experiments mounted on these pallets were exposed directly to the space environment when the shuttle's cargo bay doors were opened after the AEROSPACE VEHICLE achieved orbit around Earth. A train of pallets could also be flown without the concurrent use of the habitable module. The first Spacelab mission (called STS-9/Spacelab 1) was launched in November 1983. It was a highly successful joint NASA and ESA mission consisting of both the habitable module and an exposed instrument platform. The final Spacelab mission (called STS-55/Spacelab D-2) was launched in April 1993. It was the second flight of the German *(Deutsche)* Spacelab configuration and continued microgravity research that had started with the first German Spacelab mission (STS-61A/Spacelab D-1) flown in October 1985. NASA, other ESA countries, and Japan contributed some of the 90 experiments conducted during the Spacelab D-2 mission.

space launch vehicle(s) The expendable or reusable rocket-propelled vehicle(s) used to lift a payload or spacecraft from the surface of Earth and place it in orbit around the planet or on an interplanetary trajectory. *See also* LAUNCH VEHICLE; SPACE TRANSPORTATION SYSTEM.

spaceman A person, male or female, who travels in outer space. The term *astronaut* is preferred. *See also* ASTRONAUT; COSMONAUT.

space medicine The branch of aerospace medicine concerned specifically with the health of persons who make, or expect to make, flights beyond Earth's sensible (measurable) atmosphere into outer space.

space mine A (hypothetical) military satellite with an explosive charge (either nuclear or nonnuclear) that is designed to position itself within lethal range of a target satellite and then detonate at a preprogrammed condition or time, upon direct command, or when it is attacked.

space nuclear power (SNP) Through the cooperative efforts of the U.S. Department of Energy (DOE), formerly called the Atomic Energy Commission, and NASA, the United States has used nuclear energy in its space program to provide electrical power for many missions, including science stations on the Moon, extensive exploration missions to the outer planets—Jupiter, Saturn, Uranus, Neptune and beyond—and even to search for life on the surface of Mars.

For example, when the *Apollo 12* astronauts departed from the lunar surface on their return trip to Earth (November 1969), they left behind a nuclear-powered science station that sent information back to scientists on Earth for several years. That science station, as well as similar stations left on the Moon by the *Apollo 14* through 17 missions, operated on electrical power supplied by plutonium-238 fueled, radioisotope thermoelectric generators (RTGs). In fact, since 1961 nuclear power systems have helped assure the success of many space missions, including the *Pioneer 10* and 11 missions to Jupiter and Saturn; the *Viking 1* and 2 landers on Mars; the spectacular *Voyager 1* and 2 missions to Jupiter, Saturn, Uranus, Neptune and beyond; the Ulysses mission to the Sun's polar regions; and the Galileo mission to Jupiter (see table).

Energy supplies that are reliable, transportable, and abundant represent a very important technology in the development of an extraterrestrial civilization. Space nuclear power systems can play an ever-expanding role in supporting the advanced space exploration and settlement missions of the 21st century, including permanent lunar and Martian surface bases. Even more ambitious space activities, such as asteroid movement and mining, planetary engineering, and human expeditions to the outer regions of the solar system, will require compact energy systems (in advanced space nuclear reactor designs) at the megawatt and gigawatt levels.

Space nuclear power supplies offer several distinct advantages over the more traditional solar and chemical space power systems. These advantages include: compact size; modest mass requirements; very long operating lifetimes; the ability to operate in extremely hostile environments (for e.g., intense trapped radiation belts, the surface of Mars, the moons of the outer planets, and even interstellar space); and the ability to operate independent of distance from the Sun or orientation to the Sun.

Space nuclear power systems use the thermal energy or heat released by nuclear processes. These processes include the spontaneous (but predictable) decay of radioisotopes, the controlled splitting or fissioning of heavy atomic nuclei (such as fissile uranium-235) in a self-sustained neutron chain reaction, and (eventually) the joining together or fusing of light atomic nuclei (such as deuterium and tritium) in a controlled thermonuclear reaction. This "nuclear" heat then can converted directly or through a variety of thermodynamic (heat-engine) cycles into electric power. Until controlled thermonuclear fusion capabilities are achieved, space nuclear power applications will be based on the use of either radioisotope decay or nuclear fission reactors.

The radioisotope thermoelectric generator consists of two main functional components: the thermoelectric converter and the nuclear heat source. The radioisotope plutonium-238 has been used as the heat source in all U.S. space missions involving radioisotope power supplies. Plutonium-238 has a half-life of about 87.7 years and therefore supports a long operational life. (The half-life is the time required for one-half the number of unstable nuclei present at a given time to undergo radioactive decay.) In the nuclear decay process, plutonium-238 emits primarily alpha radiation

Summary of Space Nuclear Power Systems
Launched by the United States (1961–1997)

Power Source	Spacecraft	Mission Type	Launch Date	Status
SNAP-3A	Transit 4A	Navigational	June 29, 1961	Successfully achieved orbit
SNAP-3A	Transit 4B	Navigational	November 15, 1961	Successfully achieved orbit
SNAP-9A	Transit-5BN-1	Navigational	September 28, 1963	Successfully achieved orbit
SNAP-9A	Transit-5BN-2	Navigational	December 5, 1963	Successfully achieved orbit
SNAP-9A	Transit-5BN-3	Navigational	April 21, 1964	Mission aborted: burned up on reentry
SNAP-10A	Snapshot	Experimental	April 3, 1965	Successfully achieved orbit (reactor)
SNAP-19B2	Nimbus-B-1	Meteorological	May 18, 1968	Mission aborted: heat source retrieved
SNAP-19B3	Nimbus III	Meteorological	April 14, 1969	Successfully achieved orbit
SNAP-27	Apollo 12	Lunar	November 14, 1969	Successfully placed on lunar surface
SNAP-27	Apollo 13	Lunar	April 11, 1970	Mission aborted on way to Moon. Heat source returned to South Pacific Ocean
SNAP-27	Apollo 14	Lunar	January 31, 1971	Successfully placed on lunar surface
SNAP-27	Apollo 15	Lunar	July 26, 1971	Successfully placed on Lunar surface
SNAP-19	Pioneer 10	Planetary	March 2, 1972	Successfully operated to Jupiter and beyond; in interstellar space
SNAP-27	Apollo 16	Lunar	April 16, 1972	Successfully placed on lunar surface
Transit-RTG	"Transit" (Triad-01-IX)	Navigational	September 2, 1972	Successfully achieved orbit
SNAP-27	Apollo 17	Lunar	December 7, 1972	Successfully placed on lunar surface
SNAP-19	Pioneer 11	Planetary	April 5, 1973	Successfully operated to Jupiter, Saturn and beyond
SNAP-19	Viking 1	Mars	August 20, 1975	Successfully landed on Mars
SNAP-19	Viking 2	Mars	September 9, 1975	Successfully landed on Mars
MHW	LES 8/9	Communications	March 14, 1976	Successfully achieved orbit
MHW	Voyager 2	Planetary	August 20, 1977	Successfully operated to Jupiter, Saturn, Uranus, Neptune and beyond
MHW	Voyager 1	Planetary	September 5, 1977	Successfully operated to Jupiter, Saturn and beyond
GPHS-RTG	Galileo	Planetary	October 18, 1989	Successfully sent on interplanetary trajectory to Jupiter (1996 arrival)
GPHS-RTG	Ulysses	Solar-Polar	October 6, 1990	Successfully sent on interplanetary trajectory to explore polar regions of Sun (1994–1995)
GPHS-RTG	Cassini	Planetary	October 15, 1997	Successfully sent on interplanetary trajectory to Saturn (July 2004 arrival)

Source: NASA; Department of Energy.

that has very low penetrating power. Consequently, only light-weight shielding is required to protect the spacecraft from its nuclear emissions. A thermoelectric converter uses the "thermocouple principle" to directly convert a portion of the nuclear (decay) heat into electricity.

A nuclear reactor also can be used to provide electric power in space. The Soviet and, later, Russian space program has flown several space nuclear reactors (most recently a system called Topaz). The United States has flown only one space nuclear reactor, an experimental system called the SNAP-10A, which was launched and operated on-orbit in 1965. The objective of the SNAP-10A program was to develop a space nuclear reactor power unit capable of producing a minimum of 500 watts-electric for a period of one year while operating in space. The SNAP-10A reactor was a small zirconium hydride (ZrH) thermal reactor fueled by uranium-235. The SNAP-10A orbital test was successful, although the mission was prematurely (and safely) terminated on-orbit by the failure of an electronic component outside the reactor.

Since the United States first used nuclear power in space, great emphasis has been placed on the safety of people and the protection of the terrestrial environment. A continuing major objective in any new space nuclear power program is to avoid undue risks. In the case of radioisotope power supplies, this means designing the system to contain the radioisotope fuel under all normal and potential accident conditions. For space nuclear reactors, such as the SNAP-10A and more advanced systems, this means launching the reactor in a "cold" (nonoperating) configuration and starting up the reactor only after a safe, stable Earth orbit or interplanetary trajectory has been achieved. *See also* FISSION (NUCLEAR); FUSION (NUCLEAR); NUCLEAR-ELECTRIC PROPULSION (NEP) SYSTEM; RADIOISOTOPE THERMOELECTRIC GENERATOR (RTG); SPACE NUCLEAR PROPULSION.

space nuclear propulsion Nuclear fission reactors can be used in two basic ways to propel a space vehicle: (1) to generate electric power for an electric propulsion unit; and (2) as a thermal energy or heat source to raise a propellant (working material) to extremely high temperature for subsequent expulsion out a nozzle. In the second application, the system is often called a *nuclear rocket.*

In a nuclear rocket, chemical combustion is not required. Instead, a single propellant, usually hydrogen, is heated by the energy released in the nuclear fission process, which occurs in a controlled manner in the reactor's core. Conventional rockets, in which chemical fuels are burned, have severe limitations in the specific impulse a given propellant combination can produce. These limitations are imposed by the relatively high molecular weight of the chemical combustion products. At attainable combustion chamber temperatures, the best chemical rockets are limited to specific impulse values of about 4,300 meters per second (440 seconds). Nuclear rocket systems using fission reactions, fusion reactions, and even possibly matter-antimatter annihilation reactions (the "photon rocket") have been proposed because of their much greater propulsion performance capabilities.

Engineering developments will be needed in the 21st century to permit the use of advanced fission reactor systems, such as the gaseous-core reactor rocket or even fusion-powered systems. However, the solid-core nuclear reactor rocket is within a test-flight demonstration of engineering reality. In this nuclear rocket concept, hydrogen propellant is heated to extremely high temperatures while passing through flow channels within the solid-fuel elements of a compact nuclear reactor system that uses uranium-235 as the fuel. The high-temperature gaseous hydrogen then expands through a nozzle to produce propulsive thrust. From the mid-1950s until the early 1970s, the United States conducted a nuclear rocket program called Project Rover. The primary objective of the project was to develop a nuclear rocket for a manned mission to Mars. Unfortunately, despite the technical success of this rocket program, overall space program emphasis changed, and the nuclear rocket and the human-crewed Mars mission planning were discontinued in 1973. *See also* FISSION (NUCLEAR); FUSION (NUCLEAR); ELECTRIC PROPULSION; NERVA; NUCLEAR-ELECTRIC PROPULSION (NEP) SYSTEM; ROVER PROGRAM; SPACE NUCLEAR POWER; SPECIFIC IMPULSE.

space physics The branch of science that investigates the magnetic and electric phenomena that occur in outer space, in the upper atmosphere of planets, and on the Sun. Space physicists use balloons, rockets, satellites, and deep-space probes to study many of these phenomena in situ (i.e., in their place of origin). The scope of space physics ranges from investigating the generation and transport of energy from the Sun, to searching, past the orbit of Pluto, for a magnetic boundary (called the *heliopause*) separating our solar system from the rest of the Milky Way Galaxy. Phenomena such as aurora, trapped radiation belts, the solar wind, solar flares, and sun spots fall within the realm of space physics.

space platform An uncrewed (un-manned), free-flying orbital platform that is dedicated to a specific mission, such as commercial space activities (e.g., materials processing in space) or scientific research (e.g., the Hubble Space Telescope). This platform can be deployed and serviced or retrieved by the space shuttle. In the future, platform servicing can be accomplished by astronauts from the international space station—if the space platform orbits near the station or can be delivered to the vicinity of the space station by a space tug. *See also* HUBBLE SPACE TELESCOPE; SPACE TUG.

spaceport A spaceport is both a door-way to outer space from the surface of a planet and a port of entry from outer space to a planet's surface. At a spaceport we find the sophisticated facilities required for the assembly, testing, launching, and (in the case of reusable aerospace vehicles) landing and postflight refurbishment of space launch vehicles. Typical operations per-formed at a spaceport include the assembly of space vehicles; preflight preparation of space launch vehicles and their payloads; testing and checkout of space vehicles, spacecraft, and support equipment; coordi-nation of launch vehicle tracking and data-acquisition requirements; countdown and launch operations; and (in the case of reusable space vehicles) landing operations and refurbishment. A great variety of tech-nical and administrative activities also are needed to support the operation of a space-port. These include design engineering, safety and security, quality assurance, cryo-genic fluids management, toxic and haz-ardous materials handling, maintenance, logistics, computer operations, communi-cations, and documentation.

Expendable (one-time use) space launch vehicles now can be found at spaceport facilities around the globe. NASA's Kennedy Space Center in Florida is the spaceport for the (partially) reusable aerospace vehicle the space shuttle. In the next century, highly automated spaceports also will appear on the lunar and Martian surfaces to support permanent human bases on these alien worlds. *See also* CAPE CANAVERAL AIR FORCE STATION; KENNEDY SPACE CENTER.

space radiation environment One of the major concerns associated with the development of a permanent human pres-ence in outer space is the ionizing radia-tion environment, both natural and human-made. The natural portion of the space radiation environment consists pri-marily of Earth's trapped radiation belts (also called the Van Allen belts), solar par-ticle events (SPEs), and galactic cosmic rays (GCRs). Ionizing radiation sources associated with human activities can include: space nuclear power systems (fis-sion reactors and radioisotope), the deto-nation of nuclear explosives in the upper portion of Earth's atmosphere or in outer space (activities currently banned by inter-national treaty), space-based particle accelerators, and radioisotopes used for calibration and scientific activity.

Earth's trapped radiation environment is most intense at altitudes ranging from 1,000 kilometers (km) to 30,000 km. Peak intensities occur at about 4,000 km and 22,000 km. Below approximately 10,000 km altitude, most trapped particles are relatively low-energy electrons (typically, a few million electron volts [MeV]) and protons. In fact, below about 500 km altitude, only the trapped protons and their secondary nuclear interaction products represent a chronic ionizing radiation hazard.

Trapped electrons collide with atoms in the outer skin of a spacecraft creating penetrating X rays and gamma rays (called secondary radiations) that can cause tissue damage. Trapped energetic protons can penetrate several grams of material (typically 1–2 centimeters of aluminum are required to stop them), causing ionization of atoms as they terminate their passage in a series of nuclear collisions. Most human-occupied spacecraft missions in LOW EARTH ORBIT (LEO) are restricted to altitudes below 500 km and inclinations below about 60 degrees to avoid prolonged (chronic) exposure to this type of radiation. For orbits below 500 km altitude and inclinations less than about 60 degrees, the predominant part of an astronaut's overall radiation exposure will be due to trapped protons from the SOUTH ATLANTIC ANOMALY (SAA). The SAA is a region of Earth's inner radiation belts that dips close to the planet over the southern Atlantic Ocean southeast of the Brazilian coast. Passage through the SAA generally represents the most significant source of chronic natural space radiation for space travelers in LEO. Earth's geomagnetic field generally protects astronauts and spacecraft in LEO from cosmic ray and solar flare particles.

However, spacecraft in highly elliptical orbits around Earth will pass through the Van Allen belts each day. Furthermore, those spacecraft with a high-apogee altitude (say greater than 30,000 km) also will experience long exposures to galactic cosmic rays and solar flare environments. Similarly, astronauts traveling through interplanetary space to the Moon or Mars will be exposed to both a continuous galactic cosmic ray environment and a potential solar flare environment (i.e., a solar particle event).

A solar flare is a bright eruption from the Sun's chromosphere that may appear within minutes and then fade within an hour. Solar flares cover a wide range of intensity and size. They eject high-energy protons that represent a serious hazard to astronauts traveling beyond LEO. The SPEs associated with solar flares can last one to two days. Anomalously large solar particle events (ALSPEs), the most intense variety of solar particle event, can deliver potentially lethal doses of energetic particles—even behind modest spacecraft shielding (e.g., 1 to 2 grams per centimeter squared of aluminum). The majority of SPE particles are energetic protons, but heavier nuclei also are present.

Galactic cosmic rays (GCRs) originate outside the solar system. GCR particles are the most energetic of the three general types of natural ionizing radiation in space and contain all elements from atomic number 1 to 92. Specifically, galactic cosmic rays have the following general composition: protons (82–85%), alpha particles (12–14%), and highly ionizing heavy nuclei (1–2%), such as carbon, oxygen, neon, magnesium, silicon, and iron. The ions that are heavier than helium have been given the name HZE particles, meaning high atomic number (Z) and high energy (E). Iron (Fe) ions appear to contribute substantially to the overall HZE population. Galactic cosmic rays range in energy from 10s of MeV to 100s of GeV (a GeV is one billion [10^9] electron volts) and are very difficult to shield against. In particular, HZE particles produce high-dose ionization tracks and kill living cells as they travel through tissue.

An effective space radiation protection program for astronauts on extended missions in interplanetary space or on the lunar or Martian surface should include sufficient permanent shielding (of the spacecraft or surface base habitat modules), adequate active dosimetry, the availability of "solar storm shelters" (zones of increased shielding protection) on crewed spacecraft or on the planetary surface, and an effective solar particle event warning system that monitors the Sun for ALSPEs.

The ionizing radiation environment found in space also can harm sensitive electronic equipment (e.g., a single-event upset) and spacecraft materials. Design precautions, operational planning, localized shielding, device redundancy, and computer-memory "voting" procedures are techniques used by aerospace engineers to overcome or offset space radiation-induced problems that can occur in a spacecraft, especially one operating beyond LEO. *See also* EARTH'S TRAPPED RADIATION BELTS; RADIATION SICKNESS; SINGLE EVENT UPSET.

space resources Generally, when people think about outer space, visions of vast emptiness, devoid of anything useful, come to their minds. However, space is really a new frontier that is rich with resources, including unlimited (solar) energy, a full range of raw materials, and an environment that is both special (e.g., high vacuum, microgravity, physical isolation from the terrestrial biosphere) and reasonably predictable.

Since the start of the space age, preliminary investigations of the MOON, MARS, several asteroids and comets, and meteorites have provided tantalizing hints about the rich mineral potential of the extraterrestrial environment. For example, the U.S. Apollo expeditions to the lunar surface established that the average lunar soil contains more than 90% of the material needed to construct a complicated space industrial facility. The soil in the lunar highlands is rich in anorthosite, a mineral suitable for the extraction of aluminum, silicon, and oxygen. Other lunar soils have been found to contain ore-bearing granules of ferrous metals such as iron, nickel, titanium and chromium. Iron can be concentrated from the lunar soil (called regolith) before the raw material even is refined simply by sweeping magnets over regolith to gather the iron granules scattered within.

Remote sensing data of the lunar surface obtained by the CLEMENTINE spacecraft and LUNAR PROSPECTOR have encouraged some scientists to suggest that water ice may be trapped in perpetually shaded polar regions. If this holds true, then "ice mines" on the Moon could provide both oxygen and hydrogen—vital resources for permanent lunar settlements and space industrial facilities. The Moon would be able both to export chemical propellants for propulsion systems and to resupply materials for life support systems.

Its vast mineral-resource potential, frozen volatile reservoirs and strategic location will make Mars a critical "supply depot" for human expansion into the mineral-rich ASTEROID belt and to the giant outer planets and their fascinating collection of resource-laden moons. Smart robot explorers will assist the first human settlers on Mars, enabling these "Martians" to assess quickly and efficiently the full resource potential of their new world. As these early settlements mature, they will become economically self-sufficient by exporting propellants, life support system consumables, food, raw materials, and manufactured products to feed the next wave of human expansion to the outer regions of the solar system. Cargo spacecraft also will travel between cislunar space and Mars, carrying specialty items to eager consumer markets in both civilizations.

The ASTEROIDS especially Earth-crossing asteroids, represent another interesting category of space resources. Current remote sensing data and analysis of meteorites (many of which scientists believe originate from broken-up asteroids) indicate that carbonaceous (C-type) asteroids may contain up to 10% water, 6% carbon, significant amount of sulfur, and useful amounts of nitrogen. S-class asteroids, which are common near the inner edge of the main asteroid belt and among the Earth-crossing asteroids, may contain up to 30% free metals (alloys of iron, nickel, and cobalt, along with high concentrations of precious metals). E-class asteroids may be rich sources of titanium, magnesium, manganese, and other metals. Finally, chondrite asteroids, which are found among the Earth-crossing population, are believed to contain accessible amounts of nickel, perhaps more concentrated than the richest deposits found on Earth.

Using smart machines, possibly including self-replicating systems, space settlers in the 21st century will be able to manipulate large quantities of extraterrestrial matter and move it to wherever it is needed in the solar system. Many of these space resources will be used as the feedstock for the orbiting and planetary surface base industries that will form the basis of interplanetary trade and commerce. For example, atmospheric ("aerostat") mining stations could be set up around JUPITER and SATURN, extracting such materials as hydrogen and helium—especially helium-3, an isotope of great potential value in nuclear fusion research and applications. Similarly, Venus could be "mined" for the carbon dioxide in its atmosphere, Europa for water, and Titan for hydrocarbons. Large fleets of robot spacecraft might even be used to gather chunks of water ice from the Saturnian ring system, while a sister fleet of robot vehicles extracts metals from the main asteroid belt. Even the nuclei of selected COMETS could be intercepted and mined for frozen volatiles, including water ice.

space settlement A large extraterrestrial habitat where from 1,000 to perhaps 10,000 people would live, work, and play, while supporting space industrialization activity, such as the operation of a large space manufacturing complex or the construction of satellite power systems. One possible design is a spherical space settlement called the *Bernal sphere*. This giant spherical habitat would be approximately 2 kilometers in circumference. Up to 10,000 people would live in residences along the inner surface of the large sphere. Rotation of the settlement at about 1.9 revolutions per minute (RPM) would provide Earth-like gravity levels at the sphere's equator, but there would be essentially microgravity conditions at the poles. Because of the short distances between locations in the equatorial residential zone, passenger vehicles would not be necessary. Instead, the space settlers would travel on foot or perhaps by bicycle. The climb from the residential equatorial area up to the sphere's poles would take about 20 minutes and would lead the hiker past small villages, each at progressively lower levels of artificial gravity. A corridor at the axis

would permit residents to float safely in microgravity out to exterior facilities, such as observatories, docking ports, and industrial and agricultural areas. Ringed areas above and below the main sphere in this type of space settlement would be the external agricultural toruses. *See also* BERNAL SPHERE; SATELLITE POWER SYSTEM.

spaceship In general, any crewed vehicle or craft capable of traveling in outer space; specifically, a crewed space vehicle capable of performing an interplanetary (and eventually an interstellar) journey.

space shuttle Space shuttles (i.e., the orbiter vehicles) are the main element of America's SPACE TRANSPORTATION SYSTEM (STS) and are used for space research and applications. The shuttles are the first vehicles capable of being launched into space and returning to Earth on a routine basis.

Space shuttles are used as orbiting laboratories in which scientists and mission specialists conduct a wide variety of scientific experiments. Astronaut crews aboard space shuttles place satellites into orbit. They also rendezvous with satellites to carry out repairs and/or return a malfunctioning spacecraft to Earth for refurbishment. Recently, space shuttles have been used as part of Phase I of the International Space Station program. In this application, orbiter vehicles such as the Atlantis (OV-104) have performed important rendezvous and docking missions with the Russian *Mir* space station.

Space shuttles are true aerospace vehicles. They leave Earth and its atmosphere under rocket power provided by three liquid-fueled main engines and two solid-propellant boosters attached to an external liquid-propellant tank. After their missions in orbit end, the orbiter vehicles streak back through the atmosphere and are maneuvered to land like an airplane. The shuttles, however, are without power during reentry, and they land on runways much like a glider.

The operational space shuttle fleet consists of the: *Discovery* (OV-103), *Atlantis* (OV-104), and *Endeavour* (OV-105). The *Enterprise* (OV-101) served as a test article but never flew in space. The *Challenger* (OV-99) was lost along with its crew on January 28, 1986, during the first two minutes of the STS 51-L mission and the *Columbia* (OV-102) and its crew were lost on February 1, 2003, at the end of the STS 107 mission while returning to Earth.

space shuttle main engine (SSME) Each of three main LIQUID-PROPELLANT ROCKET ENGINES on an ORBITER vehicle that is capable of producing a THRUST of 1.67 million NEWTONS (375,000 lbf) at sea level and about 2.09 million newtons (470,000 lbf) in the VACUUM of OUTER SPACE. These engines burn for approximately eight minutes during launch ascent and together consume about 242,000 LITERS of CRYOGENIC PROPELLANTS (LIQUID HYDROGEN and LIQUID OXYGEN) per minute when all three operate at full power. The SPACE SHUTTLE's large EXTERNAL TANK provides the liquid hydrogen and liquid oxygen for these engines and is then discarded after main engine cutoff (MECO).

space sickness A form of motion sickness experienced by about 50% of astronauts or cosmonauts when they encounter the microgravity ("weightless") environment of an orbiting spacecraft a few minutes or hours after launch. Space sickness symptoms include nausea, vomiting, and general malaise. This condition is generally only temporary (typically lasting no more than a day or so) and can be treated (but not prevented) with medications. When a person enters an extended period of microgravity, the fluids shift from the lower part of the body to the upper part, causing many physical changes to occur. For example, a person's eyes will appear smaller because the face becomes fuller (i.e., puffy or swollen in appearance) and the waistline shrinks a few centimeters. Also referred to as *space adaptation syndrome.*

space simulator A device or facility that stimulates some condition or conditions found in space. It is used for testing equipment or in training programs. *See also* SIMULATOR.

space station A space station is an orbiting space system that is designed to accommodate long-term human habitation in space. The concept of people living and working in artificial habitats in outer space appeared in 19th-century science fiction literature in stories such as Edward Everett Hale's "Brick Moon" (1869) and Jules Verne's "Off on a Comet" (1878).

At the beginning of the 20th century, Konstantin Tsiolkovsky (1857–1935) provided the technical underpinnings for this concept with his truly visionary writings about the use of orbiting stations as a springboard for exploring the cosmos. Tsiolkovsky, the father of Russian astronautics, provided a more technical introduction to the space station concept in his 1895 work *Dreams of Earth and Heaven, Nature and Man.* He greatly expanded on the idea of a space station in his 1903 work entitled "The Rocket into Cosmic Space." In this technical classic, Tsiolkovsky described all the essential ingredients needed for a crewed space station including the use of solar energy, the use of rotation to provide artificial gravity and the use of a closed ecological system complete with "space greenhouse."

Throughout the first half of the 20th century the space station concept continued to technically evolve. For example, the German scientist Hermann Oberth described the potential applications of a space station in his classic treatise: *The Rocket to Interplanetary Space* (German title: *Das Rakete zu den Planetenraumen*) (1923). The suggested applications included the use of a space station as an astronomical observatory, an Earth-monitoring facility, and a scientific research platform. In 1929 an Austrian named Herman Potocnik (pen name Hermann Noordung) introduced the concept of a rotating, wheel-shaped space station. Noordung called his design "Wohnrad" ("Living Wheel"). Another Austrian, Guido von Pirquet, wrote many technical papers on space flight, including the use of a space station as a refueling node for space tugs. In the late 1920s and early 1930s, von Pirquet also suggested the use of multiple space stations at different locations in cislunar space. After World War II Dr. Wernher von Braun (with the help of space artist Chesley Bonestell) popularized the concept of a wheel-shaped space station in the United States.

The U.S. National Aeronautics and Space Administration (NASA), created in 1958, became the forum for the American space station debate. How long should such an orbiting facility last? What was its primary function? How many crew? What orbital altitude and inclination? Should it be built in space or on the ground and then deployed in space? In 1960 space station advocates from every part of the fledgling space industry gathered in Los Angeles for a Manned Space Station Symposium where they agreed that the space station was a logical goal but disagreed on what it was, where it should be located, or how it should be built.

Then, in 1961, President John F. Kennedy decided that the Moon was a worthy target of the American spirit and heritage. A lunar landing had a definite advantage over a space station: Everyone could agree on the definition of landing on the Moon, but few could agree on the definition of a space station. However, this disagreement was actually beneficial. It forced space station designers and advocates to think about what they could do, the cost of design, and what was necessary to make the project a success. What were the true requirements for a space station? How could they best be met? The space station requirements review process started informally within NASA in 1963 and has continued up to the present day. For over three decades, NASA planners and officials have asked the scientific, engineering, and business communities over and over again. What would you want? What do you need? As answers flowed in, NASA developed a variety of space station concepts to help satisfy these projected requirements. The INTERNATIONAL SPACE STATION (ISS) is the latest in this evolving series of space station concepts.

Even before the Apollo program had landed men successfully on the Moon, NASA engineers and scientists were considering the next giant step in the U.S. crewed space flight program. That next step became the simultaneous development of two complementary space technology capabilities. One was a safe, reliable transportation system that could provide routine access to space. The other was an orbital space station where human beings could live and work in space. This space station would serve as a base camp from which other, more advanced space technology developments could be initiated. This long-range strategy set the stage for two of the most significant American space activities carried out in the 1970s and 1980s: Skylab and the space shuttle (or Space Transportation System).

On May 14, 1973, the United States launched its first space station, called Skylab. It was launched by a Saturn V booster from the Apollo program. Skylab demonstrated that people could function in space for periods up to 12 weeks and, with proper exercise, could return to Earth with no ill effects.

In particular, the flight of Skylab proved that human beings could operate very effectively in a prolonged microgravity environment and that it was not essential to provide artificial gravity for people to live and work in space (at least for periods up to about six months). (Recent longer-duration flights by Russian cosmonauts and American astronauts on the MIR space station have reinforced and extended these findings.) The Skylab astronauts accomplished a wide range of emergency repairs on station equipment, including freeing a stuck solar panel array (a task that saved the entire mission), replacing rate gyros, and repairing a malfunctioning antenna. On two separate occasions the crew installed portable sun shields to replace the original equipment that was lost when Skylab was launched. These on-orbit activities clearly demonstrated the unique and valuable role people have in space.

Unfortunately, Skylab was not designed for a permanent presence in space. For example, the system was not designed to be routinely serviced on orbit—although the Skylab crews were able to perform certain repair functions. Skylab was not equipped to maintain its own orbit—a design deficiency that eventually caused its fiery demise on July 11, 1979, over the Indian Ocean and portions of western Australia. Finally, it was not designed for evolutionary growth and therefore was subject to rapid technological obsolescence. Future space station designs will take these shortcomings into account and effectively use the highly successful Skylab program experience to develop a permanent, evolutionary, and modular space station.

While the United States was concentrating on the Apollo Moon Landing program, the former Soviet Union (now called the Russian Federation) began embarking on an ambitious space station program. As early as 1962 Russian engineers described a space station comprised of modules launched separately and brought together in orbit. The world's first space station, called *Salyut-1*, was launched on April 19, 1971, by a Proton booster. (The Russian word *salyut* means "salute.") The first-generation of Russian space stations had one docking port and could not be resupplied or refueled. The stations were launched uncrewed and later occupied by crews. Two types of early Russian space stations existed: Almaz military stations and Salyut civilian stations. During the cold war, to confuse Western observers the Russians referred to both kinds of station as Salyut (see table).

The Almaz military station program was the first approved. When proposed in 1964, it had three parts: the Almaz military surveillance space station, transport logistics spacecraft for delivering military-cosmonauts and cargo, and Proton rockets for launching both. All of these spacecraft were built, but none was actually used as originally planned.

Russian engineers completed several Almaz space station hulls by 1970. The Russian leaders then ordered that the hulls be transferred to a crash program to launch a civilian space station. Work on the transport logistics spacecraft was deferred, and the Soyuz spacecraft originally built for the Russian manned Moon program was reapplied to ferry crews to the space stations.

Unfortunately, the early first-generation Russian space stations were plagued by failures. For example, the crew of *Soyuz 10,* the first spacecraft sent to *Salyut-1,* was unable to enter the station because of a docking mechanism problem. The *Soyuz 11* crew lived aboard, *Salyut-1* for three weeks but died during the return to Earth because the air escaped from their spacecraft. Then three first-generation stations failed to reach orbit or broke up in orbit before the cosmonaut crews could reach them. The second failed space station was called *Salyut-2.* It was the first Almaz military space station to fly.

However, the Russians recovered rapidly from these failures. *Salyut-3, Salyut-4,* and *Salyut-5* supported a total of five crews. In addition to military surveillance and scientific and industrial experiments, the cosmonauts performed engineering tests to help develop the second-generation stations.

The second-generation Russian space station was introduced with the launch (on September 29, 1977) and successful operation of the *Salyut-6* station. Several important design improvements appeared on this station, including the addition of a second docking port and the use of an automated Progress resupply spacecraft (a space "freighter" derived from the Soyuz spacecraft.)

With the second-generation stations, the Russian space station program evolved from short-duration to long-duration stays. Like the first-generation stations, they were launched uncrewed and their crews arrived later in a Soyuz spacecraft. Second-generation Russian stations had two docking ports. This permitted refueling and resupply by Progress spacecraft, which docked automatically at the aft port. After docking, cosmonauts on the station opened the aft port and unloaded the space freighter. Transfer of fuel to the station was accomplished automatically under supervision from ground controllers.

The availability of a second docking port also meant long-duration resident crews could receive visitors. Visiting crews often included cosmonaut-researchers from the former Soviet bloc countries or

Russian Space Station Experience			
Name	*Type*	*Launched*	*Remarks*
First-Generation Space Stations (1964–1977)			
Salyut-1	Civilian	1971	First space station
Unnamed	Civilian	1972	Failure
Salyut-2	Military	1973	First Almaz station; failure
Cosmos 557	Civilian	1973	Failure
Salyut-3	Military	1974–1975	Almaz station
Salyut-4	Civilian	1974–1977	
Salyut-5	Military	1976–1977	Last Almaz station
Second-Generation Space Stations (1977–1985)			
Salyut-6	Civilian	1977–82	Highly successful
Salyut-7	Civilian	1982–91	Last staffed in 1986
Third-Generation Space Stations (1986–1999)			
Mir	Civilian	1986–1999	First permanent space station

Source: NASA.

countries that were politically sympathetic to the former Soviet Union. For example, the Czech cosmonaut Vladimir Remek visited the *Salyut-6* station in 1978 and became the first space traveler not from either the United States or Russia.

These visiting crews helped relieve the monotony that can accompany a long stay in space. They often traded their Soyuz spacecraft for the one already docked at the station, because the Soyuz spacecraft had only a limited lifetime in orbit. The spacecraft's lifetime was gradually extended from 60 to 90 days for the early *Soyuz Ferry* to more than 180 days for the *Soyuz-TM*. By way of comparison, the Soyuz crew transfer vehicle intended for use with the international space station will have a lifetime of more than a year.

The *Salyut-6* station received 16 cosmonaut crews, including six long-duration crews. The longest stay time for a *Salyut-6* crew was 185 days. The first *Salyut-6* long-duration crew stayed in orbit for 96 days, surpassing the 84-day space endurance record that had been established in 1974 by the last *Skylab* crew. The *Salyut-6* hosted cosmonauts from Hungary, Poland, Romania, Cuba, Mongolia, Viet-nam, (East) Germany, as well as Czecho-slovakia. Twelve Progress freighter spacecraft delivered more than 20 tons of equipment, supplies, and fuel. An experimental transport logistics spacecraft called *Cosmos 1267* docked with *Salyut-6* in 1982. The transport logistics spacecraft was originally designed for the Almaz program. *Cosmos 1267* demonstrated that a large module could dock automatically with a space station—a major space technology step toward the multimodular *Mir* station and the international space station. The last cosmonaut crew left the *Salyut-6* station on April 25, 1977. The station reentered the Earth's atmosphere and was destroyed in July 1982.

The *Salyut-7* space station was launched on April 19, 1982, and was a near twin of the *Salyut-6* station. It was home to ten cosmonaut crews, including six long-duration crews. The longest crew stay time was 237 days. Guest cosmonauts from France and India worked aboard the station, as did the first Russian female space traveler (Cosmonaut Svetlana Savit-skaya) since 1963; she flew aboard the *Soyuz-T-7/Salyut-7* mission. Savitskaya also became the first woman to "walk in space" (i.e., perform an extravehicular activity) during the *Soyuz-T-12/Salyut 7* mission in 1984. Unlike the *Salyut-6* station, however, the *Salyut-7* station suffered some major technical problems. In early 1985, for example, Russian ground controllers lost contact with the then unoccupied station. In July 1985 a special crew aboard the *Soyuz-T-13* spacecraft docked with the derelict space station and made emergency repairs that extended its lifetime for another long-duration mission. The *Salyut-7* station finally was abandoned in 1986; it reentered Earth's atmosphere over Argentina in 1991.

During its lifetime on orbit, 13 Progress spacecraft delivered more than 25 tons of equipment, supplies, and fuel to *Salyut-7*. Two experimental transport logistics spacecraft, called *Cosmos 1443* and *Cosmos 1686*, docked with the station. *Cosmos 1686* was a transitional vehicle, a transport logistics spacecraft that had been redesigned to serve as an experimental space station module.

In February 1986, the Russians introduced a third-generation space station, the *Mir* ("peace") station. Design improvements included more extensive automation, more spacious crew accommodations for resident cosmonauts (and later American astronauts), and the addition of a multiport docking adapter at one end of the station. In a very real sense, *Mir* represented the world's first "permanent" space station. When docked with the Progress-M and Soyuz-TM spacecraft, this station measured more than 32.6-meters long and was about 27.4-meters wide across its modules. The orbital complex consisted of the *Mir* core module and a variety of additional modules, including the Kvant (quantum), the Kvant 2, and the Kristall modules.

The *Mir* core resembled the *Salyut-7* station but had six ports instead of two. The fore and aft ports were used primarily for docking, while the four radial ports that were located in a node at the station's front were used for berthing large modules. When launched in 1986, the core had a mass of about 20 tons. The Kvant module was added to the *Mir* core's aft port in 1987. This small, 11-ton (11,050-kg) mod-

ule contained astrophysics instruments and life support and attitude control equipment. Although Kvant blocked the core module's aft port, it had its own aft port that then served as the station's aft port.

The 18.5-ton Kvant 2 module was added in 1989. The design of this module was based on the transport logistics spacecraft originally intended for the Almaz military space station program of the early 1970s. The purpose of Kvant-2 was to provide biological research data, Earth observation data, and extravehicular activity (EVA) capability. Kvant 2 carried an EVA airlock, two solar arrays, and life support equipment.

The 19.6-ton Kristall module was added in 1990. It carried scientific equipment, retractable solar arrays, and a docking node equipped with a special ANDROGY-NOUS INTERFACE docking mechanism designed to receive spacecraft with masses of up to 100 tons. This docking unit (originally developed for the former Russian space shuttle *Buran*) was attached to the DOCKING MODULE (DM) that was being used by the American space shuttle ORBITER vehicles to link up with the *Mir* during Phase I of the *International Space Station* program.

Three more modules, all carrying American equipment, were also added to the *Mir* complex as part of the *ISS* program. These were the Spektr module, the Priroda (nature) module, and the docking module. The Spetkr module, carrying scientific equipment and solar arrays, was severely damaged on June 25, 1997, when a Progress resupply spacecraft collided with it during practice docking operations.

The Priroda module carried microgravity research and Earth observation equipment. This 19.7-ton module was the last module added to the *Mir* complex (April 1996). Unlike the other modules, however, Priroda had none of its own solar power arrays and depended on other portions of the *Mir* complex for electric power.

The docking module was delivered by the space shuttle *Atlantis* during the STS-74 mission (November 1995) and berthed at Kristall's androgynous docking port.

The Russian-built docking module became a permanent extension on *Mir* and provided better clearances for space shuttle Orbiter-*Mir* linkups.

From 1986 the *Mir* space station served as a major part of the Russian space program. However, because of active participation in the ISS program and severe budget constraints, the Russian Federation decommissioned *Mir* and abandoned the station in 1999. In March 2001, for safety reasons, the large space station was successfully deorbited and intentionally crashed into a remote area of the Pacific Ocean.

In January 1984, as part of his State of the Union address, President Ronald Reagan called for a space station program that would include participation by United States allied countries. With this presidential mandate, NASA established a Space Station Program Office in April 1984 and requested proposals from the American aerospace industry. By March 1986, the baseline design was the dual keel configuration, a rectangular framework with a truss across the middle for holding the station's living and working modules and solar arrays.

Japan, Canada, and the EUROPEAN SPACE AGENCY (ESA) each signed a bilateral memorandum of understanding in the spring of 1985 with the United States, agreeing to participate in the space station project. In 1987 the station's dual keel configuration was revised to compensate for a reduced space shuttle flight rate in the wake of the CHALLENGER ACCIDENT. The revised baseline had a single truss with the built-in option to upgrade to the dual keel design. The need for a space station "lifeboat," called the assured crew return vehicle, was also identified.

In 1988 President Reagan named the space station *Freedom*. With each annual budget cycle, *Freedom's* design underwent modifications as the U.S. Congress called for reductions in its cost. The truss was shortened, and the U.S. Habitation and Laboratory modules were reduced in size. The truss was to be launched in sections with subsystems already in place. Despite these redesign efforts, NASA and its con-

tractors were able to produce a substantial amount of HARDWARE. In 1992 the United States agreed to purchase Russian Soyuz spacecraft to serve as *Freedom's* lifeboat. This action presaged the greatly increased space cooperation between the United States and Russia that followed the COLD WAR. Another important activity, the Shuttle-*Mir* program (later called Phase I of the ISS program), was also started in 1992. The Shuttle-*Mir* program used existing assets (primarily U.S. space shuttle Orbiter vehicles and the Russian *Mir* space station) to provide the joint operational experience and to perform the joint research that would eventually lead to the successful construction and operation of the *ISS*.

In 1993 President Bill Clinton called for *Freedom* to be redesigned once again to reduce costs and to include more international involvement. The White House staff selected a design option that was called *Space Station Alpha*—a downsized configuration that would use about 75% of the hardware designs originally intended for *Freedom*. After the Russians agreed to supply major hardware elements (many of which were originally intended for a planned *Mir 2* space station), *Space Station Alpha* became officially known as the *ISS*.

The *International Space Station* program is divided into three basic phases. Phase I (an expansion of the Shuttle–Mir docking mission program) provided U.S. and Russian aerospace engineers, flight controllers, and cosmonauts and astronauts the valuable experience needed to cooperatively assemble and build the *ISS*.

Phase I officially began in 1995 and involved more than two years of continuous stays by a total of seven American astronauts aboard the Russian *Mir* space station and nine Shuttle-*Mir* docking missions. This phase of the ISS program ended in June 1998 with the successful completion of the STS-91 mission, a mission in which the space shuttle *Discovery* docked with *Mir* and "downloaded" (that is, returned to Earth) astronaut Andrew Thomas, the last American occupant of the Russian space station.

Phases II and III involve the in-orbit assembly of the station's components— Phase II the core of the *ISS* and Phase III its various scientific modules. The projected completion for the final operational configuration (shown in Figure 71) was scheduled for sometime in 2004, but this milestone will be delayed by the COLUMBIA ACCIDENT that occurred on February 1, 2003.

A very historic moment in aerospace history took place near the end of the 20th century. On December 10, 1998, STS-88 shuttle mission COMMANDER Robert Cabana and Russian cosmonaut and MISSION SPECIALIST Sergei Krikalev swung open the hatch between the shuttle *Endeavour* and the first element of the *ISS*. With this action the STS-88 astronauts completed the first steps in the orbital construction of the *ISS*. In late November 1998, a Russian PROTON rocket had successfully placed the NASA-owned, Russian-built Zarya (sunrise) control module into a perfect PARKING ORBIT. A few days later, in early December, the shuttle *Endeavour* carried the American-built Unity connecting module into orbit for RENDEZVOUS with Zarya. Astronauts Jerry Ross and James Newman then performed three arduous extra vehicular activities (totaling 21 hours and

22 minutes) to complete the initial assembly of the space station. When their space walking efforts were completed, Cabana and Krikalev were quite literally able to open the door of an important new space station era. As predicted by KONSTANTIN TSIOLKOVSKY at the start of the 20th century, humankind will soon leave the "cradle of Earth" and build permanent outposts in space on the road to the stars.

spacesuit Outer space is a very hostile environment. If astronauts/cosmonauts are to survive there, they must take part of Earth's environment with them. Air to breathe, acceptable ambient pressures, and moderate temperatures have to be contained in a shell surrounding the space traveler. This can be accomplished by providing a very large enclosed structure or habitat or, on an individual basis, by encasing the astronaut in a protective flexible capsule called the spacesuit.

Spacesuits used on previous NASA missions from the Mercury Project up through the Apollo Soyuz Test Project have provided effective protection for American astronauts. However, certain design problems have handicapped the suits. These suits were custom-fitted garments. In some suit models, more than 70

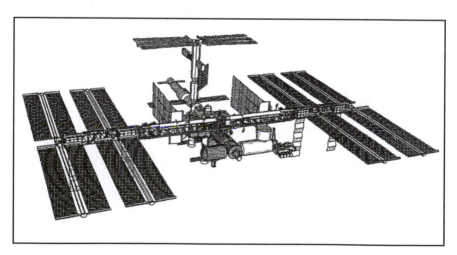

Figure 71. The fully assembled *International Space Station* (*ISS*) (artist's rendering). (*Courtesy of NASA*)

different measurements had to be taken of the astronaut in order to manufacture the spacesuit to the proper fit. As a result, a spacesuit could be worn by only one astronaut on only one mission. These early spacesuits were stiff, and even simple motions such as grasping objects quickly drained an astronaut's strength. Even donning the suit was an exhausting process that at times lasted more than an hour and required the help of an assistant.

For example, the Mercury spacesuit was a modified version of a U.S. Navy high-altitude jet aircraft pressure suit. It consisted of an inner layer of Neoprene-coated nylon fabric and a restrain outer layer of aluminized nylon. Joint mobility at the elbows and knees was provided by simple break lines sewn into the suit; but even with these break lines, it was difficult for the wearer to bend arms or legs against the force of a pressurized suit. As an elbow or knee joint was bent, the suit joints folded in on themselves, reducing suit internal volume and increasing pressure. The Mercury spacesuit was worn "soft," or unpressurized, and served only as backup for possible spacecraft cabin pressure loss—an event that never happened.

NASA spacesuit designers then followed the U.S. Air Force approach toward greater suit mobility when they developed the spacesuit for the two-man Gemini Project spacecraft. Instead of fabric-type joints used in the Mercury suit, the Gemini spacesuit had a combination of a pressure bladder and a link-net restraint layer that made the whole suit flexible when pressurized.

The gas-tight, human-shaped pressure bladder was made of Neoprene-coated nylon and covered by load-bearing link-net woven from Dacron and Teflon cords. The net layer, being slightly smaller than the pressure bladder, reduced the stiffness of the suit when pressurized and served as a type of structural shell. Improved arm and shoulder mobility resulted from the multilayer design of the Gemini suit.

Walking on the Moon's surface presented a new set of problems to spacesuit designers. Not only did the spacesuits for the "Moonwalkers" have to offer protection from jagged rocks and the intense heat of the lunar day, but the suits also had to be flexible enough to permit stooping and bending as the Apollo astronauts gathered samples from the Moon and used the lunar rover vehicle for transportation over the surface of the Moon.

The additional hazard of micrometeoroids that constantly pelt the lunar surface from deep space was met with an outer protective layer on the Apollo spacesuit. A backpack portable life support system provided oxygen for breathing, suit pressurization, and ventilation for moonwalks lasting up to seven hours.

Apollo spacesuit mobility was improved over earlier suits by use of bellowslike molded rubber joints at the shoulders, elbows, hips and knees. Modifications to the suit waist for the Apollo 15 through 17 missions provided flexibility and made it easier for astronauts to sit on the lunar rover vehicle.

From the skin out, the Apollo A7LB spacesuit began with an astronaut-worn liquid-cooling garment, similar to a pair of longjohns with a network of spaghetti-like tubing sewn onto the fabric. Cool water, circulating through the tubing, transferred metabolic heat from the astronaut's body to the backpack, where it was then radiated away to space. Next came a comfort and donning improvement layer of lightweight nylon, followed by a gas-tight pressure bladder of Neoprene-coated nylon or bellowslike molded joints components, a nylon restraint layer to prevent the bladder from ballooning, a lightweight thermal superinsulation of alternating layers of thin Kapton and glass-fiber cloth, several layers of Mylar and spacer material, and, finally, protective outer layers of Teflon-coated glass-fiber Beta cloth.

Apollo space helmets were formed from high-strength polycarbonate and were attached to the spacesuit by a pressure-sealing neckring. Unlike Mercury and Gemini helmets, which were closely fitted and moved with the astronaut's head, the Apollo helmet was fixed and the astronaut's head was free to move within it. While walking on the Moon, the Apollo crew wore an outer visor assembly over the polycarbonate helmet to shield against eye-damaging ultraviolet radiation and to maintain head and face thermal comfort.

Lunar gloves and boots completed the Apollo spacesuit. Both were designed for the rigors of exploring; the gloves also could adjust sensitive instruments. The lunar gloves consisted of integral structural restraint and pressure bladders, molded from casts of the crewperson's hands, and covered by multilayered superinsulation for thermal and abrasion protection. Thumb and fingertips were molded of silicone rubber to permit a degree of sensitivity and "feel." Pressure-sealing disconnects, similar to the helmet-to-suit connection, attached the gloves to the spacesuit arms.

The lunar boot was actually an over-shoe that the Apollo astronaut slipped on over the integral pressure boot of the spacesuit. The outer layer of the lunar boot was made from metal-woven fabric, except for the ribbed silicone rubber sole; the tongue area was made from Teflon-coated glass-fiber cloth. The boot inner layers were made from Teflon-coated glass-fiber cloth followed by 25 alternating layers of Kapton film and glass-fiber cloth to form an efficient, lightweight thermal insulation.

Modified versions of the Apollo spacesuit were used also during the Skylab Program (1973–1974) and the Apollo-Soyuz Test Project (1975).

A new spacesuit has been developed for shuttle-era astronauts that provides many improvements in comfort, convenience, and mobility over previous models. This suit, which is worn outside the orbiter during extravehicular activity (EVA), is modular and features many interchangeable parts. Torso, pants, arms, and gloves come in several different sizes and can be assembled for each mission in the proper combination to suit individual male and female astronauts. The design approach is cost-effective because the suits are reusable and not custom-fitted.

The shuttle spacesuit is called the *extravehicular mobility unit* (EMU) and consists of three main parts: liner, pressure vessel, and primary life support sys-tem (PLSS). These components are supplemented by a drink bag, communications set, helmet and visor assembly.

Containment of body wastes is a significant problem in spacesuit design. In the shuttle era EMU, the PLSS handles odors, carbon dioxide, and the containment of gases in the suit's atmosphere. The PLSS is a two-part system consisting of a backpack unit and a control and display unit located on the suit chest. A separate unit is required for urine relief. Two different urine-relief systems have been designed to accommodate both male and female astronauts. Because of the short-time durations for extravehicular activities, fecal containment is considered unnecessary.

The *manned maneuvering unit* (MMU) is a one-person, nitrogen-propelled backpack that latches to the EMU spacesuit's PLSS. Using rotational and translational hand controllers, the astronaut can fly with precision in or around the orbiter's cargo bay or to nearby free-flying payloads or structures, and can reach many otherwise inaccessible areas out-side the orbiter vehicle. Astronauts wearing MMUs have deployed, serviced, repaired, and retrieved satellite payloads.

The MMU has been called "the world's smallest reusable spacecraft." The MMU propellant (non-contaminating gaseous nitrogen stored under pressure) can be recharged from the orbiter vehicle. The reliability of the unit is guaranteed with a dual parallel system rather than a backup redundant system. In the event of a failure in one parallel system, the system would be shut down and the remaining system would be used to return the MMU to the orbiter's cargo bay. The MMU includes a 35-mm still camera that is operated by the astronaut while working in space.

Shuttle era spacesuits are pressurized at 29.6 kilopascals (4.3 pounds-force per square inch [psi]), while the shuttle cabin pressure is maintained at 101 kilopascals (14.7 psi). Because the gas in the suit is 100% oxygen (instead of 20% oxygen as is found in Earth's atmosphere), the person in the spacesuit actually has more oxygen to breathe than is available at an altitude of 3,000 meters (10,000 feet) or even at sea level without the spacesuit. However, prior to leaving the orbiter to perform tasks in space, an astronaut has to spend several hours breathing pure oxygen. This procedure (called "pre-breathing") is necessary to remove nitrogen dissolved in body fluids and thereby prevent its release as gas bubbles when pressure is reduced, a condition commonly referred to as "the bends."

In addition to new space walking tools and philosophies for astronaut-assisted assembly of the INTERNATIONAL SPACE STATION (ISS), American astronaut space walkers have an enhanced spacesuit. The shuttle spacesuit (the extravehicular mobility unit) was originally designed for sizing and maintenance between flights by skilled specialists on Earth. Such maintenance and refurbishment activities would prove difficult, if not impossible, for astronauts aboard the station.

The shuttle spacesuit has been improved for use on the ISS. It can now be stored in orbit and is certified for up to 25 extravehicular activities before it must be returned to Earth for refurbishment. The spacesuit can be adjusted in flight to fit different astronauts and can be easily cleaned and refurbished between EVAs onboard the station. The modified spacesuit has easily replaceable internal parts, reusable carbon dioxide removal cartridges, metal sizing rings that accommodate in-flight suit adjustments to fit different crewmembers, new gloves with enhanced dexterity, and a new radio with more channels to allow up to five people to communicate with one another simultaneously.

Due to orbital motion induced periods of darkness and component caused shadowing, assembly work on the space station is frequently being performed at much colder temperatures than those encountered during most of the space shuttle mission EVAs. Unlike the shuttle, the ISS cannot be turned to provide an optimum amount of sunlight to moderate temperatures during an extravehicular activity, so a variety of other enhancements now make the shuttle spacesuit more compatible for use aboard the space station. Warmth enhancements include fingertip heaters and the ability to shut off the spacesuit's cooling system. To assist assembly work in shadowed environments, the spacesuit has new helmet-mounted floodlights and spotlights. There is also a jet-pack "life jacket" (called SAFER) to allow an accidentally untethered astronaut to fly back to the station in an emergency. In 1994, as part of the STS-64 mission, astronaut Mark Lee performed an EVA during which he tested a new mobility system called the Simplified Aid for EVA Rescue (SAFER). This system is similar to, but smaller and simpler than, the MMU.

Before the arrival of the joint airlock module (called Quest), as part of the STS-104 mission to the ISS (in July 2001), space walks conducted from the space station could use only Russian spacesuits unless the space shuttle was present. The facilities of the Zvezda service module limited space station–based EVAs to only those with Russian Orlan spacesuits. The Quest module, now attached to the ISS, gives the station's occupants the capability to conduct EVAs with either Russian- or American-designed spacesuits. Prebreathing protocols and spacesuit design differences no longer limit EVA activities by astronauts and cosmonauts on the space station.

space system Generally, a system consisting of launch vehicle(s), spacecraft, and ground support equipment (GSE).

Spacetrack A global system of radar, optical, and radiometric sensors linked to a computation and analysis center in the North American Aerospace Defense Command (NORAD) combat operations center. The Spacetrack mission is detection, tracking, and cataloging of all human-made objects in orbit around Earth. It is the U.S. Air Force portion of the North American Aerospace Defense Command Space Detection and Tracking system (SPADATS). *See also* NORTH AMERICAN AEROSPACE DEFENSE COMMAND.

Space Transportation System (STS) NASA's name for the overall space shuttle program, including intergovernmental agency requirements and international and joint projects. The major components of the space shuttle system are the winged orbiter vehicle (often referred to as the shuttle or the space shuttle); the three space shuttle main engines (SSMEs); the giant external tank (ET), which feeds liquid hydrogen fuel and liquid oxygen (oxidizer) to the shuttle's three main engines; and the two solid rocket boosters (SRBs).

The orbiter is the only part of the space shuttle system that has a name in addition to a part number. The first orbiter built was the *Enterprise* (OV-101), which was designed for flight tests in the atmosphere rather than operations in space. It is now at the Smithsonian Museum at Dulles Airport outside Washington, D.C. Five operational orbiters were constructed (listed in order of completion): *Columbia* (OV-102),† *Challenger* (OV-99),* *Discovery* (OV-103), *Atlantis* (OV-104), and *Endeavour* (OV-105).

Shuttles are launched from either Pad 39A or 39B at the Kennedy Space Center, Florida. Depending on the requirements of a particular mission, a space shuttle can carry about 22,680 kilograms (kg) (50,000 pounds-mass (lbm)) of payload into LOW EARTH ORBIT (LEO). An assembled shuttle vehicle has a mass of about 2.04 million kg (4.5 million lbm) at liftoff. (See Figure 72.)

The two solid rocket boosters (SRBs) are each 45.4 meters (m) (149 feet [ft]) high and 3.7 m (12.1 ft) in diameter. Each has a mass of about 590,000 kg (1.3 million lbm). Their solid propellant consists of a mixture of powdered aluminum (fuel), ammonium perchlorate (oxidizer), and a trace of iron oxide to control the burning rate. The solid mixture is held together with a polymer binder. Each booster produces a thrust of approximately 13.8 million newtons (3.1 million pounds-force [lbf]) for the first few seconds after ignition. The thrust then gradually declines for the remainder of the two-

minute burn to avoid overstressing the flight vehicle. Together with the three main liquid-propellant engines on the orbiter, the shuttle vehicle produces a total thrust of 32.5 million newtons (7.3 million lbf) at liftoff.

Typically, the SRBs burn until the shuttle flight vehicle reaches an altitude of about 45 kilometers (km) (28 miles [mi]) and a speed of 4,970 km (3,094 mi) per hour. Then they separate and fall back into the Atlantic Ocean to be retrieved, refurbished, and prepared for another flight. After the solid rocket boosters are jettisoned, the orbiter's three main engines, fed by the huge external tank, continue to burn and provide thrust for another six minutes before they too are shut down at MECO (main engine cutoff). At this point the external tank is jettisoned and falls back to Earth, disintegrating in the atmosphere with any surviving pieces falling into remote ocean waters.

The huge external tank is 47 m (154.2 ft) long and 8.4 m (27.6 ft) in diameter. At launch, it has a total mass of about 760,250 kg (1.676 million lbm). The two inner propellant tanks contain a maximum of 1,458,400 liters (385,300 gallons) of liquid hydrogen (LH_2) and 542,650 liters (143,350 gallons) of liquid oxygen (LO_2). The external tank is the only major shuttle flight vehicle component that is expended on each launch.

The winged orbiter vehicle is both the heart and the brains of America's Space Transportation System. About the same size and mass as a commercial DC-9 aircraft, the orbiter contains the pressurized crew compartment (which can normally carry up to eight crew members), the huge cargo bay (which is 18.3 m long and 4.57 m in diameter [60 × 15 ft]), and the three main engines mounted on its aft end. The orbiter vehicle itself is 37 m (121.4 ft) long, 17 m (56 ft high), and has a wingspan of 24 m (78.7 ft). Since each of the operational vehicles varies slightly in construction, an orbiter generally has an empty mass of between 76,000 to 79,000 kg (~168,000 to 174,000 lbm).

Each of the three main engines on an orbiter vehicle is capable of producing a

† The *Columbia* and its crew were lost in a reentry accident on February 1, 2003.

* The *Challenger* and its crew were lost in a launch accident on January 28, 1986.

Figure 72. The space shuttle *Discovery* lifts off from Pad 39-B at the Kennedy Space Center to start off the STS-26 mission (September 29, 1988). *(Courtesy of NASA)*

thrust of 1.668 million newtons (375,000 lbf) at sea level and 2.09 million newtons (470,000 lbf) in the vacuum of space. These engines burn for approximately eight minutes during launch ascent and together consume about 242,250 liters (64,000 gallons) of cryogenic propellants each minute, when all three operate at full power.

An orbiter vehicle also has two smaller orbital maneuvering system (OMS) engines that operate only in space. These engines burn nitrogen tetroxide as the oxidizer and monomethyl hydrazine as the fuel. These propellants are supplied from onboard tanks carried in the two pods at the upper rear portion of the vehicle. The OMS engines are used for major maneuvers in orbit and to slow the orbiter vehicle for

reentry at the end of its mission in space. On most missions the orbiter enters an elliptical orbit, then coasts around Earth to the opposite side. The OMS engines then fire just long enough to stabilize and circularize the orbit. On some missions the OMS engines also are fired soon after the external tank separates, to place the orbiter vehicle at a desired altitude for the second OMS burn that then circularizes the orbit. Later OMS engine burns can raise or adjust the orbit to satisfy the needs of a particular mission. A shuttle flight can last from a few days to more than a week or two.

After deploying the payload spacecraft (some of which can have attached upper stages to take them to higher-altitude

NASA Space Shuttle Launches (1981–2003)

Year	Launches
1981	STS-1, STS-2
1982	STS-3, STS-4, STS-5
1983	STS-6, STS-7, STS-8, STS-9
1984	41-B, 41-C, 41-D, 41-G, 51-A
1985	51-C, 51-D, 51-B, 51-G, 51-F, 51-I, 51-J, 61-A, 61-B
1986	61-C, 51-L (*Challenger* accident)
1987	No Launches
1988	STS-26, STS-27
1989	STS-29, STS-30, STS-28, STS-34, STS-33
1990	STS-32, STS-36, STS-31, STS-41, STS-38, STS-35
1991	STS-37, STS-39, STS-40, STS-43, STS-48, STS-44
1992	STS-42, STS-45, STS-49, STS-50, STS-46, STS-47, STS-52, STS-53
1993	STS-54, STS-56, STS-55, STS-57, STS-51, STS-58, STS-61
1994	STS-60, STS-62, STS-59, STS-65, STS-64, STS-68, STS-66
1995	STS-63, STS-67, STS-71, STS-70, STS-69, STS-73, STS-74
1996	STS-72, STS-75, STS-76, STS-77, STS-78, STS-79, STS-80
1997	STS-81, STS-82, STS-83, STS-84, STS-94, STS-85, STS-86, STS-87
1998	STS-89, STS-90, STS-91, STS-95, STS-88
1999	STS-96, STS-93, STS103
2000	STS-99, STS-101, STS-106, STS-92, STS-97
2001	STS-98, STS-102, STS-100, STS-104, STS-105, STS-108
2002	STS-109, STS-110, STS-111, STS-112, STS-113
2003	STS-107 (*Columbia* accident)

Source: NASA (as of March 31, 2003).

operational orbits, such as a geostationary orbit), operating the onboard scientific instrument (e.g., Spacelab), making scientific observations of Earth or the heavens, or performing other aerospace activities, the orbiter vehicle reenters Earth's atmosphere and lands. This landing usually occurs at either the Kennedy Space Center in Florida or at Edwards Air Force Base in California (depending on weather conditions at the landing site). Unlike prior manned spacecraft, which followed a ballistic trajectory, the orbiter (now operating like an unpowered glider) has a cross-range capability (i.e., it can move to the right or left off the straight line of its reentry path) of about 2,000 km (1,250 mi). The landing speed is between 340 and 365 km (211 to 226 mi) per hour. After touchdown and rollout, the orbiter vehicle immediately is "safed" by a ground crew with special equipment. This safing operation is also the first step in preparing the orbiter for its next mission in space.

The orbiter's crew cabin has three levels. The uppermost is the flight deck, where the commander and pilot control the mission. The middeck is where the galley, toilet, sleep stations, and storage and experiment lockers are found. Also located in the middeck are the side hatch for passage to and from the orbiter vehicle before launch and after landing, the airlock hatch into the cargo bay and to outer space to support on-orbit extravehicular activities (EVAs). Below the middeck floor is a utility area for air and water tanks.

The orbiter's large cargo bay is adaptable to numerous tasks. It can carry satellites, large space platforms such as the Long-Duration Exposure Facility (LDEF), and even an entire scientific laboratory, such as the European Space Agency's Spacelab to and from LOW EARTH ORBIT. It also serves as a workstation for astronauts to repair satellites, a foundation from which to erect space structures, and a place to store and hold spacecraft that have been retrieved from orbit for return to Earth.

Mounted on the port (left) side of the orbiter's cargo bay behind the crew quarters is the remote manipulator system (RMS), which was developed and funded by the Canadian government. The RMS is a robot arm and hand with three joints similar to those found in a human being's shoulder, elbow, and wrist. There are two television cameras mounted on the RMS near the "elbow" and "wrist." These cameras provide visual information for the astronauts who are operating the RMS from the aft station on the orbiter's flight deck. The RMS is about 15 m (49 ft) in length and can move anything, from astronauts to satellites, to and from the cargo bay as well as to different points in nearby outer space.

The table provides a brief summary of all the space shuttle flights from 1981 up to the fatal COLUMBIA ACCIDENT on February 1, 2003. On that day, about 15 minutes before the anticipated end of a highly successful MICROGRAVITY research mission, the Columbia disintegrated over the western United States as it glided at high speed toward the KENNEDY SPACE CENTER landing site. The cause of the fatal accident remains under investigation. However, preliminary investigations suggest that some type of very intense heating took place inside Columbia's left wing. The accident claimed the first operational Orbiter vehicle and its seven crewmembers. Despite the tragic loss of both the Challenger in 1986 (during the STS 51-L mission) and the Columbia in 2003 (during the STS-107 mission), the space shuttle has had more than 100 successful missions and played an important role in the American space program. The remaining Orbiter vehicles—namely Discovery, Atlantis, and Endeavour—will continue to serve as the foundation of NASA's human spaceflight program and to represent the space transportation key to the successful completion and operation of the INTERNATIONAL SPACE STATION in the first decade of the 21st century.

space tribology The branch of aerospace engineering that deals with friction, lubrication, and the behavior of lubricants under the harsh environmental conditions encountered during space flight. Space tribology is actually a combination of scientific disciplines: engineering, physics, chemistry, and metallurgy, all focused and concerned with the contact of surfaces in the vacuum of outer space. The space environment often imposes special demands on the moving parts of a spacecraft. Factors such as high vacuum (i.e., very low pressure), radiation, thermal stress, and microgravity all must be considered in developing suitable approaches to reducing friction between moving components of a spacecraft. For example, new liquid lubricants, such as perfluorinated fluids, have been developed with extremely low vapor pressure and a high viscosity index. Physical vapor deposition techniques also have been used to produce thin, highly adherent films of solid lubricant on metal components. Future space missions with extended lifetimes (e.g., five to ten years) and intermittent periods of operation (as might occur in a multiple asteroid encounter mission) will place new demands on the field of space tribology. Lubricants possessing extended lifetimes and increased reliability must be developed to support these new space mission requirements. See also LUBRICATION (IN SPACE).

space tug A proposed (but as yet undeveloped) space-based reusable upper-stage vehicle that is capable of delivering, retrieving, and servicing payloads in orbits and trajectories beyond the planned operating altitude of the International Space Station. The space tug would be "hangared" in space, refurbished on orbit, and reused many times.

space vehicle In general, a vehicle or craft capable of traveling through outer space. This vehicle can be occupied by a human crew or uncrewed; capable of returning to Earth or sent on a one-way (expendable) mission. A space vehicle capable of operating both in space and in Earth's atmosphere is called an *aerospace vehicle* (as, for example, the space shuttle orbiter vehicle).

space walk Aerospace jargon used to describe an extravehicular activity (EVA). *See also* EXTRAVEHICULAR ACTIVITY.

space warfare 1. The conduct of offensive and/or defensive operations in outer space, usually resulting in the destruction or negation of an adversary's military space systems; for example, the use of antisatellite (ASAT) systems against capital military space assets. 2. The use of military space systems (e.g., surveillance satellites and secure communications satellites) to support friendly forces during hostile actions in Earth's atmosphere, on land or at sea. 3. The use of military space systems to support ballistic missile defense operations. *See also* ANTISATELLITE (ASAT) WEAPON; BALLISTIC MISSILE DEFENSE; SPACE CONTROL OPERATIONS.

space weather Variability of the near-Earth and interplanetary space environment; for example, the development of a solar flare and its potential impact on orbiting space systems.

spalling Flaking off of particles and chunks from the surface of a material as a result of localized stresses.

spare An individual part, subassembly, or assembly supplied for the maintenance or repair of systems or equipment.

Sparrow (AIM-7) The Sparrow is a radar-guided, air-to-air missile with a high-explosive warhead. The versatile missile has all-weather, all-altitude operational capability and can attack high-performance aircraft and missiles from any direction. It is a widely deployed missile used by the forces of the United States and the North Atlantic Treaty Organization (NATO).

The missile has five major sections: radome, radar guidance, warhead, flight control (autopilot plus hydraulic control system), and solid-propellant rocket motor. It has a cylindrical body with four wings at midbody and four tail fins. Although the external dimensions of the Sparrow have remained relatively unchanged from model to model, the internal components of newer missiles represent major improvements with vastly increased capabilities.

The AIM-7F joined air force inventory in 1976 as the primary medium-range, air-to-air missile for the F-15 Eagle. The AIM-7M, the only current operational version, entered service in 1982. It has improved reliability and performance over earlier models at low altitudes and in electronic counter-measures environments. It also has a significantly more lethal warhead. Today the F-4 Phantom, F-15 Eagle, and F-16 Fighting Falcon fighter aircraft carry the AIM-7M Sparrow. U.S. and NATO navies operate a surface-to-air version of this missile called the RIM-7F/M, Sea Sparrow. *See also* SEA SPARROW.

Spartan 1. A nuclear-armed, long-range (mid-course-intercept), surface-to-air guided missile that had been deployed by the United States as part of the former Safeguard ballistic missile defense system in 1975. It was designed to intercept strategic ballistic missile reentry vehicles in space. *See also* BALLISTIC MISSILE DEFENSE; SAFEGUARD. 2. A free-flying, unmanned space platform developed by NASA to support various scientific studies. A Spartan space platform is launched aboard the space shuttle and deployed from the orbiter vehicle, where it then performs a preprogrammed mission on orbit. Scientific data are collected during each mission, using a tape recorder and (in many cases) film cameras. After the Spartan platform is deployed from the orbiter, there is no command and control capability. At the end of its mission, the platform is retrieved by the orbiter, stowed in the cargo bay, and returned to Earth for data recovery. The platform then is refurbished and made available for a future mission. Power during the deployed phase of the mission is provided by onboard batteries, while platform attitude control is accomplished with pneumatic gas jets. *See also* SPACE TRANS-PORTATION SYSTEM.

special relativity Theory introduced in 1905 by the German-Swiss-American physicist Albert Einstein (1879–1955).

specific heat capacity (symbol: c) In thermodynamics, a measure of the heat capacity of a substance per unit mass. It is the quantity of thermal energy (heat) required to raise the temperature of a unit mass of a substance by one degree. In the SI UNIT system, specific heat capacity has the units: joules per kilogram-kelvin (J kg^{-1} K^{-1}).

For a simple compressible substance, the specific heat at constant volume (c_v) can be defined as: $c_V = (\partial u/\partial T)_v$, where u is internal energy and T is temperature; while the specific heat at constant pressure (c_p) can be defined as: $c_p = (\partial h/\partial T)_p$, where h is enthalpy.

specific impulse (symbol: I_{sp}) An important performance index for rocket propellants. It is defined as the thrust (or thrust force) produced by propellant combustion divided by the propellant mass flow rate. Expressed as an equation, the specific impulse is:

I_{sp} = thrust / mass flow rate = F_{thrust} / $\dot{m}$

where F_{thrust} is the thrust and m is the mass flow rate of propellant. It is also important to understand the units associated with specific impulse. In the SI UNIT system, thrust is expressed in newtons and mass flow rate in kilograms per second. Since one newton equals one kilogram-meter per second-squared (i.e., 1 N = 1 kg-m/s^2), the specific impulse in SI units becomes:

I_{sp} = newtons / (kg/sec) = meter/second (m/s). [SI units]

In the English (or traditional engineering) system of units, thrust is expressed in pounds-force (lbf), while mass flow rate of propellant in pounds-mass per second (lbm/s). Since (by definition), one pound-force is equal to one pound-mass at sea level on the surface of Earth, aerospace engineers often use the following simplifi-cation, which is, strictly speaking, valid only at sea level on Earth:

I_{sp} = lbf/(lbm/s) = seconds (s) (traditional engineering units)

See also PROPELLANT; ROCKET; ROCKET ENGINE.

specific volume (symbol: v) Volume per unit mass of a substance. The recipro-cal of density.

spectral classification The system in which STARS are given a designation. It consists of a letter and a number accord-ing to their SPECTRAL LINES, which corre-spond roughly to surface TEMPERATURE. Astronomers classify stars as O (hottest), B, A, F, G, K, and M (coolest). The numbers represent subdivisions within each major class. The SUN is a G2 star. M stars are numerous but very dim, while O and B stars are very bright but rare. Sometimes referred to as the *Harvard classification,* because astronomers at the Harvard Observatory introduced the system there in the 1890s.

spectral line A bright (or dark) line found in the spectrum of some radiant source. Bright lines indicate emission; dark lines indicate absorption.

spectrogram The photographic IMAGE of a SPECTRUM.

spectrometer An optical INSTRUMENT that splits (disperses) incoming visible LIGHT (or other ELECTROMAGNETIC RADIA-TION) from a celestial object into a SPEC-TRUM by DIFFRACTION and then measures the relative AMPLITUDES of the different WAVELENGTHS. In EMISSION SPECTROSCOPY, scientists use these data to infer the material composition and other properties of the objects that emitted the light. Using *absorp-tion spectroscopy,* scientists can infer the composition of the intervening medium that absorbed specific wavelengths of light as the radiation passed through the medium. A spectrometer is a very useful REMOTE SENS-ING tool for scientists who wish to study planetary ATMOSPHERES. Scientific SPACE-CRAFT often carry INFRARED RADIATION or ULTRAVIOLET RADIATION spectrometers.

spectroscopy The study of SPECTRAL LINES from different ATOMS and MOLECULES. Astronomers use EMISSION SPECTROSCOPY to infer the material composition of the objects that emitted the LIGHT and absorption spectroscopy to infer the composition of the intervening medium.

spectrum 1. In physics, any series of energies arranged according to frequency (or wavelength). 2. The electromagnetic spectrum; the sequence of electromagnetic waves from gamma rays (high frequency) to radio waves (low frequency). 3. The series of images produced when a beam of radiant energy is dispersed.

specular reflection Reflection in which the reflected radiation is not diffused; reflection as if from a mirror. The angle between the normal (perpendicular) to the surface and the incident beam is equal to the angle between the normal to the surface and the reflected beam. Any surface irregularities on a specular reflector must be small compared to the wavelength of the incident radiation.

speed of light (symbol: c) The speed of propagation of electromagnetic radiation (including light) through a perfect vacuum; a universal constant equal to 299,792.458 kilometers per second.

speed of sound The speed at which sound travels in a given medium under specified conditions. The speed of sound at sea level in the International Standard Atmosphere is 1,215 kilometers per hour (1,108 feet per second). Sometimes called the *acoustic velocity. See also* ATMOSPHERE.

spherical coordinates A system of coordinates defining a point on a sphere by its angular distances from a primary great circle and from a reference secondary great circle, as latitude and longitude.

spin-off A general term used to describe benefits that have derived from the (secondary) application of aerospace technologies to terrestrial problems and needs. For example, a special coating material designed to protect a reentry vehicle is also applied to fire-fighting equipment to make it more heat resistant. The manufacturer of the fire-fighting equipment improved the terrestrial product by "spinning-off" aerospace technology.

spin rocket A small rocket that imparts spin to a large rocket vehicle or spacecraft.

spin stabilization Directional stability of a missile or spacecraft obtained by the action of gyroscopic forces that result from spinning the body about its axis of symmetry.

spiral galaxy A GALAXY with spiral arms, similar to the MILKY WAY GALAXY or the ANDROMEDA GALAXY.

splashdown That portion of a human space mission in which the space capsule (reentry craft) containing the crew lands in the ocean—quite literally, "splashing down." The astronauts then are recovered by a team of helicopters, aircraft, and/or surface ships. This term was used during NASA's Mercury, Gemini, Apollo, Skylab and Apollo-Soyuz projects (1961–1975). In the space shuttle era, the orbiter vehicle returns to Earth by landing much like an aircraft. Consequently, the orbiter vehicle and its crew are said to "touch down."

The term "splashdown" also can be applied to an uncrewed, recoverable space capsule or spacecraft that has been deorbited intentionally and lands in the ocean. Once again, if the space capsule lands on the surface of Earth (under controlled circumstances), the event is described as touching down or making a soft landing. An uncontrolled or destructive landing on Earth's surface is called an impact, hard landing, or crash landing, depending on the circumstances. For example, a piece of space debris is said to "impact"; a space capsule with an undeployed or ripped parachute will either "hard land" or "crash land."

SPOT A family of EARTH OBSERVING SATELLITES developed by the French Space Agency (CNES). The spacecraft's name is an acronym for *Systeme Proba-toire d'Observation de la Terre*. *SPOT 1* was launched by an ARIANE ROCKET from Kourou, French Guiana, on February 22, 1986, and successfully placed into an 825-kilometer ALTITUDE, circular POLAR ORBIT. The *SPOT 2* SPACECRAFT was launched by an Ariane 4 rocket from Kourou on January 22, 1990, and placed in an orbit exactly opposite to the *SPOT 1* spacecraft, providing more frequent (every 13 days) repeat visits to scenes on EARTH. More recent additions to this family of remote sensing spacecraft include *SPOT 3* (launched in September 1993), *SPOT 4* (launched in March 1998), and *SPOT 5* (launched on May 4, 2002). *SPOT 5* takes a variety of relatively high-RESOLU-TION (for example, 2.5-meter panchro-matic and 5-meter multispectral) images of Earth that are sold commercially to users around the globe through the com-pany SpotImage.

Sprint A high-acceleration, short-range, nuclear-armed, surface-to-air guided mis-sile that was deployed in 1975 as part of the former Safeguard ballistic missile defense system. The Sprint system was designed to intercept strategic ballistic reentry vehicles in the atmosphere during the terminal phase of their flight. *See also* BALLISTIC MISSILE DEFENSE; SAFEGUARD.

Sputnik 1 On October 4, 1957, the for-mer Soviet Union (now called the Russian Federation) launched *Sputnik 1*, the first human-made object to be placed in orbit around Earth. A 29-meter-tall Russian A-1 rocket boosted this artificial satellite into an approximate 230 kilometer by 950 kilometer orbit. The simple, 83.5 kilogram spacecraft was essentially a hollow sphere made of steel that contained batteries and a radio transmitter to which were attached four whip antennas. As it orbited Earth, *Sputnik 1* provided scientists with information on temperatures and electron densities in Earth's upper atmosphere. It reentered the atmosphere and burned up on January 4, 1958.

The name *Sputnik* means "fellow trav-eler." Launched during the cold war under the guidance of the Russian aerospace visionary Sergei Korolov (1907–1966), Sputnik represented a technological sur-prise that sent tremors through the United States and its allies. This launch is often regarded as the birth date of the modern space age. It also marks the beginning of a heated "space race" between the United States and the former Soviet Union—a space technology race that was to culmi-nate in 1969 with an American astronaut becoming the first human to walk on the surface of the Moon.

spy satellite Popular name for a recon-naissance satellite. *See also* RECONNAIS-SANCE SATELLITE.

squib A small pyrotechnic device that may be used to fire the igniter in a rocket.

stacking Assembling the coolant tubes of a liquid propellant rocket engine's thrust chamber vertically on a mandrel that simulates the chamber/nozzle con-tour. This procedure facilitates fitting and adjusting the tubes to the required con-tour prior to brazing.

stage 1. In rocketry, an element of the missile or launch vehicle system that usually separates from the vehicle at burnout or engine cutoff. In multistage rockets, the stages are numbered chronologically in the order of burning (i.e., first stage, second stage, etc.). 2. In mechanical engineering, a set of rotor blades and stator vanes in a tur-bine or in an axial-flow compressor. 3. In fluid mechanics and thermodynamics, a step or process through which a fluid passes, especially in compression or expansion.

stage-and-a-half A liquid rocket propulsion unit of which only part falls away from the rocket vehicle during flight, as in the case of booster rockets falling away to leave the sustainer engine to con-sume the remaining fuel.

staged combustion Rocket engine cycle in which propellants are partially burned in a preburner prior to being burned in the main combustion chamber.

staging 1. In rocketry, the process during the flight of a multistage launch vehicle or missile during which a spent stage separates from the remaining vehicle and is free to decelerate or follow its own trajectory 2. In mechanical engineering, the use of two or more stages in a turbine or pump.

stagnation In thermodynamics, a condition in which a flowing fluid is brought to rest isentropically (i.e., with no change in ENTROPY).

stagnation point Point in a flow field about a body immersed in the flowing fluid at which the fluid particles have zero velocity with respect to the body.

stagnation pressure 1. Pressure that a flowing fluid would attain if it is brought to rest isentropically (i.e., with no change in ENTROPY). 2. Pressure of a flowing fluid at a point of zero fluid velocity on a body around which the fluid flows.

stagnation region The region in the vicinity of a stagnation point in a flow field about an immersed body where the fluid velocity is negligible.

stagnation temperature In thermodynamics, the temperature that a flowing fluid would attain if the fluid were brought to rest isentropically (i.e., with no change in ENTROPY).

Standard Missile (SM) A U.S. Navy surface-to-air and surface-to-surface missile that is mounted on surface ships. The Standard Missile is produced in two major types: the SM-1 MR/SM-2 (medium range) and the SM-2 (extended range). It is one of the most reliable in the navy's inventory and is used against enemy missiles, aircraft, and ships. The SM joined the fleet more than a decade ago. It replaced the Terrier and Tartar missiles and now is part of the weapons inventory of more than 100 navy ships. The SM-2 (MR) is a medium-range defense weapon for Ticonderoga-class AEGIS cruisers, Arleigh Burke–class AEGIS destroyers, and California- and Virginia-class nuclear cruisers. Oliver Hazard Perry–class frigates use the SM-1 MR.

The SM-1/SM-2 Medium Range (MR) missile has a dual-thrust solid-propellant rocket, a semiactive radar homing guidance system, and a proximity fuse, high-explosive warhead. The SM-2 Extended Range (ER) missile has a two-stage, solid-fuel rocket, an inertial/semiactive radar homing guidance system, and a proximity fuse, high-explosive warhead. The SM-1 MR missile has a range of between 27 and 37 kilometers (km) (17–23 miles [mi]); the SM-2 MR missile, between 74 and 167 km (46–104 miles); and the SM-2 ER missile, between 120 and 185 km (75–115 miles.)

star A star is essentially a self-luminous ball of very hot gas that generates energy through thermonuclear fusion reactions that take place in its core.

Stars may be classified as either "normal" or "abnormal." Normal stars, such as the Sun, shine steadily. These stars exhibit a variety of colors: red, orange, yellow, blue, and white. Most stars are smaller than the Sun, and many stars even resemble it. However, a few stars are much larger than the Sun. In addition, astronomers have observed several types of abnormal stars including giants, dwarfs, and a variety of variable stars.

Most stars can be put into one of several general spectral types called O, B, A, F, G, K, and M. (See the table.) This classification is a sequence established in order of decreasing surface temperature.

Our parent star, the Sun, is approximately 1.4 million kilometers (865,000 miles) in diameter and has an effective surface temperature of about 5,800° Kelvin. The Sun, like other stars, is a giant nuclear furnace, in which the temperature, pressure, and density are sufficient to cause light nuclei to join together, or "fuse." For example, deep inside the solar interior, hydrogen, which makes up 90% of the Sun's mass, is fused into helium atoms, releasing large amounts of energy that eventually works its way to the surface and then is radiated throughout the solar system. The Sun is currently in a state of balance, or equilibrium, between two competing forces: gravity (which wants to pull all its mass inward) and the radiation pressure and hot gas pressure resulting from the thermonuclear reactions (which push outward).

Many stars in the galaxy appear to have companions, with which they are gravitationally bound in binary, triple, or even larger systems. Compared to other stars throughout the galaxy, the Sun is slightly unusual. It does not have a known stellar companion. However, the existence of a very distant, massive, dark companion called Nemesis has been postulated by some astrophysicists in an attempt to explain an apparent "cosmic catastrophe cycle" that occurred on Earth about 65 million years ago.

Astrophysicists have discovered what appears to be the life cycle of stars. Stars originate by the condensation of enormous clouds of cosmic dust and hydrogen gas, called nebulae. Gravity is the dominant force behind the birth of a star. According to Newton's Universal Law of Gravitation, all bodies attract each other in proportion to their masses and distance apart. The dust and gas particles found in these huge interstellar clouds attract each other and gradually draw closer together. Eventually enough of these particles join together to form a central clump that is sufficiently massive to bind all the other parts of the cloud by gravitation. At this point, the edges of the cloud start to collapse inward, separating it from the remaining dust and gas in the region.

Initially the cloud contracts rapidly, because the thermal energy release related to contraction is radiated outward easily. However, when the cloud grows smaller and more dense, the heat released at the center cannot escape to the outer surface immediately. This causes a rapid rise in internal temperature, slowing down but not stopping the relentless gravitational contraction.

The actual birth of a star occurs when its interior becomes so dense and its temperature so high that thermonuclear fusion occurs. The heat released in thermonuclear fusion reactions is greater than that released through gravitational contraction, and fusion becomes the star's primary energy-producing mechanism.

Gases heated by nuclear fusion at the cloud's center begin to rise, counterbalancing the inward pull of gravity on the outer layers. The star stops collapsing and reaches a state of equilibrium between outward and inward forces. At this point, the star has become what astronomers and astrophysicists call a "main sequence star." Like the Sun, it will remain in this state of equilibrium for billions of years, until all the hydrogen fuel in its core has been converted into helium.

How long a star remains on the main sequence, burning hydrogen for its fuel, depends mostly on its mass. The Sun has an estimated main sequence lifetime of about 10 billion years, of which approximately 5 billion years have now passed. Larger stars burn their fuels faster and at much higher temperatures. These stars, therefore, have short main sequence lifetimes, sometimes as little as 1 million years. In comparison, the "red dwarf stars," which typically have less than one-tenth the mass of the Sun, burn up so slowly that trillions of years must elapse before their hydrogen supply is exhausted. When a star has used up its hydrogen fuel, it leaves the "normal" state or departs the main sequence. This happens when the core of the star has been converted from hydrogen to helium by thermonuclear reactions that have taken place.

When the hydrogen fuel in the core of a main sequence star has been consumed, the core starts to collapse. At the same time, the hydrogen fusion process moves outward from the core into the surrounding outer regions. There the process of converting hydrogen into helium continues, releasing radiant energy. But as this burning process moves into the outer regions, the star's atmosphere expands greatly, and it becomes a "red giant." The term "giant" is quite appropriate. If we put a red giant where the Sun is now, the innermost planet, Mercury, would be engulfed by it; similarly, if we put a larger "red supergiant" there, this supergiant would extend out past the orbit of Mars.

As the star's nuclear evolution continues, it might become a "variable star," pulsating in size and brightness over periods of several months to years. The visual brightness of such an "abnormal" star might now change by a factor of 100, while its total energy output varies by only a factor of 2 or 3.

As an abnormal star grows, its contracting core may become so hot that it ignites and burns nuclear fuels other than hydrogen, beginning with the helium created in millions to perhaps billions of years of main sequence burning. The subsequent behavior of such a star is complex, but in general it can be characterized as a continuing series of gravitational contractions and new nuclear reaction ignitions. Each new series of fusion reactions produces a succession of heavier elements in addition to releasing large quantities of energy. For example, the burning of helium produces carbon, the burning of carbon produces oxygen, and so forth.

Finally, when nuclear burning no longer releases enough radiant energy to support the giant star, it collapses and its dense central core becomes either a compact white dwarf or a tiny neutron star. This collapse also may trigger an explosion of the star's outer layers, which displays itself as a supernova. In exceptional cases with very massive stars, the core (or perhaps even the entire star) might become a black hole.

When a star like the Sun has burned all the nuclear fuels available, it collapses under its own gravity until the collective resistance of the electrons within it finally stops the contraction process. The "dead star" has become a "white dwarf" and may now be about the size of Earth. Its atoms are packed so tightly together that a fragment the size of a sugar cube would have a mass of thousands of kilograms. The white dwarf then cools for perhaps several billion years, going from white, to yellow, to red, and finally becomes a cold, dark sphere sometimes called a "black dwarf." (Note that the white dwarf is not experiencing nuclear burning; rather its light comes from a thin gaseous atmosphere that gradually dissipates its heat to space.) Astrophysicists estimate that there are over 10 billion white dwarf stars in the Milky Way Galaxy alone, many of which

have now become black dwarfs. This fate appears to be awaiting our own Sun and most other stars in the galaxy.

However, when a star with a mass of about 1.5 to 3 times the mass of the Sun undergoes collapse, it will contract even further and ends up as a "neutron star," with a diameter of perhaps only 20 kilometers. In neutron stars, intense gravitational forces drive electrons into atomic nuclei, forcing them to combine with protons and transforming this combination into neutrons. Atomic nuclei are, therefore, obliterated in this process, and only the collective resistance of neutrons to compression halts the collapse. At this point, the star's matter is so dense that each cubic centimeter has a mass of several billion tons.

For stars that end their life having more than a few solar masses, even the resistance of neutrons is not enough to stop the unyielding gravitational collapse. In death such massive stars may ultimately become black holes—incredibly dense point masses or singularities that are surrounded by a literal "black region" in which gravitational attraction is so strong that nothing, not even light itself, can escape.

Currently, many scientists relate the astronomical phenomena called supernovae and pulsars with neutron stars and their evolution. The final collapse of a giant star to the neutron stage may give rise to the physical conditions that cause its outer portions to explode, creating a supernova. This type of cosmic explosion releases so much energy that its debris products will temporarily outshine all the ordinary stars in the galaxy.

A regular "nova" (the Latin word for "new," the plural of which is novae) occurs more frequently and is far less violent and spectacular. One common class, called recurring novae, is due to the nuclear ignition of gas being drawn from a companion star to the surface of a white dwarf. Such binary star systems are quite common; sometimes the stars will have orbits that regularly bring them close enough for one to draw off gas from the other.

When a supernova occurs at the end of a massive star's life, the violent explosion

Stellar Spectral Classes			
Type	Description	Typical Surface Temperatures(K)	Remarks/Examples
O	very hot, large blue stars (hottest)	28,000–40,000	ultraviolet stars; very short lifetimes (3–6 million years)
B	large, hot blue stars	11,000–28,000	Rigel
A	blue-white, white stars	7,500–11,000	Vega, Sirius, Altair
F	white stars	6,000–7,500	Canopis, Polaris
G	yellow stars	5,000–6,000	the Sun
K	orange-red stars	3,500–5,000	Arcturus, Aldebaran
M	red stars (coolest)	<3,500	Antares, Betelgeuse

Source: NASA.

fills vast regions of space with matter that may radiate for hundreds or even thousands of years. The debris created by a supernova explosion eventually will cool into dust and gas, become part of a giant interstellar cloud, and perhaps once again be condensed into a star or planet. Most of the heavier elements found on Earth are thought to have originated in supernovae, because the normal thermonuclear fusion processes cannot produce such heavy elements. The violent power of a supernova explosion can, however, combine lighter elements into the heaviest elements found in nature (e.g., lead, thorium, and uranium). Consequently, both the Sun and its planets were most likely enriched by infusions of material hurled into the interstellar void by ancient supernova explosions.

Pulsars, first detected by radio astronomers in 1967, are sources of very accurately spaced bursts, or pulses, of radio signals. These radio signals are so regular, in fact, that the scientists who made the first detections were startled into thinking that they might have intercepted a radio signal from an intelligent alien civilization.

The pulsar, named because its radiowave signature regularly turns on and off, or pulses, is considered to be a rapidly spinning neutron star. One pulsar is located in the center of the Crab Nebula, where a giant cloud of gas is still glowing from a supernova explosion that occurred in the year 1054 C.E.—a spectacular celestial event observed and recorded by ancient Chinese astronomers. The discovery of this pulsar allowed scientists to understand both pulsars and supernovae.

In a supernova explosion, a massive star is literally destroyed in an instant, but the explosive debris lingers and briefly outshines everything in the galaxy. In addition to scattering material all over interstellar space, supernova explosions leave behind a dense collapsed core made of neutrons. This neutron star, with an immense magnetic field, spins many times a second, emitting beams of radio waves, X rays, and other radiations. These radiations may be focused by the pulsar's powerful magnetic field and sweep through space much like a revolving lighthouse beacon. The neutron star, the end product of a violent supernova explosion, becomes a pulsar.

Astrophysicists must develop new theories to explain how pulsars can create intense radio waves, visible light, X rays and gamma rays, all at the same time. Orbiting X-ray observatories have detected X-ray pulsars that are believed to be caused by a neutron star pulling gaseous matter from a normal companion star in a binary star system. As gas is sucked away from the normal companion to the surface of the neutron star, the gravitational attraction of the neutron star heats up the gas to millions of degrees Kelvin; this causes the gas to emit X rays.

The advent of the space age and the use of powerful orbiting observatories, such as the Hubble Space Telescope (HST), to view the universe as never before possible has greatly increased our knowledge about the many different types of stellar phenomena. Most exciting of all, perhaps, is the fact that this process of astrophysical discovery has really only just begun. *See also* ASTROPHYSICS; BLACK HOLE(S); FUSION; *HUBBLE SPACE TELESCOPE*; SUN.

star cluster   A group of STARS (numbering from a few to perhaps thousands) that were formed from a common GAS cloud and are now bound together by their mutual gravitational attraction.

***Stardust* mission**   The primary objective of NASA's Discovery class *Stardust* mission is to fly by the COMET P/Wild 2 and collect samples of dust and volatiles in the COMA of the comet. The spacecraft will then return these samples to EARTH for detailed study. The *Stardust* spacecraft was launched by a DELTA II ROCKET from CAPE CANAVERAL AIR FORCE STATION in early February 1999. Following LAUNCH, the spacecraft was placed in a HELIOCENTRIC, ELLIPTICAL ORBIT. By mid-summer 2003, it had completed its second ORBIT of the SUN. The ENCOUNTER with comet P/Wild 2 will take place in early January 2004. The spacecraft will fly past the comet at an approximate relative VELOCITY of 150 KILOMETERS per second and make a closest approach of about 150 kilometers. Comet material samples will be collected, stowed, and sealed in the special sample storage vault of the reentry CAPSULE carried onboard the *Stardust* spacecraft. Then, in mid-January 2006, the spacecraft will fly past Earth and eject the sample capsule. As the sample capsule descends through the atmosphere, it will be recovered via a mid-air snatch by U.S. Air Force airplanes somewhere over the Utah desert.

star grain   A hollow, solid-propellant rocket grain with the cross section of the hole having a multipointed star shape. *See also* ROCKET.

star probe   A specially designed and instrumented probe spacecraft that is capable of approaching within 1 million kilometers or so of the Sun's surface (photosphere). This close encounter with the Sun, the nearest star, would provide scientists with their first in-situ measurements of the physical conditions in the corona (the Sun's outer atmosphere). The challenging mission requires advanced space technologies (including propulsion, laser communications, thermal protection, etc.) and might be flown in the second or third decade of the 21st century. *See also* SUN.

starship   A space vehicle capable of traveling the great distances between star systems. Even the closest stars in the Milky Way Galaxy are often light-years apart. The term "starship" is generally used to describe an interstellar spaceship capable of carrying intelligent beings to other star systems; robot interstellar spaceships are often referred to as *interstellar probes*. *See also* INTERSTELLAR COMMUNICATION AND CONTACT; INTERSTELLAR TRAVEL.

starting pressure   In rocketry, the minimum chamber pressure required to establish shock-free flow in the exit plane of a supersonic nozzle.

star tracker   An instrument on a missile or spacecraft that locks onto a star (or pattern of "fixed" stars) and provides reference information to the missile or spacecraft's guidance system and/or attitude control system during the flight. *See also* GUIDANCE SYSTEM.

state of the art (SOA)   The level to which technology and science have been developed in any given discipline or industry at some designated cutoff time.

static   1. Involving no variation with time. 2. Involving no movement, as in *static test*. 3. Any radio interference heard as unwanted, random noises at the speaker (audio stage) of the radio receiver.

static conversion   Energy conversion in which no moving parts or equipment are involved. *See also* DIRECT CONVERSION.

static firing The firing of a rocket engine in a hold-down position. Usually done to measure thrust and to accomplish other performance tests.

static seal Device used to prevent leakage of fluid through a mechanical joint in which there is no relative motion of the mating surfaces other than that induced by changes in the operating environment.

static testing The testing of a rocket or other device in a stationary or hold-down position, either to verify structural design criteria, structural integrity, and the effects of limit loads or to measure the thrust of a rocket engine.

stationary orbit An orbit in which the satellite revolves about the primary at the same angular rate at which the primary rotates on its axis. To an observer on the surface of the primary, the satellite would appear to be stationary over a point on the primary. See also GEOSTATIONARY EARTH ORBIT; SYNCHRONOUS SATELLITE.

stationkeeping The sequence of maneuvers that maintains a space vehicle or spacecraft in a predetermined orbit.

stator 1. Part of a turbopump assembly that remains fixed or stationary relative to a rotating or moving part of the assembly. 2. In an alternating current (a.c.) generator, the part that is stationary.

statute mile A distance of 5,280 feet. One statute mile (often called simply "mile") is equal to 1.6093 kilometers. It is also equal to approximately 0.869 nautical mile.

steady flow A flow whose velocity vector components at any point in the fluid do not vary with time.

steady state Condition of a physical system in which parameters of importance (fluid velocity, temperature, pressure, etc.) do not vary significantly with time; in particular, the condition or state of rocket engine operation in which mass, momentum, volume, and pressure of the combustion products in the thrust chamber do not vary significantly with time.

steady-state universe A COSMOLOGY model (based on the PERFECT COSMOLOGICAL PRINCIPLE) suggesting that the UNIVERSE looks the same to all observers at all times.

Stefan-Boltzmann law One of the basic laws of radiation heat transfer. This law states that the amount of energy radiated per unit time from a unit surface area of an ideal BLACKBODY radiator is proportional to the fourth power of the absolute temperature of the blackbody. It can be expressed as:

$$E = \varepsilon \, \sigma \, T^4$$

where E is the emittance per unit area (watts per square meter per second [$Wm^{-2}s^{-1}$]) of the blackbody, ε is the emissivity ($\varepsilon = 1$ for a blackbody), σ is the Stefan-Boltzmann constant (5.67×10^{-8} (watts per square meter per degrees Kelvin to the fourth power [$Wm^{-2} K^{-4}$]), and T is the absolute temperature (Kelvin) of the blackbody.

stellar Of or pertaining to the stars.

stellar evolution The different phases in the lifetime of a STAR from its formation out of INTERSTELLAR GAS and dust to the time after its nuclear fuel is exhausted.

stellar guidance A system wherein a guided missile may follow a predetermined course with reference primarily to the relative position of the missile and certain preselected celestial bodies. See also GUIDANCE SYSTEM.

stellar spectrum (spectra) See SPECTRAL CLASSIFICATION.

stellar wind The flux of electromagnetic radiation and energetic particles that stream outward from a star. The stellar wind associated with the Sun is called the *solar wind. See also* SOLAR WIND.

step rocket A multistage rocket.

steradian (symbol: sr) The supplementary unit of solid angle in the international (SI) system. One steradian is equal to the solid angle subtended at the center of a sphere by an area of surface equal to the square of the radius. The surface of a sphere subtends a solid angle of 4π steradians about its center.

Stinger A shoulder-fired infrared missile that homes in on the heat (thermal energy) emitted by either jet-or propeller-driven fixed-wing aircraft or helicopters. The Stinger system employs a proportional navigation system that allows it to fly an intercept course to the target. Once the missile has traveled a safe distance from the gunner, its main engine ignites and propels it to the target.

Stirling cycle A thermodynamic cycle for a heat engine in which thermal energy (heat) is added at constant volume, followed by isothermal expansion with heat addition. The heat then is rejected at constant volume, followed by isothermal compression with heat rejection. If a regenerator is used so that the heat rejected during the constant volume process is recovered during heat addition at constant volume, then the thermodynamic efficiency of the Stirling cycle approaches (in the limit) the efficiency of the Carnot cycle. The basic principle of the Stirling engine was first proposed and patented in 1816 by Robert Stirling, a Scottish minister. *See also* CARNOT CYCLE; HEAT ENGINE; THERMODYNAMICS.

stochastic process A statistical process; a process in which there is a random variable.

stoichiometric combustion The burning of fuel and oxidizer in precisely the right proportions required for complete reaction with no excess of either reactant.

Stokes number A nondimensional number used to express the relationship between viscous force and gravity force in a fluid, as follows:

$$\text{Stokes number} = \frac{\text{viscous force}}{\text{gravity force}} = \frac{\mu V}{\rho\, g l^2}$$

where μ is absolute viscosity, V is velocity, ρ is mass density, g is the acceleration of gravity, and l is characteristic length.

storable propellant Rocket propellant (usually liquid) capable of being stored for prolonged periods of time without special temperature or pressure controls.

stowing In aerospace operations, the process of placing a payload in a retained, or attached, position in the space shuttle orbiter's cargo bay for ascent to or return from orbit.

strain In engineering, the change in the shape or volume of an object due to applied forces. There are three basic types of strain: longitudinal, volume, and shear. Longitudinal (or tensile) strain is the change in length per unit length, as occurs, for example, with the stretching of a wire. Volume (or bulk) strain involves a change in volume per unit volume, as occurs, for example, when an object is totally immersed in a liquid and experiences a hydrostatic pressure. Finally, shear strain is the angular deformation of an object without a change in its volume. Shear occurs, for example, when a rectangular block of metal is strained or distorted in such a way that two opposite faces become parallelograms, while the other two opposite do not change their shape. *See also* STRESS.

strategic defense In general, defense against longrange nuclear weapons.

stress In engineering, a force per unit area on an object that causes it to deform (i.e., experience strain). Stress can be viewed as either the system of external forces applied to deform an object or the system of internal "opposite" forces (a function of the material composition of the object) by which the object resists this deformation. The three basic types of stress are: compressive (or tensile) stress, hydrostatic pressure, and shear stress. *See also* STRAIN.

stretchout An action whereby the time for completing an aerospace activity, especially a contract, is extended beyond the time originally programmed or contracted for. Cost overruns, unanticipated technical delays, and budget cuts are frequent reasons why a stretchout is needed.

stringer A slender, lightweight, lengthwise fill-in structural member in a rocket body. A stringer reinforces and gives shape to the rocket vehicle's skin.

structured attack An attack in which the arrival of (nuclear) warheads on their diverse targets is precisely timed for maximum strategic impact. *See also* BALLISTIC MISSILE DEFENSE.

subassembly A portion of an assembly, consisting of two or more parts, that can be provisioned and replaced as an entity.

subcooled liquid In thermodynamics, a liquid whose temperature is lower than the saturation temperature for the existing pressure.

sublimation In thermodynamics, the direct transition of a material from the solid phase to the vapor phase, and vice versa, without passing through the liquid phase.

subliming ablator An ablative cooling material characterized by sublimation of the thermal protection material at the aerodynamically heated surface.

submarine-launched ballistic missile (SLBM) *See* FLEET BALLISTIC MISSILE.

submarine rocket (Subroc) Submerged, submarine-launched, surface-to-surface rocket with nuclear depth charge or homing torpedo payload, primarily used in antisubmarine applications. Also designated by the U.S. Navy as UUM-44A.

subsatellite point Intersection with Earth's (or planet's) surface of the local vertical passing through an orbiting satellite.

subsonic Of or pertaining to speeds less than the speed of sound. *See also* SPEED OF SOUND.

Sun The Sun is our parent star and the massive, luminous celestial object about which all other bodies in the solar system revolve. It provides the light and warmth upon which almost all terrestrial life depends. Its gravitational field determines the movement of the planets and other celestial bodies (e.g., comets). The Sun is a main sequence star of spectral type G-2. Like all main sequence stars, the Sun derives its abundant energy output from thermonuclear fusion reactions involving the conversion of hydrogen to helium and heavier nuclei. Photons associated with these exothermic (energy-releasing) fusion reactions diffuse outward from the Sun's core, until they reach the convective envelope. Another by-product of the thermonuclear fusion reactions is a flux of neutrinos that freely escape from the Sun.

At the center of the Sun is the core, where energy is released in thermonuclear reactions. Surrounding the core are concentric shells called the radiative zone, the convective envelope (which occurs at approximately 0.8 of the Sun's radius), the photosphere (the layer from which visible radiation emerges), the chromosphere, and, finally, the corona (the Sun's outer atmosphere). Energy is transported outward through the convective envelope by convective (mixing) motions that are organized into cells. The Sun's lower or inner atmosphere, the photosphere, is the region from which energy is radiated directly into space. Solar radiation approximates a Planck distribution (BLACKBODY source) with an effective temperature of 5,800 kelvins (K). The table provides a summary of the physical properties of the Sun.

The chromosphere, which extends for a few thousand kilometers above the photosphere, has a maximum temperature of approximately 10,000K. The corona, which extends several solar radii above the chromosphere, has temperatures of over 1 million K. These regions emit electromagnetic (EM) radiation in the ultraviolet (UV), extreme ultraviolet (EUV), and X-ray portions of the spectrum. This shorter-wavelength EM radiation although representing a relatively small portion of the Sun's total energy output, still plays a dominant role in forming planetary ionospheres and in photochemistry reactions occurring in planetary atmospheres.

Since the Sun's outer atmosphere is heated, it expands into the surrounding interplanetary medium. This continuous outflow of plasma is called the *solar wind*. It consists of protons, electrons, and alpha particles as well as small quantities of heavier ions. Typical particle velocities in the solar wind fall between 300 and 400 kilometers per second, but these velocities may get as high as 1,000 kilometers per second.

Although the total energy output of the Sun is remarkably steady, its surface displays many types of irregularities. These include sunspots, faculae, plages (bright areas), filaments, prominences, and flares. All are believed ultimately to be the result of interactions between ionized gases in the solar atmosphere and the Sun's magnetic field. Most solar activity follows the *sunspot cycle*. The number of sunspots varies, with a period of about 11 years. However, this approximately 11-year sunspot cycle is only one aspect of a more general 22-year *solar cycle* that corresponds to a reversal of the polarity patterns of the Sun's magnetic field.

Sunspots were originally observed by Galileo in 1610. They are less bright than the adjacent portions of the Sun's surface, because they are not as hot. A typical sunspot temperature might be 4,500K compared to the photosphere's temperature of 5,800K. Sunspots appear to be made up of gases boiling up from the Sun's interior. A small sunspot may be about the size of Earth, while larger ones could hold several hundred or even thousands of Earth-size planets. Extra-bright solar regions, called *plages,* often overlie sunspots. The number and size of sunspots appear to rise and fall through a fundamental 11-year cycle (or in an overall 22-year cycle, if polarity reversals in the Sun's magnetic field are considered). The greatest number occur in years when the Sun's magnetic field is the most severely twisted (called sunspot maximum). Solar physicists think that sunspot migration causes the Sun's magnetic field to reverse its direction. It then takes another 22 years for the Sun's magnetic field to return to its original configuration.

A *solar flare* is the sudden release of tremendous energy and material from the Sun. A flare may last minutes or hours, and it usually occurs in complex magnetic regions near sunspots. Exactly how or why enormous amounts of energy are liberated in solar flares is still unknown, but scientists think the process is associated

Physical and Dynamic Properties of the Sun	
Diameter	1.39×10^6 km
Mass	1.99×10^{30} kg
Distance from the Earth (average)	1.496×10^8 km [1 AU] (8.3 light min)
Luminosity	3.9×10^{26} Watts
Density (average)	1.41 g/cm^3
Equivalent blackbody temperature	5,800 kelvins (K)
Central temperature (approx.)	15,000,000 kelvins (K)
Rotation period (varies with latitude zones)	27 days (approx.)
Radiant energy output per unit surface area	6.4×10^7 W/m^2
Solar cycle (total cycle of polarity reversals of Sun's magnetic field)	22 years
Sunspot cycle	11 years (approx.)
Solar constant (at 1 AU)	1371 ± 5 W/m^2

Source: NASA.

with electrical currents generated by changing magnetic fields. The maximum number of solar flares appears to accompany the increased activity of the sunspot cycle. As a flare erupts, it discharges a large quantity of material outward from the Sun. This violent eruption also sends SHOCK WAVES through the SOLAR WIND.

Data from space-based solar observatories have indicated that *prominences* (condensed streams of ionized hydrogen atoms) appear to spring from sunspots. Their looping shape suggests that these prominences are controlled by strong magnetic fields. About 100 times as dense as the solar corona, prominences can rise at speeds of hundreds of kilometers per second. Sometimes the upper end of a prominence curves back to the Sun's surface, forming a "bridge" of hot glowing gas hundreds of thousands of kilometers long. On other occasions, the material in the prominence jets out and becomes part of the solar wind.

High-energy particles are released into heliocentric space by solar events, including very large solar flares called anomalously large solar particle events (ALSPEs). Because of their close association with infrequent large flares, these bursts of energetic particles are also relatively infrequent. However, solar flares, especially ALSPEs, represent a potential hazard to astronauts traveling in interplanetary space or working on the surface of the Moon and Mars. *See also* FUSION; SPACE RADIATION ENVIRONMENT; STAR; ULYSSES MISSION; YOHKOH.

sunlike stars Yellow, main sequence stars with; surface temperatures of 5,000–6,000K; spectral type G stars. *See also* STARS; SUN.

sunspot A relatively dark, sharply defined region on the Sun's surface, marked by an umbra (darkest region) that is approximately 2,000 kelvins (K) cooler than the effective photospheric temperature of the Sun (i.e., 5,800 K). The umbra is surrounded by a less dark, but sharply bounded region, called the penumbra. The average sunspot diameter is about 3,700 kilometers (km), but this diameter can

range up to 245,000 km. Most sunspots are found in groups of two or more, but they can occur singly. *See also* SUN.

sunspot cycle The approximately 11-year cycle in the variation of the number of SUNSPOTS. A reversal in the SUN's magnetic polarity also occurs with each successive sunspot cycle, creating a 22-year solar magnetic cycle.

sun-synchronous orbit A space-based sensor's view of Earth will depend on the characteristics of its orbit and the sensor's field of view (FOV). A sun-synchronous orbit is a special polar orbit that enables a spacecraft's sensor to maintain a fixed relation to the Sun. This feature is particularly important for environmental satellites (i.e., weather satellites) and multispectral imagery remote sensing satellites (such as the Landsat family of spacecraft). Each day a spacecraft in a sun-synchronous orbit around Earth passes over a certain area on the planet's surface at the same local time. One way to characterize sun-synchronous orbits is by the time the spacecraft cross the equator. These equator crossings (called *nodes*) occur at the same local time each day, with the descending (north-to-south) crossings occurring 12 hours (local time) from the ascending (south-to-north) crossings. In aerospace operations, the terms "A.M. polar orbiter" and "P.M. polar orbiter" are used to describe sun-synchronous satellites with morning and afternoon equator crossings, respectively.

An A.M. polar orbiter permits viewing of the land surface with adequate illumination, but before solar heating and the daily cloud buildup occur. Such a "morning platform" also provides an illumination angle that highlights geological features. The P.M. polar orbiter provides an opportunity to study the role of developed clouds in Earth's weather and climate and provides a view of the land surface after it has experienced a good deal of solar heating.

A typical morning meteorological platform might orbit at an altitude of 810 kilometers (km) (500 miles [mi]) at an inclination of 98.86 degrees and would have a period of 101 minutes. This morning platform would have its equatorial crossing time at approximately 0730 (local time). Similarly, an early-afternoon P.M. polar orbiter might orbit the Earth at an altitude of 850 km (530 mi) at an inclination of 98.70 degrees and would have a period of 102 minutes. This early-afternoon platform would cross the equator at approximately 1330 (local time). Each satellite (i.e., the morning and afternoon platform) views the same portion of Earth twice each day, therefore, the pair would provide environmental data collections with approximately six-hour gaps between each collection. In the United States, the afternoon platform is usually considered the primary meteorological mission, while the morning platform is considered to provide supplementary and backup coverage. *See also* ORBITS OF OBJECTS IN SPACE; WEATHER SATELLITE.

supergiant The largest and brightest type of STAR, with a LUMINOSITY of 10,000 to 100,000 times that of the SUN.

superheating In thermodynamics, the heating of a vapor, particularly saturated (wet) steam, to a temperature much higher than the boiling point at the existing pressure. This is done in power plants to improve efficiency and to reduce condensation in the turbines. *See also* HEAT ENGINE; RANKINE CYCLE; THERMODYNAMICS.

superior planets Planets that have orbits around the Sun that lie outside Earth's orbit. These planets include Mars, Jupiter, Saturn, Uranus, Neptune, and Pluto.

superluminal With a speed greater than the speed of light, 3×10^8 meters per second (m/s).

supernova A catastrophic stellar explosion occurring near the end of a star's life in which the star collapses and explodes, manufacturing (by nuclear transmutation) the heavy elements that it spews out into space. During a supernova explosion, the brightness of a star increases by a factor of several million times in a matter of days. *See also* STARS.

supersonic Of or pertaining to speed in excess of the speed of sound. *See also* SPEED OF SOUND.

supersonic compressor A compressor in which a supersonic velocity is imparted to the fluid relative to the rotor blades, the stator blades, or both the rotor and the stator blades, producing oblique shock waves over the blades to obtain a high-pressure rise.

supersonic diffuser A diffuser designed to reduce the velocity and increase the pressure of fluid moving at supersonic velocities.

supersonic flow In aerodynamics, flow of a fluid over a body at speeds greater than the speed of sound (i.e., greater than the acoustic velocity) and in which the shock waves start at the surface of the body. *See also* SPEED OF SOUND.

supersonic nozzle A converging-diverging nozzle designed to accelerate a fluid to supersonic speed. *See also* NOZZLE.

surface penetrator spacecraft A spacecraft probe designed to enter the surface of a celestial body, such as a comet or asteroid. The penetrator is capable of surviving a high-velocity impact and then making in-situ measurements of the penetrated surface. Data are sent back to a mothership for retransmission to scientists on Earth. *See also* PENETRATOR.

surface rover spacecraft An electrically powered robot vehicle designed to explore a planetary surface. Depending on the size of the rover and its level of sophistication, this type of mobile craft is capable of semiautonomous to fully autonomous operation. The rover could perform wide a variety of exploratory functions, including the acquisition of multispectral imagery, soil sampling and analysis, and rock inspection and collection. Data are transmitted back to Earth either directly by the rover vehicle or via a lander spacecraft or orbiting mothership. *See also* LUNAR ROVER(S); *MARS PATHFINDER*; MARS SURFACE ROVER(S).

surface tension The tendency of a liquid that has a large cohesive force to keep its surface as small as possible, forming spherical drops. Surface tension arises from intermolecular forces and is manifested in such phenomena as the absorption of liquids by porous surfaces, the rise of water (and other fluids) in a capillary tube, and the ability of liquids to "wet" a surface. *See also* HEAT PIPE.

surveillance In general, the systematic observation of places, persons, or things usually accomplished by optical, infrared, radar, or radiometric sensors. These sensors can be placed on a satellite to observe Earth or other regions of space (called *space-based surveillance*); installed on a aircraft platform (called *aerial surveillance*) to observe outer space (i.e., airborne astronomy), the surrounding atmosphere, or Earth's surface; or else situated on fixed or mobile terrestrial platforms to monitor the atmosphere and/or outer space (called *ground-based surveillance*). Surveillance includes tactical observations, strategic warning, and meteorological and environmental assessments. *See also* DEFENSE SUPPORT PROGRAM.

surveillance satellite An EARTH-orbiting MILITARY SATELLITE that watches regions of the PLANET for hostile military activities, such as BALLISTIC MISSILE launches and NUCLEAR WEAPON detonations. *See also* DEFENSE SUPPORT PROGRAM.

Surveyor Project NASA's highly successful Surveyor Project began in 1960. It consisted of seven unmanned lander spacecraft that were launched between May 1966 and January 1968, as a precursor to the human expeditions to the lunar surface in the Apollo Project. These robot lander craft were used to develop soft-landing techniques, to survey potential Apollo mission landing sites, and to improve scientific understanding of the Moon.

The *Surveyor 1* spacecraft was launched on May 30, 1966, and soft-landed in the Ocean of Storms region of the Moon. It found the bearing strength of the lunar soil was more than adequate to support the Apollo lander spacecraft

(called the lunar excursion module, or LEM). This contradicted the then-prevalent hypothesis that the Apollo LEM might sink out of sight in the fine lunar dust. The *Surveyor 1* spacecraft also telecast many pictures from the lunar surface.

The *Surveyor 3* spacecraft was launched on April 17, 1967, and soft-landed on the side of a small crater in another region of the Ocean of Storms. This robot spacecraft used a shovel attached to a mechanical "arm" to dig a trench and discovered that the load-bearing strength of the lunar soil increased with depth. It also transmitted many pictures from the lunar surface.

The *Surveyor 5* spacecraft was launched on September 8, 1967, and soft-landed in the Sea of Tranquility. An alpha-scattering device onboard this craft examined the chemical composition of the lunar soil and revealed a similarity to basalt on Earth.

The *Surveyor 6* was launched on November 7, 1967, and soft-landed in the Sinus Medii (Central Bay) region of the Moon. In addition to performing soil analysis experiments and taking many images of the lunar surface, this spacecraft also performed an extremely critical "hop experiment." NASA engineers back on Earth remotely fired the surveyor's vernier rockets to launch it briefly above the lunar surface. The spacecraft's launch did not create a dust cloud and resulted only in shallow cratering. This important demonstration indicated that the Apollo astronauts could safely lift off from the lunar surface with their rocket-propelled craft (upper portion of the LEM) when their surface exploration mission was completed.

Finally, the *Surveyor 7* spacecraft was launched on January 7, 1968, and landed in a highland area of the Moon, near the crater Tycho. Its alpha-scattering device showed that the lunar highlands contained less iron than the soil found in the mare regions (lunar plains). Numerous images of the lunar surface also were returned.

Despite the fact that the *Surveyor 2* and *4* spacecraft crashed on the Moon (rather than soft-landed and functioned), the overall Surveyor Project was extremely successful and paved the way for the Apollo surface expeditions that occurred between 1969 and 1972. *See also* APOLLO PROJECT; LANDER (SPACECRAFT); MOON.

survivability The capability of a system to avoid or withstand hostile environments without suffering irreversible impairment of its ability to accomplish its designated mission.

sustainer engine A rocket engine that maintains the velocity of a rocket vehicle once the vehicle has achieved its intended (programmed) velocity by means of a more powerful booster engine or engines (which is/are usually jettisoned). A sustainer engine also is used to provide the small amount of thrust needed to maintain the speed of a spacecraft or orbital glider that dipped into the atmosphere at perigee.

synchronous orbit An orbit over the equator in which the orbital speed of a spacecraft or satellite matches exactly the rotation of Earth on its axis, so that the spacecraft or satellite appears to stay over the same location on Earth's surface. *See also* GEOSYNCHRONOUS ORBIT.

synchronous rotation In astronomy, the rotation of a natural satellite (moon) about its primary in which the orbital period is equal to period of rotation of the satellite about its own axis. As a consequence of this condition, the satellite (moon) always presents the same side (face) to the parent body. The Moon, for example, is in synchronous rotation around Earth. (However, LIBRATION allows slightly more than one hemisphere to be seen from Earth.) Sometimes called CAPTURED ROTATION.

synchronous satellite An equatorial west-to-east satellite orbiting Earth at an altitude of approximately 35,900 kilometers (22,300 miles); at this altitude the satellite makes one revolution in 24 hours, synchronous with Earth's rotation. *See also* GEOSYNCHRONOUS ORBIT; GEOSTATIONARY EARTH ORBIT.

synchrotron radiation Electromagnetic radiation (light, ultraviolet, or X ray) produced by very energetic electrons as they spiral around lines of force in a magnetic field.

Syncom A family of NASA spacecraft launched in the 1960s to demonstrate the technologies needed to operate commercial communications satellites in geostationary Earth orbit. *Syncom 1* was launched in February 1963. After a successful launch and injection into a highly elliptical orbit, contact was lost with the spacecraft during the firing of its built-in orbit circularization rocket. *Syncom 2* was launched in July 1963 and provided important operational experience for communications satellites operating at synchronous orbit. *Syncom 3* was launched in August 1964 and successfully provided live coverage of the 1964 Olympic games from Japan. *See also* COMMUNICATIONS SATELLITE.

synoptic Pertaining to or affording an overall view.

synthetic aperture radar (SAR) A radar system that correlates the echoes of signals emitted at different points along a satellite's orbit or an aircraft's flight path. For example, a SAR system on a spacecraft illuminates its target to the side of its direction of movement and travels a known distance in orbit while the reflected, phase shift-coded pulses are returned and collected. With extensive computer processing, this procedure provides the basis for synthesizing an antenna (aperture) on the order of kilometers in size. The highest resolution achievable by such a system is theoretically equivalent to that of a single large antenna as wide as the distance between the most widely spaced points along the orbit that are used for transmitting positions. *See also* MAGELLAN MISSION; RADAR IMAGING; RADARSAT.

system integration The process of uniting the parts (components) of a spacecraft, launch vehicle, or space platform into a complete and functioning system.

T

tail 1. The tail surfaces of an aircraft, aerospace vehicle, missile, or rocket. 2. The rear portion of a body, as of an aerospace vehicle or rocket. 3. Short for the tail of a comet.

tailoring Modification of a basic solid rocket propellant by adjustment of the propellant properties to meet the requirements of a specific rocket motor.

takeoff 1. The ascent of a rocket vehicle as it departs from the launch pad at any angle. 2. The action of an aircraft as it becomes airborne. Compare this term to LIFTOFF, which refers only to the vertical ascent of a rocket or missile from its launch pad.

tandem launch The launching of two or more spacecraft or satellites using a single launch vehicle.

tank 1. A container incorporated into the structure of a liquid-propellant rocket from which a liquid propellant or propellants are fed into the combustion chamber(s). 2. A container incorporated into the structure of a nuclear rocket from which a monopropellant, such as liquid hydrogen, is fed into the nuclear reactor. 3. A ground-based or space-based container for the storage of liquid hydrogen, liquid oxygen, or other liquid propellants until they are transferred to a rocket's tanks or some other receptacle.

tankage The aggregate of the tanks carried by a liquid-propellant rocket or a nuclear rocket.

tank components 1. Devices for controlling the behavior of propellants, including positioning devices, slosh and vortex suspension devices, baffles, standpipes, and explosion devices. 2. Tank insulation.

tap A unit of impulse intensity, defined as one dyne-second per square centimeter. 1 tap = 0.1 pascal-second.

target 1. Any object, destination, point, and so on, toward which something is directed. 2. An object that reflects a sufficient amount of radiated signal to produce a return, or echo, signal on detection equipment. 3. The cooperative (usually passive) or noncooperative partner in a space rendezvous operation. 4. In military operations, a geographic area, complex, or installation planned for capture or destruction by military forces. 5. In intelligence usage, a country area, installation, or person against which intelligence operations are directed. 6. In radar, (1) generally, any discrete object that reflects or retransmits energy back to the radar equipment; (2) specifically, an object of radar search or surveillance. 7. In nuclear physics, a material subjected to particle bombardment (as in an accelerator) or neutron irradiation (as in a nuclear reactor) in order to induce a nuclear reaction.

target acquisition The detection, identification, and location of a target in sufficient detail to permit the effective employment of weapons.

target spacecraft The nonmaneuvering spacecraft in rendezvous and proximity operations.

telecommunications 1. In general, any transmission, emission, or reception of signs, signals, writings, images, sounds, or information of any nature by wire, radio, visual, or other electromagnetic or electro-optical systems. 2. In aerospace, the flow of data and information (usually by radio signals) between a spacecraft and an Earth-based communications system. A spacecraft has only a limited amount of power available to transmit a signal that sometimes must travel across millions or even billions of kilometers of space before reaching Earth. For example, an interplanetary spacecraft might have a transmitter that has no more than 20 watts of radiating power. One aerospace engineering approach is to concentrate all available power into a narrow radio beam and then to send this narrow beam in just one direction, instead of broadcasting the radio signal in all directions. Often this is accomplished by using a parabolic dish antenna on the order of 1 to 5 meters in diameter. However, even when these concentrated radio signals reach Earth, they are have very small power levels. The other portion of the telecommunications solution is to use special, large-diameter radio receivers on Earth, such as found in NASA's *Deep Space Network*. These sophisticated radio antennas are capable of detecting the very-low-power signals from distant spacecraft.

In telecommunications, the radio signal transmitted to a spacecraft is called the *uplink*. The transmission from the spacecraft to Earth is called the *downlink*. Uplink or downlink communications may consist of a pure radio-frequency (RF) tone (called a carrier), or these carriers may be modified to carry information in each direction. Commands transmitted to a spacecraft sometimes are referred to as an *upload*. Communications with a spacecraft involving only a downlink are called *one-way communications*. When an uplink signal is being received by the spacecraft at the same time that a downlink signal is being received on Earth, the telecommunications mode is often referred to as *two-way*.

Spacecraft carrier signals usually are modulated by shifting each waveform's phase slightly at a given rate. One scheme is to modulate the carrier with a

frequency, for example, near 1 megahertz (1 MHz). This 1 MHz modulation is then called a *subcarrier*. The subcarrier is modulated to carry individual phase shifts that are designated to represent binary ones (1s) and zeros (0s)—the spacecraft's telemetry data. The amount of phase shift used in modulating data onto the subcarrier is referred to as the *modulation index* and is measured in degrees. This same type of communications scheme is also on the uplink. Binary digital data modulated onto the uplink are called *command data*. They are received by the spacecraft and either acted upon immediately or stored for future use or execution. Data modulated onto the downlink are called *telemetry* and include science data from the spacecraft's instruments and spacecraft state-of-health data from sensors within the various onboard subsystems (e.g., power, propulsion, thermal control, etc).

Demodulation is the process of detecting the subcarrier and processing it separately from the carrier, detecting the individual binary phase shifts, and registering them as digital data for further processing. The device used for this is called a *modem,* which is short for *mod*ulator/*dem*odulator. These same processes of modulation and demodulation are commonly used with Earth-based computer systems and facsimile (fax) machines to transmit data back and forth over a telephone line. For example, if you have used a personal computer to "chat" over the Internet, your modem used a familiar audio frequency carrier that the telephone system could handle.

The dish-shaped *high-gain antenna* (HGA) is the type of spacecraft antenna mainly used for communications with Earth. The amount of *gain* achieved by an antenna refers to the amount of incoming radio signal power it can collect and focus into the spacecraft's receiving subsystems. In the frequency ranges used by spacecraft, the high-gain antenna incorporates a large parabolic reflector. Such an antenna may be fixed to the spacecraft bus or steerable. The larger the collecting area of the high-gain antenna, the higher the gain, and the higher the data rate it will support. However, the higher the gain, the more highly directional the antenna becomes. Therefore, when a spacecraft uses a high-gain antenna, the antenna must be pointed within a fraction of a degree of Earth for communications to occur. Once this accurate antenna-pointing is achieved, communications can take place at a high rate over the highly focused radio signal.

The low-gain antenna (LGA) provides wide-angle coverage at the expense of gain. Coverage is nearly omnidirectional, except for areas that may be shadowed by the spacecraft structure. The low-gain antenna is designed for relatively low data rates. It is useful as long as the spacecraft is relatively close to Earth (e.g., within a few astronomical units). Sometimes a spacecraft is given two low-gain antennas to provide full omnidirectional coverage, since the second LGA will avoid the spacecraft structure "blind spots" experienced by the first LGA. The low-gain antenna can be mounted on top of the high-gain antenna's subreflector.

The medium-gain antenna (MGA) represents a compromise in spacecraft engineering. It provides more gain than the low-gain antenna and has wider angle antenna-pointing accuracy requirements (typically 20 to 30 degrees) than the high-gain antenna. *See also* DEEP SPACE NETWORK.

telemetry 1. The process of making measurements at one point and transmitting the data to a distant location for evaluation and use. 2. Data modulated onto a spacecraft's communications downlink, including science data from the spacecraft's instruments and subsystem state-of-health data. *See also* TELECOMMUNICATIONS.

teleoperation The process by which a human worker (usually in a safe and comfortable environment) operates a versatile robot system that is at a distant, often hazardous location. Communications can be accomplished with radio signals, laser beams, or even "cable" (if the distance is not too great and the deployed cable does not interfere with operations or safety).

For example, an astronaut onboard the space shuttle (or space station) might "teleoperate" at a safe distance (using radio signals) a free-flying space robot retrieval system. In one space operations scenario, the teleoperated space robot rendezvouses with a damaged, potentially dangerous spacecraft. After initial rendezvous, the human operator uses instruments on the robot system to inspect the target space object. Then, if it appears that repairs can be performed safely, the human worker uses the space robot to capture it and bring it closer to the shuttle (or space station). Otherwise, again through teleoperations, the human operator uses the retrieval robot system to move the derelict spacecraft to a designated parking (storage) orbit or else put the hunk of "space junk" on a reentry trajectory that will cause it to plunge into Earth's upper atmosphere and burn up harmlessly. *See also* ROBOTICS.

telepresence The use of telecommunications, interactive displays, and a collection of sensor systems on a robot (which is at some distant location) to provide the human operator a sense of being present where the robot system actually is located. Depending on the level of sophistication in the operator's workplace as well as on the robot system, this telepresence experience can vary from a simple "steroscopic" view of the scene to a complete virtual reality activity in which sight, sound, touch, and motion are provided. Telepresence actually combines the technologies of virtual reality with robotics. Some day in the not too distant future, human controllers (on Earth), wearing sensor-laden bodysuits and three-dimensional viewer helmets, will use telepresence actually to "walk" and "work" on the Moon and other planetary bodies through their robot surrogates. Also called *virtual residency*. *See also* ROBOTICS; TELEOPERATION; VIRTUAL REALITY.

telescience A mode of scientific activity in which a distributed set of users (investigators) can interact directly with their instruments, whether in space or ground facilities, with databases, data handling and processing facilities, and with each other.

telescope In general, a device that collects electromagnetic radiation from a distant object so as to form an image of the object or to permit the radiation to be analyzed. Optical (or astronomical) telescopes can be divided into two main classes: *refracting telescopes* and *reflecting telescopes*. The oldest optical telescope is the refracting telescope, first constructed in Holland in 1608 and then developed for astronomical use by Galileo (1564–1642) in 1609. In its simplest form, the refracting telescope (or *refractor*) uses a converging lens (often called the primary lens) to gather and focus incoming light onto an eyepiece, which then magnifies the image. Galileo used a diverging lens in the eyepiece of his *Galilean telescope*, while Johannes Kepler (1571–1630) created an improved telescope in about 1611 by replacing the concave (negative) lens in the eyepiece with a convex (positive) lens. In the reflecting telescope, invented by Sir Isaac Newton (1642–1727) around 1668, a primary concave mirror located at the bottom of the telescope tube is used to collect and focus incoming light back up to a small secondary mirror, usually at a 45-degree angle to the incident light beams. This secondary mirror then reflects the incident light into an eyepiece (mounted on the side of the telescope tube) which magnifies the image. A variety of optical telescopes, consisting of various combinations of mirrors and lenses, have since been developed, including the Cassegrain and Schmidt telescopes. An optical telescope often is described in terms of its resolving power, aperture, and light-gathering power. In addition to optical telescopes, Earth-based observatories use radio telescopes and (in limited locations such as high mountaintops) infrared telescopes. However, space-based observatories (unhampered by the attenuation of Earth's atmosphere) also include infrared telescopes, gamma ray telescopes, X-ray telescopes, and ultraviolet telescopes. *See also* CHANDRA X-RAY OBSERVATORY;

COMPTON GAMMA RAY OBSERVATORY; HUBBLE SPACE TELESCOPE; INFRARED ASTRONOMICAL SATELLITE (IRA).

Television and Infrared Observation Satellite (TIROS) The TIROS series of weather satellites carried special television cameras that viewed Earth's cloud cover from a 725-kilometer (450-mile) altitude orbit. The images telemetered back to Earth provided meteorologists with a new tool—a nephanalysis, or cloud chart. On April 1, 1960, *TIROS 1,* the first true weather satellite, was launched into a near-equatorial orbit. By 1965 nine more TIROS satellites were launched. These spacecraft had progressively longer operational times and carried infrared radiometers to study Earth's heat distribution. Several were placed in polar orbits to increase cloud picture coverage. *TIROS 8* had the first Automatic Picture Transmission (APT) equipment. *TIROS 9* and *10* were test satellites of improved spacecraft configurations for the Tiros Operational Satellite (TOS) system. Operational use of the TOS satellites began in 1966. They were placed in Sun-synchronous (polar) orbits, so they could pass over the same location on Earth's surface at exactly the same time each day. This orbit enabled meteorologists to view local cloud changes on a 24-hour basis. Several Improved TOS Satellites (ITOS) were launched in the 1970s. The ITOS spacecraft served as workhorses for meteorologists of the National Oceanographic and Atmospheric Administration (NOAA), which was responsible for their operation. *See also* WEATHER SATELLITE.

temperature (symbol: T) A thermodynamic property that determines the direction of heat (thermal energy) flow. From the laws of thermodynamics, when two objects or systems are brought together, heat naturally will flow from regions of higher temperature to regions of lower temperature. In statistical mechanics, temperature can be considered as a "macroscopic" measurement of the overall kinetic energy of the individual atoms and molecules of a substance or body. *See also* ABSOLUTE TEMPERATURE.

tera- (symbol: T)   A prefix in the International System of Units meaning multiplied by 10^{12}.

terminal   1. A point at which any element in a circuit may be connected directly to one or more other elements. 2. Pertaining to a final condition or the last division of something, as *terminal guidance* or *terminal ballistics*. 3. In computer operations, an input/output (I/O) device (or station) usually consisting of a video screen or printer for output and a keyboard for input.

terminal guidance   1. The guidance applied to a missile between midcourse guidance and its arrival in the vicinity of the target. 2. With respect to an interplanetary spacecraft, the guidance and navigation performed during its approach to the target planet, which is accomplished by observing the angular position and motion and possibly the apparent size of the target body. *See also* GUIDANCE SYSTEM.

terminal phase   The final phase of a ballistic missile trajectory, lasting about a minute or less, during which warheads reenter the atmosphere and detonate at their targets. *See also* BALLISTIC MISSILE DEFENSE.

terminator   The boundary line separating the illuminated (i.e., sunlit) and dark portions of a nonluminous celestial body, like the Moon.

Terra   A Latin word meaning "Earth."

terrae   The highland regions of the Moon. Typically, the terrae are heavily cratered with large, old-appearing craters on a rugged surface with high albedo. *See also* LATIN SPACE DESIGNATIONS; MOON.

terraforming   The proposed large-scale modification or manipulation of the environment of a planet, such as Mars or Venus, to make it more suitable for human habitation. Also called *planetary engineering*.

terran   Of or relating to the planet Earth; a native of the planet Earth. *See also* TERRESTRIAL.

Terra spacecraft   The first in a new family of sophisticated NASA EARTH-OBSERVING SPACECRAFT successfully placed into POLAR ORBIT on December 18, 1999, from VANDENBERG AIR FORCE BASE, California. The five SENSORS aboard *Terra* are designed to enable scientists to examine the world's climate system. The instruments comprehensively observe and measure the changes of EARTH's landscapes, in its oceans, and within the lower ATMOSPHERE. One of the main objectives is to determine how life on Earth affects, and is affected by, changes within the climate system, with a special emphasis on better understanding the global carbon cycle. Formerly called *EOS-AM1*, this morning equator-crossing platform has a suite of sensors designed to study the diurnal properties of cloud and AEROSOL radiative fluxes. Another cluster of instruments on the spacecraft is addressing issues related to air-land exchanges of ENERGY, carbon, and water. Other spacecraft, like the *AQUA* spacecraft, are joining *Terra* in NASA's comprehensive EARTH SYSTEM SCIENCE effort during the first decade of the 21st century.

terrestrial   Of or relating to the planet Earth.

terrestrial environment   Earth's land area, including its human-made and natural surface and sub-surface features, and its interfaces and interactions with the atmosphere and the oceans. *See also* EARTH.

terrestrial planets   In addition to Earth itself, the terrestrial (or inner) planets include Mercury, Venus, and Mars. These planets are similar in their general properties and characteristics to the Earth; that is, they are small, relatively high-density bodies, composed of metals and silicates with shallow (or no) atmospheres as compared to the gaseous outer planets. *See also* EARTH; MARS; MERCURY; VENUS.

Terrier (RIM-2)   A U.S. Navy surface-to-air missile with solid-fuel rocket motor. It is equipped with radar beam rider or homing guidance and a nuclear or nonnuclear warhead.

Terrier land weapon system A surface-to-air missile system, using the Terrier RIM-2B and Terrier RIM-2C missile with ground-launching and guidance equipment, developed specifically for operations. This equipment is a lighter and land-mobile version of the U.S. Navy Terrier system.

tesla (symbol: T) The SI UNIT of magnetic flux density. It is defined as one weber of magnetic flux per square meter. The unit is named in honor of Nikola Tesla (1870–1943), a Croatian-American electrical engineer and inventor.

1 tesla = 1 weber/ (meter)^2 = 10^4 gauss

test 1. A procedure taken to determine under simulated or real conditions the capabilities, limitations, characteristics, effectiveness, reliability, or suitability of a material, device, system, or method. 2. Similarly, the procedure taken to determine the reactions, limitations, abilities, or skills of a person or crew.

test bed 1. A base, mount, or frame within or upon which a piece of equipment, such as an engine or sensor system, is secured for testing. 2. A flying test bed.

test chamber A place or room that has special characteristics where a person or device is subjected to experiment, such as an altitude chamber, wind tunnel, or acoustic chamber.

test firing The firing of a rocket engine, either live or static, with the purpose of making controlled observations of the engine or of an engine component.

test flight A flight to make controlled observations of the operation or performance of an aircraft, aerospace vehicle, rocket, or missile.

theater ballistic missile (TBM) A BALLISTIC MISSILE with limited RANGE, typically less than 400 KILOMETERS (250 mi), that is capable of carrying either a conventional HIGH EXPLOSIVE (HE), biological, chemical, or nuclear WARHEAD.

theater defense Defense against nuclear weapons on a regional level (e.g., Europe, Japan, Israel) rather than at the strategic level (i.e., globally or the United States and Russia).

Theater High Altitude Area Defense (THAAD) The Theater High Altitude Area Defense (THAAD) system is one element of the MISSILE DEFENSE AGENCY's terminal defense system against hostile BALLISTIC MISSILES launched by rogue nations in politically unstable regions of the world. The THAAD system represents a land-based upper tier BALLISTIC MISSILE DEFENSE (BMD) system that will engage short- and medium-range ballistic missiles, including those armed with WEAPONS OF MASS DESTRUCTION (WMD). THAAD is the only BMD system designed for intercepts inside and outside EARTH's ATMOSPHERE and has rapid mobility to defend against ballistic missile threats anywhere in the world. The THAAD system consists of four major components: truck-mounted launchers, interceptors, a RADAR system, and a BATTLE MANAGEMENT/COMMAND AND CONTROL SYSTEM. The KINETIC ENERGY INTERCEPTOR consists of a single-STAGE BOOSTER and a kinetic kill-vehicle that destroys its target through the FORCE of KINETIC ENERGY upon IMPACT. The first successful body-to-body intercept of a ballistic missile target was achieved on June 10, 1999.

thermal 1. Of or pertaining to heat or temperature. 2. A vertical air current caused by differential heating of the terrain.

thermal conductivity (symbol: k) An intrinsic physical property of a substance, describing its ability to conduct heat (thermal energy) as a consequence molecular motion. Typical units for thermal conductivity are joules/(second-meter-kelvin).

thermal control Regulation of the temperature of a space vehicle is a complex problem because extreme temperatures are encountered during a typical space mission. In the vacuum environment of outer space, radiative heat transfer is the only natural means to transfer thermal energy (heat) into and out of a space vehicle. In some special circumstances, a gaseous or liquid working fluid could be "dumped" from the space vehicle, to provide a temporary solution to a transient heat load—but this is extreme exception rather than the general design approach. The overall thermal energy balance for a space vehicle near a planetary body is determined by several factors: (1) thermal energy sources within the space vehicle itself; (2) direct solar radiation (the Sun has a characteristic BLACK-BODY temperature of about 5,770 kelvins (K); (3) direct thermal (infrared) radiation from the planet (e.g., Earth has an average surface temperature of about 288 K; (4) indirect (reflected) solar radiation from the planetary body (e.g., Earth reflection); and (5) vehicle emission, that is, radiation from the surface of space vehicle to the low-temperature sink of outer space. (Deep space has a temperature of about 3 K).

Under these conditions, thermally isolated portions of a space vehicle in orbit around Earth could encounter temperature variations from about 200 K, during Earth-shadowed or darkness periods, to 350 K, while operating in direct sunlight. Spacecraft materials and components can experience thermal fatigue due to repeated temperature cycling during such extremes. Consequently, aerospace engineers need to exercise great care in providing the proper thermal control for a spacecraft. Radiative heat transport is the principal mechanism for heat flow into and out of the spacecraft, while conduction heat transfer generally controls the flow of heat within the spacecraft.

There are two major approaches to spacecraft thermal control: passive and active. Passive thermal control techniques include: the use of special paints and coatings, insulation blankets, radiating fins, Sun shields, heat pipes, as well as the careful selection of the space vehicle's overall geometry (i.e., both the external and internal placement of temperature-sensitive components). Active thermal control techniques include the use of heaters (including small radioisotope sources) and coolers, louvers and shutters, or the closed-loop pumping of cryogenic materials.

An open-loop flow (or "overboard dump") of a rapidly heated working fluid might be used to satisfy a one-time or occasional special mission requirement to remove a large amount of thermal energy in a short period of time. Similarly, a sacrificial ablative surface could be used to handle a singular, large transitory external heat load. But these transitory (essentially one-shot) thermal control approaches are the exception rather than the aerospace engineering norm.

For interplanetary spacecraft, aerospace engineers often use passive thermal control techniques such as surface coatings, paint, and insulation blankets to provide an acceptable thermal environment throughout the mission. Components painted black will radiate more efficiently. Surfaces covered with white paint or white thermal blankets will reflect sunlight effectively and protect the spacecraft from excessive solar heating. Gold (i.e., gold-foil surfaces) and quartz mirror tiles also are used on the surfaces of special components. Active heating can be used to keep components within tolerable temperature limits. Resistive electric heaters, controlled either autonomously or via command, can be applied to special components to keep them above a certain minimum allowable temperature during the mission. Similarly, radioisotope heat sources (generally containing a small quantity of plutonium-238) can be installed where necessary to provide "at-risk" components with a small, essentially permanent supply of thermal energy.

thermal cycling Exposure of a component to alternating levels of relatively high and low temperatures.

thermal efficiency (symbol: η_{th}) *See* THERMODYNAMIC EFFICIENCY.

thermal equilibrium A condition that exists when energy transfer as heat between two thermodynamic systems (e.g., System 1 and System 2) is possible but none occurs. We say that System 1 and System 2 are in thermal equilibrium and that they have the same temperature. *See also* ZEROTH LAW OF THERMODYNAMICS.

thermal infrared The infrared region (IR) of the electromagnetic spectrum extending from about 3 to 14 micrometers (μm) wavelength. The thermal IR band most frequently used in remote sensing of Earth (from space) is the 8 to 14 micrometer wavelength band. This band corresponds to windows in the atmospheric absorption bands and, since Earth has an average surface temperature of about 288 kelvins (K) its thermal emission peaks near 10 micrometers wavelength.

thermal kill The destruction of a target by heating it, using directed energy (such as a high-energy laser beam), to the degree that structural components fail.

thermal protection system A system designed to protect a space vehicle from undesirable heating. Usually this is accomplished by using a system of special materials that can reject, absorb, or reradiate the unwanted thermal energy.

thermal radiation In general, the electromagnetic radiation emitted by any object as a consequence of its temperature. Thermal radiation ranges in wavelength from the longest infrared wavelengths to the shortest ultraviolet wavelengths and includes the optical (or visible) regions of the spectrum. *See also* PLANCK'S RADIATION LAW.

thermal tagging A target discrimination technique in which a high-powered laser heats up an object; a subsequent measure of its temperature then would help indicate whether the object was light (i.e., displaying a higher temperature after receiving a known heat input) or massive (i.e., exhibiting a lower temperature after receiving a similar heat input). Light objects most likely would be decoys, while the more massive objectives probably would be warhead-carrying reentry vehicles. The targets exhibiting this higher-temperature response to thermal tagging then would either be ignored or else identified for closer inspection by other missile defensive system sensors later in their trajectory, prior to the final commitment of "shoot-down" resources. *See also* BALLISTIC MISSILE DEFENSE.

thermionic conversion The direct conversion of heat (thermal energy) into electricity by evaporating electrons from a hot metal surface and condensing them on a cooler surface. The energy required to remove an electron completely from a metal is called the *work function,* which varies with the nature of the metal and its surface condition. The basic thermionic converter consists of two metals (or electrodes) with different work functions sealed in an evacuated chamber. The electrode with the larger work function is maintained at a higher temperature than the one with the smaller work function.

The emitter (or higher-temperature electrode) evaporates electrons, thereby acquiring a positive charge. The collector (or lower-temperature electrode) collects these evaporated or emitted electrons and becomes negatively charged. As a result, a voltage develops between the two electrodes, and a direct electric current will flow in an external circuit (or load) connecting them. The voltage is determined mainly by the difference in the work functions of the electrode materials.

However, the emission of electrons from the higher-temperature electrode can be inhibited as a result of the accumulation of electrons in its vicinity. This phenomenon is called the *space charge effect,* its impact can be reduced by introducing a small quantity of cesium metal into the evacuated thermionic converter chamber that contains the electrodes.

To achieve a substantial electron emission rate per unit area of emitter, the emitter temperature in a thermionic converter containing cesium must be at least 1000° Celsius (1830° Fahrenheit). Thermionic conversion has been suggested for space power generation, especially when used in conjunction with a space nuclear reactor as the heat source. *See also* DIRECT CONVERSION.

thermistor A semiconductor electronic device that uses the temperature-dependent change of resistivity of the substance. The thermistor has a very large negative temperature coefficient of resistance; that is, the electrical resistance decreases as the temperature increases. It can be used for temperature measurements or electronic circuit control.

thermocouple   A device consisting essentially of two conductors made of different metals, joined at both ends, producing a loop in which an electric current will flow when there is a difference in temperature between the two junctions. The amount of current that will flow in an attached circuit is dependent on: the temperature difference between the measurement (hot) and reference (cold) junction; the characteristics of the two different metals used; and the characteristics of the attached circuit. Depending on the different metals chosen, a thermocouple can be used as a thermometer over a certain temperature range.

thermodynamic   efficiency (symbol: η_{th})   In thermodynamics, the ratio of work done by a heat engine (W_{out}) to the total heat supplied by the thermal energy source (Q_{in}).

$$\eta_{th} = work_{out}/heat_{in}$$

Also called *thermal efficiency* and *Carnot efficiency. See also* HEAT ENGINE; THERMODYNAMICS.

thermodynamics   The branch of science that treats the relationships between thermal energy (heat) and mechanical energy. The field of thermodynamics in both physics and engineering involves the study of *systems.* A *thermodynamic system* is simply a collection of matter and space with its boundaries defined in such a way that energy transfer (as work and heat) across the boundaries can be identified and understood easily. The *surroundings* represent everything else that is not included in the thermodynamic system being studied.

Thermodynamic systems usually are placed in one of three groups: closed systems, open systems, and isolated systems. A *closed system* is a system for which only energy (but not matter) can cross the boundaries. An *open system* can experience both matter and energy transfer

across its boundaries. An *isolated system* can experience neither matter nor energy transfer across its boundaries. A *control volume* is a fixed region in space that is defined and studied as a thermodynamic system. Often the control volume is used to help in the analysis of open systems.

Steady state refers to a condition where the properties at any given point within the thermodynamic system are constant over time. Neither mass nor energy accumulate (or deplete) in a steady state system.

A *thermodynamic process* is the succession of physical states that a system passes through. Thermodynamic processes often can described by one of the following terms: cyclic, reversible, irreversible, adiabatic, or isentropic. In a *cyclic process,* the system experiences a series of states and ultimately returns to its original state. The CARNOT CYCLE and the BRAYTON CYCLE are examples. A *reversible process* can proceed in either direction and results in no change in the thermodynamic system or its surroundings. An *irreversible process* can proceed in only one direction; if the process were reversed, there would be a permanent (i.e., irreversible) change made to the system or its surroundings. An *adiabatic process* is one in which there is no flow of heat across the boundaries of the thermodynamic system. An *isentropic process* is one, in which the ENTROPY of the thermodynamic system remains constant or unchanged. An isentropic process is actually a reversible adiabatic process.

Thermodynamics is an elegant branch of physics based on two fundamental laws. The FIRST LAW OF THERMODYNAMICS states: *Energy can neither be created nor destroyed, but only altered in form.* This concept often is referred to as the *Conservation of Energy Principle.* In its simplest formulation involving a thermodynamic system, this law can be expressed as:

$$\Sigma\, energy_{in} - \Sigma\, energy_{out} = \Delta\, energy_{storage}$$

This expression often is referred to as an *energy balance* in which the energy flows into a system and out of a system are balanced with respect to any change in energy stored in the system. First-law energy balances are very useful in engineering thermodynamics.

The SECOND LAW OF THERMODYNAMICS involves the concept of entropy and is used to determine the possibility of certain processes and to establish the maximum efficiency of allowed processes. There are many statements of the second law. The statement attributed to the German physicist Rudolph Clausius (1822–1888), which was made in 1850, is considered one of the earliest: *It is impossible to construct a device that operates in a cycle and produces no effect other than the removal of heat from a body at one temperature and the absorption of an equal quantity of heat by a body at a higher temperature.*

The Second Law of Thermodynamics demonstrates that the maximum possible efficiency of a system (e.g., heat engine) is the Carnot efficiency, which is expressed as:

$$\eta_{max} = \eta_{Carnot} = (T_H - T_L)/T_H$$

where T_H is the high (absolute) temperature and T_L is the low (absolute) temperature associated with the system—that is, the temperature at which heat is added to the system (T_H) and the temperature at which heat is rejected from the system (T_L).

Another way of expressing the second law is to look at the possible changes in entropy of an isolated system. From the Second Law of Thermodynamics, we know that the entropy of an isolated system can never decrease; it can only remain constant or increase—that is,

$$\Delta \, entropy_{isolated} \geq 0$$

Thermodynamics plays a very important role in aerospace engineering, especially in defining the limits performance of space power and propulsion systems.

thermoelectric conversion The direct conversion of thermal energy (heat) into electrical energy based on the SEEBECK THERMOELECTRIC EFFECT. In 1821 the Russian-German scientist Thomas Johann Seebeck (1770–1831) was the first person to observe that if two different metals were joined at two locations that were then kept at different temperatures, an electric current will flow in a loop. Such a combination of two materials capable of producing electricity are directly as a

result of a temperature difference is called a *thermocouple.*

If the thermocouple materials A and B are joined at the hot junction and the other ends are kept cold, an electromotive force is generated between the cold ends. A direct current also will flow in a circuit (or load) connected between these cold ends. This electric current will continue to flow as long as heat is supplied at the hot junction and removed from the cold ends. For a given thermocouple (using either different metals or semiconductor materials), the voltage and electric power output are increased by increasing the temperature difference between the hot and cold junctions. The practical performance limits of a particular thermoelectric conversion device generally are set by the nature of the thermocouple materials and the (high) temperature of the available heat source. Often, as in space power applications, several thermocouples are connected in series to increase both voltage and power. *See also* SPACE NUCLEAR POWER; THERMOCOUPLE.

thermometer An instrument or device for measuring temperature.

thermonuclear Pertaining to nuclear reactions in which very high temperatures (i.e., millions of degrees Kelvin) are needed to bring about the fusion (joining) of light nuclei, such as deuterium (D) and tritium (T), with the accompanying release of energy. *See also* FUSION (NUCLEAR).

thermonuclear weapon A nuclear weapon in which very high temperatures (from a nuclear fission explosion) are used to bring about the fusion of light nuclei, such as deuterium (D) and tritium (T), which are isotopes of hydrogen, with the accompanying release of enormous quantities of energy. Sometimes referred to as a *hydrogen bomb.*

Third Law of Thermodynamics Based on the work of Hermann Nernst (1864–1941), Max Planck (1858–1947), Ludwig Boltzmann (1844–1906), Gilbert Lewis (1875–1946), Albert Einstein (1879–1955), and others in the early 1900s, the Third Law of Thermodynamics can be stated as: *The entropy of any pure substance in thermodynamic equilibrium approaches zero as the absolute temperature approaches zero.* The third law is important in that it furnishes a basis for calculating the absolute entropies of substances, either elements or compounds; these data then can be used in analyzing chemical reactions. *See also* THERMODYNAMICS.

Thor An early intermediate-range ballistic missile (IRBM) developed by the U.S. Air Force. This one-stage, liquid-propellant rocket had a range of over 2,700 kilometers (1,700 miles). It was first launched on January 25, 1957, and is now retired from service. Originally developed as a nuclear weapon delivery system, this rocket also served the U.S. Air Force and NASA as a space launch vehicle, especially when combined with a variety of upper-stage vehicles, such as the Agena.

Thousand Astronomical Unit (TAU) mission A proposed future NASA mission involving an advanced-technology robot spacecraft that would be sent on a 50-year journey into very deep space about 1,000 astronomical units (some 160 billion kilometers) away from Earth. The TAU spacecraft would feature an advanced multimegawatt nuclear reactor, ion propulsion, and a laser (optical) communications system. Initially, the TAU spacecraft would be directed for an encounter with Pluto and its moon Charon; followed by passage through the heliopause; perhaps even reaching the inner Oort cloud (the hypothetical region where comets are thought to originate) at the end of its long mission. This advanced robot spacecraft would investigate low-energy cosmic rays, low-frequency radio waves, interstellar gases, and deep-space phenomena. It also would perform high-precision astrometry (the measurement of distances between stars).

three-body problem The problem in classical celestial mechanics that treats the motion of a small body, usually of negligible mass, relative to and under the gravitational influence of two other finite point masses.

three-dimensional grain configuration In rocketry, a solid-propellant grain whose surface is described by three-dimensional analytical geometry (i.e., one that considers end effects).

throat In rocketry, the portion of a convergent/divergent nozzle at which the cross-sectional area is at a minimum. This is the region in which exhaust gases from the combustion chamber transition from subsonic flow to supersonic flow. *See also* NOZZLE.

throttling The varying of the thrust of a rocket engine during powered flight by some technique. Adjusting the propellant flow rate (e.g., tightening of fuel lines), changing of thrust chamber pressure, pulsed thrust, and variation of nozzle expansion are methods to achieve throttling.

throttling process In thermodynamics, an adiabatic process in which the ENTHALPY of a flowing fluid remains constant and no work is done. This process, usually involving a throttling device that restricts fluid flow, leads to a decrease in fluid pressure.

thrust (symbol: T) 1. In general, the forward force provided by a reaction motor. 2. In rocketry, the fundamental thrust equation is:
$$T = \dot{m}V_e + (P_e - P_a) A_e$$
where T is the thrust (force); $\dot{m}$ is the mass flow rate of propellant; V_e is the exhaust velocity at the nozzle exit; P_e is the exhaust pressure at the nozzle exit; p_e is the ambient pressure; and A_e is exit area of the nozzle. For a well-designed nozzle in which the exhaust gases are properly and completely expanded, $p_e \approx p_a$, so that the thrust becomes approximately represented by the term: $T \approx \dot{m} V_e$.

thrust barrel Structure in a rocket vehicle designed to accept the thrust load from two or more engines. Also called the *thrust structure*.

thrust chamber The place in a chemical rocket engine where the propellants are burned (or combusted) to produce the hot, high-pressure gases that are then accelerated and exhausted by the nozzle, thereby creating forward thrust. The combustion chamber and nozzle assembly. *See also* ROCKET.

thrust collector The structure attached to a rocket engine during static (i.e., captive) testing to transmit the engine thrust to thrust-measuring instruments.

thrust-time profile A plot of thrust level versus time for the duration of firing of a rocket engine or engines.

thrust terminator A device for ending the thrust in a rocket engine, either through propellant cutoff (in the case of a liquid-propellant rocket engine) or through the diversion of the flow of exhaust gases from the nozzle (in the case with a solid-propellant rocket).

thrust vector control (TVC) The intentional change in the direction of the thrust of a rocket engine to guide the vehicle. Thrust vector control can be achieved by using such devices as jet vanes and rings, which deflect the exhaust gas flow as it leaves the rocket's nozzle, or by using gimbaled nozzles and nozzle fluid injection techniques. However, interrupting the nozzle exhaust flow will introduce some degradation in the rocket engine's performance. Therefore, the gimbaled nozzle approach to thrust vector control often is the technique that imposes the minimum engine performance degradation.

time (symbol: t) In general, time is defined as the duration between two instants. The SI UNIT of time is the second, which now is precisely defined using an atomic standard. Time also may be designated as solar time, lunar time, or sidereal time, depending on the astronomical reference used. For example, *solar time* is measured by successive intervals between transits of the Sun across the meridian. *Lunar time* involves phases of Earth's Moon. The *lunar month,* for example, represents the time taken by the Moon to complete one revolution around Earth, as measured from new Moon to new Moon. The lunar month is also called the *synodic month* and is 29.5306 days long. *Sidereal time* is measured by successive transits of the vernal equinox across the local meridian. On Earth, time can be local or *universal time* (UT), which is related to Greenwich meridian.

time line In aerospace operations involving human crews, the planned schedule for astronauts during their space mission.

time tic Markings on telemetry records to indicate time intervals.

time tick A time signal consisting of one or more short audible sounds or beats.

Titan The largest Saturnian moon, approximately 5,150 kilometers in diameter. It has a sidereal period of 15.945 days and orbits at a mean distance of 1,221,850 kilometers from the planet. *See also* SATURN.

Titan (launch vehicle) The family of air force launch vehicles that was started in 1955. The Titan I missile was the first U.S. two-stage intercontinental ballistic missile (ICBM) and the first underground silo-based ICBM. The Titan I vehicle provided many structural and propulsion techniques that were later incorporated in the Titan II vehicle. Years later the Titan IV evolved from the Titan III family and is similar to the Titan 34D vehicle.

The Titan II was a liquid-propellant, two-stage, intercontinental ballistic missile that was guided to its target by an all-inertial guidance and control system. This missile, designated as LGM-25C, was equipped with a nuclear warhead and designed for deployment in hardened and dispersed underground silos. The air force built more than 140 Titan ICBMs; at one time they served as the foundation of America's nuclear deterrent force during the cold war. The Titan II vehicles also were used as space launch vehicles in

NASA's Gemini Project in the mid-1960s. Deactivation of the Titan II ICBM force began in July 1982. The last missile was taken from its silo at Little Rock Air Force Base, Arkansas, on June 23, 1987. The deactivated Titan II missiles then were placed in storage at Norton Air Force Base, California. Some of these retired ICBMs have found new use as space launch vehicles. The Titan II space launch vehicle is a modified Titan II ICBM that is designed to provide low- to medium-mass launch capability into polar LOW EARTH ORBIT (LEO). The modified Titan II is capable of lifting about 1,900 kilograms (kg) (4,200 pounds-mass [lbm]) into polar LEO. With the addition of two strap-on solid rocket motors (i.e., the graphite epoxy motors) to the first stage, the payload capability to polar LEO can be increased to 3,530 kilograms (7,800 lbm).

The versatile Titan III vehicles supported a variety of defense and civilian space launch needs. The Titan IIIA vehicle was developed to test the integrity of the inertially guided three-stage Titan IIIC liquid-propulsion system core vehicle. The Titan IIIB consists of the first two stages of the core Titan III vehicle (without the two strap-on solid rocket motors) with an Agena vehicle used as the third stage. The Titan IIIC consists of the Titan IIIA core vehicle with two solid rocket motors (sometimes called "Stage 0") attached on opposites sides of the liquid-propellant core vehicle. This configuration was developed for space launches from Complex 40 at Cape Canaveral Air Force Station (AFS), Florida. The Titan IIID vehicle, launched from Vandenberg Air Force Base, California, is essentially a Titan IIIC configuration without the Transtage and is radio-guided during launch. The Titan IIIE configuration was developed for NASA and is launched from Complex 41 at Cape Canaveral AFS. The Titan IIIE is basically a standard Titan IIID with a Centaur vehicle used as the third stage. Finally, the Titan 34D configuration uses a stretched core vehicle in conjunction with larger solid rocket motors to increase booster performance.

The Titan IV is the newest and largest unmanned space booster used by the air force. It was developed to launch the nation's largest, high-priority, high-value "shuttle-class" defense payloads. This "heavy-lift" vehicle is flexible in that it can be launched with several optional upper stage (such as the Centaur and the Inertia Upper Stage) for greater and more varied spacelift capability. The Titan IV vehicle's first stage consists of an LR87 liquid-propellant rocket that features structurally independent tanks for its fuel (aerozine 50) and oxidizer (nitrogen tetroxide). This design minimizes the hazard of the two propellants mixing if a leak should develop in either tank. Additionally, these liquid propellants can be stored at normal temperature and pressure, eliminating launch pad delays (as often encountered with the boil-off and refilling of cryogenic propellants) and giving the Titan IV vehicle the capability to meet critical defense program and planetary mission launch windows.

The Titan IV vehicle can have a length of up to 61.2 meters (204 feet) and carry up to 17,550 kg (39,000 lbm) into a 144-kilometer (km) (90-mile [mi]) altitude orbit when launched from Cape Canaveral AFS and up to 13,950 kg (31,000 lbm) into a 160-km (100-mi) altitude polar orbit when launched from Vandenberg AFB.

The Titan IVB rocket is the most recent and largest unmanned space booster used by the UNITED STATES AIR FORCE. It is a heavy-lift space launch vehicle designed to carry government payloads, such as the DEFENSE SUPPORT PROGRAM surveillance satellite or NATIONAL RECONNAISSANCE OFFICE (NRO) satellites into space from either CAPE CANAVERAL AIR FORCE STATION, Florida, or Vandenberg Air Force Base, California. The Titan IVB can place a 21,670-kilogram (47,800-lbm) payload into low Earth orbit or more than 5,760 kilograms (12,700 lbm) into geosynchronous orbit. The first Titan IVB rocket was successfully flown from Cape Canaveral Air Force Station on February 23, 1997. A powerful Titan IV-Centaur configuration successfully sent NASA's CASSINI mission to SATURN from Cape Canaveral Air Force Station on October 15, 1997.

tolerance The allowable variation in measurements within which the dimensions of an item are judged acceptable.

tolerance stackup Additive effects of all the allowable manufacturing tolerances on the final dimensions of the assembly; also called *tolerance buildup.*

Tomahawk Long-range, subsonic cruise missile used by the U.S. Navy for land attack and for antisurface warfare. Tomahawk is an all-weather submarine or ship-launched antiship or land-attack cruise missile. After launch, a solid-propellant rocket engine propels the missile until a small turbofan engine takes over for the cruise portion of the flight. This cruise missile is a highly survivable weapon. Radar detection is difficult because of its small cross section and low-altitude flight. Similarly, infrared detection is also difficult because the turbofan emits little heat.

The antiship variant of Tomahawk uses a combined active radar seeker and passive system to seek out, engage, and destroy a hostile ship at long range. Its modified Harpoon cruise missile guidance system permits the Tomahawk to be launched and fly at low altitudes in the general direction of an enemy warship to avoid radar detection. Then, at a programmed distance, the missile begins an active radar search to seek out, acquire, and hit the target ship.

The land-attack version has inertial and terrain contour matching (TERCOM) guidance. The TERCOM guidance system uses a stored map reference to compare with the actual terrain to help the missile determine its position. If necessary, a course correction is made to place the missile on course to the target.

The basic Tomahawk is 5.56 meters (m) (18.25 feet [ft]) long and has a mass of 1192 kilograms (2,650 pounds-mass [lbm]), not including the booster. It has a diameter of 51.81 centimeters (20.4 inches) and a wing span (when deployed) of 2.67 m (8.75 ft). This missile is subsonic and cruises at about 880 kilometers (km) (550 miles [mi]) per hour. It can carry a conventional or nuclear warhead. In the land-attack (conventional warhead) configuration, it has a range of 1,100 km (about 690 mi); while in the land-attack (nuclear warhead) configuration, it has a range of 2,480 km (1,555 mi). In the antiship role, it has a range of over 460 km (288 mi). This missile was first deployed in 1983.

ton (symbol: T or t) A unit of mass in both the SI and American standard system of units. 1. In the SI UNIT system, one ton (sometimes spelled tonne) is defined as 1,000 kilograms.

1 ton (tonne) = 1,000 kilograms = 2,205 pounds-mass (lbm).

Also called the *metric ton.* 2. In the American standard system of units (i.e., the foot-pound-second system), one (short) ton is defined as 2,000 pounds-mass (lbm). In the United Kingdom, an *imperial* (or *long*) *ton* contains approximately 2,240 pounds-mass (lbm).

topside sounder A spacecraft instrument that is designed to measure ion concentration in the ionosphere from above Earth's atmosphere.

torque (symbol: τ) In physics, the moment of a force about an axis; the product of a force and the distance of its line of action from the axis.

torr A unit of pressure named in honor of the Italian physicist Evangelista Torricelli (1608–1647). One standard atmosphere (on Earth) is equal to 760 torr; or 1 torr = 1 millimeter of mercury = 133.32 pascals.

torsion The state of being twisted.

total impulse (symbol: I_T) The integral of the thrust force (T) over an interval of time, t:

$$I_T = \int T \, dt$$

See also SPECIFIC IMPULSE.

touchdown The (moment of) landing of an aerospace vehicle or spacecraft on the surface of a planet or moon.

toughness In engineering, the ability of a material (especially a metal) to absorb energy and deform plastically before fracturing.

TOW missile The U.S. TOW missile (Tube-launched Optically Tracked, Wire Command-Link Guided) is a long-range, heavy antitank system designed to attack and defeat armored vehicles and other targets, such as field fortifications. The TOW 2A missile is 14.91 centimeters (cm) (5.87 inches [in]) in diameter and 128.02 cm (50.40 in) long; the TOW 2B missile is 14.9 cm (5.8 in) in diameter and 121.9 cm (48 in) long. The maximum effective range is about 3.75 kilometers (2.33 miles). The basic TOW weapon system was fielded in 1970. This system is designed to attack and defeat tanks and other armored vehicles. It is used primarily in antitank warfare and is a command to line of sight, wire-guided weapon. The system will operate in all weather conditions and on an obscured battlefield. The TOW 2 launcher is the most recent launcher upgrade. The TOW 2B missile incorporates a new fly-over, shoot-down technology. The TOW missile can be mounted on the Bradley Fighting Vehicle System (BFVS), the Improved TOW Vehicle (ITV), the High Mobility Multipurpose Wheeled Vehicle (HMMWV), and the AH-1S Cobra Helicopter. The TOW missile is also in use by more than 40 nations besides the United States as their primary heavy antiarmor weapons system.

track 1. (noun) The actual path or line of movement of an aircraft, rocket, aerospace vehicle, and the like over the surface of Earth. It is the projection of the vehicle's flight path on the planet's surface. 2. (verb) To observe or plot the path of something moving, such as an aircraft or rocket, by one means or another, as by telescope or by radar; said of persons or of electronic equipment, as, for example, "the radar *tracked* the satellite while it passed overhead."

tracking 1. The process of following the movement of a satellite, rocket or aerospace vehicle. Tracking usually is accom-plished with optical, infrared, radar, or radiosystems. 2. In BALLISTIC MISSILE DEFENSE (BMD), the monitoring of the course of a moving target. Ballistic objects may have their tracks predicted by the defensive system, using several observations and physical laws.

Tracking and Data Relay Satellite (TDRS) NASA's Tracking and Data Relay Satellite (TDRS) network provides almost full-time coverage not only for the space shuttle but also for up to 24 other orbiting spacecraft simultaneously. Services provided include communications, tracking, telemetry, and data acquisition.

The TDRS satellites orbit geosynchronously at 35,888 kilometers (22,300 miles) above Earth and look down on an orbiting space shuttle or spacecraft. For most of their orbits around Earth, these spacecraft remain in sight of one or more TDRS satellites. In the past, spacecraft could communicate with Earth only when they were in view of a ground tracking station, typically less than 15% of each orbit. The full TDRS constellation enables spacecraft to communicate with Earth for about 85 to 100% of the orbit, depending on their altitude. All space shuttle missions and nearly all other NASA spacecraft in Earth orbit require TDRS support capabilities for mission success.

Each TDRS is a three-axis stabilized satellite measuring 17.4 meters (57 feet) across when its solar panels are fully deployed. The satellite has a mass of 2,268 kilograms (5,000 pounds-mass [lbm]). The spacecraft's design uses three modules. The equipment module, forming the base of the satellite's central hexagon, houses the subsystems that actually control and operate the satellite. Attached solar power arrays generate more than 1,700 watts of electrical power. When TDRS is in the shadow of Earth, nickel-cadmium batteries supply electric power.

The communications payload module is the middle portion of the hexagon and includes the electronic equipment that regulates the flow of transmissions between the satellite's antennas and other communications functions.

Finally, the antenna module is a platform atop the hexagon holding seven antenna systems. TDRS has uplink, or forward, channels that receive transmissions (radio signals) from the ground, amplify them, and retransmit them to user spacecraft. The downlink, or return, channels receive the user spacecraft's transmissions, amplify them, and retransmit them to the ground terminal. The prime transmission link between Earth and TDRS is a 2-meter (6.7-foot) dish antenna attached by a boom to the central hexagon. This parabolic reflector operates in the Ku-band (12–14 gigahertz [GHz]) frequency to relay transmissions to and from the ground terminal. All spacecraft telemetry data downlinked by TDRS is channeled through a highly automated ground station complex at White Sands, New Mexico. This location was chosen for its clear line of sight to the satellites and dry climate. The latter minimizes rain degradation of radio signal transmission.

The TDRS spacecraft with its attached upper stage is launched aboard the space shuttle from the Kennedy Space Center in Florida and deployed from the orbiter's payload bay, nominally about six hours into the mission. To boost the TDRS into a geosynchronous orbit, the Inertial Upper Stage (or IUS) vehicle fires twice. The upper-stage vehicle's first-stage motor fires about an hour after deployment, placing the attached TDRS into an elliptical geotransfer orbit. The first stage then separates. The IUS second-stage motor fires about 12.5 hours into the mission, circularizing the orbit and shifting the flight path so that the satellite is moving above the equator. The IUS second stage and the TDRS then separate at about 13 hours after launch. Once in geosynchronous orbit its appendages, including the solar panels and parabolic antennas, are deployed. About 24 hours after launch, the satellite is ready for ground controllers to begin checkout procedures. Initially the spacecraft is positioned at an intermediate location (in geosynchronous orbit) for checkout and testing, then it is moved to its final operational location (in geosynchronous orbit).

In June 2000 NASA launched the first of three newly designed and improved TDRS spacecraft (called *TDRS-H, -I,* and *-J*) from Cape Canaveral Air Force Station to replenish the existing on-orbit fleet of spacecraft. The TDRS-H was joined in GEOSYNCHRONOUS ORBIT by *TDRS-I* when an ATLAS IIA EXPENDABLE LAUNCH VEHICLE successfully lifted that new satellite into space on September 30, 2002. The improved TDRS satellites retain and augment two large ANTENNAS that move smoothly to track satellites orbiting below, providing high data rate communications for the INTERNATIONAL SPACE STATION, the HUBBLE SPACE TELESCOPE, and other Earth orbiting spacecraft.

tracking and pointing   In ballistic missile defense (BMD), once a target is detected, it must be followed, or "tracked." When the target is tracked successfully, a high-velocity (kinetic energy weapon) interceptor, high-energy laser (HEL) beam, or neutral particle beam (NPB) is "pointed" at the target. Tracking and pointing frequently are integrated operations. *See also* BALLISTIC MISSILE DEFENSE.

trainer   In aerospace, a teaching device or facility that primarily provides a physical representation of flight hardware. The trainer also may have limited computer capabilities. Compare this term with SIMULATOR.

trajectory   In general, the path traced by any object or body moving as a result of an externally applied force, considered in three dimensions. *Trajectory* sometimes is used to mean *flight path* or *orbit,* but "orbit" usually means a closed path and "trajectory" means a path that is not closed.

trajectory correction maneuver (TCM)   Once a spacecraft's solar or planetary orbital parameters are known, spacecraft controllers on Earth can compare these values to those actually desired for the mission. To correct any discrepancy, a trajectory correction maneuver (TCM) can be planned and executed. This process involves computing the direction and magnitude of the vector required to correct to the desired trajectory. An appropriate time then is determined for making the change. The spacecraft is commanded to rotate to the attitude (in three-dimensional space) computed for implementing the change, and its thrusters are fired for the required amount of time. TCMs generally involved a velocity change (or "delta-V") on the order of meters per second or tens of meters per second.

transceiver A combination of *trans*mitter and re*ceiver* in a single housing, with some components being used by both units.

transducer General term for any device that converts one form of energy (usually in some type of signal) to another form of energy. For example, a microphone is an electroacoustic transducer in which sound waves (acoustic signals) are converted into corresponding electrical signals that then can be amplified, recorded, or transmitted to a remote location. The photocell and thermocouple are also transducers, converting light and heat (respectively) into electrical signals.

transfer orbit In interplanetary travel, an elliptical trajectory tangent to the orbits of both the departure planet and target planet (or moon). *See also* HOHMANN TRANSFER ORBIT.

transient The condition of a physical system in which the parameters of importance (e.g., temperature, pressure, fluid velocity, etc.) vary significantly with time; in particular, the condition or state of a rocket engine's operation in which the mass, momentum, volume, and pressure of the combustion products within the thrust chamber vary significantly with time.

transient period The interval from start or ignition to the time when steady-state conditions are reached, as in a rocket engine.

transit (planetary) The passage of one CELESTIAL BODY in front of another (larger-diameter) celestial body, such as VENUS across the face of the SUN.

Transit A U.S. Navy navigational satellite system that has provided passive, all-weather, worldwide position information to surface ships (military and commercial) and submarines for over 30 years.

transit time The length of time required by a rocket, missiles, or spacecraft to reach its intended target or destination.

translation Movement in a straight line without rotation.

translunar Of or pertaining to space beyond the Moon's orbit around Earth.

transmission grating A diffraction grating in which incoming signal energy is resolved into spectral components upon transmission through the grating.

transmittance (transmissivity) In radiation heat transfer, the transmittance (commonly used symbol: τ) for a body is defined as the ratio of the incident radiant energy transmitted through the body to the total radiant energy falling upon the body. The body may be *transparent* to the incident radiant energy (i.e., permitting some or all of the radiant energy to travel through the body), or else it may be *opaque* and therefore inhibit or prevent the transmission of incident radiant energy. An ideal BLACKBODY is perfectly opaque; therefore, this blackbody has a transmittance of zero, that is, $\tau_{blackbody} = 0$. *Compare with* ABSORPTANCE and REFLECTANCE.

transmitter A device for the generation of signals of any type and form that are to be transmitted. For example, in radio and radar, a transmitter is that portion of the equipment that includes electronic circuits designed to generate, amplify, and shape the radio frequency (RF) energy that is delivered to the antenna where it is radiated out into space. In aerospace applications, the spacecraft's transmitter generates a tone at a single designated frequency, typically in the S-band (about 2 gigahertz) or X-band (about 5 gigahertz) region of the electromagnetic spectrum. This tone is called the *carrier*. The carrier then can be sent from the spacecraft to Earth as it is, or it can be modulated with a data-carrying subcarrier within the transmitter. The signal generated by the spacecraft's transmitter is passed to a power amplifier, where its power typically is boosted several tens of watts. The microwave-band power amplifier can be a solid-state amplifier (SSA) or a traveling wave tube (TWT) amplifier. The output of the power amplifier then is conducted through wave guides to the spacecraft's antenna, often a high-gain antenna (HGA).

Trans-Neptunian Object (TNO) Any of the numerous small, icy celestial bodies that lie in the outer fringes of the SOLAR SYSTEM beyond NEPTUNE. TNOs include PLUTINOS and KUIPER BELT objects. *See also* QUOAOAR.

transpiration cooling A form of mass transfer cooling that involves controlled injection of a fluid mass through a porous surface. This process basically is limited by the maximum rate at which the coolant material can be pumped through the surface.

transponder A combined receiver and transmitter whose function is to transmit signals automatically when triggered by an appropriate interrogating signal.

trap A part of a solid-propellant rocket engine used to prevent the loss of unburned propellant through the nozzle.

tribology The branch of engineering science that deals with friction, lubrication, and the behavior of lubricants.

Trident missile (D-5) The U.S. Navy's Trident II (D-5) missile is the main armament aboard the Ohio-class fleet ballistic missile submarines (SSBNs), which now have replaced the aging fleet ballistic missile submarines built in the 1960s. The Trident-II (D-5) three-stage, solid-propellant rocket is the latest in a line of fleet ballistic missiles that began with the Polaris (A-1) missile. The Trident-II missile incorporates

many state-of-the-art advances in rocketry and electronics, giving this submarine-launched fleet ballistic missile greater range, payload capacity, and accuracy than its predecessors. *See also* FLEET BALLISTIC MISSILE.

triple point In thermodynamics, the temperature of a mixture of a substance that contains the liquid, gaseous, and solid phases of the material in thermal equilibrium at standard pressure. For example, by international agreement, the temperature for the *triple point of water* has been set at 273.16 kelvins (K) (491.69° Rankine), creating an easily reproducible standard for a one-point absolute temperature scale.

triplet In rocketry, an injector orifice pattern consisting of one or more sets of three orifices that produce streams converging to a point; usually fuel is injected through the outer orifices and oxidizer is injected through the central orifice.

tritium (symbol: T or $^{3}_{1}$H) The RADIOISOTOPE of HYDROGEN with two NEUTRONS and one PROTON in the NUCLEUS. It has a HALF-LIFE of 12.3 YEARS.

Triton The largest MOON of NEPTUNE. It has a diameter of approximately 2,700 kilometers (1,675 mi) and an average surface temperature of just 35 K, making it one of the coldest objects yet discovered in the SOLAR SYSTEM.

Trojan group A group of asteroids that lie near the two Lagrangian points in Jupiter's orbit around the Sun. Achilles was the first asteroid in the group to be identified (in 1906); many subsequently discovered members of this group have been named in honor of the heroes, both Greek and Trojan, of the Trojan War. *See also* ASTEROID.

truncation error In computations, the error resulting from the use of only a finite number of terms of an infinite series or from the approximation of operations in the infinitesimal calculus by operations in the calculus of finite differences.

Tsiolkovsky, Konstantin Eduardovich (1857–1935) Russian space travel pioneer who is regarded as one of the three founding fathers of ASTRONAUTICS—the other two technical visionaries being ROBERT GODDARD and HERMANN OBERTH. Tsiolkovsky was a nearly deaf schoolteacher in an obscure rural town within czarist Russia, yet, despite the physical handicap and remote location, his writings accurately projected future space technologies. In 1895 he published the book *Dreams of Earth and Sky,* in which he discussed the concept of an ARTIFICIAL SATELLITE orbiting EARTH. Many of the most important principles of astronautics appeared in the seminal 1903 work *Exploration of Space by Reactive Devices.* This book linked the use of the ROCKET to space travel and even introduced a design for a LIQUID-PROPELLANT ROCKET ENGINE, using LIQUID HYDROGEN and LIQUID OXYGEN as its chemical PROPELLANTS. His 1924 work *Cosmic Rocket Trains* introduced the concept of the MULTISTAGE ROCKET. Although he never personally constructed any of the rockets proposed in his visionary books, these works inspired many future rocket scientists, including SERGEI KOROLEV, whose powerful rockets did place the world's first artificial satellite *(SPUTNIK 1)* into ORBIT in 1957.

Tsyklon A medium-capacity Russian/Ukranian launch vehicle capable of placing 3,600-kilogram payloads into LOW EARTH ORBIT (LEO). First launched in 1977, this vehicle consists of three hypergolic-propellant stages. It is currently manufactured by NPO Yuzhnoye, Ukraine.

T-time In aerospace operations, any specific time (minus or plus) that is referenced to "launch time," or "zero," at the end of a countdown. T-time is used to refer to times during a countdown sequence that is intended to result in the ignition of a rocket propulsion unit or units to launch a missile or rocket vehicle. For example, one will hear the phrase: "T-minus 20 seconds and counting" during a countdown sequence. This refers to the point in the launch sequence that occurs 20 seconds before the rocket engines are ignited.

tube-wall construction Use of parallel metal tubes that carry coolant to or from the combustion chamber or nozzle wall.

tumble 1. To rotate end over end—said of a rocket, of an ejection capsule, or of a spent propulsion stage that has been jettisoned. 2. Of a gyro, to precess suddenly and to an extreme extent as a result of exceeding its operating limits of pitch.

Tunguska event A violent explosion that occurred in a remote part of Siberia (Russia) in late June 1908. One contemporary hypothesis suggests that this wide-area (about 80 KILOMETERS [50 mi] in diameter), destructive event was caused by the entrance of an extinct cometary NUCLEUS (about 60 METERS [200 ft] in diameter) into EARTHS ATMOSPHERE. Most of the KINETIC ENERGY of the COMET's fragments was probably dissipated through an explosive disruption of the atmosphere several kilometers above the surface of the devastated Siberian forest. Subsequent investigations (decades after the event) failed to find an IMPACT CRATER, although many square kilometers of forest were reported as being laid flat by the explosive event. Investigators reported that trees were knocked to the ground to distances of about 20 kilometers (12.5 mi) from the end point of the (projected) fireball TRAJECTORY. Other trees were snapped off or knocked over at distances as great as 40 kilometers (25 mi). The estimated ENERGY released during this event was equivalent to the detonation of a THERMONUCLEAR WEAPON with a YIELD between 12 and 20 MEGATONs. Circumstantial evidence further suggests that fires were ignited up to 15 kilometers (9.3 mi) from the endpoint of the intense burst of radiant energy. The combined environmental effects were quite similar to those expected from a large-yield nuclear detonation at a similar altitude, except, of course, without any prompt burst of NEUTRONS or GAMMA RAYS or any lingering RADIOACTIVITY. Other scientists have suggested that the Tunguska event object was a large stony METEORITE that exploded in the upper atmosphere. *See also* EXTRATERRESTRIAL CATASTROPHE THEORY.

turbine   A machine that converts the energy of a fluid stream into mechanical energy of rotation. The working fluid used to drive a turbine can be gaseous or liquid. For example, a highly compressed gas drives an *expansion turbine,* hot gas drives a *gas turbine,* steam (or other vapor) drives a *steam (or vapor) turbine,* water drives a *hydraulic turbine,* and wind spins a *wind turbine (or windmill).* Generally (except for wind and water turbines), a turbine consists of two sets of curved blades (or vanes) along side each other. One set of vanes is fixed and the other set of vanes can move. The moving vanes are spaced around the circumference of a cylinder or *rotor,* which can rotate about a central shaft. The fixed set of vanes *(the stator)* often is attached to the inside casing that encloses the rotor (or moving portion of the turbine).

In order to make efficient use of the energy of the working fluid, gas and steam turbines often have a series of successive stages. Each stage consists of a set of fixed (stator) and moving (rotor) blades. The pressure of the working fluid decreases as it passes from stage to stage. The overall diameter of each successive stage of the turbine therefore can be increased to maintain a constant torque (or rotational effect) as the working fluid expands and loses pressure (and energy).

turbopump system   In a liquid-propellant rocket, the assembly of components (e.g., propellant pumps, turbine[s], power source, etc.) designed to raise the pressure of the propellants received from the vehicle tanks and deliver them to the main thrust chamber at specified pressures and flow rates.

turbulence   A state of fluid flow in which the instantaneous velocities exhibit irregular and apparently random fluctuations so that in practice only statistical properties can be recognized and subjected to analysis. These fluctuations often constitute major deformations of the flow and are capable of transporting momentum, energy, and suspended matter at rates far in excess of the rate of transport by the molecular processes of diffusion and conduction in a nonturbulent or laminar flow.

turbulence ring   In a liquid-propellant rocket engine, the circumferential protuberance in the gas-side wall of a combustion chamber intended to generate turbulent flow and thereby enhance the mixing of burning gases.

turbulent flow   Fluid flow in which the velocity at a given point fluctuates randomly and irregularly in both magnitude and direction. The opposite of LAMINAR FLOW.

two-body problem   The problem in classical celestial mechanics that addresses the relative motion of two point masses under their mutual gravitational attraction.

two-dimensional grain configuration   In rocketry, the solid-propellant grain whose burning surface is described by two-dimensional analytical geometry (i.e., the cross section is independent of length).

two-phase flow   Simultaneous flow of gases and solid particles (e.g., condensed metal oxides); or the simultaneous flow of liquids and gases (vapors).

ullage The amount that a container, such as a fuel tank, lacks of being full.

ullage pressure Pressure in the ullage space of a container, either supplied or self-generated.

ultra- A prefix meaning "surpassing a specified limit, range, or scope" or "beyond."

ultrahigh frequency (UHF) A radio frequency in the range 0.3 gigahertz to 3.0 gigahertz.

ultrasonic Of or pertaining to frequencies above those that affect the human ear, that is, acoustic waves at frequencies greater than approximately 20,000 hertz; for example, an *ultrasonic vibrator.*

ultraviolet (UV) astronomy Astronomy based on the ultraviolet (UV, 10 to 400 nanometer wavelength) portion of the electromagnetic (EM) spectrum. Because of the strong absorption of UV radiation by Earth's atmosphere, ultraviolet astronomy must be performed using high-altitude balloons, rocket probes, and orbiting observatories. Ultraviolet data gathered from spacecraft are extremely useful in investigating interstellar and intergalactic phenomena. Observations in the ultraviolet wavelengths have shown, for example, that the very-low-density material that can be found in the interstellar medium is quite similar throughout our galaxy but that its distribution is far from homogeneous. In fact, UV data have led some astrophysicists to postulate that low-density cavities, or "bubbles," in interstellar space are caused by supernova explosions and are filled with gases that are much

hotter than the surrounding interstellar medium. Ultraviolet data gathered from space-based observatories have revealed that some stars blow off material in irregular bursts and not in a steady flow as originally thought. Ultraviolet data are also of considerable use in studying many of the phenomena that occur in distant galaxies. *See also* ASTROPHYSICS.

ultraviolet (UV) radiation That portion of the electromagnetic spectrum that lies beyond visible (violet) light and is longer in wavelength than X rays. Generally taken as electromagnetic radiation with wavelengths between 400 nanometers (just past violet light in the visible spectrum) and about 10 nanometers (the extreme ultraviolet cutoff and the beginning of X rays). *See also* ELECTROMAGNETIC SPECTRUM.

Ulysses **mission** An international space project to study the poles of the Sun and the interstellar environment above and below these solar poles. The *Ulysses* spacecraft was built by Dornier Systems of Germany for the European Space Agency (ESA), which is also responsible for in-space operations of the *Ulysses* mission. NASA provided launch support using the space shuttle *Discovery* and an upperstage configuration consisting of a two-stage inertial upper stage (IUS) rocket and a PAM-S (Payload Assist Module) configuration. In addition, the United States, through the Department of Energy, provided the radioisotope thermoelectric generator (RTG) that supplies electric power to this spacecraft. *Ulysses* is tracked and its scientific data collected by NASA's Deep Space Network (DSN). Spacecraft monitoring and control, as well as data reduction and analysis, is performed at NASA's Jet Propulsion Laboratory (JPL) by a joint ESA/JPL team.

The *Ulysses* mission, named for the legendary Greek hero in Homer's epic saga of the Trojan War who wandered into many previously unexplored areas on his return home, is a survey mission designed to examine: the properties of the SOLAR WIND, the structure of the Sun-solar wind interface, the heliospheric magnetic field, solar radio bursts and plasma waves, solar and galactic cosmic rays and the interplanetary/interstellar neutral gas and dust environment—all as a function of solar latitude.

The 370-kilogram (814-pounds-mass [lbm]) spacecraft was carried into LOW EARTH ORBIT (LEO) by *Discovery* on October 6, 1990. Then it was deployed successfully on an interplanetary trajectory that encountered Jupiter (for a gravity-assistmaneuver) in February 1992, flew past the southern polar regions of the Sun (80 degrees south solar latitude) in September 1994, and then passed over the northern polar regions of the Sun in the fall 1995, ending the prime mission of this nuclear-powered spacecraft. After completing its first solar orbit on September 29, 1995, the *Ulysses* spacecraft began its voyage back out to Jupiter, where it will loop around and return to the vicinity of the Sun in September 2000. At that time the Sun will be in a very active phase of its 11-year solar cycle, and *Ulysses* will find itself battling through the atmosphere of a star that is no longer docile.

Ulysses is the first spacecraft to explore the third dimension of space over the poles of the Sun. When the spacecraft passed over those polar regions in 1994 and 1995, scientists made some surprising new discoveries. For example, two clearly separate and distinct solar wind regimes exist, with fast wind emerging from the solar poles. Scientists also were surprised to observe how cosmic rays make their way into the solar system from galaxies beyond the MILKY WAY GALAXY (our home galaxy). The magnetic field of the Sun over its poles turned out to be very different from previous expectations, based on observations

from Earth. Finally, *Ulysses* detected a beam of particles from interstellar space that was penetrating the solar system at a velocity of about 80,000 kilometers per hour (50,000 miles per hour). (This beam velocity corresponds to about 22.22 km/s).

From September 20, 2000, until January 16, 2001, the *Ulysses* spacecraft successfully conducted its second scientific investigation of the south polar region of the Sun. After crossing the ECLIPTIC on May 25, 2001, the far-traveling ROBOT SPACECRAFT began its second pass (on August 31, 2001) over the northern polar region of the Sun. The second northern polar pass ended on December 10, 2001. When *Ulysses* made its first investigation of the Sun's poles between 1994 and 1996, the spacecraft encountered a QUIET SUN. However, during its most recent investigation of the the solar poles, *Ulysses* visited an ACTIVE SUN that had reached solar maximum. The hardy spacecraft has provided continuous observations of the Sun and the heliosphere for more than a decade and will still collect useful data for the space science community throughout the first decade of the 21st century.

This mission was originally called the *International Solar Polar Mission (ISPM)*. The original mission planned for two spacecraft, one built by NASA and the other by ESA. However, NASA canceled its spacecraft part of the original mission in 1981 and instead provided launch and tracking support for the single spacecraft built by ESA. *See also* SPACE NUCLEAR POWER; SUN.

umbilical An electrical or fluid servicing line between the ground or a tower and an upright rocket vehicle before launch. Also called *umbilical cord*.

UMR injector Injector that produces a uniform mixture ratio (UMR) and thus a combustion region with relatively uniform temperature distribution.

underexpansion A condition in the operation of a rocket nozzle in which the exit area of the nozzle is insufficient to permit proper expansion of the propellant gas. Consequently, the existing gas is at a pressure greater than the ambient pressure, and this leads to the formation of external expansion waves.

unidentified flying object (UFO) A flying object (apparently) seen in Earth's skies by an observer who cannot determine its nature. The vast majority of such "UFO" sightings can, in fact, be explained by known phenomena. However, these phenomena may be beyond the knowledge or experience of the person making the observation. Common phenomena that have given rise to UFO reports include artificial Earth satellites, aircraft, high-altitude weather balloons, certain types of clouds, and the planet Venus.

Modern interest in UFOs appears to have begun with a sighting report made by a private pilot named Kenneth Arnold. In June 1947 he reported seeing a mysterious formation of shining disks in the daytime sky near Mount Rainier in Washington State. When newspaper reporters heard of his account of "shining saucerlike disks," the popular term "flying saucer" was born.

In 1948 the United States Air Force began to investigate these UFO reports. *Project Sign* was the name given by the air force to its initial study of UFO phenomena. In the late 1940s Project Sign was replaced by *Project Grudge,* which in turn became the more familiar *Project Blue Book.* Under Project Blue Book, headquartered at Wright-Patterson Air Force Base in Ohio, the air force investigated many UFO reports from 1952 to 1969. Then on December 17, 1969 the secretary of the air force announced the termination of the project. Of a total of 12,618 sightings reported to Project Blue Book, 701 remained "unidentified."

Since the termination of this project, nothing has occurred that would support a resumption of UFO investigation by the air force. Similarly, while NASA is now the focal point for answering public inquiries to the White House concerning UFOs, it is not engaged in a research program involving these UFO phenomena or sightings.

unit The unit defines a measurement of a physical quantity, such as length, mass, or time. There are two common unit systems in use in aerospace applications today, the International System of Units (SI), which is based on the meter-kilogram-second (mks) set of fundamental units, and the American standard system of units, which is based on the foot-pound-mass-second (fps) set of fundamental units. In the standard system of units, 1 pound-mass (lbm) is defined as equaling 1 poundforce (lbf) on the surface of Earth at sea level. Derived units, such as energy, power, and force, are based on combinations of the fundamental units in accordance with physical laws, such as NEWTON'S LAWS OF MOTION.

United States Air Force Starting with the nuclear missile race during the COLD WAR, the United States Air Force, through various commands and organizations, has served as the primary agent for the space defense needs of the United States. For example, all military satellites have been launched from CAPE CANAVERAL AIR FORCE STATION, Florida, or VANDENBERG AIR FORCE BASE (AFB), California—rocket ranges owned and operated by the U.S. Air Force. The end of the cold war (circa 1989) and the dramatic changes in national defense needs at the start of the 21st century have placed additional emphasis on access to space and the use of military space systems for intelligence gathering, surveillance, secure and dependable information exchange, and the protection and enhancement of American fight forces deployed around the world.

The Air Force Space Command (AFSPC) was created on September 1, 1982. Headquartered at Peterson AFB, Colorado, it is the major command that directly discharges much of the U.S. Air Force for national defense responsibilities with respect to space systems and land-based INTERCONTINENTAL BALLISTIC MISSILES (ICBMs). Through its subordinate organization at Vandenberg AFB, for example, the AFSPC provides space war fighting forces to the U.S. STRATEGIC COMMAND and warning support to the NORTH AMERICAN AIR DEFENSE COMMAND (NORAD). AFSPC also exercises overall responsibility for the development, acquisition, operation, and maintenance of land-based ICBMs. The AFSPC's Space Warfare Center at Schriever AFB, Colorado, plays a major role in fully integrating space systems into the operational components of today's air force. The Space Warfare Center's force enhancement mission examines ways to use space systems to support fighters in the areas of navigation, weather, intelligence, communications, and theater ballistic missile warning and also how these apply to theater operations. Finally, AFSPC's Space and Missile Center at Los Angeles AFB, California, designs and acquires all U.S. Air Force and most Department of Defense space systems. It oversees launches, completes on-orbit CHECKOUTS, and then turns the various space systems over to the appropriate user agencies.

United States Naval Observatory (USNO)
The astronomical OBSERVATORY founded by the United States government in Washington, D.C., in 1844 to perform ASTROMETRY in support of timekeeping and navigation.

United States Space Command (USSPACECOM)
A unified command of the Department of Defense with headquarters at Peterson Air Force Base, Colorado Springs, Colorado. USSPACECOM was activated on September 23, 1985, to consolidate and operate military assets affecting U.S. activities in space. The command was composed of three components: the Air Force, Naval, and Army Space Commands. The USSPACECOM had three primary missions: space operations, warning, and planning for BALLISTIC MISSILE DEFENSE (BMD). The commander in chief, USSPACECOM, was directly responsible to the president through the secretary of defense and the Joint Chiefs of Staff. USSPACECOM exercises operational command of assigned U.S. military space assets through the air force, navy, and army component space commands. Merged into the U.S. Strategic Command (USSTRATCOM) on October 1, 2002.

United States Strategic Command (USSTRATCOM)
The United States Strategic Command (USSTRATCOM) was formed on October 1, 2002, with the merger of the UNITED STATES SPACE COMMAND (USSPACECOM) and USSTRATCOM—the strategic forces organization within the Department of Defense. The merger was part of an ongoing initiative within the Department of Defense to transform the American military into a 21st-century fighting machine. The blending of the two previously existing organizations improved combat effectiveness and accelerated the process of information collection and assessment needed for strategic decision making. As a result, the merged command is now responsible for both early warning of and defense against missile attacks as well as long-range strategic attacks.

The newly created United States Strategic Command is headquartered at Offutt Air Force Base, Nebraska—the well-known headquarters of the U.S. Air Force's Strategic Air Command during the COLD WAR. USSTRATCOM is one of nine unified commands within the Department of Defense. The command serves as the command and control center for U.S. strategic (nuclear) forces and controls military space operations, computer network operations, information operations, strategic warning and intelligence, and global strategic planning. Finally, the command is charged with deterring and defending against the proliferation of WEAPONS OF MASS DESTRUCTION.

USSTRATCOM coordinates the use of the Department of Defense's military space forces in providing missile warning, communications, navigation, weather, imagery, and signals intelligence. It provides space support to deployed American military forces worldwide and also defends the information technology structure within the Department of Defense.

The command's space missions consist of space support, force enhancement, space control, and force application. All military SATELLITES are launched from CAPE CANAVERAL AIR FORCE STATION, Florida, or VANDENBERG AIR FORCE BASE, California. Once the satellites reach their final orbits and begin operating, U.S. Army, Navy, and/or Air Force personnel track and "fly" the spacecraft and operate their specialized payloads through a worldwide network of ground stations. USSTRATCOM has several military space components—the Army Space Command (ARSPACE), in Arlington, Virginia; the Naval Space Command (NAVSPACE) in Dahlgren, Virginia; and Space Air Force (SPACEAF) at Vandenberg Air Force Base in California.

universal time (UT) The worldwide civil time standard, equivalent to GREENWICH MEAN TIME (GMT).

universal time coordinated (UTC) The worldwide scientific standard of timekeeping, based on carefully maintained atomic clocks. It is kept accurate to within microseconds. The addition (or subtraction) of "leap seconds" as necessary at two opportunities every year keeps UTC in step with Earth's rotation. Its reference point is Greenwich, England. When it is midnight there on Earth's prime meridian, it is midnight (00:00:00.000000) UTC, often referred to as "all balls" in aerospace jargon.

universe Everything that came into being at the moment of the big bang, and everything that has evolved from that initial mass of energy; everything that we can (in principle) observe. All energy (radiation), all matter, and the space that contains them. *See also* BIG BANG THEORY.

unmanned Without human crew; *unpersoned* or *uncrewed* are more contemporary terms.

unmanned aerial vehicle (UAV) A robot AIRCRAFT flown and controlled through TELEOPERATION by a distant human operator. Also called a REMOTELY PILOTED VEHICLE (RPV).

upcomer Nozzle tube in which coolant flows in a direction opposite to that of the exhaust gas flow.

uplink The TELEMETRY SIGNAL sent from a ground station to a SPACECRAFT or PLANETARY PROBE.

uplink data Information that is passed from a ground station on Earth to a spacecraft, space probe, or space platform. *See also* TELECOMMUNICATIONS.

upper atmosphere The general term applied to outer layers of Earth's atmosphere above the troposphere. It includes the stratosphere, mesosphere, thermosphere, and exosphere. Sounding rockets and Earth-observing satellites often are used to gather scientific data about this region. *See also* ATMOSPHERE; UPPER ATMOSPHERE RESEARCH SATELLITE.

upper stage In general, the second, third, or later stage in a multistage rocket. Often getting into LOW EARTH ORBIT (LEO) is only part of the effort against the planet's gravitational forces for geosynchronous satellites and interplanetary spacecraft. Once propelled into LEO by a reusable aerospace vehicle, such as the space shuttle, or an expendable launch vehicle, such as the Delta or Titan, these payloads depend on an upper-stage vehicle to boost them on the next phase of their journey. Upper-stage vehicles carried by the space shuttle are solid-propellant rocket vehicles that provide the extra thrust, or "kick," needed to move spacecraft into higher orbits or interplanetary trajectories. Solid-propellant or liquid-propellant rocket vehicles are used as upper stages with expendable launch vehicles.

The upper-stage vehicle, its payload, and any required supporting hardware (e.g., a special cradle attached to the shuttle's cargo bay or an adapter mounted to the first or second stage of an expendable launch vehicle) are carried into LEO by the launch vehicle. Once in LEO, the upper-stage vehicle and its attached payload separate from the launch vehicle and fire the rocket engine (or engines) necessary to deliver the payload to its final destination. *See also* AGENA; CENTAUR; INERTIAL, UPPER STAGE (IUS); PAYLOAD ASSIST MODULE.

Uranian Of or relating to the planet Uranus. *See also* URANUS.

Uranus Unknown to ancient astronomers, the planet Uranus was discovered by Sir William Herschel in 1781. Initially called *Georgium Sidus* (George's star, after England's King George III) and *Herschel* (after its discoverer), in the 19th century the seventh planet from the Sun was finally named Uranus after the ancient Greek god of the sky and father of the Titan Cronos (Saturn in Roman mythology).

At nearly 3 billion kilometers from the Sun, Uranus is too distant from Earth to permit telescopic imaging of its features by ground-based telescopes. Because of the methane in its upper atmosphere, the planet appears as only a blue-green disk or blob in the most powerful of terrestrial telescopes. On January 24, 1986, a revolution took place in our understanding and knowledge about this planet, as the *Voyager 2* encountered the Uranian system at a relative veloc-

ity of over 14 kilometers per second. What we know about Uranus today is largely the result of that spectacular encounter.

Uranus has one particularly interesting property—its axis of rotation lies in the plane of its orbit rather than vertical to the orbital plane, as occurs with the other planets. Because of this curious situation, Uranus moves around the Sun like a barrel rolling along on its side rather than like a top spinning on its end. In other words, Uranus is tipped over on its side, with its orbiting moons and rings creating the appearance of a giant bull's-eye. The northern and southern polar regions are alternatively exposed to sunlight or to the darkness of space during the planet's 84-year-long orbit around the Sun. At its closest approach, *Voyager 2* came within

Table 1. Selected Physical and Dynamic Properties of the Planet Uranus	
Diameter (equatorial)	51,120 km
Mass (estimated)	8.7×10^{25} kg
"Surface" gravity	8.69 m/s^2
Mean density (estimated)	1.3 g/cm^3
Albedo (visual)	0.66
Temperature (blackbody)	58 kelvins (K)
Magnetic field	Yes, intermediate strength (field titled 60° with respect to axis of rotation)
Atmosphere	Hydrogen (~ 83%) helium (~15%) methane (~2%)
"Surface" features	Bland and featureless (except for some discrete methane clouds)
Escape velocity	21.3 km/s
Radiation belts	Yes (intensity similar to those at Saturn)
Rotation period	17.24 hours
Eccentricity	0.047
Mean orbital velocity	6.8 km/s
Sidereal year (a Uranian year)	84 years
Inclination of planet's equator to its orbit around the Sun	97.9°
Number of (known) natural satellites	21
Rings	Yes (11)
Average distance from Sun	2.871×10^9 km (19.19 AU) [159.4 light-min]
Solar flux at average distance from Sun	3.7 W/m^2 (approx.)

Source: Based on NASA data.

Table 2. Selected Physical and Dynamic Property Data
for Major Uranian Moons

Name	Diameter (km)	Period (day)	Distance from Center of Uranus (km)	Visual Albedo	Average Density (g/cm³)
Miranda	470	1.414	129,800	0.27	1.3
Ariel	1,160	2.520	191,200	0.34	1.7
Umbriel	1,170	4,144	266,000	0.18	1.5
Titania	1,580	8,706	435,800	0.27	1.7
Oberon	1,520	13,463	582,600	0.24	1.6

Source: Based on NASA data.

81,500 kilometers of the Uranian cloud-tops. The spacecraft telemetered back to Earth thousands of spectacular images and large quantities of other scientific data about the planet, its moons, rings, atmosphere, and interior.

The upper atmosphere of Uranus consists mainly of hydrogen (H_2) (~83 percent) and helium (He) (approximately 15 percent) with small amounts of methane (CH_4) (about 2 percent), water vapor (H_2O), and ammonia (NH_3). The methane in the upper atmosphere of Uranus and its preferential absorption of red light gives the planet its overall blue-green color.

Table 1 presents selected physical and dynamic property data for Uranus, while Table 2 describes some of the physical features of the major Uranian moons. The large Uranian moons appear to be about 50% water ice, 20% carbon- and nitrogen-based materials, and 30% rock. Their surfaces, almost uniformly dark gray in color, display varying degrees of geologic history. Very ancient, heavily cratered surfaces are apparent on some of the moons, while others show strong evidence of internal geologic activity. Miranda, the innermost of the five large Uranian moons, is considered by scientists to be one of the strangest bodies yet observed in the solar system. *Voyager 2* images revealed an unusual world consisting of huge fault canyons as deep as 20 kilometers terraced layers, and a mixture of old and young surfaces. *See also* VOYAGER.

vacuum The absence of gas or a region in which there is a very low gas pressure. This is a relative term. For example, a *soft vacuum* (or *low vacuum*) has a pressure of about 0.01 pascal (i.e., 10^{-2} pascal); a *hard vacuum* (or *high vacuum*) typically has a pressure between 10^{-2} and 10^{-7} pascal; while pressures below 10^{-7} pascal are referred to as an *ultrahard* (or *ultrahigh*) vacuum.

Valles Marineris An extensive canyon system on MARS near the PLANET's EQUATOR, discovered in 1971 by NASA's MARINER 9 SPACECRAFT.

valve Mechanical device by which the flow of fluid may be started, stopped, or regulated by a movable part that opens, closes, or partially obstructs a passageway in a containing structure, called the *valve housing.*

Van Allen radiation belts A doughnut-shaped zone of high-intensity particulate radiation around Earth from an altitude of about 320 to 32,400 kilometers (200 to 20,000 miles) above the magnetic equator. The radiation of the Van Allen belts is composed of protons and electrons temporarily trapped in Earth's magnetic field. Spacecraft and their occupants or sensitive equipment orbiting within the belts or passing through them must be protected against this ionizing radiation. The existence of the belts was first confirmed by instruments placed on EXPLORER I (the first U.S. Earth satellite) by Dr. James Van Allen (1914–), after whom the region is now named. *See also* EARTH'S TRAPPED RADIATION BELTS.

Vandenberg Air Force Base (VAFB) Vandenberg Air Force Base is located 89 kilometers (55 miles) north of Santa Barbara near Lompoc, California. It is the site of all military, NASA, and commercial space launches accomplished on the West Coast of the United States. The base, named in honor of General Hoyt S. Vandenberg (Air Force Chief of Staff from 1948 to 1953), also provides launch facilities for the testing of intercontinental ballistic missiles. The first missile was launched from Vandenberg in 1958 and the world's first polar-orbiting satellite was launched from there aboard a Thor/Agena launch vehicle in 1959. *See also* CAPE CANAVERAL AIR FORCE STATION.

Vanguard Project An early U.S. space project involving a new three-stage rocket *(Vanguard rocket)* and a series of scientific spacecraft *(Vanguard satellites.)* The U.S. Navy successfully launched the first satellite in this project, called *Vanguard 1,* on March 17, 1958. This tiny Earth-orbiting satellite provided information leading in 1959 to the identification of the slight, but significant, pear shape of the Earth. An earlier, widely publicized launch attempt had ended in dramatic failure, when the Vanguard rocket exploded just as it lifted off the launch pad. The *Vanguard 2* satellite was launched on February 17, 1959. It transmitted the world's first picture of Earth's cloud cover (as observed from a satellite), but a wobble in the spacecraft due to an inadvertent bump by its launch vehicle resulted in a less than satisfactory picture quality. Finally, the *Vanguard 3* satellite was launched on September 18, 1959. This early spacecraft deepened our understanding of the space environment, including the VAN ALLEN RADIATION BELT and micro-meteoroids.

vapor The gaseous phase of a substance; in thermodynamics, this term often is used interchangeably with *gas.*

vaporization In thermodynamics, the transition of a material from the liquid phase to the gaseous (or vapor) phase, generally as a result of heating or pressure change.

vapor pressure 1. The pressure exerted by the atoms or molecules of a given vapor. For a pure substance confined within a container, it is the vapor's pressure on the walls of its containing vessel; for a vapor (or gas) mixed with other vapors (or gases), it is that particular vapor's contribution to the total pressure (i.e., its partial pressure). In meteorology, the term "vapor pressure" generally is used to denote the partial pressure of water vapor in the atmosphere. 2. The sum of the partial pressures of all the vapors in a system.

vapor turbine A turbine in which part of the thermal energy (heat) supplied by a vapor is converted into mechanical work of rotation. The *steam turbine* is a common type of vapor turbine. Sometimes called a *condensing turbine. See also* RANKINE CYCLE; TURBINE.

variable-area exhaust nozzle An exhaust nozzle on a jet engine or a rocket engine that has an exhaust opening that can be varied in area by means of some mechanical device. *See also* NOZZLE.

variable star A STAR that does not shine steadily but whose brightness (LUMINOSITY) changes over a short period of time.

vector Any physical quantity, such as force, velocity, or acceleration, that has both magnitude and direction at each point in space, as opposed to a *scalar,* which has magnitude only.

vector steering A steering method for rockets and spacecraft in which one or more thrust chambers are gimbal-mounted ("gimbaled") so that the direction of the thrust (i.e., the thrust vector) may be tilted in relation to the vehicle's center of gravity to produce a turning movement.

Vega Twin Russian spacecraft that were launched in December 1984 and performed a probe/flyby mission to Venus and then continued on for an encounter with Comet Halley. These spacecraft were modified versions of the Venera spacecraft. In June 1985 the *Vega 1* and 2 spacecraft flew by Venus. Each spacecraft dropped off a probe package (consisting of an instrumented balloon and a lander) and then proceeded, using a gravity assist from Venus, to encounter Comet Halley. *Vega 1* flew past Comet Halley on March 6, 1986, coming within 9,000 kilometers of its nucleus. During its closest approach, the *Vega 1* spacecraft acquired hundreds of images of the comet, as well as other important scientific data. Similarly, *Vega 2* encountered Comet Halley on March 9, 1986 and also returned many images and important data. The Russians quickly provided these images and data to scientists at the European Space Agency (ESA), so they could make last-minute adjustments in the trajectory of the Giotto spacecraft prior to its encounter with Comet Halley on March 14, 1986. The *Vega* spacecraft are now in orbit around the Sun. *See also* COMET; *GIOTTO* SPACECRAFT; VENERA; VENUS.

vehicle In general, an aerospace structure, machine, or device (e.g., a rocket) that is designed to carry a payload through the atmosphere and/or space; more specifically a rocket vehicle. *See also* LAUNCH VEHICLE.

vehicle tank Tank that serves both as a primary integral structure of a rocket vehicle (or spacecraft) and as a container of pressurized propellants.

Vela A family of research and development spacecraft launched by the United States in the 1960s and early 1970s to detect nuclear detonations in the atmosphere down to Earth's surface, or in outer space at distances of more than 160 million kilometers (km) (100 million miles

[mi]). These spacecraft were jointly developed by the Department of Defense and the Atomic Energy Commission (now Department of Energy) and were placed in pairs, 180 degrees apart in very high altitude (about 115,000 km [71,500 mi]) orbits around Earth. The first pair of Vela spacecraft, called *Vela 1A* and *Vela 1B*, were launched successfully on October 17, 1963. The last pair of these highly successful, 26-sided (polyhedron-shaped) spacecraft, called *Vela 6A* and *Vela 6B*, were launched successfully on April 8, 1970. It is interesting to note that the United States, the former Soviet Union, and the United Kingdom signed the *Limited Nuclear Test Ban Treaty* in October 1963. This treaty prohibits the signatories from testing nuclear weapons in Earth's atmosphere, under water, or in outer space. In addition to supporting important U.S. government nuclear test monitoring objectives, the Vela satellites also supported a modest revolution in astrophysics. Between 1969 and 1972, the Vela satellites detected 16 very short bursts of gamma ray photons with energies of 0.2 to 1.5 million electron volts. These mysterious cosmic gamma ray bursts lasted from less than a tenth of a second to about 30 seconds. Although the Vela instruments were not designed primarily for astrophysical research, simultaneous observations by several spacecraft started astrophysicists on their contemporary hunt for "gamma ray bursters." *See also* GAMMA RAY ASTRONOMY.

velocity   A vector quantity that describes the rate of change of position. Velocity has both magnitude (speed) and direction, and it is expressed in terms of units of length per unit of time (e.g., meters per second).

velocity of light (symbol: c)   *See* SPEED OF LIGHT.

Venera   A family of mostly successful robotic space missions flown by Russia (i.e., the former Soviet Union) to the planet Venus between 1961 and 1984. These missions included orbiters, landers, and atmospheric probes. In October 1967, for example, the *Venera 4* spacecraft placed a landing capsule/probe into the Venusian atmosphere, collecting data that indicated the planet's atmosphere was from 90 to 95% carbon dioxide (CO_2). This 380-kilogram (kg) probe descended by parachute for about 94 minutes when data transmissions ceased at an altitude of some 25 kilometers (km). In December 1970 the 495-kg atmospheric probe from the *Venera 7* spacecraft reached the surface of Venus and transmitted data for about 20 minutes. The *Venera 8* spacecraft launched a probe into the Venusian atmosphere in July 1972. This probe, with an improved communications system, successfully landed on the surface and survived for about 50 minutes in the infernolike conditions found on the surface of Venus.

In October 1975 the *Venera 9* and *10* spacecraft sent probes to the surface of Venus that landed successfully and transmitted the first black-and-white images of the planet's rock-strewn surface. *Venera 11* and *12*, launched in 1978, were also landerprobe/flyby missions with improved sensors.

In the 1980s, the Russians launched four more sophisticated Venera spacecraft to Venus. The *Venera 13* and *14* spacecraft were 5,000-kg flyby/lander configurations sent to the planet at the end of 1981. *Venera 13* landed on Venus on March 1, 1982, and its identical companion, *Venera 14,* touched down on March 5, 1982. These landers returned black-and-white and color images of the Venusian surface. They also performed the first soil analysis of the Venusian surface.

The final pair of Venera spacecraft (called *Venera 15* and *16*), were launched in June 1983. These spacecraft did not carry landers but rather performed mapping missions of the planet's cloud-enshrouded surface, using their synthetic aperture radar (SAR) systems. The spacecraft orbited for about one year and produced detailed radar images of the planet's surface at between one and two kilometers resolution. *See also* VENUS.

vengeance weapon 2 (Vergeltungwaffe 2)   *See* V-2 ROCKET.

vent-and-relief valve Specialized version of a relief valve wherein the valve assembly acts as an outlet for ullage vapor during the filling of a tank (e.g., propellant tank) and then performs as a relief valve during operations.

venturi tube A short tube having a smaller diameter in the middle than at the ends. This device is used for measuring the quantity of fluid flowing through a pipe. When a fluid flows through such a tube, its pressure decreases and its velocity increases as the diameter becomes smaller. The amount of pressure decrease is proportional to the speed of the flow and the amount of flow restriction.

vent valve Pressure-relieving valve that is operated on external command, as contrasted to a relief valve, which opens automatically when pressure reaches a given level.

Venus Venus is the second planet out from the Sun. Because the planet appears to observers on Earth, as either an evening or a morning star, it is often called the Evening Star or the Morning Star. Venus is named after the Roman goddess of love and beauty. Among the planets in our solar system, it is the only one named after a female mythological deity. It is also called an inferior planet, because it revolves around the Sun within the orbit of Earth. The planet maintains an average distance of about 0.723 astronomical unit (AU) (~108 million kilometers) from the Sun. The table provides physical and dynamic data for Earth's nearest planetary neighbor. At closest approach, Venus is approximately 42 million kilometers from Earth.

Since the 1960s, visits by numerous American and Russian spacecraft have now dispelled the pre-space age romantic fantasies that this cloudenshrouded planet was a prehistoric world that mirrored a younger Earth. Except for a few physical similarities of size and gravity, Earth and Venus are very different worlds. For example, the surface temperature on Venus approaches 500° Celsius (773° Kelvin), its atmospheric pressure is more than 90 times that of Earth, it has no surface water, and its dense, hostile atmosphere with sulfuric acid clouds and an overabundance of carbon dioxide (about 96%) represents a runaway greenhouse of disastrous proportions.

Cloud-penetrating synthetic aperture radar (SAR) systems carried by the American *Pioneer* Venus orbiter and Magellan spacecraft and the Russian *Venera 15* and *16* spacecraft have provided a detailed characterization of the previously unobservable Venusian surface. These radar imagery data, especially the high-resolution images collected by the Magellan mission, are now challenging planetary scientists to explain some of the interesting findings.

Scientists generally divide the surface of Venus into three classes of terrain: highlands (or *tesserae*) (10% of surface); rolling uplands (70%); and lowland plains (20%). One very prominent highland region is called Ishtar Terra, after Ishtar, the ancient Babylonian goddess of love. This region is about the size of Australia and stands several kilometers above the average planetary radius. Ishtar Terra contains the highest peaks yet discovered on Venus, including Maxwell Montes, which is often regarded as the single most impressive topographical feature on the planet.

The relatively fresh appearance and small number of impact craters suggest to planetary scientists that the surface as a whole is young (by geologic standards). It also appears that Venus experienced a dramatic resurfacing event some 300 to 500 million years ago. But whether this resurfacing phenomenon (which covered most of the planet with lava) was caused by a brief chain of catastrophic events or by the influences of low-level volcanism operating over longer periods of time is currently the subject of scientific investigation and debate. The fractured Venusian highlands, such as Ishtar Terra and Aphrodite Terra, represent the planet's older surface material not covered by younger flows of lava. Maat Mons is the largest volcano on Venus, and it appears to have experienced a recent surge of volcanic activity. A volcanic feature unique to the planet are the *coronae*—large (often hundreds of kilometers across), heav-

Physical and Dynamic Properties of Venus	
Diameter (equatorial)	12,100 km
Mass	4.87×10^{24} kg
Density (mean)	5.25 g/cm^3
Surface gravity	8.88 m/sec^2
Escape velocity	10.4 km/sec
Albedo (over visible spectrum)	0.7–0.8
Surface temperature (approx.)	750 K (477°C)
Atmospheric pressure (at surface)	9600 kPa (~1,400 psi)
Atmosphere	CO_2 (96.4%),
(main components)	N_2 (3.4%)
(minor components)	Sulfur dioxide (150 ppm), argon (70 ppm), water vapor (20 ppm)
Surface wind speeds	0.3–1.0 m/sec
Surface materials	Basaltic rock and altered materials
Magnetic field	Negligible
Radiation belts	None
Number of natural satellites	None
Average distance from Sun	1.082×10^8 km (0.723 AU)
Solar flux (at top of atmosphere)	2,620 W/m^2
Rotation period (a Venusian "day")	243 days (retrograde)
Eccentricity	0.007
Mean orbital velocity	35.0 km/sec
Sidereal year (period of one revolution around Sun)	224.7 days
Earth-to-Venus distances	
Maximum	2.59×10^8 km (1.73 AU)
Minimum	0.42×10^8 km (0.28 AU)

Source: NASA.

ily fractured circular regions that sometimes are surrounded by a trench. The channels, or *canali*, on Venus are long, riverlike surface features where lava once flowed.

Yet despite all the interesting facts that have been revealed about Venus by the American and Russian space missions, equally interesting questions also have been raised by scientists as they examined these new data. Questions such as "What was Venus like before the resurfacing event?" await resolution in the 21st century, when even more rugged probes and landers again visit Earth's nearest planetary neighbor. *See also* MAGELLAN MISSION; PIONEER-VENUS; VENERA.

Venusian   Of or pertaining to the planet Venus.

vernal equinox   The spring EQUINOX, which occurs on or about March 21.

vernier engine   A rocket engine of small thrust used primarily to obtain a fine adjustment in the velocity and trajectory or in the attitude of a rocket or aerospace vehicle.

Very Large Array (VLA)   The Very Large Array (VLA) is a spatially extended radiotelescope facility at Socorro, New Mexico. It consists of 27 antennas, each 25 meters (82 feet) in diameter, that are configured in a giant "Y" arrangement on railroad tracks over a 20-kilometer (12.5-mile) distance. The VLA is operated by the National Radio Astronomy Observatory and sponsored by the National Science Foundation.

very large scale integration (VLSI)

An integrated circuit containing more than 64,000 transistors. *See also* INTEGRATED CIRCUIT.

Viking Project

The Viking Project was the culmination of an initial series of American missions to explore Mars in the 1960s and 1970s. This series of interplanetary missions began in 1964 with *Mariner 4,* continued with the *Mariner 6* and 7 flyby missions in 1969, and then the *Mariner 9* orbital mission in 1971 and 1972.

Viking was designed to orbit Mars and to land and operate on the surface of the Red Planet. Two identical spacecraft, each consisting of a lander and an orbiter, were built.

The orbiters carried the following scientific instruments:

1. A pair of cameras with 1,500-millimeter focal length that performed systematic searches for landing sites, then looked at and mapped almost 100% of the Martian surface. Cameras onboard the *Viking 1* and *Viking 2* orbiters took more than 51,000 photographs of Mars.
2. A Mars atmospheric water detector that mapped the Martian atmosphere for water vapor and tracked seasonal changes in the amount of water vapor.
3. An infrared thermal mapper that measured the temperatures of the surface, polar caps, and clouds; it also mapped seasonal changes. In addition, although the Viking orbiter radios were not considered scientific instruments, they were used as such. By measuring the distortion of radio signals as these signals traveled from the Viking orbiter spacecraft to Earth, scientists were able to measure the density of the Martian atmosphere.

The Viking landers carried the following instruments: (See Figure 73.)

1. The biology instrument, consisting of three separate experiments designed to detect evidence of microbial life in the Martian soil.
2. A gas chromatograph/mass spectrometer (GCMS) that searched the Martian soil for complex organic molecules.
3. An X-ray fluorescence spectrometer that analyzed samples of the Martian soil to determine its elemental composition.
4. A meteorology instrument that measured air temperature and wind speed and direction at the landing sites.
5. A pair of slow-scan cameras that were mounted about 1 meter (3.28 feet) apart on the top of each lander. These cameras provided black-and-white, color, and stereo photographs of the Martian surface.
6. A seismometer had been designed to record any "Marsquakes" that might occur on the Red Planet. Such information would have helped planetary scientists determine the nature of Mars's internal structure. Unfortunately, the seismometer on Lander 1 did not function after landing and the instrument on Lander 2 observed no clear signs of internal (tectonic) activity.
7. An upper-atmosphere mass spectrometer that conducted its primary measurements as each lander plunged through the Martian atmosphere on its way to the landing site. This instrument made the lander's first important scientific discovery—the presence of nitrogen in the Martian atmosphere.
8. A retarding potential analyzer that measured the Martian ionosphere, again during entry operations.
9. Accelerometers, a stagnation pressure instrument, and a recovery temperature instrument that helped determine the structure of the lower Martian atmosphere as the landers approached the surface.
10. A surface sampler boom that employed its collector head to scoop up small quantities of Martian soil to feed the biology, organic chemistry, and inorganic chemistry instruments. It also provided clues to the soil's physical properties. Magnets attached to the sampler, for example, provided information on the soil's iron content.

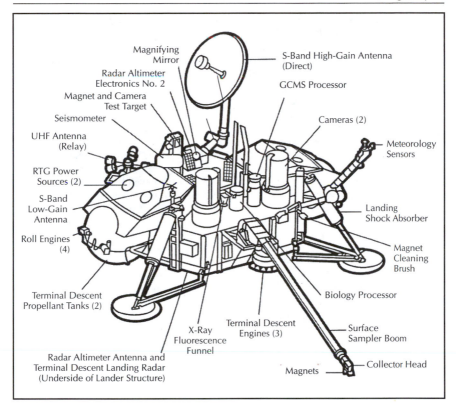

Figure 73. The Viking lander spacecraft and its complement of instruments.

11. The lander radios also were used to conduct scientific experiments. Physicists were able to refine their estimates of Mars's orbit by measuring the time for radio signals to travel between Mars and Earth. The great accuracy of these radio-wave measurements also allowed scientists to confirm portions of Einstein's General Theory of Relativity.

Both Viking missions were launched from Cape Canaveral, Florida. *Viking 1* was launched on August 20, 1975, and *Viking 2* on September 9, 1975. The landers were sterilized before launch to prevent contamination of Mars by terrestrial microorganisms. These spacecraft spent nearly a year in transit to the Red Planet.

Viking 1 achieved Mars orbit on June 19, 1976; and *Viking 2* began orbiting Mars on August 7, 1976. The *Viking 1* lander accomplished the first soft landing on Mars on July 20, 1976, on the western slope of Chryse Planitia (the Plains of Gold) at 22.46 degrees north latitude, 48.01 degrees west longitude. The *Viking 2* lander touched down successfully on September 3, 1976, at Utopia Planitia located at 47.96 degrees north latitude, 225.77 degrees west longitude.

The Viking mission was planned to continue for 90 days after landing. Each orbiter and lander, however, operated far beyond its design lifetime. For example, the *Viking 1* orbiter exceeded four years of active flight operations in orbit around Mars.

The Viking Project's primary mission ended on November 15, 1976, just 11 days before Mars passed behind the Sun (an astronomical event called a "superior conjunction"). After conjunction, in mid-December 1976, telemetry and command operations were reestablished and extended mission operations began.

The *Viking 2* orbiter mission ended on July 25, 1978, due to exhaustion of attitude-control system gas. The *Viking 1* orbiter spacecraft also began to run low on attitude-control system gas, but through careful planning it was possible to continue collecting scientific data (at a reduced level) for another two years. Finally, with its control gas supply exhausted, the *Viking 1* orbiter's electrical power was commanded off on August 7, 1980.

The last data from the *Viking 2* lander were received on April 11, 1980. The *Viking 1* lander made its final transmission on November 11, 1982. After over six months of effort to regain contact with the *Viking 1* lander, the Viking mission came to an end on May 23, 1983.

With the single exception of the seismic instruments, the entire complement of scientific instruments of the Viking Project acquired far more data about Mars than ever anticipated. The primary objective of the lander was to determine whether (microbial) life currently exists on Mars. The evidence provided by the landers is still subject to debate, although most scientists feel these results are strongly indicative that life does *not now* exist on Mars. However, recent analyses of Martian meteorites have renewed interest in this very important question, and Mars is once again the target of intense scientific investigation by even more sophisticated scientific spacecraft. *See also* MARINER SPACECRAFT; MARS; *MARS GLOBAL SURVEYOR* MISSION; *MARS PATHFINDER*; MARS SAMPLE RETURN MISSION; MARTIAN METEORITES.

virtual reality (VR) A computer-generated artificial reality that captures and displays in varying degrees of detail the essence or effect of physical reality (i.e., the "real-world" scene, event, or process) being modeled or studied. With the aid of a data glove, headphones, and/or head-mounted stereoscopic display, a person is projected into the three-dimensional world created by the computer.

A virtual reality system generally has several integral parts. There is always a computerized description (i.e., the "database") of the scene or event to be studied or manipulated. It can be a physical place, such as a planet's surface made from digitized images sent back by robot space probes. It can even be more abstract, such as a description of the ozone levels at various heights in Earth's atmosphere or the astrophysical processes occurring inside a pulsar or a black hole.

VR systems also use a special helmet or headset ("goggles") to supply the sights and sounds of the artificial, computer-generated environment. Video displays are coordinated to produce a three-dimensional effect. Headphones make sounds appears to come from any direction. Special sensors track head motions, so that the visual and audio images shift in response.

Most VR systems also include a glove with special electronic sensors. This "data" glove lets a person interact with the virtual world through hand gestures. He or she can move or touch objects in the computer-generated visual display, and these objects then respond as they would in the physical world. Advanced versions of such gloves also provide artificial "tactile" sensations so that an object "feels like the real thing" being touched or manipulated (e.g., smooth or rough, hard or soft, cold or warm, light or heavy, flexible or stiff, etc.).

The field of virtual reality is quite new, and rapid advances should be anticipated over the next decade, as computer techniques, visual displays, and sensory feedback systems (e.g., advanced data gloves) continue to improve in their ability to project and model the real world. VR systems have many potential roles in aerospace. For example, sophisticated virtual reality systems will let scientists "walk on another world" while working safely here on Earth. Future mission planners will identify the best routes (based on safety, resource consumption, and mission objectives) for both robots and humans to explore the surface of the Moon and Mars, before the new missions are even launched. Astronauts will use VR training systems regularly to try out space maintenance and repair tasks and perfect their skills long before they lift off on an actual mission. Aerospace engineers will use VR system as an indispensable design tool to fully examine and test new aerospace hardware, long before any "metal is bent" in building even a prototype model of the item.

viscosity A measure of the internal friction or flow resistance of a fluid when it is subjected to shear stress. The *dynamic viscosity* is defined as the force that must be applied per unit area to permit adjacent layers of fluid to move with unit velocity relative to each other. The dynamic viscosity is sometimes expressed in *poise* (centimeter-gram-second [c.g.s.] system) or in pascal-seconds (SI unit system). One poise is equal to 0.1 newton-second per square meter (i.e., 1 poise = 10^{-1} N s m^{-2}). The *kinematic viscosity* is defined as the dynamic viscosity divided by the fluid's density. The kinematic viscosity can be expressed in *stokes* (centimeter-gram-second [c.g.s.] system) or in square meters per second (SI unit system). One stoke is equal to 10^{-4} m^2 s^{-1}. In general, the viscosity of a liquid usually decreases as the temperature is increased; the viscosity of a gas increases as the temperature increases. *See also* SI UNITS.

viscous fluid A fluid whose molecular viscosity is sufficiently large to make the viscous forces a significant part of the total force field in the fluid.

void In rocketry, an air bubble in a cured propellant grain or in a rocket motor insulation.

volcano A vent in the crust of a PLANET or MOON from which molten lava, GASes, and other pyroclastic materials flow.

volt (symbol: V) The SI UNIT of electric potential difference and electromotive

force. One volt is equal to the difference of electric potential between two points of a conductor carrying a constant current of 1 ampere when the power dissipated between these points equals 1 watt. This unit is named after the Italian scientist Count Alessandro Volta (1745–1827), who performed pioneering work involving electricity and electric cells.

volume (symbol: V) The space occupied by a solid object or a mass of fluid (liquid or confined gas).

Von Braun, Wernher (1912–1977) German-American ROCKET engineer and space travel advocate who developed the V-2 ROCKET during World War II for the German Army and then assisted the COLD WAR era U.S. space program in the development of both military rockets and civilian LAUNCH VEHICLES. Inspired by HERMANN J. OBERTH's vision of rockets for INTERPLANETARY travel, von Braun devoted his professional life to the development of ever more powerful LIQUID-PROPELLANT ROCKET ENGINES. In the mid-1950s, a professional friendship with Walt Disney, the entertainment genius, allowed von Braun to communicate his dream of space travel to millions of Americans. Von Braun actually fulfilled a significant portion of these dreams by developing the mighty SATURN V launch vehicle that successfully sent the first human beings to the MOON during the NASA's APOLLO PROJECT.

Voskhod An early Russian three-person spacecraft that evolved from the Vostok spacecraft. *Voskhod 1* was launched on October 12, 1964 and carried the first three-person crew into space. The cosmonauts, Vladimir Komarov, Konstantin Feoktistov, and Boris Yegorov, flew on a one-day duration Earth orbital mission. *Voskhod 2* was launched on March 18, 1965 and carried a crew of two cosmonauts, including Alexei Leonov, who performed the world's first "spacewalk" (about 10 minutes in duration) during the orbital mission. *Voskhod* means "sunrise" in Russian.

Vostok   The first Russian manned spacecraft. This spacecraft was occupied by a single cosmonaut and consisted of a spherical cabin (about 2.3 meters in diameter) that was attached to a biconical instrument module. *Vostok 1* was launched on April 12, 1961, carrying Cosmonaut Yuri Gagarin, the first human to fly in space. Gagarin's flight made one orbit of Earth and lasted about 108 minutes.

Voyager   Once every 176 years the giant outer planets—Jupiter, Saturn, Uranus, and Neptune—align themselves in such a pattern that a spacecraft launched from Earth to Jupiter at just the right time might be able to visit the other three planets on the same mission, using a technique called *gravity assist*. NASA space scientists named this multiple giant planet encounter mission the "Grand Tour" and took advantage of a unique celestial alignment opportunity in 1977 by launching two sophisticated spacecraft, called *Voyager 1* and 2. (See Figure 74.)

Each Voyager spacecraft has a mass of 825 kilograms (1,815 pounds-mass [lbm]) and carries a complement of scientific instruments to investigate the outer planets and their many moons and intriguing ring systems. These instruments, provided electric power by a long-lived nuclear system called a radioisotope thermoelectric generator (RTG), recorded spectacular closeup images of the giant outer planets and their interesting moon systems, explored complex ring systems, and measured properties of the interplanetary medium.

Taking advantage of the 1977 Grand Tour launch window, the *Voyager 2* spacecraft lifted off from Cape Canaveral, Florida on August 20, 1977, onboard a Titan-Centaur rocket. (NASA called the first Voyager spacecraft launched *Voyager 2,* because the second Voyager spacecraft to be launched eventually would overtake it and become *Voyager 1*.) *Voyager 1* was launched on September 5, 1977. This spacecraft followed the same trajectory as its *Voyager 2* twin and overtook its sister ship just after entering the asteroid belt in mid-December 1977.

Voyager 1 made its closest approach to Jupiter on March 5, 1979, and then used Jupiter's gravity to swing itself to

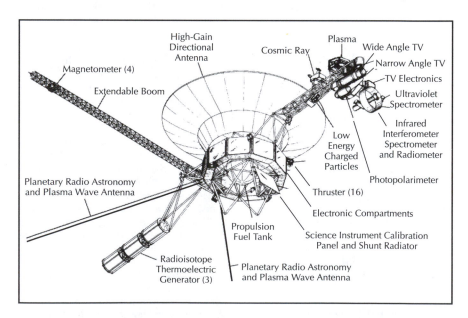

Figure 74. The Voyager spacecraft and its complement of sophisticated instruments.

Saturn. On November 12, 1980, *Voyager 1* successfully encountered the Saturnian system and then was flung up out of the ecliptic plane on an interstellar trajectory. The *Voyager 2* spacecraft successfully encountered the Jovian system on July 9, 1979 (closest approach), and then used the gravity assist technique to follow *Voyager 1* to Saturn. On August 25, 1981, *Voyager 2* encountered Saturn and then went on to successfully encounter both Uranus (January 24, 1986) and Neptune (August 25, 1989). Space scientists consider the end of *Voyager 2*'s encounter of the Neptunian system as the end of a truly extraordinary epoch in planetary exploration. In the first 12 years since they were launched from Cape Canaveral, these incredible spacecraft contributed more to our understanding of the giant outer planets of our solar system than was accomplished in over three millennia of Earth-based observations. Following its encounter with the Neptunian system, *Voyager 2* also was placed on an interstellar trajectory and (like its *Voyager 1* twin) now continues to travel outward from the Sun.

As the influence of the Sun's magnetic field and SOLAR WIND grow weaker, both Voyager spacecraft eventually will pass out of the heliosphere and into the interstellar medium. Through NASA's Voyager Interstellar Mission (VIM) (which began officially on January 1, 1990), the two Voyager spacecraft will continue to be tracked on their outward journey. The two major objectives of the VIM are: an investigation of the interplanetary and interstellar media, and a characterization of the interaction between the two; and a continuation of the successful Voyager program of ultraviolet astronomy. During the VIM, the spacecraft will search for the heliopause (the outermost extent of the solar wind, beyond which lies interstellar space). It is hoped that one Voyager spacecraft will still be functioning when it penetrates the heliopause and will provide scientists with the first true sampling of the interstellar environment. Barring a catastrophic failure on board either Voyager spacecraft, their nuclear power systems

should provide useful levels of electric power until at least 2015.

Since both Voyager spacecraft eventually would journey beyond the solar system, their designers placed a special interstellar message (a record entitled: "The Sounds of Earth") on each in the hope that perhaps millions of years from now, some intelligent alien race will find either spacecraft drifting quietly through the interstellar void. If they are able to decipher the instructions for using this record, they will learn about our contemporary terrestrial civilization and the men and women who sent Voyager on its stellar journey. *See also* JUPITER; NEPTUNE; ORBITS OF OBJECTS IN SPACE; SATURN; URANUS.

Vulcan The hypothetical (but nonexistent) PLANET that some 19th-century astronomers believed existed in the extremely hot ORBIT between MERCURY and the SUN.

V-2 rocket The German V-2 rocket of World War II is the "grandfather" of many

Figure 75. A reassembled German V-2 rocket lifting off its launch pad at the U.S. Army's White Sands Missile Range in southern New Mexico, circa 1947. *(Courtesy of U.S. Army)*

of the large missiles flown by the United States. Based on the findings made by the American rocket scientist Dr. Robert H. Goddard (1882–1945) in the 1920s and 1930s, a team of German rocket scientists developed this weapon system at Peenemunde, Germany, and terrorized Allied populations of Europe and Britain until the end of the war. The German program began in early 1940; the first V-2 rocket was launched on July 6, 1942. In the closing days of World War II, the United States captured many German rocket scientists, including Dr. Wernher Von Braun (1912–1977), as part of Operation Paper Clip, and shipped these scientists plus numerous captured V-2 rocket compo-

nents to the United States. These V-2 rockets were then assembled and tested at the White Sands Missile Range in New Mexico as part of an emerging U.S. missile program. (See Figure 75.)

The V-2 rocket itself had a liquid-propellant engine that burned alcohol and liquid oxygen. It was about 14 meters (46 feet) long and 1.66 meters (5.45 feet) in diameter. After World War II, the United States (using American and captured German rocket personnel) assembled and tested 67 V-2 rockets in New Mexico between 1946 and 1952. This experience provided the technical heritage for many important American rockets, including the Redstone, Jupiter, and Saturn vehicles.

W

warhead That part of a missile or rocket that contains either the nuclear or thermonuclear system, high-explosive system, chemical or biological agents, or inert materials intended to inflict damage upon the enemy.

watt (symbol: W) The SI UNIT of power (i.e., work per unit time). One watt is defined as 1 joule (J) per second, that is 1 W = 1 J/s. In electrical engineering, 1 watt corresponds to the product of 1 ampere (A) times 1 volt (V), that is, 1 W = 1 A × 1 V. This represents the rate of electric energy dissipation in a circuit in which a current of 1 ampere is flowing through a voltage difference of 1 volt. This unit is named in honor of James Watt (1736–1819), the Scottish engineer who developed the steam engine.

wave A periodic disturbance that is propagated in a medium in such a manner that at any point in the medium, the quantity serving as a measure of the disturbance is a function of time, while at any instant the displacement at a point is a function of the position of the point. At each spatial point there is an oscillation. The number of oscillations that occur per unit time is the *frequency* (symbol: v). The distance between one wave crest to the next wave crest (or one trough to the next trough) is called the *wavelength* (symbol: λ). *See also* WAVELENGTH.

wavelength (symbol: λ) In general, the mean distance between maxima (or minima) of a periodic pattern. Specifically, the least distance between particles moving in the same phase of oscillation in a wave disturbance. The wavelength is measured along the direction of propagation of the wave, usually from the midpoint of a crest (or trough) to the midpoint of the next crest (or trough). The wavelength (λ) is related to the frequency (v) and phase speed (c) (i.e., speed of propagation of the wave disturbance) by the simple formula: λ = c/v. The reciprocal of the wavelength is called the *wave number.*

weapon debris (nuclear) The residue of a nuclear weapon after it has exploded; that is, the materials used for the casing and other components of the weapon, plus unexpended plutonium or uranium, together with the fission products created in the fission chain reaction and other radioisotopes formed by various phenomena, including neutron capture (activation).

weapon engagement zone In aerospace defense, airspace of defined dimensions within which the responsibility for engagement normally rests with a particular weapon system. Also called: *missile engagement zone* and *short-range air defense engagement zone.*

weapons of mass destruction (WMD) In arms control usage, weapons that are capable of a high order of destruction and/or of being used in such a manner as to destroy large numbers of people. These types of weapons can be nuclear, chemical, biological, or radiological. The term generally excludes the means of transporting or propelling the weapon where such means is a separable and divisible part of the weapon (e.g., a guided missile).

weapon system A delivery vehicle and weapon combination including all related equipment, materials, services, and personnel required so that the system becomes self-sufficient in its intended operational environment.

weather satellite   One of the first applications of data and images supplied by Earth-orbiting satellites was to improve the understanding and prediction of weather. The weather satellite (also known as the *meteorological satellite* or the *environmental satellite*) is an uncrewed spacecraft that carries a variety of sensors. Two types of weather satellites generally are encountered: geostationary and polar-orbiting weather satellites, which are named for their types of orbits.

Today weather satellites are used to observe and measure a wide range of atmospheric properties and processes to support increasingly more sophisticated weather warning and forecasting activities. Imaging instruments provide detailed pictures of clouds and cloud motions as well as measurements of sea-surface temperature. Sounders collect data in several infrared or microwave spectral bands that are processed to provide profiles of temperature and moisture as a function of altitude. Radar altimeters, scatterometers, and imagers (i.e., synthetic aperture radar, or SAR) can measure ocean currents, sea-surface winds, and the structure of snow and ice cover.

In the United States, several federal agencies have charters for monitoring and forecasting weather. The National Weather Service (NWS) of the National Oceanic and Atmospheric Administration (NOAA) has the primary responsibility for providing severe storm and flood warnings as well as short and medium-range weather forecasts. The Federal Aviation Administration (FAA) provides specialized forecasts and warnings for aircraft. The Defense Meteorological Satellite Program (DMSP) within the Department of Defense (DoD) formerly supported the specialized needs of the military and intelligence communities, which emphasize global capabilities to monitor clouds and visibility in support of combat and reconnaissance activities and to monitor sea-surface conditions in support of naval operations. In response to a 1994 presidential directive,

NOAA has assumed operational responsibility for DMSP.

Global change research strives to monitor and understand the processes of natural and anthropogenic (people-caused) changes in Earth's physical, biological, and human environments. Weather satellites support this research by providing measurements of stratospheric ozone and ozone-depleting chemicals; by providing long-term scientific records of Earth's climate; by monitoring Earth's radiation balance and the concentrations of greenhouse gases and aerosols; by monitoring ocean temperatures, currents, and biological productivity; by monitoring the volume of ice sheets and glaciers; and by monitoring land use and vegetation. These variables provide important information concerning the complex processes and interactions of global environment change, including climate change. *See also* DEFENSE METEOROLOGICAL SATELLITE PROGRAM (DMSP); GEOSTATIONARY OPERATIONAL ENVIRONMENTAL SATELLITE (GOES); GLOBAL CHANGE; POLAR-ORBITING OPERATIONAL ENVIRONMENTAL SATELLITE (POES).

web   The minimum thickness of a solid-propellant grain from the initial ignition surface to the insulated case wall or to the intersection of another burning surface at the time when the burning surface undergoes a major change; for an end-burning grain, the length of the grain. *See also* ROCKET; SOLID-PROPELLANT ROCKET.

weber (symbol: Wb)   The SI UNIT of magnetic flux. It is defined as the flux that produces an electromotive force of 1 volt when, linking a circuit of one turn, the flux is reduced to zero at a uniform rate in 1 second. Named in honor of Wilhelm Weber, a German physicist (1804–1891) who studied magnetism. 1 weber = 10^8 maxwells.

weight (symbol: w)   1. The force with which a body is attracted toward Earth by gravity. 2. The product of the mass (m) of a body and the gravitational acceleration (g) acting on the body, w = mg. For example, a 1 kilogram (1 kg) mass on the surface of Earth would experience a downward force or "weight" of approximately 9.8 newtons.

In a dynamic situation, the weight can be a multiple of that under resting conditions. Weight also varies on other planets in accordance with their value of gravitational acceleration.

weightlessness The condition of free fall or zero-g, in which objects inside an orbiting, unaccelerated spacecraft are "weightless"—even though the objects and the spacecraft are still under the influence of the celestial object's gravity; the condition in which no acceleration, whether of gravity or other force, can be detected by an observer within the system in question. *See also* G; MICROGRAVITY.

welding The process of joining two or more pieces of metal by applying thermal energy (heat), pressure, or both, with or without filler material to produce a localized union through fusion or recrystallization across the interface.

wet To come in contact with and flow across (a surface, body, or area)—said of a fluid.

wet emplacement A launch pad that provides a deluge of water for cooling the flame bucket and other equipment during the launch of a missile or aerospace vehicle. *Compare with* DRY EMPLACEMENT and FLAME DEFLECTOR.

white dwarf (star) A compact STAR at the end of its life cycle. Once a star of one SOLAR MASS or less exhausts its nuclear fuel, it collapses under GRAVITY into a very dense object about the size of EARTH.

white room A clean, dust-free room that is used for the assembly, calibration, and (if necessary) repair of delicate spacecraft components and devices, such as gyros and sensor systems. *See also* CLEAN ROOM.

White Sands Missile Range (WSMR)

White Sands Missile Range is a multiservice test range whose main function is the support of missile development and test programs for the U.S. Army, Navy, Air Force, NASA, and other government agencies and private firms. WSMR is under operational control of the U.S. Army Test and Evaluation Command (TECOM), Aberdeen Proving Ground, Maryland. The missile range is located in the Tularosa Basin of south-central New Mexico. The range boundaries extend almost 160 kilometers (km) (100 miles [mi]) north to south by about 65 km (40 mi) east to west.

During World War II, the U.S. Army Ordnance Corps recognized the possibilities of rocket warfare and sponsored research and development in methods of missile guidance. The missile range was established on July 9, 1945 as White Sands Proving Ground (the named later changed to White Sands Missile Range) to serve as America's testing range for the development of rocket technology and missile weapons. Following World War II, captured German V-2 rockets were assembled and tested at White Sands. Other rockets such as Nike, Corporal, Lance, and Viking also experienced developmental testing at the range. Launch Complex 33 (LC-33) at WSMR now has been designated as a National Historic Landmark, based on the fact that it was the country's first major rocket launch facility. It is also interesting to note the world's first atomic bomb, the TRINITY shot, was exploded on July 16, 1945 near the northern boundary of the complex.

During the early part of the space shuttle program, astronauts Jack Lousma and C. Gordon Fullerton landed the orbiter *Columbia* at the Northrup Strip in the Alkali Flats area of the WSMR to successfully end the STS-3 mission (March 30, 1982). Because heavy rains had made the primary shuttle landing site at Edwards Air Force Base, California, unacceptably wet, NASA chose the alternate site to support *Columbia's* return to Earth. After this event, the Northrup Strip was renamed the White Sands Space Harbor. Although STS-3 has been the only shuttle mission (to date) to land at the range, WSMR continues to support ongoing shuttle missions, including support as a contingency or end-of-mission (EOM) landing site.

wick A group or braid of thin fibers that transports (or quite literally "sucks up") a liquid, if the adhesive force between the fiber and the liquid is greater than the liquid's cohesive force.

Wien's displacement law In radiation transport, the statement that the wavelength of maximum radiation intensity (λ_{max}) for a BLACKBODY is inversely proportional to the absolute temperature of the radiating blackbody:
$$\lambda_{max}T = b \text{ (a constant)}$$
If the absolute temperature (T) is in degrees Kelvin and the maximum wavelength (λ_{max}) is in centimeters, then the constant (b) has the value: 0.2898 cm$^{-\circ}$ K.

Classification of Wind Tunnels (According to Airflow Speed in Test Section)	
Tunnel Class	*Free-stream Mach-number[a] Range*
Low speed	0 to 0.5
High speed	0.5 to 0.9
Transonic	0.7 to 1.4
Supersonic	1.4 to 5.0
Hypersonic	>5.0

[a] Mach number = $\dfrac{\text{Local stream velocity}}{\text{Local velocity of sound}}$

Source: NASA.

"wilco" Aerospace language for "I will comply."

WIMP A hypothetical PARTICLE, called the *w*eakly *i*nteracting *m*assive *p*article, thought by some scientists to pervade the UNIVERSE as the hard to observe DARK MATTER. *See also* MACHO.

window In general, a gap in a linear continuum. An *atmospheric window*, for example, is a range of wavelengths in the electromagnetic spectrum to which the atmosphere is transparent; a *launch window* is the time interval during which conditions are favorable for launching an aerospace vehicle or spacecraft on a specific mission.

wind tunnel A ground facility that supports aerodynamic testing of aircraft, missiles, propulsion systems, or their components under simulated flight conditions. The concept of a wind tunnel is quite simple: Instead of flying the object (or a precise scale model of the object) through the air at the desired test speed and attitude, it is supported in the test attitude within the wind tunnel, and the air flows past the object at the test speed. Wind tunnels exist in a wide variety of shapes and sizes. One wind tunnel design parameter of particular interest is the velocity of airflow, usually given in free-stream Mach number (M) range. (The Mach number is the ratio of the free stream velocity to the local velocity of sound.) The table provides a general classification of wind tunnels according to the speed of the airflow in the test section. The test section in a wind tunnel usually includes windows and appropriate illumination for viewing the test object. Today no aircraft, spacecraft, or space launch or reentry vehicle is built or committed to flight until after its design and components have been thoroughly tested in wind tunnels. *See also* MACH NUMBER.

wing The aerodynamic lifting surface that provides conventional lift and control for an aircraft, winged missile, or aerospace vehicle.

work (symbol: W) In physics and thermodynamics, work (W) is defined as the energy (E) expended by a force (F) acting though a distance (s). Mathematically,
$$\text{Work} = \int F \cdot d\,s$$
In the SI unit system, work is expressed in joules (J). When a force of 1 newton (N) moves through a distance of 1 meter, 1 joule of work is performed; that is, 1 joule = 1 newton-meter.

working fluid A fluid (gas or liquid) used as the medium for the transfer of energy from one part of a system to another part. *See also* BRAYTON CYCLE; CARNOT CYCLE; RANKINE CYCLE.

X

X In aerospace engineering, a prefix used to designate an experimental missile, aerospace vehicle, or rocket. For example, the *X-15 experimental rocket plane.*

X-1 The rocket-powered research aircraft, patterned on the lines of a 50-caliber machine-gun bullet, that was the first human-crewed vehicle to fly faster than the speed of sound. The speed of sound in air varies with altitude. On October 14, 1947 the Bell X-1, named "Glamorous Glennis" and piloted by Captain Charles "Chuck" Yeager, was carried aloft under the bomb bay of a B-29 bomber and then released. The pilot ignited the aircraft's rocket engine, climbed, and accelerated, reaching Mach 1.06, or 1,127 kilometers (km) (700 miles [mi]), per hour as it flew over Edwards Air Force Base in California at an altitude of 13.1 km (8.1 mi). At this altitude, the speed of sound (or Mach 1.00) is 1,078.7 km (670 mi) per hour. The rocket-powered experimental aircraft, having used up all of its propellant, then glided to a landing on its tricycle gear at Muroc Dry Lake in the Mojave Desert. *See also* x-15.

X-15 The North American X-15 was a rocket-powered experimental aircraft 15.24 meters (m) (50 feet [ft]) long with a wingspan of 6.71 m (22 ft). It was a missile-shaped vehicle with an unusual wedge-shaped vertical tail, thin stubby wings, and unique fairings that extended along the side of the fuselage. The X-15 had an empty mass of 6,340 kilograms (kg) (14,000 pounds-mass [lbm]) and a launch mass of 15,420 kg (34,000-lbm). The vehicle's pilot-controlled rocket engine was capable of developing 253,500 newtons (57,000 pounds-force [lbf]) of thrust.

The X-15 research aircraft helped bridge the gap between human flight within the atmosphere and human flight in space. It was developed and flown in the 1960s to provide in-flight information and data on aerodynamics, structures, flight controls, and the physiological aspects of high-speed, high-altitude flight. For flight in the dense air of the lower ("aircraft-usable") portions of the atmosphere, the X-15 employed conventional aerodynamic controls. However, for flight in the thin upper portions of Earth's atmospheric envelope, the X-15 used a ballistic control system. Eight hydrogen peroxide-fueled thruster rockets, located on the nose of the aircraft, provided pitch and yaw control.

Because of its large fuel consumption, the X-15 was air launched from a B-52 aircraft (i.e., a "mothership") at an altitude of about 13,700 m (45,000 ft) and a speed of about 805 kilometers (km) (500 miles [mi]) per hour. Then the pilot ignited the rocket engine, which provided thrust for the first 80 to 120 seconds of flight, depending on the type mission being flown. The remainder of the normal 10- to 11-minute duration flight was powerless and ended with a 322-km (200-mi) per hour glide landing at Edwards Air Force Base in California. Generally, one of two types of X-15 flight profiles was used: a high-altitude flight plan that called for the pilot to maintain a steep rate of climb; or a speed profile that called for the pilot to push over and maintain a level altitude.

First flown in 1959, the three X-15 aircraft made a total of 199 flights. The X-15 flew more than six times the speed of sound and reached a maximum altitude of 107.8 km (67 mi) and a maximum speed of 7,273 km (4,520 mi) per hour. The final flight occurred on October 24, 1968. It is interesting to note that Apollo Astronaut Neil Armstrong (the first human to walk on the Moon) was one of the pilots who flew the X-15 aircraft.

X-20 The name given to an early U.S. Air Force space plane concept whose mission was to explore human flight in the hypersonic and orbital regimes. However, the proposed vehicle did not have a significant payload capacity, and therefore it would not have been suitable in space logistics missions (i.e., in "payload-hauling" missions). The X-20, or DYNA-SOAR, space plane was to be placed into orbit by a Titan booster. When its orbital mission was completed, the military pilot was to control this sleek glider through atmospheric reentry and then land it on a runway, much like a conventional jet fighter. Despite its innovative nature, the project was canceled in December 1963, before the prototype vehicle was ever built.

X-33 NASA's X-33 REUSABLE LAUNCH VEHICLE (RLV) prototype program attempted to simultaneously demonstrate several major advancements in SPACE LAUNCH VEHICLE technology that would increase safety and reliability while lowering the cost of placing a PAYLOAD into LOW EARTH ORBIT by an ORDER OF MAGNITUDE or so—that is, from about $20,000 per KILOGRAM to perhaps as little as $2,000 per kilogram. The X-33 vehicle was a half-scale prototype of the conceptual SINGLE-STAGE-TO-ORBIT (SSTO) RLV, called *VentureStar,* being pursued by the Lockheed-Martin Company as a commercial development. However, after a series of disappointing technical setbacks and serious cost overruns, NASA abruptly cancelled the X-33 program in March 2001.

X-34 The vehicle being developed under NASA's X-34 program served as a suborbital flying laboratory for several space technology advancements applicable to future REUSABLE LAUNCH VEHICLES. The program involved a partnership between NASA and the Orbital Sciences Corporation of Dulles, Virginia. Despite

success in X-34 vehicle development and testing, NASA decided to cancel the program early in 2001 due to funding constraints and other factors.

X-37 NASA's X-37 vehicle is an advanced technology demonstrator that is helping define the future of reusable space transportation systems. It is a cooperative program involving NASA, the UNITED STATES AIR FORCE, and the Boeing Company. The X-37 is a REUSABLE LAUNCH VEHICLE designed to operate in both the orbital and REENTRY phases of space flight. The robotic space plane can be ferried into ORBIT by the SPACE SHUTTLE or by an EXPENDABLE LAUNCH VEHICLE. The program is demonstrating new AIRFRAME, AVIONICS, and operational technologies relevant to future SPACECRAFT and LAUNCH VEHICLE designs. The shape of the X-37 vehicle is a 120 percent scale derivative of the U.S. Air Force X-40A SPACE VEHICLE, also designed by the Boeing Company.

X ray A penetrating form of electromagnetic radiation of very short wavelength (approximately 0.01 to 10 nanometers or 0.1 to 100 angstroms) and high photon energy (approximately 100 electron volts to some 100 kiloelectron volts). X rays are emitted when either the inner orbital electrons of an excited atom return to their normal energy states (these photons are called *characteristic X rays*) or when a fast-moving charged particle (generally an electron) loses energy in the form of photons upon being accelerated and deflected by the electric field surrounding the nucleus of a high atomic number element (this process is called *bremsstrahlung,* or "braking radiation"). Unlike gamma rays, X rays are nonnuclear in origin. *See also* ADVANCED X-RAY ASTROPHYSICS FACILITY; X-RAY ASTRONOMY.

X-ray astronomy Since Earth's atmosphere absorbs most of the X rays coming from celestial phenomena, astronomers must use orbiting space platforms to study these interesting emissions, which usually are associated with very energetic, violent processes occurring in the universe. In fact, X-ray emissions carry detailed information about the temperature, density, age, and other physical conditions of the celestial objects that have produced them. X-ray astronomy is the most advanced of the three general disciplines associated with high-energy astrophysics: X-ray, gamma ray, and cosmic ray astronomy. The observation of X-ray emissions has been very valuable in the study of high-energy events, such as mass transfer in binary star systems, the interaction of supernovae remnants with interstellar gas, and the functioning of quasars.

Space-based X-ray observatories provide data allowing scientists to study and understand: stellar structure and evolution, including binary star systems, supernovae remnants, pulsars and BLACK HOLE candidates; large-scale galactic phenomena, including the interstellar medium itself and soft X-ray emissions of local galaxies; the nature of active galaxies, including the spectral characteristics and time variation of X-ray emissions from the central (or nuclear) regions of such galaxies; and rich clusters of galaxies, including their associated X-ray emissions. X-ray emissions also help scientists observe and study solar flares occurring on our parent star, the Sun. *See also* CHANDRA X-RAY OBSERVATORY.

X-ray binary The most often encountered type of luminous galactic X-ray source. It is a close binary star system in which material from a large, normal star flows (under gravitational forces) onto a compact stellar companion, such as a neutron star or a black hole (for the most luminous X-ray sources) or perhaps a white dwarf (for less luminous sources). *See also* X-RAY ASTRONOMY.

X-ray Timing Explorer (XTE) *See* ROSSI X-RAY TIMING EXPLORER.

Y   In aerospace engineering, the symbol for prototype. For example, a prototype guided missile would be designated as YGM. Prototype vehicles usually are manufactured in limited quantities and are used to support operational and preproduction tests.

YAL-1A Attack Laser   *See* AIRBORNE LASER.

yard (symbol: yd)   A fundamental unit of length in the standard (foot-pound-second) system of units. Since 1963, 1 yard has been defined as exactly 0.9144 meter.

1 yard = 3 feet = 0.9144 meter

yaw   The rotation or oscillation of an aircraft, missile, or aerospace vehicle about its vertical axis so as to cause the longitudinal axis of the vehicle to deviate from the flight line or heading in its horizontal plane. *See also* PITCH; ROLL.

year   A period of one revolution of Earth around the Sun. The choice of a reference point determines the exact length of a year. The *calendar year* (based on the Gregorian calendar) is made up of an average of 365.2425 mean solar days. For convenience, the calendar year is taken as three successive years of 365 days followed by one "leap" year consisting of 366 days. A *solar day* is defined as the time interval between two successive passages of the Sun across the meridian; while a *mean solar day* of 24 hours is considered as the average value of the solar day for one year. Scientists and astronomers also use other, more specialized definitions. For example, the period of revolution from perihelion to perihelion (point of Earth's closest approach to the Sun) is called the *anomalistic year* and cor-

responds to approximately 365.259 mean solar days. The *tropical year* is defined as the average period of time between successive passages of the Sun across the vernal equinox (i.e., return of the Sun to the first point in Aries) and corresponds to approximately 365.242 mean solar days. Finally, the *sidereal year* is based on the return passage of certain "fixed" stars and is approximately 365.256 mean solar days.

yield   The total energy released in a nuclear explosion. It is usually expressed in equivalent tons of TNT (the quantity of trinitrotohene required to produce a corresponding amount of energy). *Low yield* is generally considered to be less than 20 kilotons; low-intermediate yield, from 20 to 200 kilotons; intermediate yield, from 200 kilotons to 1 megaton. There is no standardized term to describe yields from 1 megaton upward.

YM   A prefix often used in aerospace engineering to designate a prototype missile.

Yohkoh   A solar X-ray observation satellite launched by the Japanese Institute of Space and Astronautical Sciences (ISAS) on August 30, 1991. The main objective of this satellite is to study the high-energy radiations from solar flares (i.e., hard and soft X rays and energetic neutrons) as well as quiet Sun structures and presolar flare conditions. *Yohkoh*, which means "sunbeam" in Japanese, is a three-axis stabilized observatory-type satellite in a nearly circular Earth orbit, carrying four instruments: two imagers and two spectrometers. The imaging instruments (a hard X-ray telescope [20–80 kiloelectron volts [keV] energy range] and a soft X-ray telescope [0.1–4 keV energy range]) have almost full-Sun fields of view to avoid missing any flares on the visible disk of the Sun. This mission is a cooperative mission of Japan, the United States, and the United Kingdom. For example, *Yohkok*'s soft X-ray telescope (SXT) was developed for NASA by the Lockheed Palo Alto Research Laboratory, in partnership with the National Astronomical Observatory of Japan and the Institute for Astronomy of the University of Tokyo. *See also* SUN.

young stellar object (YSO) Any celestial object in an early stage of STAR formation, from a PROTOSTAR to a MAIN-SEQUENCE STAR.

Z

Z The symbol for *atomic number,* or the number of protons (or electrons) in the atoms of a given element.

Zenit A three-stage Russian launch vehicle capable of placing about 13,700 kilograms in LOW EARTH ORBIT (LEO). It was first launched in 1985.

zenith That point on the celestial sphere vertically overhead. The point 180 degrees from the zenith is called the *nadir.*

zero-g The condition of free fall and weightlessness. When there are no forces on objects in an orbiting spacecraft, they are said to be in zero-g. *See also* G; MICROGRAVITY.

zero-gravity (zero-g) aircraft An aircraft that flies a special parabolic trajectory to create low-gravity conditions (typically 0.01 g) for short periods of time (10–30 seconds), where 1 g here represents the acceleration due to gravity at Earth's surface (9.8 meters per second per second). For example, a modified KC-135 aircraft can simulate up to 40 periods of low-gravity for 25-second intervals during one flight. This aircraft accommodates a variety of experiments and often is used to support crew training and to refine space-flight experiment equipment and techniques. The KC-135, like other "zero-gravity" research aircraft, obtains weightlessness by flying a parabolic trajectory. The plane climbs rapidly at a 45-degree angle (pull-up), slows as it traces a parabola (pushover), and then descends at a 45-degree angle (pull-out). The forces of acceleration and deceleration produce twice the normal gravity during the pull-up and pull-out legs of the flight, while the brief pushover

at the top of the parabola produces less than 1% of Earth's sea-level gravity. *See also* MICROGRAVITY.

zero-length launching A technique in which the first motion of the missile or aircraft removes it from the launcher.

Zeroth Law of Thermodynamics Two systems, each in thermal equilibrium with a third system, are in thermal equilibrium with each other. This statement is frequently called the Zeroth Law of Thermodynamics and is actually an implicit part of the concept of temperature. It is important in the field of thermometry and in the establishment of empirical temperature scales. *See also* FIRST LAW OF THERMODYNAMICS; SECOND LAW OF THERMODYNAMICS; THERMODYNAMICS.

zodiac From ancient Greek meaning "circle of figures." Early astronomers described the band in the sky about 9° on each side of the ECLIPTIC, which they divided into 30° intervals—each representing a sign of the zodiac. Within their GEOCENTRIC COSMOLOGY, the SUN appeared to enter a different CONSTELLATION of the zodiac each month. So the signs of the zodiac helped them mark the annual REVOLUTION of EARTH around the SUN. These twelve constellations (or signs) are Aries (ram), Taurus (bull), Gemini (twins), Cancer (crab), Leo (lion), Virgo (maiden), Libra (scales), Scorpius (scorpion), Sagittarius (archer), Capricornus (goat), Aquarius (water-bearer), and Pisces (fish). Although the signs of the zodiac originally (more than 2,500 years ago) corresponded in position to the twelve constellations just named, because of the phenomenon of PRECESSION the zodiacal signs do not presently coincide with these constellations. For example, when people today say the Sun enters Aries at the VERNAL EQUINOX, it has, in fact, shifted forward (from ancient times) and is now actually in the constellation Pisces.

zodiacal light A faint cone of light extending upward from the horizon in the direction of the ecliptic. Zodiacal light is seen from the tropical latitudes for a few hours after sunset or before sunrise. It is due to sunlight being reflected by tiny pieces of interplanetary dust in orbit around the Sun.

Zond A family of Russian spacecraft that explored the Moon, Mars, Venus, and interplanetary space in the 1960s. *Zond 1,* a partial success, was launched on April 2, 1964 toward Venus but stopped transmitting about two months before it flew by the planet at a distance of 100,000 kilometers (km). *Zond 3,* launched as a test interplanetary probe on July 18, 1965, flew past the Moon at a distance of 9,200 km and obtained photographs of the lunar farside. The last Zond mission, *Zond 8,* was launched on October 20, 1970. It circled the Moon at a distance of 1,100 km and returned to Earth, splashing down in the Indian Ocean. The Zond lunar circumnavigation missions (i.e., *Zond 5, 6, 7* and *8*) are considered by some space analysts as precursors to Russian manned missions in response to the U.S. Apollo program, during the cold war "race to the Moon" in the late 1960s.

zulu time (symbol: Z) Time based on Greenwich mean time (GMT). *See* GREENWICH MEAN TIME.

APPENDIXES

Appendix I: Special Reference List

Special Units for Astronomical Investigations

Astronomical unit *(AU)*: The mean distance from Earth to the Sun—approximately 1.495979×10^{11} m

Light-year *(ly)*: The distance light travels in 1 year's time—approximately 9.46055×10^{15} m

Parsec *(pa)*: The parallax shift of 1 second of arc (3.26 light-years)—approximately 3.085768×10^{16} m

Speed of light *(c)*: 2.9979×10^8 m/s

Source: NASA.

International System (SI) Units and Their Conversion Factors

Quantity	Name of Unit	Symbol	Conversion Factor
distance	meter	m	1 km = 0.621 mi.
			1 m = 3.28 ft.
			1 cm = 0.394 in.
			1 mm = 0.039 in.
			1 μm = 3.9×10^{-5} in. = 10^4 Å
			1 nm = 10 Å
mass	kilogram	kg	1 tonne = 1.102 tons
			1 kg = 2.20 lb.
			1 g = 0.0022 lb. = 0.035 oz.
			1 mg = 2.20×10^{-6} lb. = 3.5×10^{-5} oz.
time	second	s	1 yr. = 3.156×10^7 s
			1 day = 8.64×10^4 s
			1 hr. = 3,600 s
temperature	kelvin	K	273 K = 0°C = 32°F
			373 K = 100°C = 212°F
area	square meter	m^2	1 m^2 = 10^4 cm^2 = 10.8 ft.^{2}
volume	cubic meter	m^3	1 m^3 = 10^6 cm^3 = 35 ft.^{3}
frequency	hertz	Hz	1 Hz = 1 cycle/s
			1 kHz = 1,000 cycles/s
			1 MHz = 10^6 cycles/s
density	kilogram per cubic meter	kg/m^3	1 kg/m^3 = 0.001 g/cm^3
			1 g/cm^3 = density of water
speed, velocity	meter per second	m/s	1 m/s = 3.28 ft./s
			1 km/s = 2,240 mi./hr.
force	newton	N	1 N = 10^5 dynes = 0.224 lbf
pressure	newton per square meter	N/m^2	1 N/m^2 = 1.45×10^{-4} lb./in.^{2}
energy	joule	J	1 J = 0.239 cal
photon energy	electronvolt	eV	1 eV = 1.60×10^{-19} J; 1 J = 10^7 erg
power	watt	W	1 W = 1 J/s
atomic mass	atomic mass unit	amu	1 amu = 1.66×10^{-27} kg
wavelength of light	angstrom	Å	1 Å = 0.1 nm = 10^{-10} m
acceleration of gravity	g	g	1 g = 9.8 m/s^2

Source: NASA.

Recommended SI Unit Prefixes

Submultiple	Prefix	Symbol	Multiple	Prefix	Symbol
10^{-1}	deci-	d	10^{1}	deca-	da
10^{-2}	centi-	c	10^{2}	hecto-	h
10^{-3}	milli-	m	10^{3}	kilo-	k
10^{-6}	micro-	μ	10^{6}	mega-	M
10^{-9}	nano-	n	10^{9}	giga-	G
10^{-12}	pico-	p	10^{12}	tera-	T
10^{-15}	femto-	f	10^{15}	peta-	P
10^{-18}	atto-	a	10^{18}	exa-	E
10^{-21}	zepto-	z	10^{21}	zetta-	Z
10^{-24}	yocto-	y	10^{24}	yotta-	Y

Common Metric/English Conversion Factors
(for Space Technology Activities)

	Multiply	By	To Obtain
length	inches	2.54	centimeters
	centimeters	0.3937	inches
	feet	0.3048	meters
	meters	3.281	feet
	miles	1.6093	kilometers
	kilometers	0.6214	miles
	kilometers	0.54	nautical miles
	nautical miles	1.852	kilometers
	kilometers	3281	feet
	feet	0.0003048	kilometers
weight and mass	ounces	28.350	grams
	grams	0.0353	ounces
	pounds	0.4536	kilograms
	kilograms	2.205	pounds
	tons	0.9072	metric tons
	metric tons	1.102	tons
liquid measure	fluid ounces	0.0296	liters
	gallons	3.7854	liters
	liters	0.2642	gallons
	liters	33.8140	fluid ounces
temperature	degrees Fahrenheit plus 459.67	0.5555	kelvins
	degrees Celsius plus 273.16	1.0	kelvins
	kelvins	1.80	degrees Fahrenheit minus 459.67
	kelvins	1.0	degrees Celsius minus 273.16
	degrees Fahrenheit minus 32	0.5555	degrees Celsius
	degrees Celsius	1.80	degrees Fahrenheit plus 32
thrust (force)	pounds force	4.448	newtons
Pressure	newtons	0.225	pounds force
	millimeters mercury	133.32	pascals (newtons per square meter)
	pounds per square inch	6.895	kilopascals (1,000 pascals)
	pascals	0.0075	millimeters mercury at 0°C
	kilopascals	0.1450	pounds per square inch

Source: NASA.

Appendix II: Journeys into Cyberspace

In recent years, numerous websites concerning space exploration and technology have appeared on the Internet. Visits to such sites can provide information about the status of ongoing missions, such as the Cassini mission to Saturn. This book can serve as an important companion, as you explore a new website and encounter a word, phrase, or form of space technology that is unfamiliar and not fully discussed within the particular site. To help enrich the content of this book and make your space technology-related travels in cyberspace more enjoyable and productive, the following is a selected list of websites that are recommended for your viewing. From these sites you will also be able to link to many other exciting space-related locations on the Web. (Note: This is obviously a partial list of the many space-related websites now available. Every effort has been made to ensure the accuracy of the information provided at the time of publication. Any inconvenience you may experience due to the dynamic nature of the Internet is regretted.)

Selected Space Organization Home Pages
European Space Agency (ESA) http://www.esrin.esa.it/
National Aeronautics and Space Administration (NASA) http://www.nasa.gov
National Oceanic and Atmospheric Administration (NOAA) http://www.noaa.gov
National Reconnaissance Office (NRO) http://www.nro.odci.gov/
U.S. Air Force Space Command (AFSPC) http://www.spacecom.af.mil/hqafspc/
United States Air Force (USAF)—Air Force Link http://www.af.mil/news/
United States Air Force—Patrick AFB/Cape Canaveral AFS http://www.patrickaf.mil/
United States Strategic Command (USSTRATCOM) http://www.spacecom.mil/

Selected Space Missions
Cassini mission (Saturn) http://www.jpl.nasa.gov/cassini/
Galileo mission (Jupiter) http://www.jpl.nasa.gov/galileo/

Other Interesting Space and Astronomy Sites
Contemporary Aerospace News and Reports http://www.space.com
Lunar and Planetary Institute (LPI) http://cass.jsc.nasa.gov/
NASA's Space Science News http://science.nasa.gov/
National Air and Space Museum (Smithsonian Institution)
    http://ceps.nasm.edu/NASMpage.html
National Space Science Data Center (NSSDC) (numerous space
    missions—multinational) http://nssdc.gsfc.nasa.gov/planetary/
The Planetary Society http://planetary.org